METHODS IN MOLECULAR BIOLOGY

Series Editor
John M. Walker
School of Life Sciences
University of Hertfordshire
Hatfield, Hertfordshire, AL10 9AB, UK

For further volumes:
http://www.springer.com/series/7651

Chloroplast Biotechnology

Methods and Protocols

Edited by

Pal Maliga

Waksman Institute of Microbiology, Rutgers University, Piscataway, NJ, USA

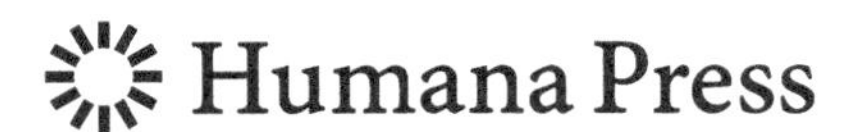

Editor
Pal Maliga
Waksman Institute of Microbiology
Rutgers University
Piscataway, NJ, USA

ISSN 1064-3745 ISSN 1940-6029 (electronic)
ISBN 978-1-4939-5870-2 ISBN 978-1-62703-995-6 (eBook)
DOI 10.1007/978-1-62703-995-6
Springer New York Heidelberg Dordrecht London

Printed on acid-free paper

Humana Press is a brand of Springer
Springer is part of Springer Science+Business Media (www.springer.com)

Dedication

I dedicate this book to Zora, my wife, for her love, support, and seminal contributions to helping create this technology.

Preface

Plastids of higher plants are semi-autonomous organelles with a small, highly polyploid genome, their own transcription-translation machinery and metabolism suitable for engineering by targeted transformation. Tractable metabolic pathways include photosynthesis, starch, amino acid, lipid, and pigment biosynthesis. Transformation of the plastid genome was achieved in 1988 for *Chlamydomonas reinhardtii*, a unicellular alga, and 1990 in tobacco, the first of several higher plants accessible with this technology. This volume presents the relevant background for understanding and applying Chloroplast Biotechnology and step-by-step instruction by experts to facilitate entry into plastome engineering of plants. Detailed chapters include protocols that have been successfully applied to tobacco, crop plants (tomato, petunia, potato, eggplant, lettuce, soybean, cabbage, sugar beet and alfalfa), Chlamydomonas and avascular plants Bryophytes (*Physcomitrella patens, Marchantia polymorpha*). This book will be useful for all interested in fundamental chloroplast molecular biology as well as applications in agriculture, industrial biotechnology and healthcare.

Piscataway, NJ, USA *Pal Maliga*

Contents

Contributors

ELENA MARTIN AVILA • *Faculty of Life Sciences, The University of Manchester, Manchester, UK*
KAILASH C. BANSAL • *National Bureau of Plant Genetic Resources (ICAR), New Delhi, India*
PIERLUIGI BARONE • *Department of Crop Sciences, University of Illinois, Urbana, IL, USA*
DANIEL BARRERA • *California Center for Algae Biotechnology, La Jolla, CA, USA; Division of Molecular Biology, The University of California San Diego, La Jolla, CA, USA*
MICHELE BELLUCCI • *Istituto di Genetica Vegetale, Consiglio Nazionale delle Ricerche (CNR), Perugia, Italy*
RALPH BOCK • *Max-Planck-Institut für Molekulare Pflanzenphysiologie, Potsdam-Golm, Germany*
TEODORO CARDI • *Consiglio per la Ricerca e la Sperimentazione in Agricoltura – Centro di Ricerca per l'Orticoltura (CRA-ORT), Pontecagnano, (SA), Italy*
SHOTA CHIYODA • *Graduate School of Biostudies, Kyoto University, Kyoto, Japan*
WAN-RU CHU • *Department of Horticulture, National Chung Hsing University, Taichung, Taiwan, ROC*
ANIL DAY • *Faculty of Life Sciences, The University of Manchester, Manchester, UK*
ARELI HERRERA DÍAZ • *Department of Biology I – Botany, Ludwig-Maximilians-University of Munich, Planegg-Martinsried, Germany*
MANUEL DUBALD • *BioScience, Bayer CropScience, Morrisville, NC, USA*
KEES M.P. VAN DUN • *Rijk Zwaan Breeding B.V., Fijnaart, The Netherlands*
CHLOE ECONOMOU • *Institute for Structural and Molecular Biology, University College London, London, UK*
MARK EDWARDS • *Southern Cross Plant Science, Centre for Plant Conservation Genetics, Southern Cross University, Lismore, NSW, Australia*
DANIELA GARGANO • *CNB-CSIC, Spanish National Biotechnology Centre, Madrid, Spain*
JAVIER GIMPEL • *California Center for Algae Biotechnology, La Jolla, CA, USA; Division of Molecular Biology, The University of California San Diego, La Jolla, CA, USA*
JOHN C. GRAY • *Department of Plant Sciences, University of Cambridge, Cambridge, UK*
MAUREEN R. HANSON • *Department of Molecular Biology and Genetics, Cornell University, Ithaca, NY, USA*
ROBERT J. HENRY • *Queensland Alliance for Agriculture and Food Innovation, University of Queensland, Brisbane, QLD, Australia*
BRIDGET V. HOGG • *School of Biology & Environmental Science, University College Dublin, Dublin, Ireland*
ROBERT K. JANSEN • *Integrative Biology, University of Texas at Austin, Austin, TX, USA; Genomics and Biotechnology Section, Department of Biological Sciences, Faculty of Science, King Abdulaziz University, Jeddah, Saudi Arabia*
TAKAYUKI KOHCHI • *Graduate School of Biostudies, Kyoto University, Kyoto, Japan*
HANS-ULRICH KOOP • *Department of Biology I – Botany, Ludwig-Maximilians-University of Munich, Planegg-Martinsried, Germany*

CILIA L.C. LELIVELT • *Rijk Zwaan Breeding B.V., Fijnaart, The Netherlands*
CHUNJING LIN • *Agro-Biotechnology Research Institute, Jilin Academy of Agricultural Sciences, Changchun, P. R. China*
GERHARD LINK • *Plant Cell Physiology and Molecular Biology, University of Bochum, Bochum, Germany*
CHENG-WEI LIU • *Department of Post-Modern Agriculture, Ming Dao University, Chang Hua, Taiwan, ROC*
YANZHI LIU • *Agro-Biotechnology Research Institute, Jilin Academy of Agricultural Sciences, Changchun, P. R. China*
PANAGIOTIS MADESIS • *Faculty of Life Sciences, The University of Manchester, Manchester, UK*
PAL MALIGA • *Waksman Institute of Microbiology, Rutgers University, Piscataway, NJ, USA*
FRANCESCA DE MARCHIS • *Istituto di Genetica Vegetale, Consiglio Nazionale delle Ricerche (CNR), Perugia, Italy*
STEPHEN MAYFIELD • *California Center for Algae Biotechnology, La Jolla, CA, USA; Division of Molecular Biology, The University of California San Diego, La Jolla, CA, USA*
MATTHEW S. MCCABE • *Animal Bioscience Centre, Teagasc, Grange, Dunsany, Co Meath, Ireland*
ELISABETH A. MUDD • *Faculty of Life Sciences, The University of Manchester, Manchester, UK*
CATHERINE J. NOCK • *Southern Cross Plant Science, Centre for Plant Conservation Genetics, Southern Cross University, Lismore, NSW, Australia*
JACQUELINE M. NUGENT • *Department of Biology, National University of Ireland, Maynooth, Co. Kildare, Ireland*
JENNIFER ORTELT • *Plant Cell Physiology and Molecular Biology, University of Bochum, Bochum, Germany*
BERNARD PELISSIER • *Bayer CropScience, Lyon, France*
SAUL PURTON • *Institute for Structural and Molecular Biology, University College London, London, UK*
SILVIA RAMUNDO • *Department of Molecular Biology, University of Geneva, Geneva, Switzerland; Department of Plant Biology, University of Geneva, Geneva, Switzerland*
NICOLE RICE • *Southern Cross Plant Science, Centre for Plant Conservation Genetics, Southern Cross University, Lismore, NSW, Australia*
JEAN-DAVID ROCHAIX • *Department of Molecular Biology, University of Geneva, Geneva, Switzerland; Department of Plant Biology, University of Geneva, Geneva, Switzerland*
STEPHANIE RUF • *Max-Planck-Institut für Molekulare Pflanzenphysiologie, Potsdam-Golm, Germany*
TRACEY A. RUHLMAN • *Integrative Biology, University of Texas at Austin, Austin, TX, USA*
AMIRALI SATTARZADEH • *Department of Molecular Biology and Genetics, Cornell University, Ithaca, NY, USA*
NUNZIA SCOTTI • *CNR-IBBR, National Research Council of Italy, Institute of Biosciences and BioResources, Portici, Italy*
ROBERT E. SHARWOOD • *Hawkesbury Institute for the Environment, University of Western Sydney, Penrith, NSW, Australia*
AJAY K. SINGH • *National Institute of Abiotic Stress Management (ICAR), Malegaon, Baramati, Pune, India*
C. BASTIAAN DE SNOO • *Rijk Zwaan Breeding B.V., Fijnaart, The Netherlands*
JEFFREY M. STAUB • *Monsanto Company, St. Louis, MO, USA*
MAMORU SUGITA • *Center for Gene Research, Nagoya University, Nagoya, Japan*
MASAHIRO SUGIURA • *Center for Gene Research, Nagoya University, Nagoya, Japan*

RAYMOND SURZYCKI • *Solarvest BioEnergy Inc., Bloomington, IN, USA*
JOANNA SZAUB • *Institute for Structural and Molecular Biology, University College London, London, UK*
SITHICHOKE TANGPHATSORNRUANG • *National Center for Genetic Engineering and Biotechnology, Klong Luang, Pathumthani, Thailand*
GHISLAINE TISSOT • *Bayer CropScience, Lyon, France*
MENQ-JIAU TSENG • *Department of Horticulture, National Chung Hsing University, Taichung, Taiwan, ROC*
TARINEE TUNGSUCHAT-HUANG • *Waksman Institute of Microbiology, Rutgers University, Piscataway, NJ, USA*
VLADIMIR T. VALKOV • *CNR-IBBR, National Research Council of Italy, Institute of Biosciences and BioResources, Naples, Italy*
YUNPENG WANG • *Agro-Biotechnology Research Institute, Jilin Academy of Agricultural Sciences, Changchun, P. R. China*
THANYANAN WANNATHONG • *Department of Biology, Silpakorn University, Nakornpathom, Thailand*
ZHENGYI WEI • *Agro-Biotechnology Research Institute, Jilin Academy of Agricultural Sciences, Changchun, P. R. China*
ANDREAS WEIHE • *Institut für Biologie/Genetik, Humboldt-Universität zu Berlin, Berlin, Germany*
SPENCER M. WHITNEY • *College of Medicine, Biology and Ecology, Research School of Biology, The Australian National University, Canberra, ACT, Australia*
JACK M. WIDHOLM • *Department of Crop Sciences, University of Illinois, Urbana, IL, USA*
SHAOCHEN XING • *Agro-Biotechnology Research Institute, Jilin Academy of Agricultural Sciences, Changchun, P. R. China*
KATSUYUKI T. YAMATO • *Faculty of Biology-Oriented Science and Technology (BOST), Kinki University, Wakayama, Kyoto, Japan*
MING-TE YANG • *Institute of Molecular Biology, National Chung Hsing University, Taichung, Taiwan, ROC*
XING-HAI ZHANG • *Department of Crop Sciences, University of Illinois, Urbana, IL, USA*

Part I

Background

Chapter 1

The Plastid Genomes of Flowering Plants

Tracey A. Ruhlman and Robert K. Jansen

Abstract

The plastid genome (plastome) has proved a valuable source of data for evaluating evolutionary relationships among angiosperms. Through basic and applied approaches, plastid transformation technology offers the potential to understand and improve plant productivity, providing food, fiber, energy and medicines to meet the needs of a burgeoning global population. The growing genomic resources available to both phylogenetic and biotechnological investigations are allowing novel insights and expanding the scope of plastome research to encompass new species. In this chapter we present an overview of some of the seminal and contemporary research that has contributed to our current understanding of plastome evolution and attempt to highlight the relationship between evolutionary mechanisms and tools of plastid genetic engineering.

Key words Angiosperm, DNA recombination, DNA replication, DNA repair, Genome evolution, Inheritance, Intergenic region, Inverted repeat, Plastome, Phylogeny

1 Introduction

The most notable defining feature of the plant cell is the presence of plastids, the bioenergetic organelles responsible for photosynthesis and myriad metabolic activities. Contemporary plastids carry the remnant genome (plastome) of their evolutionary ancestor, a photosynthetic bacterium believed to be related to extant cyanobacteria. Over eons, the coding capacity of the plastome has been greatly reduced relative to its progenitor such that a very small fraction of the ancestral gene complement remains [1]. The majority of plastid proteins are encoded in the nucleus and posttranslationally imported; the expression of plastome sequences is controlled by imported nuclear factors.

The evolutionary trajectory, from free-living organism to endosymbiont, to organelle, that has shaped the plastome may be ongoing [2–4]. A survey of the plastome sequences in publicly available databases reveals that despite the prevalence of highly conserved sequence and organization in the majority of flowering plants sampled to date, a salient fraction show marked variation in

Pal Maliga (ed.), *Chloroplast Biotechnology: Methods and Protocols*, Methods in Molecular Biology, vol. 1132,
DOI 10.1007/978-1-62703-995-6_1, © Springer Science+Business Media New York 2014

their rates of sequence evolution and genomic architecture [5]. Although some examples of divergence are subtle, others are conspicuous and lead one to wonder: By what mechanisms do these changes arise, and how are plants able to tolerate changes that appear disruptive?

In this chapter we outline some general features of the typical angiosperm plastome including its structure, organization and gene content. We consider cases where genes otherwise found in the plastome are disrupted or missing and how these changes, along with genomic characters such as rearrangements, are used not only to infer phylogenetic relationships but also to extend our understanding of how organelle genomes change through evolutionary time. Further, we discuss mechanisms that may be influencing genomic stability and consider how these same activities are inherently involved in the introduction of exogenous DNA sequences via plastid transformation.

2 Characteristics of the Angiosperm Plastome

Complete plastome sequences are represented schematically as circular maps (Fig. 1). Early studies concerned with the architectural features of plastomes used denaturation mapping and restriction enzyme digestion of DNA molecules isolated from purified plastids to characterize plastome size and structure. Prior to 1990 plastid DNA (ptDNA) from a diverse range of angiosperm species, representative gymnosperms and ferns, and a number of photosynthetic green algae, had already been scrutinized [6]. Among angiosperms the findings were largely in agreement and provided the framework for our current understanding of plastome structure. With the development of technology facilitating the direct sequencing of complete plastomes many of the seminal predictions have been confirmed providing a reasonably clear picture of the typical angiosperm plastome. The circular maps presented in the literature represent a single monomer, though a number of studies have identified alternative forms including multimeric circles as well as linear and branched molecules (discussed below). The monomer is highly gene dense relative to nuclear or mitochondrial genomes, with 120–130 genes packed into 120–170 kb. Nonetheless, the gene space accounts for only ~50 % of total nucleotide sequence with the remainder comprising introns, regulatory regions and intergenic spacers [7]. Plastomes are also highly AT rich; overall GC content is typically on the order of 30–40 % and in some regions that do not encode proteins AT content exceeds 80 % [8]. The proportion of GC, which is higher in protein-coding regions, varies across plastomes by location, codon position and functional group. For example, genes encoding photosynthetic functions have the highest GC content while the NAD(P)H genes have the lowest.

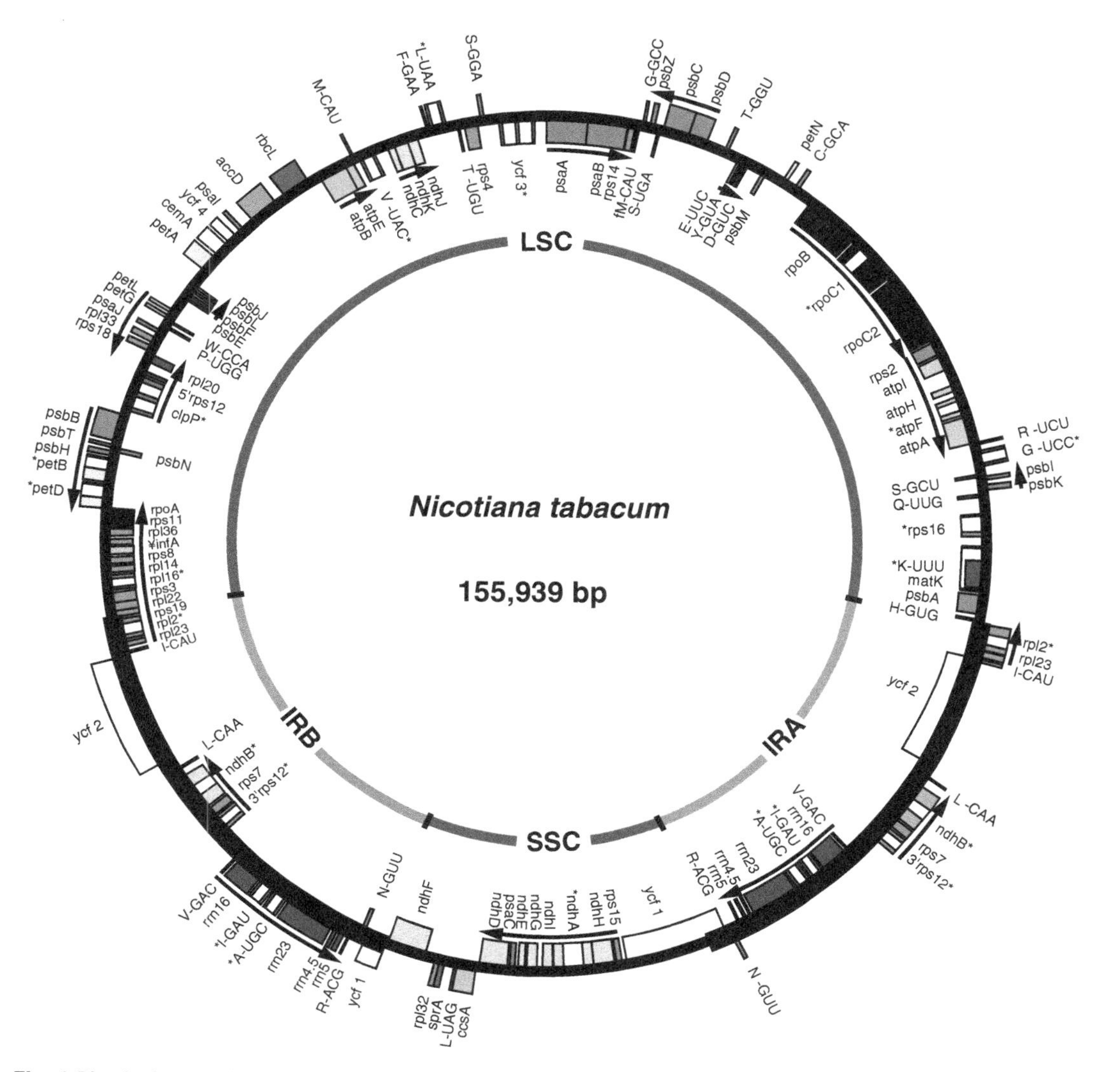

Fig. 1 Physical map of the *Nicotiana tabacum* plastid genome (NC_001879). The *N. tabacum* plastid genome is representative of the ancestral gene order that is seen among unrearranged angiosperms. Genes annotated inside the *circle* are transcribed clockwise, while those outside are transcribed counterclockwise. *Arrows* indicate polycistronic transcription units. Introns are annotated as *open boxes*, and genes containing introns are marked with *asterisks*

Like those of its prokaryotic predecessor, a number of plastid genes are cotranscribed from operons. Polycistronic precursors often contain transcripts encoding proteins involved in similar functions or subunits of higher order complexes, reflecting the need for concerted regulation of expression. Some plastid polycistronic RNAs, however, are multifunctional such as the ubiquitous ribosomal operon containing tRNA and rRNA sequences. Others, like the *rpl23* gene cluster (syn. S10 operon), encode polypeptides belonging to different functional complexes.

3 Gene Content

The gene complement of the vast majority of angiosperm plastomes is highly conserved and the arrangement of sequences collinear (Fig. 2). Genes have been grouped into three general classes: those of the genetic system, photosynthetic components and "others." A comprehensive catalogue of the typical plastome gene content is presented by Bock [9]. Genes that do not fit well into either of the major functional categories include *clpP*, *accD*, *ccsA* and *cemA*.

Encoding a subunit of plastid-localized ATP-dependent caseinolytic protease (Clp) is *clpP*. Apart from this subunit, all other polypeptides that comprise the Clp holoenzyme and its paralogs are imported from the cytoplasm; as many as 20 plastid-targeted Clp polypeptides are predicted for *Arabidopsis* [10, 11].

Among the metabolic activities found in plastids is fatty acid synthesis. The *accD* gene encodes the beta-carboxyl transferase subunit of acetyl-CoA carboxylase (ACCase), which assembles with three nuclear-encoded subunits and is essential for leaf development in *Nicotiana tabacum* (tobacco; [12, 13]). The functional copy of *accD* has been lost from angiosperm plastomes at least seven times [5] (Fig. 3). In grasses, the plastid-encoded product has been substituted by a single-subunit eukaryotic ACCase [14], and a nuclear copy of the prokaryotic *accD*, likely transferred from the plastid, has been identified in *Trifolium* [15].

Required for heme attachment to cytochrome, *ccsA*-encoded cytochrome c biogenesis protein (CcsA) associates with a number of nuclear proteins forming the thylakoid-bound system II cytochrome assembly machinery [16, 17]. Another integral membrane protein, the product of *cemA*, is thought to interact with heme molecules due to homology with characterized heme attachment domains [18, 19]. At present no functional analyses have been conducted in higher plants, but localization studies confirm the product of *cemA* as a polytopic protein of the inner envelope membrane. Inferences have been drawn from analyses in *Chlamydomonas* where the *cemA* product is two times larger than its angiosperm homolog [20]; these studies suggest a role in proton extrusion and promotion of efficient inorganic carbon uptake into plastids

Plastid protein-coding genes are usually named as an abbreviation of the encoded protein's function. Predicted open reading frames (ORFs) receive the designation *ycf* (hypothetical chloroplast reading frames) until some function can be ascribed to their product. Like those described above, most *ycfs* have been renamed as their role in plastid function has been elucidated. The genes for the conserved photosystem I (PSI) assembly proteins *ycf3* and *ycf4* are yet to be renamed, and *ycf15* is no longer considered to be a protein-coding gene [21–23]. Short, non-conserved ORFs are observed in all plastomes but are presumed to be nonfunctional

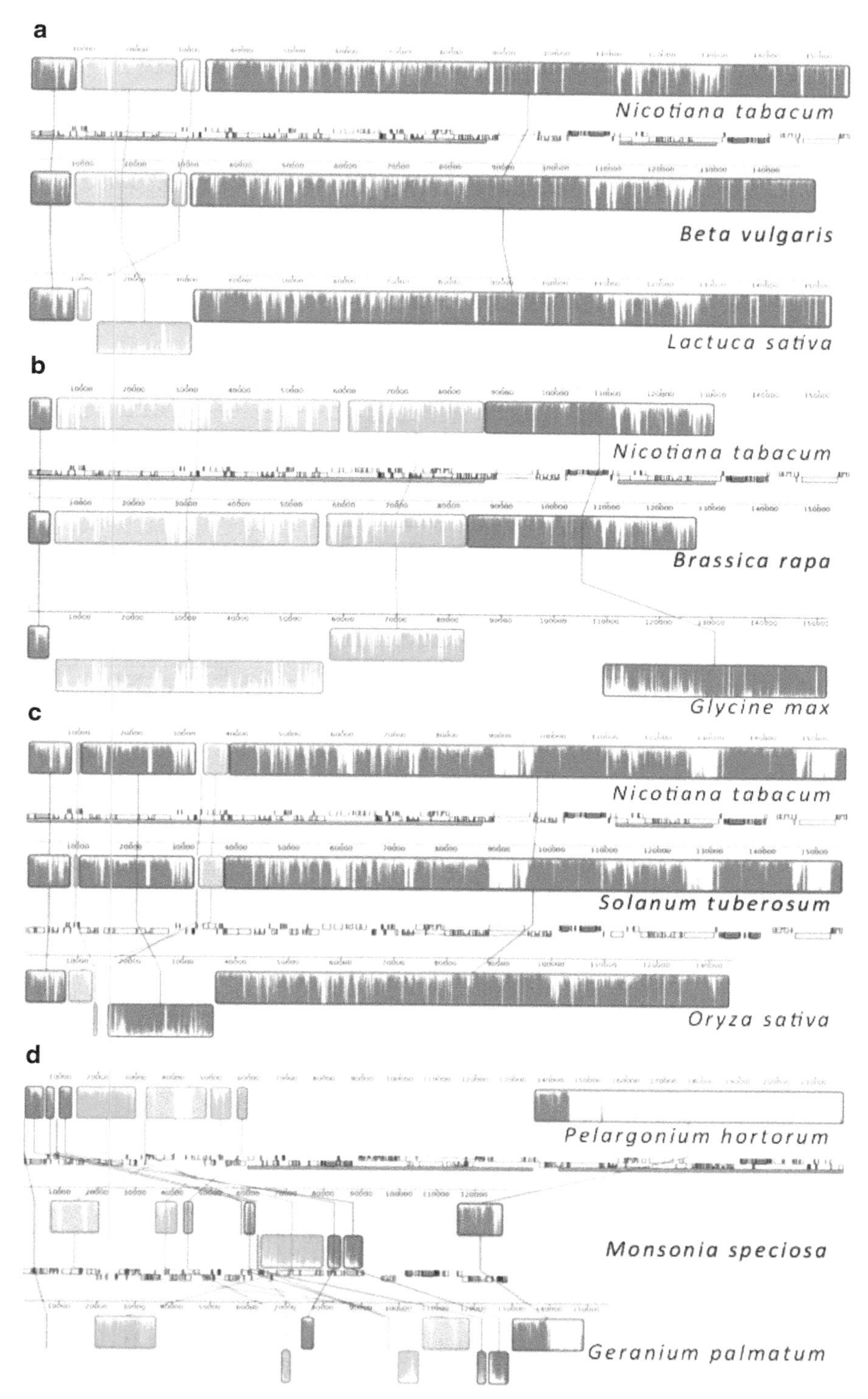

Fig. 2 Gene order comparison of highly conserved and rearranged plastid genomes. Whole-plastid genome sequences were downloaded from Genbank for *N. tabacum* (NC_001879), *B. vulgaris* (EF534108.1), *L. sativa* (NC_007578), *Brassica rapa* (DQ231548.1), *Glycine max* (NC_007942), *Solanum tuberosum* (NC_008096), *O. sativa* (NC_001320), *P. hortorum* (NC_008454), *M. speciosa* (NC_014582), and *G. palmatum* (NC_014573). Alignments were performed in Geneious Pro [222] with the mauveAligner algorithm [223], which aligns syntenic blocks of genes and predicts inversions relative to a reference genome. *N. tabacum* was set as the reference genome in (**a**)–(**c**). In (**d**), *P. hortorum* was set as the reference

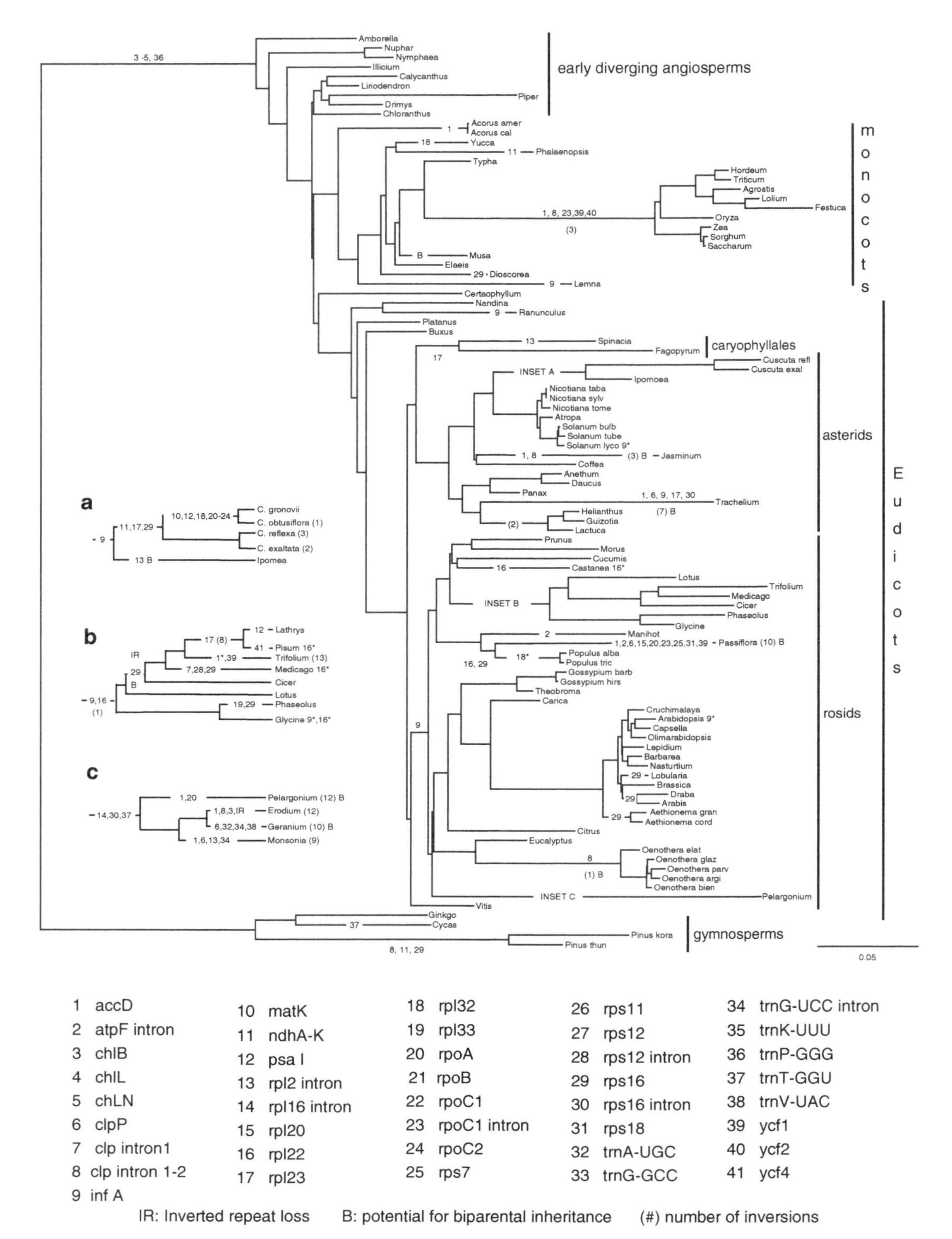

Fig. 3 Angiosperm phylogeny. The large maximum likelihood phylogram was constructed from 97 taxa based on 81 plastid gene sequences and is adapted from Jansen et al. [83]; scale bar indicates the increment of 0.05 substitutions per site. *Inset trees* are cladograms; *branches* do not represent evolutionary distance. Gene and intron losses are represented by *numbers* (see key) plotted on branches and are based on sequence homology or divergence, relative to the ancestral angiosperm plastome, as reported in the literature (see references in main text) or examined by the authors. *Asterisk* indicates reported cases of nuclear transfer, i.e., cDNA of nuclear origin detected or fusion product imported into plastids

due to the lack of conservation across closely related species and biochemical evidence for their expression [9, 24]. Although many *ycfs* have been characterized and renamed, as yet no definitive function has been attributed to the two largest ORFs present in plastomes, *ycf1* and *ycf2*; therefore, they remain grouped with the "others" (*see* **Note 1**). Both *ycf1* and *ycf2* have been designated as essential genes because their targeted disruption results in unstable mutant phenotypes [25]. However *ycf1* appears to be lost in four distinct lineages. In *Passiflora* [5] and *Trifolium* [26] the *ycf1* gene appears to be pseudogenized, and within the Geraniaceae (*Erodium*, *Geranium* and *Monsonia*; [27]) both *ycf1* and *ycf2* are lost as in Poaceae plastomes [28–30].

Greater than one-third of the typical plastome comprises protein-coding genes involved in photosynthesis and related reactions. Apart from *ndhB* all of these sequences are contained within the single-copy regions; an additional ten single-copy genes encode subunits of the NAD(P)H dehydrogenase. There are 22 genes for subunits of photosystems I and II (*psa* and *psb*); 6 genes each encode subunits of the cytochrome b6f complex (*pet*) and the F_0F_1 ATP synthase (*atp*). With the exception of the RuBisCo large subunit (*rbcL*) all of the genes in this class encode subunits of membrane complexes that are assembled together with nuclear-encoded proteins. The gene for the small subunit of RuBisCo (*rbcS*) is likewise found in the nuclear genome, but the holoenzyme assembles in the plastid stroma and does no associate with membranes [31].

The genetic system genes are the largest functional class including 30 tRNAs, 4 rRNAs, 21 proteins that associate with the large or the small ribosomal subunits, 4 subuints of the plastid-encoded RNA polymerase (PEP), intron-encoded maturase K (*matK*; group II intron splicing factor) and translation initiation factor 1 (*infA*).

4 The Large Inverted Repeat

4.1 Genes Are Duplicated in the Inverted Repeat

Most angiosperm plastomes retain a large (~25 kb) inverted repeat (IR), the two copies designated IR_A and IR_B. Genes contained within the IR are duplicated as IR_A and IR_B are thought to be perfectly identical in nucleotide sequence. Usually the genes of the plastid ribosomal operon (*rrn16*, *rrn23*, *rrn4.5* and *rrn5*) and seven tRNA genes are duplicated in the IR. In addition to three complete sequences for ribosomal proteins, the IR also contains *ycf2* and exons two and three of the *trans*-spliced *rps12* gene. The two IR copies are separated by unequally sized single-copy domains, referred to as the large (~80 kb) and small (~20 kb) single-copy regions, LSC and SSC, respectively. This arrangement is seen in *N. tabacum* (Fig. 1) and is thought to be the ancestral form for angiosperms [21]. The widespread occurrence of the IR across not only land plants but also in chlorophyte lineages suggests that this is an ancient feature. The publication of the *Amborella* plastome [22] has confirmed the IR's presence in the most basal angiosperm. Speculation as to the origin of the IR, and its possible role in plastome stability, has fostered decades of research and hypotheses. Kolodner and Tewari [32] detected the presence of the large palindromic sequences in *Lactuca sativa* (lettuce), *Spinacia oleracea* (spinach) and *Zea mays* (corn) using intramolecular homoduplex formation visualized by electron microscopy and reported the absence of this feature in *Pisum sativum* (pea). Subsequent analyses including complete plastome sequences have confirmed the loss of one IR copy in a lineage of legumes [33, 34].

4.2 The IR and Recombination

Long before sequencing of complete plastomes became routine, restriction endonuclease mapping was widely used to characterize and compare the structure of angiosperm plastomes [6]. Intramolecular recombination between the IR copies was suggested as a mechanism to prevent the divergence of the two copies from each other and has the potential to reverse the polarity of the segment between the two copies (i.e., the phenomenon later described by Palmer as flip-flop recombination; [32, 35]). The observed difference in conformation of circular dimers (head to tail in pea, head to head 70–80 % in *L. sativa* and *S. oleracea*) led to the suggestion that the inverted sequences were also involved in intermolecular recombination among the many DNA copies present in plastids.

4.3 Speculation Regarding the Origin of the IR

Based on restriction mapping in *Nicotiana* species, a dimeric circle comprising two complete plastome copies, joined together in a head-to-head (and tail-to-tail) fashion, was proposed by Tassopolo and Kung [36] as a possible origin of the IR. Repeated deletions in an identified "hot spot" mapped to a region downstream of JL_A (junction of LSC and IR_A) extending some distance into the LSC region eventually yielded the IR currently observed. More likely the widely observed variation at IR single-copy junctions results from both expansion and contraction of the repeats [37]. In the case of *Nicotiana acuminata*, the abundant plastome sequences now available suggest that a large insertion, not a deletion, in the "hot spot" region caused plastome expansion in this lineage.

In 2010 [38] phylogenetic analyses of the cyanobacterial ancestor of plastids suggested that the order Chroococcales includes the closest extant relative. This placement was refuted in a recent analysis [39], which suggested a more ancient cyanobacterial origin of plastids. Cyanobacteria, including some of the previously suggested close relatives, have genomes that contain at least duplicated copies of a ribosomal operon (dispersed and/or in reverse orientation). As the rRNAs are the most commonly observed IR genes in plant plastids, a possible origin of the plant IR may lie in a very distant ancestor.

Whatever its origin, the presence of the IR is deemed ancient and certainly predates the branch eventually leading to angiosperms. Across most angiosperms gene content and position of the IR are highly conserved. Usually around 25 kb, variation in the size of the IR, which can range from a little as 7 and up to ~76 kb, often accounts for plastome size variation overall [27]. The most dramatically wide variation is seen in a small but fascinating minority whose study may yield insights into the mechanisms involved in expansion and contraction of the repeats.

4.4 IR Persistence and Loss

Despite its persistence in the vast majority of angiosperm plastomes, the IR does not appear to be essential as one copy has been lost several times in eudicots, at least two times in rosids and possibly twice in asterids based on available data. Within Fabaceae, this molecular synapomorphy defines the monophyletic inverted repeat-lacking clade (IRLC; [40]) that includes *Trifolium*, *Pisum*, *Cicer* and *Medicago*, a designation that is strongly supported by independent phylogenetic analyses of molecular data. Furthermore IR loss within this clade is unambiguously confirmed by assembled and annotated complete plastome sequences [15, 26, 41]. Although not essential, IR loss has had an impact on plastome evolution in the IRLC. Acceleration in the rate of synonymous nucleotide substitution in formerly IR genes and changes in gene order are observed in legume plastomes where the IR is lost [42–45].

Two independent IR losses are reported within the Geraniaceae. Although no supporting data are available, Palmer and Downie [46] inferred IR loss in *Monsonia* (formerly *Sarcocaulon*) and *Erodium* based on hybridization of small probes with sequences surrounding IR single-copy junctions. Recently published complete plastome sequences [27] from three species in the family found the IR missing in *Erodium texanum*, while *Monsonia speciosa* contained both copies of the IR, albeit in a highly reduced form. Thus far all *Erodium* species sequenced show loss of the IR [47]. Plastome sequences have been generated for *M. vanderietieae*, the species reported earlier as lacking the IR. A number of clades within Geraniaceae exhibit unprecedented reorganization of plastome sequences making the assembly of finished genomes a challenging endeavor. Also reconfigurations in gene order can obscure data interpretation when hybridization approaches are employed for IR detection. Among the data from the *M. vanderietieae* draft plastid genome, two large repeats of at least 3 kb have been detected [48] suggesting that the IR may not be completely lost, rather just hiding. Certainly finding differential loss of the IR or for that matter any genomic feature with noteworthy variance at the species level is tantalizing as it suggests a recent evolutionary event. Nonetheless the current data suggest that in this case both of the sequenced *Monsonia* species retain a similarly reduced version of the IR, suggesting that there has only been a single IR loss in Geraniaceae.

Among the asterid nonphotosynthetic parasites there are two reported cases of IR loss. The Orobanchaceae holoparasite *Conopholis americana* was predicted by one group to contain [49] and by another to have lost the IR [46]. Recently assembled plastome sequences confirm the *C. americana* loss; however, a definitive loss cannot as yet be confirmed for *Striga asiatica*, also in the Orobanchaceae. Again assembling complete sequences for species that contain highly reorganized plastomes makes assignment of particular characters challenging. If data analysis ultimately yields resolution in favor of IR loss in *Striga*, this will represent a second, independent loss within the family [50] (*see* **Note 2**). IR loss is not a universal feature among the holoparasites however. *Epifagus virginiana*, for example, has experienced extreme reduction in the single-copy regions due to the loss of the photosynthetic genes while retaining an IR of ~25 kb [51]. The reduced IR (9,767 bp) found in the subterranean orchid *Rhizanthella gardneri* plastome (~60 kb) lacks the rRNA genes. This is the primary reason for the size difference between these two small angiosperm plastomes [52]. Within the same family, in the *Neottia nidus-avis* plastome, the IR is thought to have undergone expansion relative to other orchids contributing to its larger plastome of ~92 kb [53].

4.5 The IR and Plastome Stability

The presence of the IR may play a role in structural stability of the plastome and certainly preserves the homogeneity of the sequences encoded in each copy. Differential rates of nucleotide substitution between genes of the single-copy regions and those of the IR [29, 30, 45, 54, 55], along with the thinking that both copies display identical sequence, suggest that an efficient gene conversion mechanism is at work. Such a mechanism has been invoked to explain copy correction in plastid transformants when point mutations or foreign sequence are introduced into the IR [56–58].

Following a mutation or a transgenic integration event, although replication and partitioning may contribute, homoplasmy is likely to be driven predominantly by recombination processes. As illustrated by Khakhlova and Bock [59], gene conversion proceeds very rapidly and early in regenerating callus and shoots under selection. While biased gene conversion may favor wild-type alleles where mutations are neutral [55], deleterious mutations may be removed or resistance alleles (i.e., transgenic resistance alleles) fixed under selective conditions yielding homoplasmy [59, 60].

The differential rate of nucleotide substitution between the single-copy regions and IR is well established [30, 45, 54, 55]. In plastomes containing an IR, IR genes experience significantly lower rates of substitution with some variability depending on the species and genes examined. Comparison of synonymous substitution rates (d_s) of six formerly IR-located genes in IRLC plastomes with those in IR-containing relatives revealed that d_s is higher in the IRLC genes. In the IR-containing plastomes single-copy region genes display d_s that is 2.3-fold higher than IR genes, while in the IRLC, d_s is uniform across the entire plastome.

The homogenizing effect of gene conversion in the IR is demonstrated both theoretically and experimentally. It is not known if there is a significant difference in the frequency of IR recombination within or between plastome molecules, nor is there much information about intra- or intermolecular recombination between single-copy regions. Small inversions mediated by intramolecular recombination events have played a major role in the evolution of pseudogene diversity and can be useful for increasing phylogenetic resolution at the species level [61–64]. One could speculate that simple proximity facilitates more rapid and frequent recombination between IR copies in the same molecule, but given the various hypotheses regarding the physical form of ptDNA previous notions of what constitutes intra- versus intermolecular may have to be reconsidered. We know that transplastomic plants have been generated targeting insertion to the LSC and that these lines reach homoplasmy [58]. The very few reports that describe gene conversion with regard to single-copy genes in wild-type plants are based on examination of sequence data alone. Maintenance of an *rpl23* pseudogene situated in the LSC by gene conversion with the functional allele present in the IR has been suggested for representatives of the grass family [65]. In this study, limited nonparallel sampling for the functional copies and pseudogenes makes the proposal of maintenance for this pseudogene, by any mechanism, highly speculative. The polyploid state of the plastome would provide ample template for intermolecular gene conversion to restore the wild-type sequence following a mutation and may be one of the ways plastomes have escaped Muller's ratchet [66]. In any case we can be certain that intermolecular recombination occurs in plastids as evidenced by the transplastomic plants that arise through site-specific integration of foreign sequences by homologous recombination with the flanking sequence on the transformation vector.

Presence of an IR does not ensure genome stability. The IR, while nearly omnipresent, undergoes seemingly constant expansion and contraction through evolutionary time [37]. This can be observed not only in numerous examinations of the junction sites [30, 53, 67–70] but also in the plastomes where the IR is massively expanded or drastically reduced [27, 52, 71, 72]. Within the family Oleaceae in the Jasmineae tribe IR length is somewhat expanded relative to most angiosperms (29 kb), and moderate rearrangement is seen. Arrays of repeats, duplicated sequences inserted into coding and noncoding regions, and gene and intron loss are reported [73]. Two overlapping inversions resulted in the relocation of the *ycf4–psaI* region in Jasmineae plastomes, while this feature and other peculiarities are not seen in the tribe that includes cultivated olives (Oleaceae; [74]). The *ycf4–psaI* region was recently described as hypermutable among IRLC members [15].

An extreme locus-specific elevation in nucleotide substitution coupled with length mutations in the same region led the authors to speculate that repeated DNA breakage and, presumably error-prone, repair was involved.

5 Synteny Across Plastomes

Plastid genes tend to be collinear across a broad range of angiosperms, and simple, phylogenetically informative inversions can be useful for the resolution of relationships, especially at deep nodes in the tree of life [75–77]. There are however several cases where numerous overlapping inversions, insertions and deletions (indels), and expansions and contractions of the IR result in such reconfiguration of gene order as to confound prediction of evolutionary relationships.

5.1 Genomic Rearrangements

In the Campanulaceae, Cosner et al. [78] used restriction site/gene mapping to evaluate the distribution of gene order changes among 18 members of this family relative to *N. tabacum*. Across the 18 taxa, so many gene order changes were observed that inference of the individual events within the family was highly problematic. However, despite this complexity, the resulting gene order phylogenies exhibited low levels of homoplasy and were congruent with independent trees generated from DNA sequences for the same taxa.

Nowhere among photosynthetic angiosperm plastomes are genomic rearrangement and substitution rate acceleration more dramatically illustrated than in the family Geraniaceae. As mentioned, genera within the Geraniaceae contain both the largest (*Pelargonium*) and smallest (*Monsonia*) IRs of any reported to date as well as an IR-lacking lineage (*Erodium*). Massive accumulation of dispersed repeats, often associated with changes in gene order, is observed in *Geranium*, *Erodium* and *Pelargonium*. Also highly unusual is the disruption of two conserved operons within the family and the presence of rRNA genes outside the IR in *Monsonia*. The presence of large repeats (>100 bp) in these three genera is unprecedented. Sequence homology indicates that some of the large repeats have arisen from full or partial duplication of genes. These duplicated sequences, along with sequences of protein and tRNA genes commonly present in plastomes, are predicted to be pseudogenes [27]. Members of the long branch clade (LBC) of *Erodium*, unlike sister species *E. texanum*, are further lacking functional genes encoding all subunits of NAD(P)H [79].

Estimating the minimum number of events required to arrive at the observed configuration, Chumley and colleagues [72] proposed 8 IR boundary changes, 12 inversions and several insertions of duplicated sequence for the *Pelargonium hortorum* plastome. Indels can arise readily in the face of IR expansion and contraction. As a change occurs at an IR boundary, that change will be reflected as an indel in the other IR copy through copy correction by gene conversion. However attractive it may be to attempt, reconstructing evolutionary history based on the number of rearrangement events required to convert one genome to another would be risky when considering plastomes like those in the Geraniaceae. In this case it may be impossible to accurately estimate the number of genomic changes or the order in which they occurred. Perhaps a more reliable approach to reconstructing the evolutionary history of genomic changes would be to compare genomes of very closely related species or even different populations of the same species in the lineages that exhibit such a high degree of rearrangement (*see* **Note 3**).

As it turns out when species that are thought to be closely related display substitution rate acceleration and structural changes in plastomes accurate phylogenetic tree topologies remain obscure and may suggest that such lineages have undergone recent and rapid divergence. Although not as extreme as seen in Geraniaceae, genomic rearrangement and elevated nonsynonymous substitution rates have been documented in *Silene* plastomes [70]. Four species were recently examined and while two, *S. latifolia* and *S. vulgaris*, appeared to retain ancestral plastome characteristics, two others, *S. noctiflora* and *S. conica*, were estimated to have experienced four and one repeat-mediated inversions, respectively. In all four species *infA* and *rpl23* are pseudogenized and in *S. noctiflora* and *S. conica matK*, *rpoA* and *accD* have diverged and may also represent pseudogenes. These two species have lost the introns in *rpoC1* and both of the *clpP* introns, while *S. noctiflora* has lost the *rpl16* intron and *S. conica* has lost the intron of *atpF*. All of the intron losses noted in *Silene* plastomes have been documented in other angiosperm lineages as well (Fig. 3).

A positive correlation has been noted between nucleotide substitutions and genomic rearrangements ([5, 27, 70]; Fig. 3). At first glance this seems fairly intuitive. For example, in plastomes such as those described, a fully or a partially duplicated gene sequence is inserted at some alternate locus (a genomic rearrangement). It may be separated physically from the sequences that regulate its expression; it may be truncated; for whatever reason it is nonfunctional. While the functional copy is maintained, over time the duplicated sequence experiences an accelerated rate of nucleotide substitution because there is reduced selective constraint on its divergence. This source of substitution rate acceleration is not usually examined however. Comparisons are made based on intact genes and duplicated sequences are not included in such analyses suggesting other explanations for the observed correlation. Aberrant DNA repair pathways, such as those involved in double-strand break (DSB) repair and recombination-mediated replication are implicated in genomic rearrangement and could further contribute to divergence in nucleotide sequences.

6 Plastome Gene Loss

6.1 Defining Gene Losses

The *functional* loss of plastome-encoded sequences occurs by various means. Authors have often described this situation with little regard to the mechanics that yield a loss of function. There are numerous examples where plastome gene losses are reported in angiosperms, and in only a few cases, functional replacement by nuclear homolog is suggested or demonstrated (Fig. 3) [15, 80–83]. Gene loss may include pseudogenization due to nonsynonymous substitution resulting in amino acid changes that impair the functional capacity of a protein, introduction of stop codon or elimination of start or stop codons. Descriptions of gene loss based on substitution by sequence analysis typically do not consider changes in regulatory regions such as promoters and 5′ or 3′ untranslated regions (UTRs), or intron sequences, that could potentially disrupt expression of the gene product yielding a functional loss. Apart from pseudogenization of coding regions by nucleotide substitution there is gene loss due to indels. Insertions into coding and regulatory regions or deletions that compromise regulatory function, introduce frameshift mutations or, in the case of deletions, remove entire coding regions constitute another class of gene loss. Among the Campanulaceae, the Geraniaceae and in *Trifolium*, genome rearrangements have resulted in disruption of operons that are highly conserved in other angiosperms [26, 27, 72, 84]. By their definition the genes of an operon are cotranscribed from a promoter situated upstream of the most distal gene at the 5′ end. While the coding regions of the "stranded" genes appear to be functional by sequence inspection, whether and how these genes are transcribed and regulated remain a mystery. Interestingly, a recent analysis of transcriptional start sites using differential RNA sequencing has identified numerous examples where plastid genes exhibited independence from their canonical operon promoters [85]. In any case, assignment of functional status to plastid-encoded, or plastid-targeted, genes must be made with a caveat; conclusions reached through sequence inspection should be supported by biochemical evidence.

6.2 Old and New Gene Loss

Notwithstanding the preceding discussion, recent gene losses reinforce the notion that the plastome is a dynamic molecule, still following its evolutionary trajectory while adhering to the constraints of its environment. Recent events include the loss of *ndh* genes in *Erodium*, more than a dozen independent ribosomal protein gene losses and, in one clade of the parasitic genus *Cuscuta*, the loss of *psaI*, *matK* and the entire gene suite for subunits of the plastid-encoded RNA polymerase. In addition to protein-coding genes, recent losses are also reported for tRNA genes. The loss of the *ndh* genes in Orchidaceae [86, 87]; *accD*, *ycf1* and *ycf2* in Poaceae; and *infA* in all but one rosid represent older events but still occurred long after the origin of angiosperms (Fig. 3). The early and rapid transfer of genes to the nucleus following endosymbiosis [1] reduced the coding capacity of the plastome to less than 5 % of the ancestral genome by the time the angiosperm lineage originated. The analysis of nuclear genomes from evolutionarily distant *Arabidopsis* and *Oryza* reveals that large tracts of ptDNA have been integrated into these genomes. One insert in *Oryza sativa* (rice) represents a nearly complete copy of the plastome possessing very high identity (>99 %) to the sequences of the plastid genome [88–90]. With the raw material present in the nucleus, is it possible for any transferred gene sequence to eventually gain the appropriate features and functionally to replace its plastid counterpart? Lloyd and Timmis [91] have recently observed three distinct mechanisms by which a gene of plastid origin may become activated as a nuclear integrant and suggest that such phenomena are more common than previously thought. Nonetheless, the early processes involved in acquisition of sequence elements conferring transcriptional activity and plastid localization are likely to be stochastic and therefore rare events.

6.3 Gene Retention

The persistence of organellar genomes has inspired a number of theoretical hypotheses (reviewed in refs. 92–94). Recent and/or repeated functional transfer events, that is, replacement of a plastome-encoded gene product with an imported, nuclear-encoded gene product, seem to favor ribosomal proteins or proteins whose activity is not intimately linked to photosynthesis. According to the *co*-location for *r*edox *r*egulation (CORR) hypothesis proposed by Allen [95] the nonrandom sample of genes that are ubiquitously retained in photosynthetic plastomes link supply and demand at the level of regulation of gene expression within plastids. More precisely, this core set of genes provides plastids with a mechanism to respond rapidly and directly to the local redox environment. The CORR hypothesis does not address the need for import of many more nuclear-encoded gene products to execute transcription, translation, photosynthesis and related reactions. The absence of *rpoA* from the moss plastomes where the α-subunit of PEP is imported from the cytoplasm [96] weakens the so-called autonomy of plastids in terms of gene expression, a central feature of the CORR theory. More recently Wright et al. [97] have proposed that organellar genomes could themselves serve as long-term redox damage sensors by providing a tonic, or constitutive, retrograde signal. Sensing oxidative damage to ptDNA molecules as a means to activate control mechanisms could, theoretically, be perpetuated through any subset of genes as damage and repair pathways should not preferentially favor the specific subset retained in contemporary plastomes. Retention of the *rbcL* gene in plastids defies at least one assumption of hydrophobicity hypotheses; it is not a membrane protein with hydrophobic domains. Transgenic experiments have shown that the large subunit can be imported by plastids from the cytoplasm following nuclear expression [98]. The concept of ordered assembly, outlined by Daley and Whelan [94], offers another plausible motivation for retention of photosynthetic genes in plastomes. The study of multisubunit protein complexes, such as photosystem II, has shown that assembly proceeds in an ordered manner, with the sequential addition of specific scaffolding and functional subunits [99]. Again the problem that arises is that plastomes do not encode the entire complement of genes required for photosystem assembly [100]. In this case perhaps retention of at least some of the subunit genes in the plastome assures plastid-localized assembly.

It is quite possible that there are a variety of reasons, several of which have been proposed, why some genes as opposed to others presently remain plastid encoded. The reasons may be different for different genes or subsets of genes and the underlying mechanisms combinatorial. Given the small window we have opened into the evolution of plastomes it is likely that we have but scratched the surface and continued study will reveal further intricacies governing the subcellular localization of genes. With the availability of complete nuclear genome sequences come new opportunities to employ bioinformatic approaches to study the fate of plastid genes in the nucleus.

7 Plastid Genes in the Nucleus

7.1 Transcriptional Activation in the Nucleus

Using data from the most current annotation of the 12 *O. sativa* chromosomes and the TIGR Database of Rice Transcript Assemblies, Akbarova et al. [101] identified cDNAs of nuclear origin that include the complete sequences of five plastid genes: *atpI*, *psbJ*, *psbL*, *rpl14* and *rps7*. This finding may be indicative of an intermediate stage in the evolution of the *O. sativa* plastome, which currently contains apparently transcriptionally active copies of these five genes. This situation fits perfectly with the paradigm for plastid genes functionally transferred to the nucleus, as suggested by Timmis et al. [2]. Theory predicts that there would be a period during which both the nuclear- and plastid-encoded products were active in plastids, a period of functional redundancy. During this time selection would act to favor one copy over the other; perhaps one copy would be lost or potentially undergo mutation and adopt another functional role in the cell. If the plastid copy were lost this would constitute a functional nuclear transfer event. Such an intermediate state of functional redundancy is supported by the finding that while plastid-encoded *rps16* continues to be transcribed and spliced in species representing at least four genera, a mitochondrially derived *rps16* gene now located in the nucleus possesses a dual-targeting sequence capable of delivering its product into plastids [102].

Far outside of plants, in the photosynthetic protist *Paulinella*, examination of 3,000 expressed sequence tags revealed that the *psaI* gene has been relocated to the nucleus in a recent transfer event. Here the chromatophore (analogous to the much older plastid) copy of *psaI* is easily detectable but has been interrupted by two stop codons within the former coding region [103], while ten other genes for PSI subunits are encoded in the *Paulinella* plastome. Although unable to locate the nuclear copies, Magee et al. [15] report the loss of the PSI genes *ycf4* and *psaI* from *Pisum* and *Lathyrus*, respectively. Until now the only reported loss of photosynthesis-related genes among angiosperms had been in members of the nonphotosynthetic heterotrophs of Orchidaceae and *Cuscuta* and *E. virginiana*.

7.2 Promiscuous ptDNA and Experimental Artifacts

From inspection of the available angiosperm nuclear genomes it is now clear that both short and long tracts of nuclear ptDNA, or NUPTs, are present and retain high identity to plastome sequences [87]. A recent investigation revealed that intact ORFs of plastid origin could be maintained in the nucleus within NUPTs for several million years [104].

In addition ptDNA insertions are found abundantly in mitochondrial DNA (mtDNA) [105–109]. These transferred sequences present a problem that must be recognized and managed when analyzing amplified ptDNA for phylogenetic reconstructions. Duplications such as those uncovered in *O. sativa*, where 53 plastid genes were present in multiple copies throughout the 12 chromosomes [101], could result in PCR amplification of nontarget sequences. This phenomenon has been evaluated with regard to spurious amplification of mtDNA from nuclear inserts (NUMTs) and a recent comment by Arthofer et al. [90] offers suggestions about how artifactual phylogenetic inferences may be avoided when ptDNA sequences are considered. Careful consideration of the potential to amplify spurious sequences should inform primer design, and novel approaches, such as the application of methylation-sensitive PCR to distinguish nuclear sequences from organelle as described by Kumar and Bendich [110], may alleviate some of these issues.

So-called promiscuous DNAs can confound molecular analysis in transplastomic experiments as well. The plastome, which persists in all plastid differentiation types, may be present in up to 100 identical copies per plastid in leaf cells [111]. Mesophyll cells of a mature leaf may carry up to 100 chloroplasts with the result that the plastome can comprise a significant portion (up to ~20 %; [112, 113]) of cellular DNA content. For the establishment of stable transplastomic lines homoplasmy, the iteration of the transgene throughout all copies of the plastome, must be achieved. In the case of deletion studies, it is desirable to demonstrate homoplasmy of the transplastome as evidence that the mutation involved a nonessential gene [12, 25, 114, 115]. The mixed genotype is unstable and this condition, called heteroplasmy, does not persist in the absence of selection, resolving into one state (wild type) or the other (transformed) over time [116]. Culturing on nonlethal selective media can facilitate the transformation of all copies. Plastids carrying a resistance marker, and in turn the cells that harbor these plastids, are preferentially maintained as plastome molecules are divided up between daughter plastids and subsequently as plastids are partitioned between daughter cells at mitosis [117, 118]. The same applies to the maintenance of a wild-type allele when the experimental design includes disruption of an essential gene.

The homoplasmic state of putative transformants is typically evaluated by Southern analysis of restriction fragments that differ between the wild type and transplastome, PCR amplification across the insertion/deletion site or by germination of seed from self-fertilized transplastomic plants. Given the high identity and length of NUPTs, hybridization experiments, and PCR in particular due to its sensitivity, could lead to an incorrect assignment of heteroplasmy as the NUPTs will more likely produce a result corresponding to the wild-type plastome [119, 120]. Analysis of transplastomic *N. tabacum* carrying a mutation in the *psbZ* gene found persistent heteroplasmy even when cultured in sucrose media, thereby alleviating the demand for photoautotrophy. This led investigators to propose an essential, nonphotosynthetic function [121, 122]. By analyzing ptDNA purified by pulsed-field gel electrophoresis, Swiatek and co-workers [123] subsequently demonstrated homoplasmy of the transplastome for mutant line for *psbZ* and *cemA* but not the essential *ycf1* mutant. In this study total genomic samples and ptDNA prepared by gradient centrifugation were shown to contain promiscuous ptDNA sufficient to produce misleading PCR results. Furthermore quantitative evaluation of template abundance responsible for promiscuous *cemA* signals suggested that multiple copies may be present in the nucleus and/or mitochondria [123].

8 Recombination and the Plastome

The generation of novel plastome genotypes by transformation initially relies on integration of foreign sequence by intermolecular homologous recombination (HR; [57, 124]). Mechanistically similar to gene conversion, HR and repair pathways participate in the subsequent events that yield homoplasmic transplastomic cells and eventually stable transplastomic plants. Intra- or intermolecular recombination between repeated sequences, both in wild-type plastomes [12, 27, 63, 84, 125, 126] and as a result of transplastomic experiments [114, 127, 128], can generate inversions when repeats are palindromic or deletions when direct. Furthermore the role of HR proteins in damage repair may be compromised when foreign DNA is introduced, and through associated tissue culture and selective pressure, as these manipulations can place additional stress on recombination machinery leading to unintended events [129, 130]. Numerous recombination products have been observed during the development of transplastomic lines [128, 131] and likely vastly more remain either unexplored or unreported as such recombination products are undesirable to biotechnologists.

8.1 DNA Replication, Repair and Recombination Activities

Among the DNA repair and recombination genes identified in the nuclear genomes of *Oryza* and *Arabidopsis* TargetP [132] prediction suggests that 19 and 17 %, respectively, are targeted to plastids [133]. As yet only a small number of such activities have been explored.

8.1.1 RecA Protein

Early work in *P. sativum* demonstrated DNA strand transfer activity for plastid-localized RecA [134, 135]. Further study implicated RecA in recombination-mediated repair of damaged ptDNA [136]. Using *Arabidopsis* mutant lines generated by T-DNA insertion Rowen et al. [137] now show that reduced RecA1 (AT1G79050) activity leads to a destabilization and reduction in ptDNA. The reduction in plastome copy number in mutant lines relative to wild type suggests that RecA1 may participate in recombination-mediated replication.

8.2 DNA Replication, Physical Form and Copy Number

It is likely that ptDNA replicates by a number of mechanisms acting concomitantly and/or sequentially along a developmental continuum. Proposed models include displacement-loop and rolling circle (θ and σ; [138]) replication of circular templates and recombination-dependent mechanisms acting on circular or linear templates [139, 140]. There is strong experimental evidence for multiple replication origins mapping to both the IR and single-copy regions [141–148]. Several different groups using distinct approaches have demonstrated that ptDNA is present in vivo in many forms. These include monomeric and multimeric circles, linear molecules or highly branched forms and condensed, high molecular weight DNA complexes [138, 140, 149–152]. Given the range of potential templates available to the replication machinery it is possible that different mechanisms predominate in replication of each of the different forms of ptDNA.

Replication is thought to be most active in plastids of meristematic tissues and leaf primordia, when cells and their organelles are growing and dividing most rapidly. At these early developmental stages, branched linear molecules were reported to comprise the most abundant topological form of ptDNA in some species, with reducing complexity as leaves develop [140, 152]. Plastome copy number reaches a maximum in young leaves and appears to be reduced approximately threefold in terms of genomes per plastid in mature leaves well before senescence [140, 152, 153]. Given the abundance of plastids in the cells of mature leaves, this still constitutes a considerable amount of DNA on a per cell basis. Estimation of ptDNA copy number in plastids of maturing leaves of *Medicago*, *Pisum*, *Zea* and *Arabidopsis* [152, 154–156] suggested that a large proportion of mature leaf plastids were completely devoid of ptDNA, having lost earlier populations through degradation. In their examination of isolated plastids of three maize cultivars, Zheng et al. [156] report that during greening of dark-grown seedlings, 6 h of exposure to light induced ptDNA instability resulting in the loss of half of the ptDNA that was present before transfer to the light. By 24 h of light exposure ptDNA abundance was equal to that of the mature plastids of light-grown plants, which were found to contain tenfold fewer plastome copies than etioplasts. The authors propose that the high plastome copy numbers seen prior to the initiation of photosynthesis are needed to facilitate accumulation of plastid-encoded proteins. Once photosynthesis is initiated, this demand is ablated and copy number declines. While this group observed disagreement in copy number estimates between DAPI staining and quantitative PCR, they suggest that the detection of NUPTs, both in PCR-based and hybridization approaches, artificially inflates plastome copy number estimates.

At least three studies challenge the notion that plastome abundance declines as cells mature in *Arabidopsis* and report an absence of significant variation in plastome abundance during development [157–159]. Further analysis in *Beta vulgaris* (beet) supports the majority of *Arabidopsis* findings and demonstrates the constancy of plastome copy number as leaves mature [113]. Methodological differences may indeed be responsible for the inconsistencies; regardless, three independent labs using different methods report agreement of their results for *Arabidopsis*.

8.3 Maintaining Plastome Integrity

Recombination-mediated replication is likely to play an integral role in the maintenance of plastome integrity. In the potentially genotoxic environment of the plastid, ptDNA damage is bound to arise. Collapsed replication forks, DNA lesions, DSB and slipped-strand mispairing threaten plastome stability and may set up scenarios under which error-prone DNA repair can occur. Aberrant replication and repair mechanisms can ultimately lead to reconfiguration of plastomes. Very little is known about the nuclear-encoded proteins that function in plastids to ameliorate these phenomena, but a few recent studies are shedding some light on the mechanisms involved in securing ptDNA.

8.3.1 Plastid-Targeted Whirlies

Like RecA1, the plastid-targeted Whirly proteins are characterized as single-stranded DNA-binding proteins and are proposed to contribute to plastome stability by prohibiting illegitimate recombination events [160]. *Arabidopsis* plants in which one or both known plastid-targeted Whirlies are knocked out have been used to investigate the role and mechanism of these proteins in plastome maintenance. Outward- and inward-facing PCR amplification and sequencing identified 40 reorganized ptDNA molecules in the double-knockout lines including deletions and circularized and/or head-to-tail concatemers delimited by direct repeats of 10–18 bp. In the lines where only one of the two Whirlies was disrupted rearrangements were less frequent. Surprisingly, even analysis of wild-type plants revealed two rearrangement events. Extension of this approach to maize Whirly mutants confirmed that these proteins are similarly involved in plastome maintenance in this system. Analysis of the short direct repeats associated with illegitimate recombination events in *Arabidopsis* mutants identified an overrepresentation of A/T homopolymeric regions, known to cause slipped-strand-mediated replication stalling and DSB [161]. In some lines repeats bordering the rearranged regions contained some mismatches suggesting that recombination events did not require perfect homology [160]. Continued study of Whirlies has revealed that a specific motif conserved across this protein family is integral to the assembly of higher order homomeric structures. While previous experiments had demonstrated that the phenotype of the *Arabidopsis why1 why3* mutant lines could be complemented by transformation with the wild-type *WHY1*, transformation with a *why1* construct carrying the mutation for a specific lysine residue shown to abolish higher order assembly in *why2* lines failed to recover Whirly-1 activity [162].

8.3.2 Double-Strand Break Repair in Plastids

The frequency and resolution of DSBs in plastomes are likely influenced by several protein factors including RecA, Whirlies, DNA polymerase Ib [163] and the homolog of bacterial *MutS*, MSH1 [164]. To study how plastids of wild-type *Arabidopsis* cope with a DSB scenario, Kwon et al. [165] created an inducible system in which the I-CreII endonuclease of *Chlamydomonas reinhardtii* is targeted to plastid where it cleaves the endogenous *psbA* gene. Amplicon sequencing of nested PCR products from across the I-CreII cleavage site showed deletions mapping to either side of the cleavage site. Evaluation of the junction sequences indicated that DSBs had occurred and were predominantly repaired by microhomology-mediated end joining (MMEJ), but in a few cases data was suggestive of nonhomologous end joining (NHEJ). Perfectly homologous repeats of 6–12 bp, or imperfect 10–16 bp repeats, were sufficient to direct DSB repair by MMEJ. No rearrangements were detected in this study, likely due to the presence of factors (i.e., Whirlies) that minimize activation of a microhomology-mediated break-induced replication (MMBIR) repair pathway. Wild-type *Arabidopsis* plants were resistant to ptDNA rearrangements compared to the Whirly double-knockout plants, which accrue a heterogeneous pool of ptDNA molecules when treated with DNA-damaging agents. Because MMEJ can only account for deletions following a repair event [166] the investigators proposed that an MMBIR response yielded the varied and complex ptDNA molecules observed [160, 167]. This response, which relies on template switching after replication fork collapse, could utilize microhomologous regions within the same molecule or on two different molecules producing the diversity of rearranged products observed in these studies [168].

Repair pathways mediated by microhomologous regions in the plastome represent an error-prone mechanism to overcome severe DNA damage. NHEJ can resolve a DSB with high fidelity or alter DNA, with the result being tightly linked to the nature of the break and the activity of attendant protein factors [169, 170]. A major DNA repair pathway in the eukaryotic nucleus and also active in bacteria and plant mitochondria, NHEJ was not thought to occur in angiosperm plastids. The absence of promiscuous DNA from plastomes supports the assumption that NHEJ does not occur in this compartment. To our knowledge there is but one reported case of identifiable extraplastidal DNA: the *Daucus carota* (carrot) plastome apparently carries a fragment with high homology to a mitochondrial gene (cox1; [107]). Extensive BLAST searching with novel ptDNA sequences from *Trifolium* failed to produce a single hit to indicate a horizontal gene transfer event into plastids [26]. Investigating transposition of the IS 150 element in transplastomic *N. tabacum*, Kohl and Bock [171] were unable to detect religated, IS 150 lacking plastomes, again suggesting the absence of NHEJ in plastids. Observing DSBs at high frequency at both ends of IS element, these authors proposed two possible pathways to explain the lack of detection of religated molecules: the damaged plastomes were degraded or repaired via homologous recombination with IS 150-containing plastomes. More recently two independent investigations have reported NHEJ products in *Arabidopsis*. In one case repair of DSBs by NHEJ following I-CreII activity was detected at low frequency [165], and in another NHEJ repair events represented 17 % of the rearranged products in Whirly knockout lines [167]. The finding that NHEJ, although likely a quantitatively minor pathway, can occur in plastids makes elucidating the nature of linear ptDNA ends all the more compelling.

9 Plastome Evolution: Phylogenetic and Biotechnological Considerations

The processes of DNA replication, recombination and repair in plastids [130, 172] have shaped the evolution of a conservative genome in most angiosperms. Given the highly orchestrated nature of these activities, perturbances to one aspect of the system can be ramified in unpredictable ways. Large- and small-scale rearrangements, indels and pseudogenization have diversified plastomes and consideration of such diversity is imperative to experimental design for phylogenetic reconstruction as well as for biotechnological applications. Although gene order and protein-coding sequences are well conserved across most angiosperms, pertinent differences are observed across families and genera. More subtle differences can be detected at and below the species level facilitating resolution over short evolutionary time scales [64, 173].

9.1 Targeting Plastome Insertion

To achieve efficient foreign sequence integration by homologous recombination transplastomic approaches have used endogenous plastome sequences to target insertions. Studies demonstrate a positive correlation between the rate of recombination and both the length and degree of sequence homology in prokaryotes [174, 175]. Decreasing identity between transformation vectors and target plastome regions, i.e., using *N. tabacum* flanking sequences to transform different species, yields much lower transformation efficiencies [176–180]. Both intra- and interfamilial variation in target sequences negatively impacted homologous recombination events and concomitantly transformation efficiency.

Between *L. sativa* and *N. tabacum* in which the ribosomal operon region has been targeted for transgene integration, gene order is conserved but large and small deletions within the introns of *trnI* and *trnA* unique to *Nicotiana* restrict the length of nearly identical sequence to less than 600 bp across the integration site. Although the minimum flanking sequence length has not been defined empirically, as little as a 400 bp on either side of the expression cassette appears to be sufficient to obtain transformation at a reasonable frequency [181]. Targeting sequences generally extend from 1 to 1.5 kb on either size of the expression cassette [182, 183].

9.2 Phylogenetic Utility of "Noncoding" Sequences

The level of conservation observed for plastome protein-coding sequences is so great as to limit their phylogenetic utility when sampling few genes or taxa, especially at lower taxonomic levels. Alternatively, concatenated sequences for all protein-coding regions plus the four ribosomal RNAs have been employed to gain resolution among early diverging lineages and several major angiosperm clades ([5] and references therein; [184, 185]). For evaluating recent divergence such as interspecific hybridization or cultivars/haplotypes within species, amplified noncoding sequences have been used. Introns and intergenic regions, including spacers and regulatory sequences, yield some phylogenetically informative sites but also demonstrate the need to recognize the functional nature of regulatory elements when analyzing DNA sequence [61–64, 186, 187]. More recently investigators have explored the use of whole-plastome sequences to evaluate relationships at and below the species level. Studies such as those of wild *Oryza* species [188] and individuals of *Jacobaea vulgaris* [189] have found that most phylogenetically informative sites occurred in intergenic regions.

9.2.1 Small Inversions in Regulatory Elements

Indeed the intergenic regions used for phylogenetic studies include various elements associated with transcription and translation of plastid genes. The observation of small inversions predicted to arise from recombination between palindromes in both 5′ and 3′ UTR sequences is not surprising. The mRNA-binding proteins involved in both activation and degradation of plastid transcripts are likely highly sensitive to cognate structures. The orientation of the loop sequence may be free from directional selection and thus has no evolutionary significance (feature would be freely variable within groups), as long as the overall structure is preserved. Like the IR-mediated interconversion of single-copy regions shown by Palmer [34] to exist in equimolar proportions within an individual, such small inversions could be present as a mixed population of isomers. A mutation that alters the thermodynamic stability of the structure such that one orientation was favored over another could lead to the fixation of that allele over time.

Oppositely, the very same kinds of nucleic acid–protein interactions should deter mutations that disrupt conserved stem and loop structures in UTRs, ribosome-binding sites, conserved nucleotides involved in intron structure and processing or promoter elements. While the plastid gene expression system appears to be quite robust in its ability to utilize heterologous regulatory elements [58, 180, 190] specific stem and loop interactions are observed [191–195] and predicted structures for some 5′ regions, such as the *psbA* 5′ UTR, are distinct between angiosperm families [180]. Similar structures localized in the 3′ UTRs of plastid transcripts, unlike those of Eubacteria, do not act as terminators of transcription but rather participate in polycistron processing, 3′ end maturation and transcript stability [193, 195–197]. The structural and functional constraints placed on sequence evolution in noncoding regions should be widely recognized and incorporated into phylogenetic approaches [61, 198].

9.2.2 Nucleotide Substitution in Noncoding Regions

In addition to rearrangements, the diversity of angiosperm plastomes is reflected in nucleotide substitution rates [30, 86, 199–201]. Rate comparisons may consider synonymous or nonsynonymous substitutions, or a ratio of the two, for protein-coding genes. In regions that do not code for amino acid sequences, base substitutions have, at least for phylogenetic purposes, often been deemed neutral and compared to synonymous coding substitutions. An evaluation of 34 single-copy introns and intergenic regions across a small but taxonomically disparate set of angiosperms found that nucleotide substitutions accounted for approximately 72 % of the phylogenetically informative characters identified [202]. However, where functional sequences are concerned base substitutions may not be neutral. On closer inspection regions of extreme sequence variability tend to localize in areas with the least structural and functional constraint [203, 204], an observation consistent with directional selection on regulatory sequences. Analysis of nucleotide substitution rates for protein-coding genes allows unambiguous assignment in terms of conservative and non-conservative change; nonetheless, this approach typically does not consider protein domain functions, such as participation higher order intramolecular structure, interaction with other polypeptide subunits and enzymatic capacity. Thus some residues or groups of residues may be under localized selection pressures, a point that could be overlooked when evaluating rate variation at the genomic scale, across genes or taxa.

10 Plastome Inheritance

A cursory review of available plastome sequences could miss the diversity that becomes apparent with critical examination. Against the background of apparently restrictive conservation, the unusual, the unexplained, begs investigation. Plastids and their genomes are inherited in a maternal fashion in approximately 80 % of angiosperms [205–208]. It is intriguing that a number of the most bizarre plastomes exhibit biparental inheritance (Fig. 3). Given the highly active recombination system functioning in plastids could there be a relationship between the presence of two plastome genotypes and genomic instability? Extensive recombination demonstrated in somatic hybrids and male sterile lines of *Nicotiana* [209–211] and through the incorporation of foreign sequences using plastid transformation strategies leaves little room for the notion that different plastome genotypes cannot recombine. Further evidence for ptDNA recombination has been documented in interspecific somatic hybrids of *Solanum* [212] and *Brassica* [213] and in wild populations of lodgepole pine [214]. Likely the greatest impediment to recombination is a physical one; the presence of the plastid envelopes presents a formidable barrier. The mechanical forces applied during plastid transformation overcome this barrier, but in the absence of such interventions how would different genotypes achieve the necessary proximity? Plastid fusion has been reported in *Hosta* mutants [215] suggesting the possibility for the exchange of genetic material. Also, the presence of stroma-filled tubules, or stromules, connecting two or more plastids constitutes a possible pathway for exchange [216–219].

Biparental inheritance is thought to be a derived trait in angiosperms and has arisen independently in several lineages [205, 208, 220, 221]. An intriguing hypothesis that addresses the possible correlation between plastome instability and emergence of biparental inheritance of ptDNA has been proposed. Zhang and Sodmergen [221] suggest that, rather than contributing to disruptive change, the presence of paternal plastids and their plastomes provides resolution to plastid-genome incompatibility resulting from deleterious mutations in the maternal plastids. The potential for biparental inheritance (PBPI) is identified using cytological evidence, the detection of plastid DNA signals in male gametic cells [205, 207]. While species known to exhibit biparental inheritance are always positive for PBPI, genetic inheritance is rarely demonstrated in the majority of PBPI species. This suggests that although the potential exists for paternal plastids and their DNA to be transmitted the occurrence is low relative to maternal transmission and would be easy to miss were there no defect to favor retention of the male parent plastid and its haplotype. The "paternal rescue" hypothesis suggests that PBPI can overcome defects that may render maternal plastids incompatible with the nuclear genome through stable transmission of paternal plastids to progeny but does not address a mechanism. Are maternal, defective plastids excluded in toto, or can we imagine a scenario where two plastid haplotypes, maternal and paternal, recombine (i.e., undergo gene conversion) to restore a wild-type phenotype?

11 Conclusion

Like any endeavor involving empirical investigation, studies of plastome evolution, however enlightening, leave unresolved questions and newly envisioned directions for research. The dizzying advance of DNA sequencing technology promises databases replete with information. Integrating bioinformatic, genomic and biochemical data will be paramount in the design and execution of experiments aimed at elucidating the role of nuclear-encoded proteins in plastome maintenance. Of particular interest will be projects that target DNA replication, recombination and repair pathways in those families where the plastome is, or once was, destabilized. Plastid biotechnology, aided by available genomic resources, should be extended to include important agronomic and model species. Plastome modification as a tool to address basic research questions has yielded previously elusive insights, and we enthusiastically encourage novel applications of this technology to progress our understanding of plastome evolution in form and function.

12 Notes

1. Using protein A-tagged Tic20-I for co-purifiation, a component of the plastid inner membrane translocon (TIC), Tic214, was identified as the product of plastid-encoded *ycf1*.
2. In addition to *C. americana*, IR loss has been confirmed for *Phelipanche ramosa* (61,709 kb). The nearest relative examined in this study, *P. purpurea*, was found to have a small IR,

perhaps accounting for the 1182 bp difference in plastome size between the congeners. These findings suggest at least two independent losses within the Orobanchaceae. If data analysis ultimately yields resolution in favor of IR loss in Striga, this will represent a third, independent loss within the family.

3. In an effort to elucidate the ancestral plastome organization for Geraniaceae, [226] employed increased sampling of complete plastomes to include all genera within Geraniaceae and representatives of three additional Geraniales families. Whereas previous taxon sampling did not allow reconstruction of the evolutionary events leading to contemporary plastomes [72] the increase sampling permitted a strongly supported reconstruction.

Acknowledgments

Support was provided by the National Science Foundation (DEB-0717372 to R.K.J. and IOS-1027259 to R.K.J. and T.A.R.) and the Fred G. Gloeckner Foundation (to T.A.R. and R.K.J.). The authors thank J. Chris Blazier for comments on an earlier version of the manuscript.

References

1. Martin W, Rujan T, Richly E, Hansen A, Cornelsen S, Lins T, Leister D, Stoebe B, Hasegawa M, Penny D (2002) Evolutionary analysis of *Arabidopsis*, cyanobacterial, and chloroplast genomes reveals plastid phylogeny and thousands of cyanobacterial genes in the nucleus. Proc Natl Acad Sci USA 99:12246–12251
2. Timmis JN, Ayliffe MA, Huang CY, Martin W (2004) Endosymbiotic gene transfer: organelle genomes forge eukaryotic chromosomes. Nat Rev Genet 5:123–135
3. Stegemann S, Hartmann S, Ruf S, Bock R (2003) High-frequency gene transfer from the chloroplast genome to the nucleus. Proc Natl Acad Sci USA 100:8828–8833
4. Sheppard AE, Madesis P, Lloyd AH, Day A, Ayliffe MA, Timmis JN (2011) Introducing an RNA editing requirement into a plastid-localised transgene reduces but does not eliminate functional gene transfer to the nucleus. Plant Mol Biol 76:299–309
5. Jansen RK, Cai Z, Raubeson LA, Daniell H, Depamphilis CW, Leebens-Mack J, Müller KF, Guisinger-Bellian M, Haberle RC, Hansen AK, Chumley TW, Lee S, Peery R, McNeal JR, Kuehl JV, Boore JL (2007) Analysis of 81 genes from 64 plastid genomes resolves relationships in angiosperms and identifies genome-scale evolutionary patterns. Proc Natl Acad Sci USA 104: 19369–19374
6. Palmer JD (1991) Plastid chromosomes: structure and evolution. In: Bogorad L, Vasil IK (eds) The molecular biology of plastids cell culture and somatic cell genetics of plants, vol 7a. Springer, Vienna, pp 5–53
7. Wu C, Lai Y, Lin C, Wang Y, Chaw S (2009) Evolution of reduced and compact chloroplast genomes (cpDNAs) in gnetophytes: selection toward a lower-cost strategy. Mol Phylogenet Evol 52:115–124
8. Cai Z, Penaflor C, Kuehl J, Leebens-Mack J, Carlson J, dePamphilis C, Boore J, Jansen R (2006) Complete plastid genome sequences of *Drimys*, *Liriodendron*, and *Piper*: implications for the phylogenetic relationships of magnoliids. BMC Evol Biol 6:77
9. Bock R (2007) Structure, function, and inheritance of plastid genomes. In: Bock R (ed) Cell and molecular biology of plastids, vol 19. Springer, Berlin, pp 29–63
10. Clarke AK, MacDonald TM, Sjogren LLE (2005) The ATP-dependent Clp protease in chloroplasts of higher plants. Physiol Plant 123:406–412

11. Kuroda H, Maliga P (2003) The plastid clpP1 protease gene is essential for plant development. Nature 425:86–89
12. Kode V, Mudd EA, Iamtham S, Day A (2005) The tobacco plastid *accD* gene is essential and is required for leaf development. Plant J 44:237–244
13. Cronan JE, Waldrop GL (2002) Multi-subunit acetyl-CoA carboxylases. Prog Lipid Res 41:407–435
14. Konishi T, Shinohara K, Yamada K, Sasaki Y (1996) Acetyl-CoA carboxylase in higher plants: most plants other than Gramineae have both the prokaryotic and the eukaryotic forms of this enzyme. Plant Cell Physiol 37:117–122
15. Magee AM, Aspinall S, Rice DW, Cusack BP, Semon M, Perry AS, Stefanovic S, Milbourne D, Barth S, Palmer JD, Gray JC, Kavanagh TA, Wolfe KH (2010) Localized hypermutation and associated gene losses in legume chloroplast genomes. Genome Res 20:1700–1710
16. Page MLD, Hamel PP, Gabilly ST, Zegzouti H, Perea JV, Alonso JM, Ecker JR, Theg SM, Christensen SK, Merchant S (2004) A homolog of prokaryotic thiol disulfide transporter CcdA is required for the assembly of the cytochrome b6f complex in *Arabidopsis* chloroplasts. J Biol Chem 279:32474–32482
17. Hamel PP, Dreyfuss BW, Xie Z, Gabilly ST, Merchant S (2003) Essential histidine and tryptophan residues in CcsA, a system II polytopic cytochrome c biogenesis protein. J Biol Chem 278:2593–2603
18. Sasaki Y, Sekiguchi K, Nagano Y, Matsuno R (1993) Chloroplast envelope protein encoded by chloroplast genome. FEBS Lett 316: 93–98
19. Willey DL, Gray JC (1990) An open reading frame encoding a putative haem-binding polypeptide is cotranscribed with the pea chloroplast gene for apocytochrome f. Plant Mol Biol 15:347–356
20. Rolland N, Dorne AJ, Amoroso G, Sültemeyer DF, Joyard J, Rochaix JD (1997) Disruption of the plastid ycf10 open reading frame affects uptake of inorganic carbon in the chloroplast of *Chlamydomonas*. EMBO J 16:6713–6726
21. Raubeson LA, Peery R, Chumley TW, Dziubek C, Fourcade HM, Boore JL, Jansen RK (2007) Comparative chloroplast genomics: analyses including new sequences from the angiosperms *Nuphar advena* and *Ranunculus macranthus*. BMC Genomics 8:174
22. Goremykin VV, Hirsch-Ernst KI, Wölfl S, Hellwig FH (2003) Analysis of the *Amborella trichopoda* chloroplast genome sequence suggests that *Amborella* is not a basal angiosperm. Mol Biol Evol 20:1499–1505
23. Schmitz-Linneweber C, Maier RM, Alcaraz J, Cottet A, Herrmann RG, Mache R (2001) The plastid chromosome of spinach (*Spinacia oleracea*): complete nucleotide sequence and gene organization. Plant Mol Biol 45:307–315
24. Kahlau S, Aspinall S, Gray JC, Bock R (2006) Sequence of the tomato chloroplast DNA and evolutionary comparison of Solanaceous plastid genomes. J Mol Evol 63:194–207
25. Drescher A, Ruf S, Calsa T, Carrer H, Bock R (2000) The two largest chloroplast genome-encoded open reading frames of higher plants are essential genes. Plant J 22:97–104
26. Cai Z, Guisinger M, Kim H, Ruck E, Blazier JC, McMurtry V, Kuehl JV, Boore J, Jansen RK (2008) Extensive reorganization of the plastid genome of *Trifolium subterraneum* (Fabaceae) is associated with numerous repeated sequences and novel DNA insertions. J Mol Evol 67:696–704
27. Guisinger MM, Kuehl JV, Boore JL, Jansen RK (2011) Extreme reconfiguration of plastid genomes in the angiosperm family Geraniaceae: rearrangements, repeats, and codon usage. Mol Biol Evol 28:583–600
28. Hiratsuka J, Shimada H, Whittier R, Ishibashi T, Sakamoto M, Mori M, Kondo C, Honji Y, Sun CR, Meng BY (1989) The complete sequence of the rice (*Oryza sativa*) chloroplast genome: intermolecular recombination between distinct tRNA genes accounts for a major plastid DNA inversion during the evolution of the cereals. Mol Gen Genet 217:185–194
29. Maier RM, Neckermann K, Igloi GL, Kössel H (1995) Complete sequence of the maize chloroplast genome: gene content, hotspots of divergence and fine tuning of genetic information by transcript editing. J Mol Biol 251:614–628
30. Guisinger MM, Chumley TW, Kuehl JV, Boore JL, Jansen RK (2010) Implications of the plastid genome sequence of *Typha* (Typhaceae, Poales) for understanding genome evolution in Poaceae. J Mol Evol 70:149–166
31. Portis AR, Parry MAJ (2007) Discoveries in Rubisco (Ribulose 1,5-bisphosphate carboxylase/oxygenase): a historical perspective. Photosynth Res 94:121–143
32. Kolodner R, Tewari KK (1979) Inverted repeats in chloroplast DNA from higher plants. Proc Natl Acad Sci USA 76:41–45
33. Palmer JD, Thompson WF (1981) Rearrangements in the chloroplast genomes

of mung bean and pea. Proc Natl Acad Sci USA 78:5533–5537

34. Palmer JD (1983) Chloroplast DNA exists in two orientations. Nature 301:92–93
35. Stein DB, Palmer JD, Thompson WF (1986) Structural evolution and flip-flop recombination of chloroplast DNA in the fern genus *Osmunda*. Curr Genet 10:835–841
36. Tassopulu D, Kung SD (1984) *Nicotiana* chloroplast genome. 6. Deletion and hot spot—a proposed origin of the inverted repeats. Theor Appl Genet 67:185–193
37. Goulding SE, Wolfe KH, Olmstead RG, Morden CW (1996) Ebb and flow of the chloroplast inverted repeat. Mol Gen Genet 252:195–206
38. Falcon LI, Magallon S, Castillo A (2010) Dating the cyanobacterial ancestor of the chloroplast. ISME J 4:777–783
39. Criscuolo A, Gribaldo S (2011) Large-scale phylogenomic analyses indicate a deep origin of primary plastids within cyanobacteria. Mol Biol Evol 28:3019–3032
40. Wojciechowski MF, Lavin M, Sanderson MJ (2004) A phylogeny of legumes (Leguminosae) based on analysis of the plastid *matK* gene resolves many well-supported subclades within the family. Am J Bot 91:1846–1862
41. Jansen RK, Wojciechowski MF, Sanniyasi E, Lee S, Daniell H (2008) Complete plastid genome sequence of the chickpea (*Cicer arietinum*) and the phylogenetic distribution of *rps12* and *clpP* intron losses among legumes (Leguminosae). Mol Phylogenet Evol 48: 1204–1217
42. Palmer JD, Thompson WF (1982) Chloroplast DNA rearrangements are more frequent when a large inverted repeat sequence is lost. Cell 29:537–550
43. Palmer JD, Osorio B, Aldrich J, Thompson WF (1987) Chloroplast DNA evolution among legumes: loss of a large inverted repeat occurred prior to other sequence rearrangements. Curr Genet 11:275–286
44. Wolfe KH (1988) The site of deletion of the inverted repeat in pea chloroplast DNA contains duplicated gene fragments. Curr Genet 13:97–99
45. Perry AS, Wolfe KH (2002) Nucleotide substitution rates in legume chloroplast DNA depend on the presence of the inverted repeat. J Mol Evol 55:501–508
46. Downie S, Palmer J (1992) Use of chloroplast DNA rearrangements in reconstructing plant phylogeny. In: Soltis PS, Soltis DE, Doyle JJ (eds) Molecular systematics of plants. Chapman & Hall, New York, pp 14–35
47. Blazier JC Personal communication
48. Guisinger MM Personal communication
49. Wimpee CF, Wrobel RL, Garvin DK (1991) A divergent plastid genome in *Conopholis americana*, an achlorophyllous parasitic plant. Plant Mol Biol 17:161–166
50. DePamphilis CW Unpublished results, Pennsylvania State University
51. Wolfe KH, Morden CW, Palmer JD (1992) Function and evolution of a minimal plastid genome from a nonphotosynthetic parasitic plant. Proc Natl Acad Sci USA 89: 10648–10652
52. Delannoy E, Fujii S, Colas des Francs-Small C, Brundrett M, Small I (2011) Rampant gene loss in the underground orchid *Rhizanthella gardneri* highlights evolutionary constraints on plastid genomes. Mol Biol Evol 28:2077–2086
53. Logacheva MD, Schelkunov MI, Penin AA (2011) Sequencing and analysis of plastid genome in mycoheterotrophic orchid *Neottia nidus-avis*. Genome Biol Evol 3:1296–1303
54. Wolfe KH, Li W, Sharp PM (1987) Rates of nucleotide substitution vary greatly among plant mitochondrial, chloroplast, and nuclear DNAs. Proc Natl Acad Sci USA 84:9054–9058
55. Birky CW Jr, Walsh JB (1992) Biased gene conversion, copy number, and apparent mutation rate differences within chloroplast and bacterial genomes. Genetics 130:677–683
56. Svab Z, Hajdukiewicz P, Maliga P (1990) Stable transformation of plastids in higher plants. Proc Natl Acad Sci USA 87: 8526–8530
57. Maliga P (1993) Towards plastid transformation in flowering plants. Trends Biotechnol 11:101–107
58. Daniell H, Kumar S, Dufourmantel N (2005) Breakthrough in chloroplast genetic engineering of agronomically important crops. Trends Biotechnol 23:238–245
59. Khakhlova O, Bock R (2006) Elimination of deleterious mutations in plastid genomes by gene conversion. Plant J 46:85–94
60. Tungsuchat-Huang T, Sinagawa-García SR, Paredes-López O, Maliga P (2010) Study of plastid genome stability in tobacco reveals that the loss of marker genes is more likely by gene conversion than by recombination between 34-bp *loxP* repeats. Plant Physiol 153:252–259
61. Kelchner SA (2000) The evolution of non-coding chloroplast DNA and its application in

plant systematics. Ann Mo Bot Gard 87: 482–498
62. Kim K, Lee H (2005) Widespread occurrence of small inversions in the chloroplast genomes of land plants. Mol Cells 19:104–113
63. Bain JF, Jansen RK (2006) A chloroplast DNA hairpin structure provides useful phylogenetic data within tribe Senecioneae (Asteraceae). Can J Bot 84:862–868
64. Ansell SW, Schneider H, Pedersen N, Grundmann M, Russell SJ, Vogel JC (2007) Recombination diversifies chloroplast *trnF* pseudogenes in *Arabidopsis lyrata*. J Evol Biol 20:2400–2411
65. Morton BR, Clegg MT (1993) A chloroplast DNA mutational hotspot and gene conversion in a noncoding region near *rbcL* in the grass family (Poaceae). Curr Genet 24:357–365
66. Muller HJ (1964) The relation of recombination to mutational advance. Mutat Res 106:2–9
67. Plunkett GM, Downie SR (2000) Expansion and contraction of the chloroplast inverted repeat in Apiaceae subfamily Apioideae. Syst Bot 25:648–667
68. Wang R, Cheng C, Chang C, Wu C, Su T, Chaw S (2008) Dynamics and evolution of the inverted repeat-large single copy junctions in the chloroplast genomes of monocots. BMC Evol Biol 8:36
69. Davis JI, Soreng RJ (2010) Migration of endpoints of two genes relative to boundaries between regions of the plastid genome in the grass family (Poaceae). Am J Bot 97:874–892
70. Sloan DB, Alverson AJ, Wu M, Palmer JD, Taylor DR (2012) Recent acceleration of plastid sequence and structural evolution coincides with extreme mitochondrial divergence in the angiosperm genus *Silene*. Genome Biol Evol 4:294. doi:10.1093/1093/gbe/evs006
71. Palmer JD, Nugent JM, Herbon LA (1987) Unusual structure of geranium chloroplast DNA: a triple-sized inverted repeat, extensive gene duplications, multiple inversions, and two repeat families. Proc Natl Acad Sci USA 84:769–773
72. Chumley TW, Palmer JD, Mower JP, Fourcade HM, Calie PJ, Boore JL, Jansen RK (2006) The complete chloroplast genome sequence of *Pelargonium × hortorum*: organization and evolution of the largest and most highly rearranged chloroplast genome of land plants. Mol Biol Evol 23:2175–2190
73. Lee H, Jansen RK, Chumley TW, Kim K (2007) Gene relocations within chloroplast genomes of *Jasminum* and *Menodora* (Oleaceae) are due to multiple, overlapping inversions. Mol Biol Evol 24:1161–1180
74. Mariotti R, Cultrera NGM, Díez CM, Baldoni L, Rubini A (2010) Identification of new polymorphic regions and differentiation of cultivated olives (*Olea europaea* L.) through plastome sequence comparison. BMC Plant Biol 10:211
75. Jansen RK, Palmer JD (1987) A chloroplast DNA inversion marks an ancient evolutionary split in the sunflower family (Asteraceae). Proc Natl Acad Sci USA 84:5818–5822
76. Raubeson LA, Jansen RK (1992) Chloroplast DNA evidence on the ancient evolutionary split in vascular land plants. Science 255:1697–1699
77. Stein DB, Conant DS, Ahearn ME, Jordan ET, Kirch SA, Hasebe M, Iwatsuki K, Tan MK, Thomson JA (1992) Structural rearrangements of the chloroplast genome provide an important phylogenetic link in ferns. Proc Natl Acad Sci USA 89:1856–1860
78. Cosner ME, Raubeson LA, Jansen RK (2004) Chloroplast DNA rearrangements in Campanulaceae: phylogenetic utility of highly rearranged genomes. BMC Evol Biol 4:27
79. Blazier C, Guisinger M, Jansen RK (2011) Recent loss of plastid-encoded *ndh* genes within *Erodium* (Geraniaceae). Plant Mol Biol 76:263–272
80. Gantt JS, Baldauf SL, Calie PJ, Weeden NF, Palmer JD (1991) Transfer of *rpl22* to the nucleus greatly preceded its loss from the chloroplast and involved the gain of an intron. EMBO J 10:3073–3078
81. Millen RS, Olmstead RG, Adams KL, Palmer JD, Lao NT, Heggie L, Kavanagh TA, Hibberd JM, Gray JC, Morden CW, Calie PJ, Jermiin LS, Wolfe KH (2001) Many parallel losses of *infA* from chloroplast DNA during angiosperm evolution with multiple independent transfers to the nucleus. Plant Cell 13:645–658
82. Ueda M, Fujimoto M, Arimura S, Murata J, Tsutsumi N, Kadowaki K (2007) Loss of the *rpl32* gene from the chloroplast genome and subsequent acquisition of a preexisting transit peptide within the nuclear gene in *Populus*. Gene 402:51–56
83. Jansen RK, Saski C, Lee S, Hansen AK, Daniell H (2011) Complete plastid genome sequences of three rosids (*Castanea*, *Prunus*, *Theobroma*): evidence for at least two independent transfers of *rpl22* to the nucleus. Mol Biol Evol 28:835–847
84. Haberle RC, Fourcade HM, Boore JL, Jansen RK (2008) Extensive rearrangements in the chloroplast genome of *Trachelium caeruleum* are associated with repeats and tRNA genes. J Mol Evol 66:350–361

85. Petya Zhelyazkova P, Sharma CM, Förstner KU, Liere K, Vogel J, Börner T (2012) The primary transcriptome of barley chloroplasts: numerous noncoding RNAs and the dominating role of the plastid-encoded RNA polymerase. Plant Cell 24:123–136
86. Chang CC, Lin HC, Lin IP, Chow TY, Chen HH, Chen WH, Cheng CH, Lin CY, Liu SM, Chang CC, Chaw SM (2006) The chloroplast genome of *Phalaenopsis aphrodite* (Orchidaceae): comparative analysis of evolutionary rate with that of grasses and its phylogenetic implications. Mol Biol Evol 12:279–291
87. Wu FH, Chan MT, Liao DC, Hsu CT, Lee YW, Daniell H, Duvall MR, Lin CS (2010) Complete chloroplast genome of *Oncidium* Gower Ramsey and evaluation of molecular markers for identification and breeding in Oncidiinae. BMC Plant Biol 10:68
88. Shahmuradov IA, Akbarova YY, Solovyev VV, Aliyev JA (2003) Abundance of plastid DNA insertions in nuclear genomes of rice and *Arabidopsis*. Plant Mol Biol 52:923–934
89. Matsuo M, Ito Y, Yamauchi R, Obokata J (2005) The rice nuclear genome continuously integrates, shuffles, and eliminates the chloroplast genome to cause chloroplast-nuclear DNA flux. Plant Cell 17:665–675
90. Arthofer W, Schuler S, Steiner FM, Schlick-Steiner BC (2010) Chloroplast DNA-based studies in molecular ecology may be compromised by nuclear-encoded plastid sequence. Mol Ecol 19:3853–3856
91. Lloyd AH, Timmis JN (2011) The origin and characterization of new nuclear genes originating from a cytoplasmic organellar genome. Mol Biol Evol 28:2019–2028
92. Allen JF (2003) The function of genomes in bioenergetic organelles. Philos Trans R Soc Lond B Biol Sci 358:19–38
93. Allen JF, Puthiyaveetil S, Strom J, Allen CA (2005) Energy transduction anchors genes in organelles. Bioessays 27:426–435
94. Daley DO, Whelan J (2005) Why genes persist in organelle genomes. Genome Biol 6:110
95. Allen JF (1993) Redox control of gene expression and the function of chloroplast genomes—an hypothesis. Photosynth Res 36:95–102
96. Sugiura C, Kobayashi Y, Aoki S, Sugita C, Sugita M (2003) Complete chloroplast DNA sequence of the moss *Physcomitrella patens*: evidence for the loss and relocation of *rpoA* from the chloroplast to the nucleus. Nucleic Acids Res 31:5324–5331
97. Wright AF, Murphy MP, Turnbull DM (2009) Do organellar genomes function as long-term redox damage sensors? Trends Genet 25:253–261
98. Kanevski I, Maliga P (1994) Relocation of the plastid *rbcL* gene to the nucleus yields functional ribulose-1,5-bisphosphate carboxylase in tobacco chloroplasts. Proc Natl Acad Sci USA 91:1969–1973
99. Zerges W (2002) Does complexity constrain organelle evolution? Trends Plant Sci 7:175–182
100. Dekker JP, Boekema EJ (2005) Supramolecular organization of thylakoid membrane proteins in green plants. Biochim Biophys Acta 1706:12–39
101. Akbarova YY, Solovyev VV, Shahmuradov IA (2010) Possible functional and evolutionary role of plastid DNA inserted in rice genome. Appl Comput Math 9:19–33
102. Ueda M, Nishikawa T, Fujimoto M, Takanashi H, Arimura S, Tsutsumi N, Kadowaki K (2008) Substitution of the gene for chloroplast RPS16 was assisted by generation of a dual targeting signal. Mol Biol Evol 25: 1566–1575
103. Reyes-Prieto A, Yoon HS, Moustafa A, Yang EC, Andersen RA, Boo SM, Nakayama T, Ishida K, Bhattacharya D (2010) Differential gene retention in plastids of common recent origin. Mol Biol Evol 27:1530–1537
104. Rousseau-Gueutin M, Ayliffe MA, Timmis JN (2011) Conservation of plastid sequences in the plant nuclear genome for millions of years facilitates endosymbiotic evolution. Plant Physiol 157:2181–2193
105. Stern DB, Lonsdale DM (1982) Mitochondrial and chloroplast genomes of maize have a 12-kilobase DNA sequence in common. Nature 299:698–702
106. Nakazono M, Hira A (1993) Identification of the entire set of transferred chloroplast DNA sequences in the mitochondrial genome of rice. Mol Gen Genet 236:341–346
107. Goremykin VV, Salamini F, Velasco R, Viola R (2009) Mitochondrial DNA of *Vitis vinifera* and the issue of rampant horizontal gene transfer. Mol Biol Evol 26:99–110
108. Alverson AJ, Wei XX, Rice DW, Stern DB, Barry K, Palmer JD (2010) Insights into the evolution of mitochondrial genome size from complete sequences of *Citrullus lanatus* and *Cucurbita pepo* (Cucurbitaceae). Mol Biol Evol 27:1436–1448
109. Sloan DB, Alverson AJ, Chuckalovcak JP, Wu M, McCauley DE, Palmer JD, Taylor DR (2012) Rapid evolution of enormous, multichromosomal genomes in flowering plant mitochondria with exceptionally high mutation rates. PLoS Biol 10:e1001241. doi:10.1371/journal.pbio.1001241

110. Kumar RA, Bendich AJ (2011) Distinguishing authentic mitochondrial and plastid DNAs from similar DNA sequences in the nucleus using the polymerase chain reaction. Curr Genet 57:287–295
111. Maier RM, Schmitz-Linneweber C (2004) Plastid genomes. In: Daniell H, Chase C (eds) Molecular biology and biotechnology of plant organelles: chloroplasts and mitochondria. Springer, Dordrecht, pp 115–150
112. Bendich AJ (1987) Why do chloroplasts and mitochondria contain so many copies of their genome? Bioessays 6:279–282
113. Rauwolf U, Golczyk H, Greiner S, Herrmann RG (2010) Variable amounts of DNA related to the size of chloroplasts III. Biochemical determinations of DNA amounts per organelle. Mol Genet Genomics 283:35–47
114. Rogalski M, Ruf S, Bock R (2006) Tobacco plastid ribosomal protein S18 is essential for cell survival. Nucleic Acids Res 34: 4537–4545
115. Fleischmann TT, Scharff LB, Alkatib S, Hasdorf S, Schottler MA, Bock R (2011) Nonessential plastid-encoded ribosomal proteins in tobacco: a developmental role for plastid translation and implications for reductive genome evolution. Plant Cell 23: 3137–3155
116. Maliga P, Carrer H, Kanevski I, Staub J, Svab Z (1993) Plastid engineering in land plants: a conservative genome is open to change. Philos Trans R Soc Lond B Biol Sci 342:203–208
117. Maliga P, Sz-Breznovits A, Marton L, Joo F (1975) Non-mendelian streptomycin-resistant tobacco mutant with altered chloroplasts and mitochondria. Nature 255:401–402
118. Moller S (2005) Pastid division in higher plants. In: Moller SG (ed) Plastids. Blackwell, Oxford, pp 126–152
119. McCabe MS, Klaas M, Gonzalez-Rabade N, Poage M, Badillo-Corona JA, Zhou F, Karcher D, Bock R, Gray JC, Dix PJ (2008) Plastid transformation of high-biomass tobacco variety Maryland Mammoth for production of human immunodeficiency virus type 1 (HIV-1) p24 antigen. Plant Biotechnol J 6:914–929
120. Zhou F, Badillo-Corona JA, Karcher D, Gonzalez-Rabade N, Piepenburg K, Borchers AI, Maloney AP, Kavanagh TA, Gray JC, Bock R (2008) High-level expression of human immunodeficiency virus antigens from the tobacco and tomato plastid genomes. Plant Biotechnol J 6:897–913
121. Maenpaa P, Gonzalez EB, Chen L, Khan MS, Gray JC, Aro EM (2000) The *ycf9* (orf62) gene in the plant chloroplast genome encodes a hydrophobic protein of stromal thylakoid membranes. J Exp Bot 51:375–382
122. Baena-Gonzalez E, Gray JC, Tyystjarvi E, Aro EM, Maenpaa P (2001) Abnormal regulation of photosynthetic electron transport in a chloroplast *ycf9* inactivation mutant. J Biol Chem 276:20795–20802
123. Swiatek M, Greiner S, Kemp S, Drescher A, Koop H, Herrmann RG, Maier RM (2003) PCR analysis of pulsed-field gel electrophoresis-purified plastid DNA, a sensitive tool to judge the hetero-/homoplastomic status of plastid transformants. Curr Genet 43:45–53
124. Klaus SMJ, Huang F, Golds TJ, Koop H (2004) Generation of marker-free plastid transformants using a transiently cointegrated selection gene. Nat Biotechnol 22:225–229
125. Kelchner SA, Wendel JF (1996) Hairpins create minute inversions in non-coding regions of chloroplast DNA. Curr Genet 30:259–262
126. Catalano SA, Saidman BO, Vilardi JC (2009) Evolution of small inversions in chloroplast genome: a case study from a recurrent inversion in angiosperms. Cladistics 25:93–104
127. Iamtham S, Day A (2000) Removal of antibiotic resistance genes from transgenic tobacco plastids. Nat Biotechnol 18:1172–1176
128. Gray BN, Ahner BA, Hanson MR (2009) Extensive homologous recombination between introduced and native regulatory plastid DNA elements in transplastomic plants. Transgenic Res 18:559–572
129. Agrawal AF, Hadany L, Otto SP (2005) The evolution of plastic recombination. Genetics 171:803–812
130. Maréchal A, Brisson N (2010) Recombination and the maintenance of plant organelle genome stability. New Phytol 186:299–317
131. Kavanagh TA, Thanh ND, Lao NT, McGrath N, Peter SO, Horváth EM, Dix PJ, Medgyesy P (1999) Homeologous plastid DNA transformation in tobacco is mediated by multiple recombination events. Genetics 152:1111–1122
132. Emanuelsson O, Nielsen H, Brunak S, von Heijne G (2000) Predicting subcellular localization of proteins based on their N-terminal amino acid sequence. J Mol Biol 300:1005–1016
133. Singh S, Roy S, Choudhury S, Sengupta D (2010) DNA repair and recombination in higher plants: insights from comparative genomics of Arabidopsis and rice. BMC Genomics 11:443
134. Cerutti H, Ibrahim HZ, Jagendorf AT (1993) Treatment of pea (*Pisum sativum* L.) protoplasts with DNA-damaging agents induces a 39-kilodalton chloroplast protein immuno-

logically related to *Escherichia coli* RecA. Plant Physiol 102:155–163
135. Cerutti H, Jagendorf AT (1993) DNA strand-transfer activity in pea (*Pisum sativum* L.) chloroplasts. Plant Physiol 102:145–153
136. Cerutti H, Johnson A, Boynton J, Gillham N (1995) Inhibition of chloroplast DNA recombination and repair by dominant negative mutants of *Escherichia coli* RecA. Mol Cell Biol 15:3003–3011
137. Rowan BA, Oldenburg DJ, Bendich AJ (2010) RecA maintains the integrity of chloroplast DNA molecules in *Arabidopsis*. J Exp Bot 61:2575–2588
138. Kolodner RD, Tewari KK (1975) Chloroplast DNA from higher plants replicates by both the Cairns and the rolling circle mechanism. Nature 256:708–711
139. Krishnan NM, Rao BJ (2009) A comparative approach to elucidate chloroplast genome replication. BMC Genomics 10:237
140. Oldenburg DJ, Bendich AJ (2004) Most chloroplast DNA of maize seedlings in linear molecules with defined ends and branched forms. J Mol Biol 335:953–970
141. Meeker R, Nielsen B, Tewari KK (1988) Localization of replication origins in pea chloroplast DNA. Mol Cell Biol 8:1216–1223
142. Lu Z, Kunnimalaiyaan M, Nielsen BL (1996) Characterization of replication origins flanking the 23S rRNA gene in tobacco chloroplast DNA. Plant Mol Biol 32:693–706
143. Takeda Y, Hirokawa H, Nagata T (1992) The replication origin of proplastid DNA in cultured cells of tobacco. Mol Gen Genet 232:191–198
144. Kunnimalaiyaan M, Shi F, Nielsen BL (1997) Analysis of the tobacco chloroplast DNA replication origin (ori B) downstream of the 23 S rRNA gene. J Mol Biol 268:273–283
145. Kunnimalaiyaan M, Nielsen BL (1997) Fine mapping of replication origins (ori A and ori B) in *Nicotiana tabacum* chloroplast DNA. Nucleic Acids Res 25:3681–3686
146. Mühlbauer SK, Lössl A, Tzekova L, Zou Z, Koop H (2002) Functional analysis of plastid DNA replication origins in tobacco by targeted inactivation. Plant J 32:175–184
147. Wang Y, Saitoh Y, Sato T, Hidaka S, Tsutsumi K (2003) Comparison of plastid DNA replication in different cells and tissues of the rice plant. Plant Mol Biol 52:905–913
148. Scharff LB, Koop H (2007) Targeted inactivation of the tobacco plastome origins of replication A and B. Plant J 50:782–794
149. Deng XW, Wing RA, Gruissem W (1989) The chloroplast genome exists in multimeric forms. Proc Natl Acad Sci USA 86:4156–4160
150. Lilly JW, Havey MJ, Jackson SA, Jiang J (2001) Cytogenomic analyses reveal the structural plasticity of the chloroplast genome in higher plants. Plant Cell 13:245–254
151. Scharff LB, Koop H (2006) Linear molecules of tobacco ptDNA end at known replication origins and additional loci. Plant Mol Biol 62:611–621
152. Shaver JM, Oldenburg DJ, Bendich AJ (2006) Changes in chloroplast DNA during development in tobacco, *Medicago truncatula*, pea, and maize. Planta 224:72–82
153. Baumgartner BJ, Rapp JC, Mullet JE (1989) Plastid transcription activity and DNA copy number increase early in barley chloroplast development. Plant Physiol 89:1011–1018
154. Rowan B, Oldenburg D, Bendich A (2004) The demise of chloroplast DNA in *Arabidopsis*. Curr Genet 46:176–181
155. Rowan B, Oldenburg D, Bendich A (2009) A multiple-method approach reveals a declining amount of chloroplast DNA during development in *Arabidopsis*. BMC Plant Biol 9:3
156. Zheng Q, Oldenburg DJ, Bendich AJ (2011) Independent effects of leaf growth and light on the development of the plastid and its DNA content in *Zea* species. J Exp Bot 62:2715–2730
157. Li W, Ruf S, Bock R (2006) Constancy of organellar genome copy numbers during leaf development and senescence in higher plants. Mol Genet Genomics 275:185–192
158. Zoschke R, Liere K, Börner T (2007) From seedling to mature plant: *Arabidopsis* plastidial genome copy number, RNA accumulation and transcription are differentially regulated during leaf development. Plant J 50:710–722
159. Evans IM, Rus AM, Belanger EM, Kimoto M, Brusslan JA (2010) Dismantling of *Arabidopsis thaliana* mesophyll cell chloroplasts during natural leaf senescence. Plant Biol 12:1–12
160. Maréchal A, Parent J, Véronneau-Lafortune F, Joyeux A, Lang BF, Brisson N (2009) Whirly proteins maintain plastid genome stability in *Arabidopsis*. Proc Natl Acad Sci USA 106:14693–14698
161. Mirkin EV, Mirkin SM (2007) Replication fork stalling at natural impediments. Microbiol Mol Biol Rev 71:13–35
162. Cappadocia L, Parent JS, Zampini E, Lepage E, Sygusch J, Brisson N (2012) A conserved lysine residue of plant Whirly proteins is necessary for higher order protein assembly and protection against DNA damage. Nucleic Acids Res 40:258–269

163. Parent JS, Lepage E, Brisson N (2011) Divergent roles for the two PolI-like organelle DNA polymerases of *Arabidopsis*. Plant Physiol 156:254–262
164. Xu YZ, Arrieta-Montiel MP, Virdi KS, de Paula WBM, Wildhalm JR, Basset GJ, Davila JI, Elthon TE, Elowsky CG, Sato SJ, Clemente TE, Mackenzie SA (2011) MutS HOMOLOG1 is a nucleoid protein that alters mitochondrial and plastid properties and plant response to high light. Plant Cell 23:3428–3441
165. Kwon T, Huq E, Herrin DL (2010) Microhomology-mediated and nonhomologous repair of a double-strand break in the chloroplast genome of *Arabidopsis*. Proc Natl Acad Sci USA 107:13954–13959
166. McVey M, Lee S (2008) MMEJ repair of double-strand breaks (director's cut): deleted sequences and alternative endings. Trends Genet 24:529–538
167. Cappadocia L, Marechal A, Parent J, Lepage E, Sygusch J, Brisson N (2010) Crystal structures of DNA-Whirly complexes and their role in *Arabidopsis* organelle genome repair. Plant Cell 22:1849–1867
168. Hastings PJ, Ira G, Lupski JR (2009) A microhomology-mediated break-induced replication model for the origin of human copy number variation. PLoS Genet 5:e1000327
169. Shuman S, Glickman MS (2007) Bacterial DNA repair by non-homologous end joining. Nat Rev Microbiol 5:852–861
170. Lieber MR (2010) The mechanism of double-strand DNA break repair by the nonhomologous DNA end-joining pathway. Annu Rev Biochem 79:181–211
171. Kohl S, Bock R (2009) Transposition of a bacterial insertion sequence in chloroplasts. Plant J 58:423–436
172. Day A, Madesis P (2007) DNA replication, recombination, and repair in plastids. In: Bock R (ed) Cell and molecular biology of plastids, vol 19. Springer, Berlin, pp 65–119
173. Pleines T, Jakob SS, Blattner FR (2008) Application of non-coding DNA regions in intraspecific analyses. Plant Syst Evol 282:281–294
174. Shen P, Huang HV (1986) Homologous recombination in *Escherichia coli*: dependence on substrate length and homology. Genetics 112:441–457
175. Fujitani Y, Yamamoto K, Kobayashi I (1995) Dependence of frequency of homologous recombination on the homology length. Genetics 140:797–809
176. Sidorov VA, Kasten D, Pang SZ, Hajdukiewicz PT, Staub JM, Nehra NS (1999) Technical advance: stable chloroplast transformation in potato: use of green fluorescent protein as a plastid marker. Plant J 19:209–216
177. Ruf S, Hermann M, Berger IJ, Carrer H, Bock R (2001) Stable genetic transformation of tomato plastids and expression of a foreign protein in fruit. Nat Biotechnol 19:870–875
178. Zubko MK, Zubko EI, Zuilen KV, Meyer P, Day A (2004) Stable transformation of petunia plastids. Transgenic Res 13:523–530
179. Nguyen TT, Nugent G, Cardi T, Dix PJ (2005) Generation of homoplasmic plastid transformants of a commercial cultivar of potato (*Solanum tuberosum* L.). Plant Sci 168:1495–1500
180. Ruhlman T, Verma D, Samson N, Daniell H (2010) The role of heterologous chloroplast sequence elements in transgene integration and expression. Plant Physiol 152:2088–2104
181. Bock R (2001) Transgenic plastids in basic research and plant biotechnology. J Mol Biol 312:425–438
182. Ruhlman T, Ahangari R, Devine A, Samsam M, Daniell H (2007) Expression of cholera toxin B–proinsulin fusion protein in lettuce and tobacco chloroplasts—oral administration protects against development of insulitis in non-obese diabetic mice. Plant Biotechnol J 5:495–510
183. Verma D, Samson NP, Koya V, Daniell H (2008) A protocol for expression of foreign genes in chloroplasts. Nat Protoc 3:739–758
184. Moore MJ, Bell CD, Soltis PS, Soltis DE (2007) Using plastid genome-scale data to resolve enigmatic relationships among basal angiosperms. Proc Natl Acad Sci USA 104:19363–19368
185. Moore MJ, Soltis PS, Bell CD, Burleigh JG, Soltis DE (2010) Phylogenetic analysis of 83 plastid genes further resolves the early diversification of eudicots. Proc Natl Acad Sci USA 107:4623–4628
186. Swangpol S, Volkaert H, Sotto RC, Seelanan T (2007) Utility of selected non-coding chloroplast DNA sequences for lineage assessment of *Musa* interspecific hybrids. J Biochem Mol Biol 40:577–587
187. Whitlock BA, Hale AM, Groff PA (2010) Intraspecific inversions pose a challenge for the *trnH-psbA* plant DNA barcode. PLoS One 5:e11533
188. Waters DLE, Nock CJ, Ishikawa R, Rice N, Henry RJ (2011) Chloroplast genome sequence confirms distinctness of Australian and Asian wild rice. Ecol Evol 2:211–217

189. Doorduin L, Gravendeel B, Lammers Y, Ariyurek Y, Chin-A-Woeng T, Vrieling K (2011) The complete chloroplast genome of 17 individuals of pest species *Jacobaea vulgaris*: SNPs, microsatellites and barcoding markers for population and phylogenetic studies. DNA Res 18:93–108
190. Tangphatsornruang S, Birch-Machin I, Newell CA, Gray JC (2011) The effect of different 3′ untranslated regions on the accumulation and stability of transcripts of a *gfp* transgene in chloroplasts of transplastomic tobacco. Plant Mol Biol 76:385–396
191. Shen Y, Danon A, Christopher DA (2001) RNA binding-proteins interact specifically with the *Arabidopsis* chloroplast *psbA* mRNA 5′ untranslated region in a redox-dependent manner. Plant Cell Physiol 42:1071–1078
192. Nickelsen J (2003) Chloroplast RNA-binding proteins. Curr Genet 43:392–399
193. Pfalz J, Bayraktar OA, Prikryl J, Barkan A (2009) Site-specific binding of a PPR protein defines and stabilizes 5′ and 3′ mRNA termini in chloroplasts. EMBO J 28:2042–2052
194. Stern DB, Goldschmidt-Clermont M, Hanson MR (2010) Chloroplast RNA Metabolism. Annu Rev Plant Biol 61:125–155
195. Zhelyazkova P, Hammani K, Rojas M, Voelker R, Vargas-Suarez M, Borner T, Barkan A (2011) Protein-mediated protection as the predominant mechanism for defining processed mRNA termini in land plant chloroplasts. Nucleic Acids Res 40:3092. doi:10.1093/nar/gkr1137
196. Stern DB, Gruissem W (1987) Control of plastid gene expression: 3′ inverted repeats act as mRNA processing and stabilizing elements, but do not terminate transcription. Cell 51:1145–1157
197. Loza-Tavera H, Vargas-Suárez M, Díaz-Mireles E, Torres-Márquez M, González de la Vara L, Moreno-Sánchez R, Gruissem W (2006) Phosphorylation of the spinach chloroplast 24 kDa RNA-binding protein (24RNP) increases its binding to *petD* and *psbA* 3′ untranslated regions. Biochimie 88:1217–1228
198. Borsch T, Quandt D (2009) Mutational dynamics and phylogenetic utility of noncoding chloroplast DNA. Plant Syst Evol 282:169–199
199. Matsuoka Y, Yamazaki Y, Ogihara Y, Tsunewaki K (2002) Whole chloroplast genome comparison of rice, maize, and wheat: implications for chloroplast gene diversification and phylogeny of cereals. Mol Biol Evol 19:2084–2091
200. Guisinger MM, Kuehl JV, Boore JL, Jansen RK (2008) Genome-wide analyses of Geraniaceae plastid DNA reveal unprecedented patterns of increased nucleotide substitutions. Proc Natl Acad Sci USA 105:18424–18429
201. Zhong B, Yonezawa T, Zhong Y, Hasegawa M (2009) Episodic evolution and adaptation of chloroplast genomes in ancestral grasses. PLoS One 4:e5297
202. Shaw J, Lickey EB, Schilling EE, Small RL (2007) Comparison of whole chloroplast genome sequences to choose noncoding regions for phylogenetic studies in angiosperms: the tortoise and the hare III. Am J Bot 94:275–288
203. Borsch T, Hilu KW, Quandt D, Wilde V, Neinhuis C, Barthlott W (2003) Noncoding plastid *trnT-trnF* sequences reveal a well resolved phylogeny of basal angiosperms. J Evol Biol 16:558–576
204. Storchova H, Olson MS (2007) The architecture of the chloroplast psbA-trnH non-coding region in angiosperms. Plant Syst Evol 268:235–256
205. Corriveau JL, Coleman AW (1988) Rapid screening method to detect potential biparental inheritance of plastid DNA and results for over 200 angiosperm species. Am J Bot 75:1443–1458
206. Harris SA, Ingram R (1991) Chloroplast DNA and biosystematics: the effects of intraspecific diversity and plastid transmission. Taxon 40:393–412
207. Zhang Q, Liu Y, Sodmergen (2003) Examination of the cytoplasmic DNA in male reproductive cells to determine the potential for cytoplasmic inheritance in 295 angiosperm species. Plant Cell Physiol 44:941–951
208. Hu Y, Zhang Q, Rao G, Sodmergen (2008) Occurrence of plastids in the sperm cells of Caprifoliaceae: biparental plastid inheritance in angiosperms is unilaterally derived from maternal inheritance. Plant Cell Physiol 49:958–968
209. Medgyesy P, Fejes E, Maliga P (1985) Interspecific chloroplast recombination in a *Nicotiana* somatic hybrid. Proc Natl Acad Sci USA 82:6960–6964
210. Thanh ND, Medgyesy P (1989) Limited chloroplast gene transfer via recombination overcomes plastome-genome incompatibility between *Nicotiana tabacum* and *Solanum tuberosum*. Plant Mol Biol 12:87–93
211. Kung SD, Zhu YS, Chen K, Shen GF, Sisson VA (1981) *Nicotiana* chloroplast genome. Mol Gen Genet 183:20–24

212. Bidani A, Nouri-Ellouz O, Lakhoua L, Sihachakr D, Cheniclet C, Mahjoub A, Drira N, Gargouri-Bouzid R (2007) Interspecific potato somatic hybrids between *Solanum berthaultii* and *Solanum tuberosum* L. showed recombinant plastome and improved tolerance to salinity. Plant Cell Tissue Organ Cult 91:179–189
213. Yadav P, Bhat SR, Prakash S, Mishra LC, Chopra VL (2009) Resynthesized *Brassica juncea* lines with novel organellar genome constitution obtained through protoplast fusion. J Genet 88:109–112
214. Marshall HD, Newton C, Ritland K (2001) Sequence-repeat polymorphisms exhibit the signature of recombination in lodgepole pine chloroplast DNA. Mol Biol Evol 18: 2136–2138
215. Vaughn KC (1981) Plastid fusion as an agent to arrest sorting out. Curr Genet 3:243–245
216. Gray JC, Hibberd JM, Linley PJ, Uijtewaal B (1999) GFP movement between chloroplasts. Nat Biotechnol 17:906–909
217. Kwok EY, Hanson MR (2004) GFP-labelled Rubisco and Aspartate aminotransferase are present in plastid stromules and traffic between plastids. J Exp Bot 55:595–604
218. Hanson MR, Sattarzadeh A (2008) Dynamic morphology of plastids and stromules in angiosperm plants. Plant Cell Environ 31:646–657
219. Gray JC, Hansen MR, Shaw DJ, Graham K, Dale R, Smallman P, Natesan SKA, Newell CA (2012) Plastid stromules are induced by stress treatments acting through abscisic acid. Plant J 69:387–398
220. Birky CW (1995) Uniparental inheritance of mitochondrial and chloroplast genes: mechanisms and evolution. Proc Natl Acad Sci USA 92:11331–11338
221. Zhang Q, Sodmergen (2010) Why does biparental plastid inheritance revive in angiosperms? J Plant Res 123:201–206
222. Drummond AJ, Ashton B, Buxton S, Cheung M, Cooper A, Heled J, Kearse M, Moir R, Stones-Havas S, Sturrock S, Thierer T, Wilson A (2010) Geneious v5.1. Available from http://www.geneious.com
223. Darling AE, Mau B, Perna NT (2010) Progressive mauve: multiple genome alignment with gene gain, loss, and rearrangement. PLoS One 5:e11147
224. Kikuchi S, Bedard J, Hirano M, Hirabayashi Y, Oishi M, Imai M, Takase M, Ide T, Nakai M (2013) Uncovering the protein translocon at the chloroplast inner envelope membrane. Science 339:571–574
225. Wicke S, Müller KF, de Pamphilis CW, Quandt D, Wickett NJ, Zhang Y, Renner SS, Schneeweiss GM (2013) Mechanisms of functional and physical genome reduction in photosynthetic and nonphotosynthetic parasitic plants of the broomrape family. Plant Cell 25:3711–3725
226. Weng M-L, Blazier JC, Govindu M, Jansen RK (2014) Reconstruction of the ancestral plastid fenome in Geraniaceae reveals a correlation between genome rearrangements, repeats and nucleotide substitution rates. Mol Biol Evol. doi:10.1093/molbev/mst257

Chapter 2

Next-Generation Technologies to Determine Plastid Genome Sequences

Robert J. Henry, Nicole Rice, Mark Edwards, and Catherine J. Nock

Abstract

Sequencing of chloroplast genomes is a key tool for analysis of chloroplasts and the impact of manipulation of chloroplast genomes by biotechnology. Advances in genome sequencing allow the complete sequencing of the chloroplast genome and assessment of variation in the chloroplast genome sequences within a plant. Isolation of chloroplast DNA has been a traditional starting point in these analyses, but the capacity of current sequencing technologies allows effective analysis of the chloroplast genome sequence by shotgun sequencing of a preparation of total DNA from the plant. Chloroplast insertions in the nuclear genome can be distinguished by their much lower copy number. Short-read sequences are best assembled by alignment with a reference chloroplast genome.

Key words DNA sequencing, Genome assembly, Next-generation sequencing, NGS, Plastid genome

1 Introduction

Traditional Sanger sequencing has routinely been applied to the analysis of specific chloroplast sequences following PCR amplification. The whole chloroplast sequence provides a more complete analysis but was beyond the scope of these techniques. Next-generation sequencing allows the generation of far greater volumes of sequence data at low cost making routine whole chloroplast genome sequencing feasible for the first time. Sequencing of the whole chloroplast genome has usually involved isolation of chloroplasts and preparation of DNA from purified chloroplasts [1]. Obtaining highly purified chloroplast DNA samples has been difficult, especially for some species. An alternative simplified strategy to obtain a whole chloroplast genome sequence has been to use whole chloroplast genome amplification to prepare samples for sequencing.

An efficient strategy avoiding both chloroplast isolation and chloroplast genome amplification is extraction of chloroplast genome sequences from shotgun whole genome sequences of total plant DNA [2]. This is a simple and cost-effective option

Pal Maliga (ed.), *Chloroplast Biotechnology: Methods and Protocols*, Methods in Molecular Biology, vol. 1132,
DOI 10.1007/978-1-62703-995-6_2, © Springer Science+Business Media New York 2014

producing reliable data on chloroplast sequence and variants within the sample.

The chloroplast genome has been sequenced for a growing number of plant species [3]. As of August 2012, the NCBI Organelle Genome Resource contained 285 records of complete eukaryota plastid genomes. These sequences provide reference genome sequences allowing analysis of variants and modified chloroplast genomes by re-sequencing using efficient next-generation technologies. Longer sequence reads may allow de novo assembly of the chloroplast genome as sequencing technology improves [4].

2 Materials

2.1 Sample Preparation

1. QIAGEN TissueLyser.
2. 96 deep well collection plate.

2.2 DNA Preparation

1. Qiagen DNeasy 96 Plant Kit.

2.3 DNA Quality Analysis/Assurance

1. Agarose gel electrophoresis kit.

2.4 Quantification of dsDNA

1. Quant-iT™ PicoGreen(R) dsDNA Reagent and Kits.
2. Qubit® 2.0 Fluorometer.

2.5 DNA Sequencing

1. Covaris S2 DNA shearing device.
2. Agilent 2100 Bioanalyzer and DNA 1000 chip.
3. Agarose gel electrophoresis kit.
4. QIAquick Gel Extraction kit.
5. QIAquick PCR Purification Kit.
6. Illumina Genome Analyzer (GAIIx).
7. Illumina software Pipeline 1.4.

2.6 Data Analyses

1. Computer with internet access.
2. CLC Genomics Workbench v.3.6.5 software package (http://www.clcbio.com).
3. DNA sequence editing program (e.g., Sequencher, http://www.genecodes.com).

3 Methods

3.1 Sample Preparation

Harvested leaf tissue directly into the sample tubes of a 96 deep well collection plate and grind in a TissueLyser as per the recommendations in the Qiagen DNeasy 96 Plant Kit (http://www.qiagen.com/literature/handbooks/literature.aspx?id=1000061) (*see* **Note 1**).

3.2 DNA Preparation

A Qiagen DNeasy 96 Plant Kit yields very high-quality DNA; however, the yield is very species specific. The method should be tested for the species you are working with first. For whole genome shotgun sequencing, you may need to increase the number of replicate DNA extractions for each sample to achieve the total mass of DNA required. For species in the genus *Oryza* [2], the manufacturer's method was followed (http://www.qiagen.com/literature/handbooks/literature.aspx?id=1000061), and four replicates of each sample were required.

3.3 DNA Quality Analysis/Assurance

Extracted DNA samples should be quantified and the quality checked using an agarose gel checking for excessive smearing or fragmentation. High-quality starting material is important for library generation. The DNA sample should be pure, having an OD 260/280 ratio of between 1.8 and 2. It is important to quantitate the input material carefully, preferably using fluorescence-based quantitation methods (such as Invitrogen PicoGreen or Qubit™ Quantitation Fluorometer assays).

3.4 Quantification of dsDNA Using the Invitrogen PicoGreen Reagent Kit

For quantification, the following protocol has been adapted from the procedure supplied with the Molecular Probes PicoGreen dsDNA quantitation kit. The PicoGreen reagent is used as an ultrasensitive fluorescent stain for quantitating double-stranded DNA (dsDNA) in solution. The protocol described here is suitable for quantitating dsDNA concentrations ranging from 1 ng/μL to 200 ng/μL of DNA sample. For further information on limitations and compounds that may interfere with the assay, please refer to Molecular Probes product information sheets available from the following: http://probes.invitrogen.com/.

1. Preparation of 1× TE buffer. The kit is supplied with 20× TE buffer. To prepare a 1 mL working solution of 1× TE buffer, mix 50 μL 20× TE buffer with 950 μL sterile Milli-Q water. To calculate the total volume of 1× TE required, use the following equation:

$$\text{Total volume } 1\times\text{TE} = 200\ \mu\text{L} \times \left(\text{Sa}_n + \left[\text{Pl}_n \times \left(\text{St}_n + 2\right)\right]\right) + 1{,}500\ \mu\text{L}$$

where St_n = total number of points within the standard curve including the standard blank, Sa_n = total number of samples, and Pl_n = total number of plates to be run consecutively.

Example, for 50 samples, the total volume of 1× TE required = 200 μL 1× TE working reagent × (50 samples + [1 plate × (7 standards + 2 waste allowance)] + 1,500 μL) for making up λ DNA working stock solutions = 13.4 mL.

Table 1
High-range λ dsDNA standard curve

TE (μL)	2 ng/μL λ DNA (μL)	dsDNA[a] (ng)
100	0	Blank
100	0	0
90	10	20
80	20	40
70	30	60
50	50	100
0	100	200

[a]Quantity of dsDNA added to 200 μL assay volume

Table 2
Low-range λ dsDNA standard curve

TE (μL)	200 pg/μL λ DNA (μL)	dsDNA[a] (ng)
100	0	Blank
100	0	0
90	10	2
80	20	4
70	30	6
50	50	10
0	100	20

[a]Quantity of dsDNA added to 200 μL assay volume

2. Preparation of λ DNA working stock solutions. The lambda DNA stock supplied with the kit (component C) contains 100 μg λ dsDNA/mL.

   To make a 2 ng/μL λ DNA working stock for a high-range standard curve (Table 1), combine 10 μL of the stock λ dsDNA with 490 μL 1× TE.

   To make a 200 pg/μL λ DNA working stock for the low-range standard curve (Table 2), carry out a two-step dilution as follows:

   Step 1—prepare a 2 ng/μL λ DNA solution by combining 10 μL of the stock λ dsDNA with 490 μL 1× TE (final concentration 1 ng/μL λ DNA).

   Step 2—prepare a 200 pg/μL λ DNA working stock by combining 100 μL of the 2 ng/μL λ DNA solution with 900 μL 1× TE.

3. Preparation of PicoGreen working solution. The PicoGreen working solution is prepared by making a 200-fold dilution of the concentrated stock (supplied with kit) in 1× TE. To make 1 mL of PicoGreen working solution, mix 5 μL concentrated stock with 995 μL 1× TE buffer (*see* **Note 2**).

   To calculate the total volume of PicoGreen working solution required, use the following equation:

$$\text{Total volume PicoGreen working solution} = 100\ \mu\text{L} \times \left(\text{Sa}_n + \left[\text{Pl}_n \times \left(\text{St}_n + 2\right)\right]\right)$$

   where Sa_n = total number of samples, St_n = total number of points within the standard curve including the standard blank, and Pl_n = total number of plates to be run consecutively.

   Example, for 50 samples, the total volume of PicoGreen working solution = 100 μL PicoGreen working reagent × (50 samples + [1 plate × (7 standards + 2 waste allowance)]) = 5,950 μL.

4. To prepare a dsDNA standard curve, dispense the appropriate volumes (see Tables 1 and 2 above) of 1× TE and λ DNA working stock into each microplate well.
5. To prepare dsDNA samples, add 2 μL DNA sample and 98 μL TE to each well (*see* **Note 3**).
6. Aliquot 100 μL PicoGreen working solution into each microplate well (standards and samples).
7. Incubate plate for 5 min at RT protected from light.
8. Measure the fluorescence in a spectrophotometer, with fluorescence and microtiter plate capability. Filters with the following wavelength parameters should be used: excitation is 485/20 nm and emission is 528/20 nm.
9. To determine the concentration of dsDNA in 1 μL of the sample, divide the result from the assay by the volume of sample added to the assay. For example, if the DNA mass shown in the assay results is 100 ng and 2 μL was used in the assay, 100 ng/2 = 50 ng dsDNA/μL sample.

3.5 DNA Sequencing

Several platforms for high-throughput sequencing can be applied to chloroplast sequencing. These include those produced by Applied Biosystems (solid), Illumina, and 454 Roche. The protocol presented here utilizes an Illumina sequencing platform. Modifications to sample preparation and data handling would be necessary for other sequencing platforms. A method using the Roche GS FLX platform is presented by Yang et al. [5].

General steps for short-read NGS platforms involve:

1. Random shearing of DNA, either via sonication or nebulization

2. Ligation of universal adapters at both ends of the DNA fragments
3. Immobilization and amplification (to improve signal intensity) of the adapter-flanked fragments

This generates clustered amplicons to serve as templates for the sequencing reactions. Through alternating cycles of base incorporation and image capture, these platforms produce highly accurate short DNA sequences which range in size from 25 to 150 bases.

For Illumina sequencing, around 1–3 μg of total DNA was sheared and prepared following the manufacturer's instructions (Illumina sample preparation for paired-end sequencing) with the following modifications. DNA was sheared using the adaptive focused acoustics method using a Covaris S2 device as follows: duty cycle 10 %, intensity 5, and cycles per burst 200 for 180 s at 6 °C. DNA quantity and quality of fragmentation were assessed with a DNA 1000 chip on an Agilent 2100 Bioanalyzer. Ligation products were purified by agarose gel electrophoresis (2 % agarose, 120 V for 120 min). A narrow size range of predominantly 300-bp fragments was excised from the gel and the products isolated with a QIAquick Gel Extraction kit without heating (alternatively Invitrogen E-Gels SizeSelect™ agarose gels are also effective). PCR products were further purified with a QIAquick PCR Purification Kit and quantified again using a DNA 1000 chip. Approximately 4 pmol per individual and 3 pmol of the PhiX control lane were sequenced for 36×2 cycles on an Illumina Genome Analyzer (GAIIx) following the manufacturer's instructions. Base calling was performed with Illumina software Pipeline 1.4. Multiplexed sequencing using indexing may also be considered as an option for obtaining chloroplast sequences.

3.6 Data Analysis

1. Import FASTQ files [6] from Illumina paired-end sequencing run to CLC Genomics Workbench (*see* **Note 4**).
2. Trim paired-end and single-read sequences to remove low-quality data and adaptor sequences (trim sequence option). A quality score *Q* of 20 corresponds to an expected base call error of $P=0.01$. Filter to remove sequence reads below a specified length after trimming.
3. Identify suitable reference chloroplast genome sequence from a close relative (ideally the same species or genus) for read mapping. Where a close relative is unavailable or the target species is unknown, de novo assembly can be used to identify the closest published chloroplast genome sequence for reference mapping (*see* **Note 5**).
4. Export reference sequence from DNA sequence database (e.g., GenBank) and import into CLC Genomics Workbench.

5. Assemble trimmed paired-end and single-read sequences by mapping against reference sequence (map reads to reference option).
6. Settings for short and long read parameters, such as mismatch and deletion/insertion cost, will depend on how closely related the target and reference sequences are. To facilitate genome assembly, less stringent setting is required for more distantly related species.
7. Set match mode to random to enable assembly of both inverted repeat regions and repetitive elements.
8. Set conflict resolution mode to vote majority to avoid SNP calling in the consensus sequencing resulting from alignment of nuclear and mitochondrial reads to the chloroplast reference.
9. Examine the consensus sequence produced by read mapping to identify gaps—regions with zero coverage (*see* **Note 6**).
10. Export the consensus sequence including gaps in FASTA format.
11. Import consensus sequence to a sequence editing program (e.g., Sequencher, http://www.genecodes.com). Where gaps cannot be resolved as indels, report gaps as N rather than dash (−) in the sequence.
12. Annotate final chloroplast genome sequence using the program DOGMA [7].

4 Notes

1. Plant tissue was collected preferably as fresh leaf tissue. DNA from collections used for other analyses or from DNA banks may be used as a source of DNA if the quality and quantity of DNA are found to be acceptable for analysis. Samples should be processed while still fresh or extracted from carefully dried or frozen material.
2. The PicoGreen working solution must be used on the day it is prepared, preferably within 4 h. Do not make up in a glass container, as there is a risk that the PicoGreen reagent will bind to the glass.
3. Dilution can be adjusted if DNA concentration is expected to be lower or higher. For example, in high DNA concentration 1 μL + 99 μL TE or 5 μL + 95 μL TE, if DNA concentration is expected to be very low or if the sample volume needs to be conserved, the sample can be diluted to a lower concentration before being assayed.

4. Numerous bioinformatics programs are available for assembly of NGS sequence reads [8]. The method presented in this chapter used the commercial software package CLC Genomics Workbench v.3.6.5 (http://www.clcbio.com).
5. When de novo assembly is required to identify a reference chloroplast genome sequence, sort the resulting contigs by length. Select the longest contigs with the highest median coverage. These will invariably include the chloroplast sequence due to the abundance of chloroplast reads in extracts from total DNA. Perform BLAST searches (http://blast.ncbi.nlm.nih.gov/Blast.cgi) on these contigs to identify the closest available chloroplast genome sequence.
6. Gaps may be due to insertion/deletions, highly divergent sequence regions (particularly where a distant reference genome is used) or inversions. Sanger sequencing can be used to confirm sequence reads in these regions.

Acknowledgments

The authors thank the Australian Plant DNA Bank and Southern Cross Plant Genomics for assistance with samples and sequencing.

References

1. Atherton RA, McComish BJ, Shepherd LD, Berry LA, Albert NW, Lockhart PJ (2010) Whole genome sequencing of enriched chloroplast DNA using the Illumina GAII platform. Plant Methods 6:22
2. Nock CJ, Waters DLE, Edwards MA, Bowen S, Rice N, Cordeiro GM, Henry RJ (2010) Chloroplast genome sequence from total DNA for plant identification. Plant Biotechnol J 9:328–333
3. Li X, Gao H, Song JY, Wang JY, Henry R, Wu HZ, Hu ZG, Yao H, Luo HM, Luo K, Pan HL, Chen AL (2012) Complete chloroplast genome sequence of Magnolia grandiflora and comparative analysis with related species. Sci China Ser C 56:189–198
4. McPherson H, van der Merwe M, Delaney SK, Edwards MA, Henry RJ, McIntosh E, Rymer PD, Milner ML, Siow J, Rossetto M (2013) Capturing chloroplast variation for molecular ecology studies: a simple next generation sequencing approach applied to a rainforest tree. BMC Ecol 13:8
5. Yang M, Zhang X, Liu G, Chen K, Yun Q, Zhao D, Al-Mssallem IS, Yu J (2010) The complete chloroplast genome sequence of the date palm (Phoenix dactylifera L.). PLoS One 5:e12762
6. Cock PJA, Fields CJ, Goto N, Heuer ML, Rice PM (2009) The Sanger FASTQ file format for sequences with quality scores, and the Solexa/Illumina FASTQ variants. Nucleic Acids Res 38:1767–1771
7. Wyman SK, Jansen RK, Boore JL (2004) Automatic annotation of organellar genomes with DOGMA. Bioinformatics 20:3252–3255
8. Miller JR, Koren S, Sutton G (2010) Assembly algorithms for next-generation sequencing data. Genomics 95:315–327

Chapter 3

Plastid Gene Transcription: Promoters and RNA Polymerases

Jennifer Ortelt and Gerhard Link

Abstract

Chloroplasts, the sites of photosynthesis and sources of reducing power, are at the core of the success story that sets apart autotrophic plants from most other living organisms. Along with their fellow organelles (e.g., amylo-, chromo-, etio-, and leucoplasts), they form a group of intracellular biosynthetic machines collectively known as plastids. These plant cell constituents have their own genome (plastome), their own (70S) ribosomes, and complete enzymatic equipment covering the full range from DNA replication via transcription and RNA processive modification to translation. Plastid RNA synthesis (gene transcription) involves the collaborative activity of two distinct types of RNA polymerases that differ in their phylogenetic origin as well as their architecture and mode of function. The existence of multiple plastid RNA polymerases is reflected by distinctive sets of regulatory DNA elements and protein factors. This complexity of the plastid transcription apparatus thus provides ample room for regulatory effects at many levels within and beyond transcription. Research in this field offers insight into the various ways in which plastid genes, both singly and groupwise, can be regulated according to the needs of the entire cell. Furthermore, it opens up strategies that allow to alter these processes in order to optimize the expression of desired gene products.

Key words Chloroplast, DNA-binding proteins, Gene expression, NEP, PEP, Promoters, RNA polymerase, Sigma factors, Transcriptional regulation

1 Introduction

Compared to prokaryotic cells, those of eukaryotic organisms generally have a tendency to be more highly compartmentalized and specialized, giving rise to numerous cell types throughout the life cycle and in response to variable environmental cues an organism may be exposed to. Complex intracellular structures, including organelles, and their functional interactions permit effective regulation and coordination of biochemical processes within and between cells. In a typical plant cell, the genetic material is sorted among the three DNA containing types of organelles, i.e., the nucleus (genome), the mitochondria (chondriome), and the plastids (plastome). Regarding their evolutionary origin, the latter two can be viewed as relics of ancient prokaryotes that have invaded

Pal Maliga (ed.), *Chloroplast Biotechnology: Methods and Protocols*, Methods in Molecular Biology, vol. 1132,
DOI 10.1007/978-1-62703-995-6_3, © Springer Science+Business Media New York 2014

a eukaryotic cell and—after appropriate long-term adjustments involving gene transfer and novel signaling mechanisms—have turned into stably transmitted endosymbionts of modern eukaryotic cells (endosymbiont hypothesis) [1, 2]. While mitochondria may have originated from purple bacteria-like ancestors [3, 4], today's plastids seem to be derived from ancient cyanobacteria [5]. Both types of organelles have retained their own DNA and a fully functional enzymatic apparatus for DNA replication and all aspects of gene expression; nevertheless, most genes for plastid (and mitochondrial) proteins reside in the nucleus, which has led to the term "genetic semiautonomy" to describe the intracellular situation of mitochondria and plastids. Indeed, of the typically 2,500–5,000 organellar proteins, at least 90 % are nuclear coded [6, 7], translated (as a precursor) on cytoplasmic polyribosomes, posttranslationally cleaved, and imported into the organelle [8, 9]. Among these nuclear-coded proteins are many that can be traced back to cyanobacteria [10], indicating considerable intracellular gene migration during evolution. Other nuclear-coded proteins, however, share only regional homology or none at all with their prokaryotic counterparts, suggesting that eukaryotic nuclear modules or entire proteins have entered the plastid compartment. Considering this quantitatively large impact of nuclear genes on plastid structure and function, a pertinent question then is to ask why plastids (and mitochondria) have retained their own DNA at all. To address this, it seems reasonable therefore to begin with a short overview on chloroplast DNA itself.

Today's view of the plastome is that it consists of multiple (circular and probably identical) double-stranded DNA molecules that are (depending on the plant species) 120–160 kb in size [11]. Each DNA circle has a coding capacity of approximately 100–150 genes, i.e., less than 5 % of a typical cyanobacterial genome [12], which again emphasizes the magnitude of gene migration to the nucleus (*see* above).

Apart from the genes for rRNAs, tRNAs, and other (small regulatory) RNAs, the plastome contains structural genes for proteins. These are known to be involved in photosynthesis (approximately 40) and carbon assimilation, in general housekeeping functions (very few), as well as in various steps of plastid gene expression ranging from transcription via RNA processing to translation (approximately 60) [12, 13]. Interestingly, many of the multimeric protein complexes found in plastids are hybrids in terms of the coding site and site of synthesis of their subunits. This is, e.g., true for the textbook example Rubisco (ribulose-1,5-bisphosphate-carboxylase/oxygenase), which is a hexadecamer consisting of 8 (chloroplast-coded) large and 8 (nuclear-coded) small subunits. The same dual principle, i.e., mixed coding sites, is evident for the photosynthesis apparatus, with a tendency that intrinsic proteins such as the PSI and PSII reaction center proteins are chloroplast-own gene products, while peripheral proteins such as those of the antenna complexes tend to be nuclear coded. This principle seems to hold true in other, perhaps less obvious, cases as well, including plastidic fatty acid biosynthesis: only a single subunit (*accD* gene product) of the participating enzyme acetyl-CoA carboxylase is chloroplast encoded, while the others are nuclear gene products. Last but not least, the plastid gene expression machinery itself is built up according to the same principle at probably all levels. Intensely studied examples include the organellar ribosomes, RNA splicing, and DNA transcription.

This then relates back to the question of why a number of prokaryotic genes have been preserved within the plastid compartment at all and why considerable effort is being spent by the organelle to replicate and express these genes. An appealing idea, though not experimentally proven, suggests that it is the close physical proximity between photosynthesis and plastid gene expression, which makes it possible to quickly replenish photosynthetic proteins that undergo degradation and turnover to highly variable extents [14].

DNA transcription in bacteria, in contrast to eukaryotes, is carried out by just one single type of DNA-dependent RNA polymerase. The basal entity, also named core enzyme, is composed of two alpha (α) and one each of beta (β) and beta prime (β') subunits ($\alpha_2\beta\beta'$ structure). This complex, which sometimes contains an additional subunit (omega, ω), is capable of synthesizing RNA molecules of considerable length but lacks specificity to initiate RNA synthesis at well-defined promoter sequences. To achieve this critical task, another protein termed sigma (σ) is required. This regulatory protein factor complements the core enzyme and confers specific promoter binding and transcription initiation on the (transiently formed) holoenzyme, before it is released (and ready for a new initiation cycle). Most often, bacterial cells contain more than one single type of sigma factor. The multiple members of this protein family are known to have distinct promoter preferences; they exist in variable amounts and can differentially interact with other proteins (anti-sigma factors). Collectively, these features greatly expand the flexibility of transcriptional regulation in prokaryotic cells.

Despite the universal occurrence of the multi-subunit RNA polymerase in bacteria, it is not the only transcription enzyme that can be isolated from prokaryotic organisms. Bacterial cells infected by T3, T7, or SP6 phages accumulate a single-subunit polymerase encoded by a phage gene. This latter enzyme does not require a sigma factor for faithful transcription initiation from its cognate (specialized) promoter sequences. Similar phage-type RNA polymerases have since been described also for mitochondria of eukaryotic cells [15, 16].

How does the transcription apparatus of plastids look like, and how does it function? Results from the 1960s indicate already that chloroplast DNA can be transcribed and translated in vivo [17]. Not much later, highly purified preparations of soluble multi-subunit RNA polymerase were obtained [18]. The polypeptide patterns as well as N-terminal protein sequences [19], along with the complete nucleotide sequence of chloroplast DNA [20], suggested the existence of a plant bacterial-type multi-subunit enzyme to be active in chloroplasts and other plastid types. On the other hand, heat bleaching and inhibitors affecting 70S translation provided initial evidence that plastid gene expression might be controlled by nuclear gene products or even a complete nuclear-coded RNA polymerase [21]. This was substantiated in work using parasitic plants defective in the chloroplast genome [22] as well as by using naturally occurring [23] or site-directed plastome mutants [24]. Genetic and molecular data of these mutants, along with the detailed knowledge of the chloroplast genome, provided direct evidence for continued transcription of chloroplast genes in the absence of bacterial-type plastid RNA polymerase. Together, these and other data discussed below have strengthened the view that the plastome is transcribed by at least two different types of plastid DNA-dependent RNA polymerases: one being encoded by the nucleus (NEP, nucleus-encoded polymerase) and the other by the plastome (PEP, plastid-encoded polymerase) [25].

Although the existence of multiple plastid RNA polymerase formally resembles the situation found in the nucleus of eukaryotic cells, the circular superhelical structure of chloroplast DNA, its transcription into polycistronic RNAs, and the subsequent translation on 70S ribosomes all are features more typical of prokaryotic than eukaryotic (nucleo-cytosolic) gene expression. The two different polymerase forms originate from different intracellular sites and likely differ in their phylogenetic origin. Furthermore, they also seem to function sequentially on a timescale: NEP is thought to act early by transcribing plastome-localized genes for PEP subunits, while the latter enzyme plays its major role later during plastid development and in fully functional chloroplasts [26–28]. This does not exclude, however, NEP as an important player in mature plastids. Indeed, the nuclear-coded polymerase seems to have a rescue function in situations where PEP is defective or completely absent.

The existence of these two principally different types of RNA polymerases in one organelle has so far been demonstrated in only a few higher plant species: Arabidopsis, tobacco, maize, mustard, spinach, and wheat. Division of labor among these polymerases regarding plastid gene expression becomes evident by distinct patterns of plastid transcripts. For instance, transcript levels of photosynthetic genes are usually low in proplastids, reaching their maximum levels in later plastid stages. On the other hand, housekeeping genes appear to be transcribed more constitutively during plastid development [29]. We will first cover the NEP system and then focus on the regulatory network of plastid transcription centered around PEP, with emphasis on higher plant systems. For further details, including the many data beyond the scope of this chapter, we alert readers interested in other aspects of chloroplast transcription to various more comprehensive reviews [25, 30–32].

2 NEP: *N*uclear-*E*ncoded *P*olymerase

In the early 1990s plastid gene expression was observed in the parasitic plant *Epifagus virginiana* [22] although this organism is unable to synthesize the bacterial-like plastid-encoded PEP due to the loss of *rpoB* and *rpoC* genes for two of its subunits [33]. These observations pointed to a nucleus-encoded RNA polymerase activity in plastids. Similar conclusions were reached in work with the 70S ribosome-deficient *albostrians* mutant of barley [23]. Experiments with plastome knockout mutants from tobacco lacking individual PEP subunits further strengthened the concept of a second, nuclear-coded, RNA polymerase in plastids [24, 34]. First direct evidence for this enzyme was provided in biochemical work showing a single catalytic polypeptide of approx. 110 kDa in spinach chloroplast [35]. This was later confirmed and extended for wheat chloroplast using both peptide and cDNA sequencing [36]. Nuclear gene(s) and/or cDNA sequences for T7-type RNA polymerase were isolated and sequenced for a number of plant species including the plant model Arabidopsis [37, 38]. Furthermore, bacterially overexpressed tobacco NEP was analyzed for its transcriptional properties in vitro [39].

The activity of the nucleus-encoded RNA polymerase varies according to developmental stage and plant tissue. In proplastids NEP seems to be the predominant transcription enzyme [27], while during chloroplast differentiation PEP is thought to take over functionally [40]. The relative impact of NEP decreases [34, 41], but it continues to drive transcription of housekeeping genes even in mature plastids [29, 34].

The nuclear-encoded plastid RNA polymerase (NEP) is closely related in sequence to both mitochondrial and T3/T7 phage polymerase enzymes [42], and therefore this gene was assigned *RpoT*. Sequencing revealed that a family of *RpoT* genes exists in most plant species that were examined. For instance, in monocotyledonous species two *RpoT* genes, *RpoTm* and *RpoTp*, have been shown to code for a mitochondrial and a plastid-localized RNA polymerase, respectively [43, 44] (Table 1). In the dicotyledonous plants Arabidopsis [37, 38] and *Nicotiana sylvestris* [45, 46], even three *RpoT* genes were found. These were related to a plastid (RpoTp), a mitochondrial (RpoTm), and an enzyme dual-targeted to both organelles (RpoTmp). Even six *RpoT* genes were described for *Nicotiana tabacum*, which can be explained by its amphidiploid

Table 1
Plastid and mitochondrial RNA polymerases in plant species

Group	Species	Gene(s)	Name(s)	Localization	References
Green alga	Cr	1	RpoTm	mt	[146]
	Ot	1	RpoTm	mt	[168]
Mosses	Pp	2	RpoT1,2	mt, both	[169, 170]
Lycophytes	Sm	1	RpoTm	mt	[171]
Monocots	Os	2	RpoTm, p	mt, cp	[44]
	Hv	2	RpoTm, p	mt, cp	[40]
	Ta	2	RpoTm, p	mt, cp	[36]
	Zm	2	RpoTm, p	mt, cp	[43]
Dicots	At	3	RpoTm, p, mp	mt, cp, both	[47]
	Ns	3	RpoTm, p, mp	mt, cp, both	[45, 46]
	Nt		2× RpoTm, p, mp	mt, cp, both	[38]

Abbreviations: At, *Arabidopsis thaliana*; Cr, *Chlamydomonas reinhardtii*; Hv, *Hordeum vulgare*; Ns, *Nicotiana sylvestris*; Nt, *Nicotiana tabacum*; Pp, *Physcomitrella patens*; Os, *Oryza sativa*; Ot, *Ostreococcus tauri*; Sm, *Selaginella moellendorffii*; Ta, *Triticum aestivum*; Zm, *Zea mays*

genome [47]. Finally, biochemical evidence for the existence of two different NEP (RpoT) types named NEP-1 and NEP-2 was obtained for spinach plastids [48].

What might be the reason of more than a single nuclear-coded polymerase co-targeted to each of these organelles? In the case of mitochondria, existing data point to RpoTm as the major RNA polymerase responsible for transcription of most if not all mitochondrial genes, while RpoTmp seems to have an accessory role with a more restricted set of transcribed genes [49]. Interestingly, RpoTmp seems to function in a gene, rather than promoter, specific manner, suggesting that so far unknown regulatory elements may play a role [49]. A somewhat analogous situation applies in the chloroplast, where RpoTp is the major enzyme and RpoTmp serves accessory, probably not yet fully resolved, roles. So far, only two promoters were shown to be specifically used by RpoTmp, *Prrn16* [50] and *PclpP58* [51]. In any case, based on their knockout mutant phenotype, both polymerases are crucial for proper plastid development. A simultaneous loss of both RpoTm and RpoTmp in a double mutant line cannot be counteracted by the plastid transcription apparatus; these seedlings are lethal [49, 52].

3 PEP: *P*lastid-*E*ncoded *P*olymerase

Evidence for RNA-synthesizing activity in isolated chloroplast dates back to the 1960s [17]. Soluble DNA-dependent RNA polymerase was first characterized from maize [53] and highly purified soon afterwards [18]. This enzyme preparation, as well as those obtained from pea [54] and spinach [55], was found to resemble bacterial RNA polymerase in both architecture and some functional aspects [56]. The plastid polymerase is built up basically by five catalytic subunits, which subsequently were shown to reveal high sequence similarity with the bacterial α, β, and β′ core subunits. The putative plant core enzyme $\alpha_2\beta\beta'\beta''$ is encoded by the plastid genes *rpoA*, *rpoB*, *rpoC1*, and *rpoC2*, respectively. The gene products of *rpoC1/2* (β′ and β″) can be aligned with the N- and the C-terminal portions of the bacterial β′ subunit [57]. Exchange of the bacterial β′ against the plant β′ and β″ subunits is possible and results in an active bacterial enzyme [58]. On the other hand, replacement of the bacterial by the plastid α-subunit (*rpoA* gene product) did not prove successful [59], reflecting the lower degree of sequence similarity as compared to β′/β″.

The plant *rpoA*, *rpoB*, and *rpoC1/2* genes are usually located on the plastome [60], being detectable on neither the genome nor the chondriome. However, notable exceptions are known, where *rpoA* genes were identified in the nucleus. These include mosses [61, 62], algae [63], and the malaria parasite *Plasmodium falciparum* [64]. Even more extreme, the freshwater protozoa *Reclinomonas* contain all the *rpo* genes for core subunits on its mitochondrial genome [65].

It has long been known that the bacterial core enzyme needs to be complemented by a regulatory protein factor named sigma (σ) to yield the initiation-competent holoenzyme [66]. Both functional and crystallization studies have provided extensive details of the mechanisms involved [67–69]. It therefore is conceivable to assume that also the plant PEP enzyme is composed of a core surrounded by regulatory factor(s). Initial attempts to address this question led to conflicting answers: the available full chloroplast DNA sequences did not reveal anything resembling a typical bacterial sigma (*rpoD*) gene. On the other hand, immunological [70] and biochemical evidence [71] was obtained for plastid proteins with sigma-like characteristics. Almost 10 years later, this question was directly answered by the cloning and functional characterization of nuclear genes for "true" plastid sigma factors [72, 73] (*see* Subheading 3.1).

We would like to emphasize that PEP, the "classical" soluble form of the plastid multi-subunit RNA polymerase (sRNAP, s*oluble* RNA p*olymerase*), is only one of several related enzymatically active preparations. Others include the so-called TAC (*transcriptionally* a*ctive* c*hromosome*) [74] representing large membrane-associated DNA-/RNA-binding complexes with up to 100 proteins [75] and various crude plastid extracts for functional in vitro studies [76–78]. To even more closely reflect the in vivo situation, *in-organellar* run-on transcription experiments using complete chloroplasts have been devised [79, 80] and since then successfully used in many subsequent studies. Together, these various strategies help complement the picture obtained by chloroplast transformation and in vivo gene expression experiments [81].

Furthermore, PEP itself can be obtained in different forms depending on the plastid type from which it is isolated. For instance, chloroplast and etioplast PEP preparations from mustard (*Sinapis alba*) were found to differ with regard to size, subunit architecture, and function [82, 83]. The subunit pattern of the etioplasts enzyme (PEP-B) resembles that of *E. coli* RNA polymerase more closely than the PEP-A form of chloroplast [82]. The latter contains at least 12–15 accessory proteins, some of which were identified as CSP41 and other RNA-binding proteins as well as Fe-superoxide dismutase 3, an annexin-like protein, and a cpCK2-like kinase [83, 84]. More recently, using higher-resolution separation and mass spectrometry, a total of 53 proteins were identified in this enzyme preparation [85]. Similar results were obtained in work with PEP fractions from Nicotiana tabacum, which were purified using an elegant (α-subunit) tagging and affinity purification approach [86]. It is conceivable that attachment of accessory proteins interferes with basic properties of the enzyme, as is indicated by findings that the etioplast PEP-B form is sensitive to the bacterial transcription inhibitor rifampicin, while the chloroplast PEP-A form is not [82]. Similar conclusions, i.e., evidence for functionally different PEP forms, were reached in studies on PEP promoter usage in monocotyledonous (wheat) leaves representing a developmental gradient from young cells at the base to mature cells at the tip [87].

3.1 Sigma Factors

In 1969 an activity that confers specific DNA binding and transcription initiation to *E. coli* core RNA polymerase was discovered and named σ (sigma) factor [66]. Based on its apparent molecular size of approx. 70 kDa, this factor was termed SIG70. Multiple factors with similar regulatory properties were subsequently found in *E. coli* as well as in nearly all other bacteria that were investigated [88, 89]. Based on sequence and function, bacterial sigma factors in general are assigned to one out of two independent families named according to their first representatives, i.e., SIG70 and SIG54, respectively. The factors of the SIG70 family are further subdivided into four groups, with those of group 1 considered to be essential, while members of group 2–4 seem to be nonessential and are designated "alternative" (specialized) factors [90]. Typical SIG70-type factors are composed of four conserved regions which together enable the basic sigma functions required for promoter and core binding [91]. Attachment of sigma factor to the polymerase leads to higher binding affinity to the gene promoter, resulting in enhanced transcription [92]. Crystal structures that provided a further refined picture were obtained, e.g., for most regions of *E. coli* SIG70 [93], complete factors from *T. aquaticus* and *T. thermophilus* [94, 95], and *E. coli* sigma E bound to a domain of its anti-sigma factor RseA [96].

Despite various indications for the existence of "sigma-like" proteins in plastids (for review *see* ref. 97), it took almost 30 years since the initial discovery [66] that complete sigma sequences from a photosynthetic organism (the red alga *Cyanidium caldarium*) were obtained [72, 73]. Soon afterwards, sequences from a number of higher plant species were published, including *Arabidopsis thaliana* [98, 99], rice [100], mustard [101], maize [102], wheat [103], and tobacco [104].

From these and other subsequent data (for review *see* ref. 105), it readily became obvious that the plant factors share much of their C-terminal portion with typical bacterial members of the SIG70 family [73]. This portion includes the conserved regions for basic sigma function except subregion 1.1, which has not demonstrated for plant proteins so far but is also known to be absent in many bacterial sigmas [90]. On the other hand, the known plastid factors all seem to differ from their bacterial counterparts in having an additional unique N-terminal portion. Recently, considerable interest has emerged to understand the function of this unconserved region and its potential impact on the regulation of plastid gene transcription.

While the green alga *Chlamydomonas reinhardtii* seems to have only one single sigma factor [106], families with up to 7 members (*Populus trichocarpa*) have been identified in higher plants so far. The typical number is six factors, as is the case, e.g., in *Arabidopsis thaliana* (AtSIG1-6) [107, 108]. Considering only the conserved region (CR), it has been possible to assign orthologs for one and the same sigma factor from different species [109]. The family members differ in their expression, both RNA and protein level, depending on plant tissue, developmental stage, and environmental cues. Work with (knockout) mutants, sense/antisense constructs, and in vitro transcription studies collectively indicate that individual members of the sigma family may reveal distinct differences in promoter preference and/or efficiency of transcription [105, 110]. To fully answer the question of specialized versus redundant function(s) of individual factors, more detailed mutational and functional analyses will be required.

3.1.1 *Arabidopsis thaliana*

The Arabidopsis sigma family comprising of AtSIG1-6 has already been extensively studied and therefore may serve as a reference to which the situation in other plant species can be compared. The derived Arabidopsis sequences obtained from cDNA and genomic clones show similarities within the C-terminal conserved region (CR, about 303–321 aa in length for individual family members). In contrast, strong differences were observed within the N-terminal unconserved region (UCR, ranging from 116 aa for AtSIG4 up to 257 aa for AtSIG2). Interestingly no apparent ortholog to the *AtSig4* gene has been published so far in any other plant species. However, recent database screening has provided hits for putative SIG4 candidates in at least several other dicotyledonous species (J. Schweer, unpublished observations).

Multiple sequence alignment defines *AtSig2* to be the family member most closely related to bacterial *rpoD* for SIG70, i.e., the primary housekeeping sigma factor. Nevertheless, the (nonlethal) phenotype of *AtSig2* knockout lines does not support the idea that this factor is essential and therefore might be a functional equivalent of SIG70. By similar arguments, due to lack of a lethal phenotype, none of the other members of the Arabidopsis family seems to qualify as a (essential) primary factor as defined for prokaryotic systems.

One way to approach the functional role of individual members of the sigma family is to look at their transcript and protein levels throughout development. All Arabidopsis sigma proteins seem to exist already at young stage (day 4 after germination), except for AtSIG1 and AtSIG5 which can first be detected at day 8 [31, 111, 112]. Based on results obtained with very early stages of seedling development, it has been suggested that AtSIG3 exists already in ungerminated seeds and AtSIG2 soon thereafter (day 2) [113].

What about the function of the individual members of the Arabidopsis sigma family? Answers have been provided primarily by work with knockout mutants. In this regard, AtSIG1 has proved to be the factor perhaps most difficult to tackle because of lack of phenotype of existing *sig1* mutant lines. However, important clues to its function came from work on an interacting protein named SIB1 (sigma-binding protein1) [114]. More recently it was shown that SIB1 is involved in responses mediated by salicylic acid and jasmonic acid as well as by infection with *Pseudomonas syringae* [115]. In vitro experiments have shown an increased promoter-binding activity in the case of SIG1-SIB1 interaction [116]. Together these results support the idea that SIG1 is specifically involved in plastid transcription in defense-regulated situations, while under normal conditions its role is redundant to those of other members of the sigma family.

AtSIG2 has been assigned a specific role in the formation of tRNA transcripts involved in protein and tetrapyrrole synthesis [117]. This is indicated by decreased chlorophyll content during cotyledon stage of *sig2* mutants as well as by reduced accumulation of transcripts representing photosynthetic protein and *trn* genes [108, 118, 119].

Recent data suggested that both factors SIG3 and SIG4 may serve highly specialized rather than general roles, because transcription of one single gene each (*psbN* and *ndhF*, respectively) seems to be preferentially affected in the mutant lines [120, 121]. However, unless interacting factor(s) would play a role, the constitutive expression of SIG3 during cotyledon development would argue against this. Furthermore, the increased SIG3 protein level in knockout mutants lacking other sigma factors suggests a more redundant role for AtSIG3 implicating relatively low promoter specificity. This is directly supported by the results of in vitro transcription experiments [118].

AtSIG5 has been shown to be involved in plastid gene expression under diverse stress conditions including osmolarity, temperature, and high light stress [117]. Furthermore this factor was demonstrated to use the specific *b*lue-*l*ight-*r*esponsive *p*romoter (BLRP) of the *psbD* gene (*see* Subheading 4.2) [113, 122].

Sigma factor 6 reveals a dual temporal role, with an early component at a distinct time point and a second, persistent, function throughout entire cotyledon development [111, 123]. Furthermore it has been shown to have promoter selectivity. Unlike any other member of the sigma family, it is involved in transcription of the *atpB/E* operon [124]. This selective behavior is likely related to specific elements within the unconserved region [125], recently identified to be phosphorylation sites [32]. Interestingly, an interacting factor, a pentatricopeptide repeat protein (PPR), termed DELAYED GREENING 1 (DG1), has been identified for AtSIG6 [126]. UCR interaction can be expected to result in better stabilization of promoter binding or binding to polymerase, which seems to regulate PEP-dependent transcription in plastids. Taken together, AtSIG6 reveals features not yet observed to the same extents for other members of the Arabidopsis sigma family, i.e., both development dependency and promoter selectivity.

3.1.2 A Look to Sigma Factors Beyond Arabidopsis

While for Arabidopsis the sigma factors and their genes have been characterized quite extensively already, with the exception of maize [127] and rice [100, 128], the information from other plant species is still incomplete. For instance, in mustard [101, 129], tobacco [104], and *Physcomitrella patens* [130], some but not all individual members have been studied. Nevertheless, these and other findings together aid in functional assignment (Fig. 1) in a similar way as has become known for diverse bacterial species. However, sigma factor classification in bacteria versus plants also reveals substantial differences. In plants no primary factor with an essential role for general cell activity and survival was identified. Sigma factors SIG2 and SIG6 (AtSIG2 and AtSIG6; ZmSIG2A, ZmSIG2B, and ZmSIG6; OsSIG2A and OsSIG2B) are predominantly involved in plastid transcription during early seedling development, i.e., in juvenile chloroplasts. In contrast, a second plant sigma group, Sig1 and SIG5 (AtSIG1 and AtSIG5, ZmSIG1A and ZmSIG1B, NtSIGA1 and NtSIGA2, PpSIG1 and PpSIG5, OsSIG1, TaSIGA), seems to participate in plastid gene expression in more mature plastids. SIG5 has been shown to be involved particularly under stress conditions, in response to blue light and (along with SIG1) during circadian rhythms [103, 113, 128, 131]. Similar considerations across species are not yet possible to the

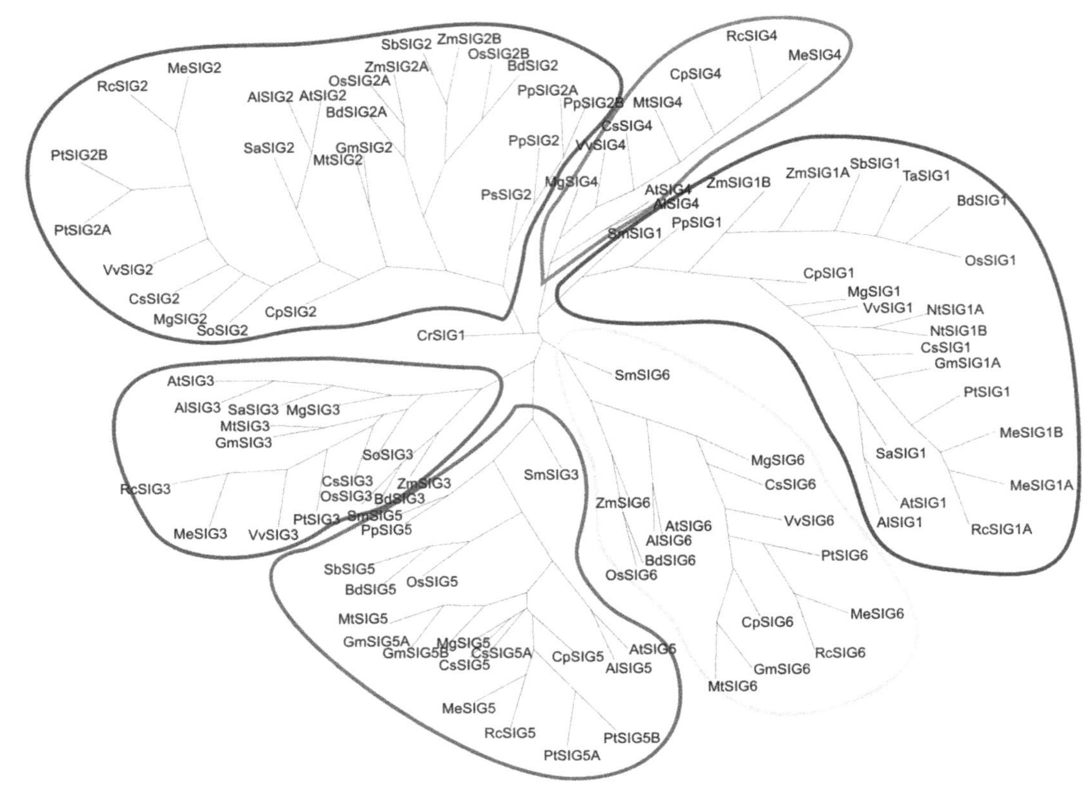

Fig. 1 Possible phylogenetic relationships among plant sigma factors. Derived sigma protein sequences were aligned and data presented as a phylogenetic tree based on similarity levels. A number of unknown proteins taken from the Phytozome collection (http://www.phytozome.net) share (at least regional) sequence similarity with characterized plant sigma factors and thus were tentatively assigned SIG. Although orthologs of Arabidopsis AtSIG4 have not yet been functionally identified in other plants, it is interesting to note that unknown Phytozome proteins from several dicotyledonous species reveal significant SIG4 similarity

same extent for sigma factors of the SIG3 and SIG4 subgroup. Available data for AtSIG3 and AtSIG4 as well as ZmSIG3 suggest rather a general redundant role.

3.2 PTK: Plastid Transcription Kinase

Soon after discovery that chloroplasts contain sigma-like proteins [71], it became clear that, in contrast to the situation in bacteria [132], the plant factors can be regulated via their phosphorylation state [133, 134]. Subsequent work led to the detection and characterization of a PEP-associated Ser/Thr protein kinase, named plastid transcription kinase (PTK), which itself was found to respond to phosphorylation and SH-group redox state in vitro [135, 136]. PTK was renamed later cpCK2 because of its striking resemblance to nucleo-cytosolic members of the CK2 group of eukaryotic Ser/Thr kinases [84]. The latter are involved in numerous signaling chains in eukaryotic cells. They are considered unique in terms of their acidic (rather than basic) phosphoacceptor motif

S/TxxE/D, where E/D at the $n+3$ position can be substituted by phosphoserine, and their ability to use GTP as donor [137].

Reversible phosphorylation of proteins is a well-known mechanism for posttranslational regulation that, via changes in charge and conformation, accounts for activation or inactivation and can result in altered substrate specificity [138]. That cpCK2 can use plant sigma factors as authentic phosphorylation substrates were first shown for mustard (*Sinapis alba*) [84] and recently for Arabidopsis [139]. Both AtSIG1 and AtSIG6 were demonstrated to respond to changes in phosphorylation state resulting in altered promoter selectivity [140]. Likewise, *in-organello* run-on transcription using isolated chloroplasts suggested that AtSIG1 regulates *psaA* transcription for PSI components in a phosphorylation-dependent manner, although the kinase activity involved was not investigated [141]. Interestingly, one of the most sensitive phosphorylation sites on the AtSIG6 sequence (at Ser174) is not a priori a typical cpCK2 phosphoacceptor motif (*see* above). It only conforms to the consensus sequence upon conversion of the Ser residue at the $n+3$ position into phosphoserine, suggesting the possibility that a second ("pathfinder") kinase might be involved in vivo. This would further increase the flexibility of transcriptional regulation via sigma factor phosphorylation.

4 Plastid Promoters

Although the details have emerged to be quite complex, a "typical" eukaryotic nuclear promoter transcribed by Pol II usually centers around the TATA box and often contains variable sets of upstream elements. Transcription initiation depends on correct assembly of the components of the initiator complex at the TATA box, followed by recruitment of Pol II and additional sets of both general and specific transcription factors [142].

Likewise, despite many variable details, "typical" bacterial promoters usually contain two conserved sequence motifs commonly known as −35 region (TGACA) and −10 (TATAAT) region termed according to their most upstream (5′-) position in front of the transcription start site. Unlike eukaryotic transcription, the holoenzyme (core + sigma) directly assembles at the promoter and initiates RNA synthesis [132].

Considering the prokaryotic origin and semiautonomous character of plastids, what might the "typical" plastid promoter be? Plastome sequences for many plant species reveal prokaryotic-type −35 and −10 motifs, while sequences reminiscent of a eukaryotic TATA box are relatively rare. Furthermore, in view of the two different types of RNA polymerases, PEP and NEP, it would seem conceivable that both −35/−10 and T3/T7-like promoter types may exist. Considerable efforts were made to group plastid genes according to their putative promoters and RNA polymerase involved: genes exclusively transcribed by PEP were named class I, those transcribed by NEP as class II, and others transcribed by both polymerase types as class III genes [24, 34]. However, based on the results of more recent studies, this static classification does not seem to fully account for the flexibility of plastid transcription. Often, plastid genes have more than one single promoter (Fig. 1a) and are differentially transcribed in a development- and environment-dependent manner [39]. This becomes evident in PEP- (tobacco *Δrpo*) or sigma factor-defective plant systems, where transcription was shown to start from different promoter(s) and sometimes with different time course as compared to the wild-type situation. These data suggest that "silent" promoters exist, which can be activated under appropriate conditions and vice versa. Genes with only one single promoter are rare on the plastome. Two representatives are *rbcL* and *psbA* (Fig. 2b), each of which is transcribed exclusively from one PEP promoter [29, 143]. Regions containing multiple promoters, for both PEP and NEP, include the operons *psbD/C*, *rrn16*, and *atpB/E* as well as the monocistronic genes *ycf1* and *clpP* (Fig. 2a).

4.1 NEP Promoters

Identification of NEP promoters has initially been guided by searches for conserved sequence elements in front of genes transcribed by NEP, followed by in vitro transcription experiments using mutagenized templates. Despite a certain degree of variation, for both different genes from the same species and the same gene from different species, three types of NEP promoters were defined [144–146]. Type Ia promoters contain the so-called YRTA box in close vicinity to the transcription start site (usually with the 5′-Y between positions −15 and −5). Type Ib promoters have an additional upstream element named GAA box for enhanced DNA-binding affinity, which is located 18–20 nt in front of the YRTA motif [147]. The third NEP promoter version, type II, is set apart because of its "nonconsensus" architecture [148, 149]. For instance, the *clpP-53* promoter from tobacco consists of a long (28 nt) sequence stretch that is completely unrelated to the type I promoter boxes and, furthermore, extends into the transcribed region (−5 to +25). Regions that are conserved in both sequence and position were found to be present in other higher plant species including Arabidopsis, rice, and spinach [150] (Fig. 2c). Transcription units that seem to be exclusively transcribed by NEP include the single genes *accD* and *ycf2* and the *rpoB* operons [34, 39].

4.2 PEP Promoters

As might be expected from the similarity between the PEP core and bacterial multi-subunit RNA polymerase, PEP promoters usually contain both a typical −35 (TTGACA) and −10 (TATAAT) region [31]. Despite this general tendency pointing to a conserved basic architecture, there are a number of specialized and more restricted features which were detected in some but not all

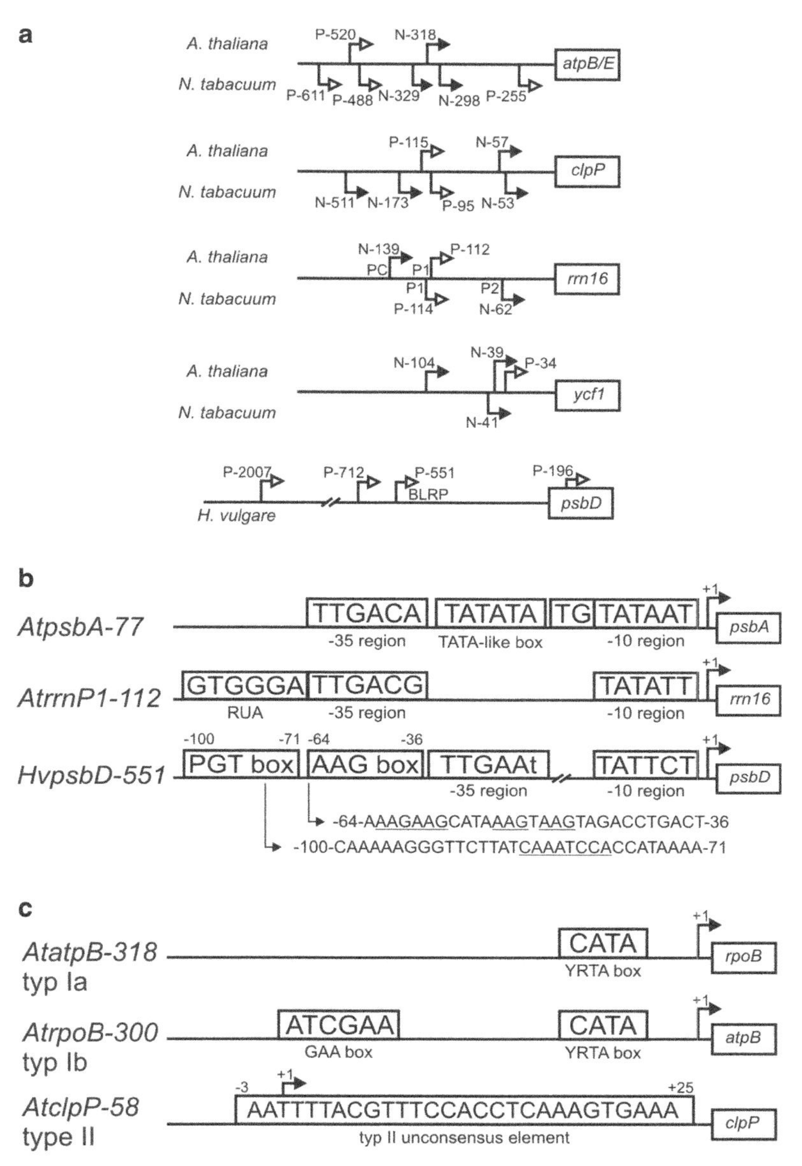

Fig. 2 Variation in plastid promoter arrangement and architecture. (**a**) Multiple promoters in front of selected chloroplast transcription units. Shown are (*from top to bottom*) promoters upstream of the *atpB/E* operon, *clpP*, the *rrn16* operon, *ycf1* (all both from Arabidopsis and tobacco), as well as *psbD* (from barley). NEP promoters are denoted by *filled arrows* and the prefix "N-" and PEP promoters by *open arrows* and the prefix "P-." (**b**) Variable PEP promoter architecture exemplified for the −35/−10 promoters in front of the Arabidopsis *psbA* gene and *rrn16* operon and the blue-light-responsive promoter of barley *psbD*. The "+1" *arrows* indicate the transcription start site. Features shown include the TATA-like box and the extended −10 region (*psbA*), the upstream RUA element (*rrn16*), as well as the PGT and AAG boxes (*psbD*). The sequences of the latter, including their critical internal regions (*underlined*), are given on separate lines. (**c**) Architecture of the three principal types of NEP promoters, drawn for Arabidopsis *rpo*B (type Ia), *atpB* (type Ib), and *clpP* (type II) (for details, *see* text)

chloroplast promoters. These include the extended −10 region (TGn) as well as the TATA boxlike (TATATA) motif, both of which seem to be part of the PEP promoter in front of the *psbA* gene for the D1 reaction center protein, i.e., the probably most efficiently transcribed protein-coding plastid gene (Fig. 2b) [97, 151]. Other more restricted promoter elements are the PGT and AAG boxes upstream of the −35 region of one of the multiple *psbD* promoters in several higher plants [152–154] (Fig. 2b).

PEP promoters with additional *cis* elements are also known for *psaA* (region U/region D) [155], *rbcL* (CDF1) [143], and *rrnP1* (RUA region, *see* below) [156] (Fig. 2b). For instance, the *rbcL* gene is transcribed from one single PEP promoter containing conserved −35 and −10 elements in several species including tobacco, maize, spinach, pea, barley, and Arabidopsis. In addition, a relatively long DNA region spanning position −102 to −16 seems to be recognized by a trimeric protein named *c*hloroplast *DNA-binding f*actor 1 (CDF1), which has been implicated in light- and organ-specific regulation.

4.2.1 A Closer View of the rrn Promoters

Multiple promoters have been detected in front of the *rrn* operon comprising the genes *rrn16*, *trnI*, *trnA*, *rrn23*, *rrn4.5*, and *rrn5*. The region responsible for transcription initiation has been dissected in much detail particularly in spinach, Arabidopsis, tobacco, barley, and mustard. This work has led to the detection of only one single promoter in spinach and mustard, which was identified as a NEP type II promoter termed PC [48, 157, 158]. In contrast, the single promoter was shown to be a PEP promoter named P1 in the case of barley [159]. Both these promoters were found to coexist, and to be utilized, in Arabidopsis [149]. Finally, in tobacco the P1 PEP promoter is accompanied by a more downstream NEP type I promoter (P2) [24, 160] (Fig. 2a).

The *rrnP1* PEP promoter consists of typical −35 and −10 regions and an additional *cis* element termed RUA (r*R*NA operon *u*pstream *a*ctivator), an activator sequence well conserved in higher plants [156]. It was suggested that this RUA may replace, or modify, the function of −10 region, which does not seem to play an essential role for the *rrnP1* promoter activity.

The *rrnPC* promoter was shown to be used in a development-dependent or organ-specific manner, as it is involved in transcription in leaves and roots but not in cotyledons [149]. This "nonconsensus" (type II) NEP promoter lacks the YRTA and GAA box revealed by type I promoters. In spinach, a sequence 28 nt upstream of the transcription start site was identified to serve as a recognition site for CDF2 (chloroplast DNA-binding factor 2) [48, 161]. At the time these data were obtained, it was not yet known which NEP polymerase (RPOTmp or RPOTp) transcribes the operon from the *rrnPC* promoter. Recent studies using primer extension experiments show that it is specifically RPOTmp which recognizes this promoter in Arabidopsis [50].

4.2.2 BLRP, the Blue-Light-Responsive Promoter of psbD

In barley, the *psbD/C* operon can be transcribed from 4 PEP promoters (*P-2007*, *P-712*, *P-551*, *P-196*), one of which *P-551* is exceptional both in terms of its architecture and (blue-light-responsive) regulation. This unique PEP promoter has a typical −10 region (TATTCT) but a less conserved and nonessential −35 region (TTGAAT) [162]. Further upstream are the two light-responsive sequence elements named AAG repeat (at position −64 to −36) and PGT box (at −100 to −71). The latter is recognized by the transcription factor PGTF1 in its unphosphorylated form in the light. In the dark this factor is phosphorylated by an ADP-dependent kinase, resulting in inhibition of promoter binding and transcription. The AAG repeat box interacts with a multiprotein complex named AGF-binding factor, which leads to increased transcription activity [162, 163]. Interestingly, there seem to be differences as to the AAG repeat numbers in monocotyledonous (two repeats) versus dicotyledonous (one repeat) species, with possible implications for the distribution of blue-light regulation of chloroplast transcription in higher plants.

5 Summary and Perspectives

In this short overview chapter, we hope to have addressed major topics in the field of chloroplast transcription that have emerged over the past decades and years. We are aware, however, that attempting to do so is a difficult task because of the need for a strict focus, which unevitably means omission of numerous relevant and often particularly exciting findings.

What are the major coordinates in the field of plastid transcription? Compared to the pioneering phase when the first chloroplast DNA sequences were cloned and characterized [164], there has been a dramatic progress in our knowledge on chloroplast genes, promoters, and other regulatory elements (Table 2). Likewise, achievements of similar importance have become the basis for our current view of organellar gene expression, ranging from transcription via RNA processing and modification to translation and beyond. With regard to transcription, it appears most notable that RNA synthesis in plastids is not a matter of one single polymerase and its specificity factors as in bacteria, but it rather represents a multi-polymerase/multifactor situation. Of the numerous components of the chloroplast transcription apparatus, only a minority (PEP core subunits) is encoded by the organellar genome, while the great majority (NEPs, PEP factors) consists of nuclear gene products.

While not too long ago, a widely held view was that the control of plastid gene expression would be predominantly at the translational level; since then a paradigm change has taken place. Recent results have made it increasingly clear that multiple

Table 2
Some early milestones in chloroplast transcription research[a]

Year	Discovery	References
1964	RNA synthesis in isolated chloroplasts	[17]
1974	Bacterial-type multi-subunit RNA polymerase	[18]
1976	Transcriptionally active chromosome (TAC)	[74]
1977	DNA restriction mapping, gene cloning, and sequencing	[20, 164]
1983	Plastid transcription extracts/promoter strength assays	[77]
1987	*In-organello* run-on transcription in isolated chloroplasts	[79, 80]
1990	Stable chloroplast transformation in higher plants	[81]
1993	2nd (phage T3/T7-type) plastid RNA polymerase	[35]
1996	Nuclear genes for plastid sigma factors	[72, 73]
1997	Phosphorylation and redox control of plastid transcription	[136, 163]
2000	Multiple accessory proteins of RNA polymerase	[83]

[a]Findings (until the year 2000) that have served as initial focal points for subsequent work on a broader scale and/or using higher-resolution techniques

regulatory levels, including transcription, together determine the final amounts and activities of plastid gene products. For instance, sigma factor mutant and complementation experiments have demonstrated the impact of these transcription factors for plastid development and function. This does not exclude a—equally important—role of other regulators, many of which act posttranscriptionally, as is known, e.g., for members of the helical PPR protein family [165]. Indeed, a member of this protein family has recently been shown to functionally interact with one sigma factor, indicating a possible cross-connection between transcription and posttranscriptional events [126].

Individual chloroplast promoters have been shown to reveal differential activity during development and/or in environmentally controlled situations (e.g., *psbD* BLR promoter [162], *psbA* light dependency [166], *rbcL* light/organ specificity [143]). It can be anticipated that differential promoter activity may be a more general feature in chloroplast molecular biology. Further progress in this direction clearly depends on more broad-scale investigations on the arrangement, architecture, and activity of upstream control regions covering the full set of chloroplast transcription units from a given plant species. To take this even further, it should again be emphasized that the plastome is organized into discrete units consisting of DNA/RNA/protein complexes (nucleoids, collectively known as TAC). It would be instructive to learn whether or not these complexes are equivalent, i.e., whether differential transcription does occur within this population. Towards this end, high-resolution in situ techniques for nucleic acid and protein quantification will be helpful, as will small-scale microarrays. Complementary biochemical subfractionation techniques of nucleoids have been described already for bacterial systems [167]. Together, this would open up new avenues for reaching a full understanding of the plastid transcriptional "landscape" in different developmental stages, in response to environmental cues, and different plant species.

Significant achievements towards these goals have recently been made, with regard to both regulatory protein contacts and the full set of transcription initiation sites. For instance, using chloroplast chromatin immunoprecipitation (cpChIP) assays, preferential association of the PEP alpha-subunit with PEP promoters in front of photosynthesis genes was demonstrated. Furthermore, using the same technique with antibodies against the PEP-associated putative DNA-binding protein pTAC3, the authors presented evidence suggesting light-induced association of the transcription complex with cognate PEP promoters [172]. In a second notable strategy, advantage was taken of the property of terminator exonuclease TEX (epicenter) to discriminate between primary and processed RNA 5′ ends. Differential RNA sequencing (dRNA-seq) then allowed the authors to define a wide complement of chloroplast initiation sites as well as alternative, differentially transcribed, and potentially regulatory RNA species [173].

Acknowledgements

We thank all colleagues who made possible sharing data and exchange of views for making this review better, and we apologize that, for reasons of space, not all important work could be included. Research from our own group was funded by the DFG (LI261/21 and SFB480). Particular thanks are due to Brigitte Link for her strong commitment and expert technical assistance throughout this work.

References

1. Mereschkowsky C (1905) Über die Natur und den Ursprung der Chromatophoren im Pflanzenreiche. Biol Centralbl 25:593–604, 689–691
2. Herrmann RG (1997) Eukaryotism, towards a new interpretation. In: Schenk HEA, Herrmann RG, Jeon KW, Müller NE, Schwemmler W (eds) Eukaryotism and symbiosis. Springer, Berlin Heidelberg, 73–118
3. Gray MW (1999) Evolution of organellar genomes. Curr Opin Genet Dev 9: 678–687
4. Emelyanov VV (2003) Mitochondrial connection to the origin of the eukaryotic cell. Eur J Biochem 270:1599–1618
5. McFadden GI (2001) Chloroplast origin and integration. Plant Physiol 125:50–53
6. Leister D (2003) Chloroplast research in the genomic age. Trends Genet 19:47–56
7. Baginsky S, Siddique A, Gruissem W (2004) Proteome analysis of tobacco bright yellow-2 (BY-2) cell culture plastids as a model for undifferentiated heterotrophic plastids. J Proteome Res 3:1128–1137

8. Cline K, Henry R (1996) Import and routing of nucleus-encoded chloroplast proteins. Annu Rev Cell Dev Biol 12:1–26
9. The Arabidopsis Genome Initiative (2000) Analysis of the genome sequence of the flowering plant *Arabidopsis thaliana*. Nature 408:796–815
10. Martin W, Rujan T, Richly E, Hansen A, Cornelsen S, Lins T, Leister D, Stoebe B, Hasegawa M, Penny D (2002) Evolutionary analysis of *Arabidopsis*, cyanobacterial, and chloroplast genomes reveals plastid phylogeny and thousands of cyanobacterial genes in the nucleus. Proc Natl Acad Sci U S A 99:12246–12251
11. Palmer JD (1985) Comparative organization of chloroplast genomes. Annu Rev Genet 19:325–354
12. Sugiura M (1992) The chloroplast genome. Plant Mol Biol 19:149–168
13. Tsudzuki J, Ito S, Tsudzuki T, Wakasugi T, Sugiura M (1994) A new gene encoding tRNA(Pro) (GGG) is present in the chloroplast genome of black pine: a compilation of 32 tRNA genes from black pine chloroplasts. Curr Genet 26:153–158
14. Allen JF (1993) Control of gene expression by redox potential and the requirement for chloroplast and mitochondrial genomes. J Theor Biol 165:609–631
15. Clayton DA (1987) Nuclear gene products that function in mitochondrial DNA replication. Philos Trans R Soc Lond B Biol Sci 317:473–482
16. Kelly JL, Greenleaf AL, Lehman IR (1986) Isolation of the nuclear gene encoding a subunit of the yeast mitochondrial RNA polymerase. J Biol Chem 261:10348–10351
17. Kirk JT (1964) Studies on RNA synthesis in chloroplast preparations. Biochem Biophys Res Commun 16:233–238
18. Smith HJ, Bogorad L (1974) The polypeptide subunit structure of the DNA-dependent RNA polymerase of *Zea mays* chloroplasts. Proc Natl Acad Sci U S A 71:4839–4842
19. Hu J, Bogorad L (1990) Maize chloroplast RNA polymerase: the 180-, 120-, and 38-kilodalton polypeptides are encoded in chloroplast genes. Proc Natl Acad Sci U S A 87:1531–1535
20. Shinozaki K, Ohme M, Tanaka M, Wakasugi T, Hayashida N, Matsubayashi T, Zaita N, Chunwongse J, Obokata J, Yamaguchi-Shinozaki K, Ohto C, Torazawa K, Meng BY, Sugita M, Deno H, Kamogashira T, Yamada K, Kusuda J, Takaiwa F, Kato A, Tohdoh N, Shimada H, Sugiura M (1986) The complete nucleotide sequence of the tobacco chloroplast genome: its gene organization and expression. EMBO J 5:2043–2049
21. Bunger W, Feierabend J (1980) Capacity for RNA synthesis in 70S ribosome-deficient plastids of heat-bleached rye leaves. Planta 149:163–169
22. dePamphilis CW, Palmer JD (1990) Loss of photosynthetic and chlororespiratory genes from the plastid genome of a parasitic flowering plant. Nature 348:337–339
23. Hess WR, Prombona A, Fieder B, Subramanian AR, Borner T (1993) Chloroplast rps15 and the rpoB/C1/C2 gene cluster are strongly transcribed in ribosome-deficient plastids: evidence for a functioning non-chloroplast-encoded RNA polymerase. EMBO J 12:563–571
24. Allison LA, Simon LD, Maliga P (1996) Deletion of rpoB reveals a second distinct transcription system in plastids of higher plants. EMBO J 15:2802–2809
25. Maliga P (1998) Two plastid RNA polymerases of higher plants: an evolving story. Trends Plant Sci 3:4–6
26. Serino G, Maliga P (1998) RNA polymerase subunits encoded by the plastid rpo genes are not shared with the nucleus-encoded plastid enzyme. Plant Physiol 117:1165–1170
27. Hess WR, Borner T (1999) Organellar RNA polymerases of higher plants. Int Rev Cytol 190:1–59
28. Liere K, Maliga P (1999) In vitro characterization of the tobacco rpoB promoter reveals a core sequence motif conserved between phage-type plastid and plant mitochondrial promoters. EMBO J 18:249–257
29. Baumgartner BJ, Rapp JC, Mullet JE (1993) Plastid genes encoding the transcription/translation apparatus are differentially transcribed early in barley (Hordeum vulgare) chloroplast development (evidence for selective stabilization of psbA mRNA). Plant Physiol 101:781–791
30. Allison LA (2000) The role of sigma factors in plastid transcription. Biochimie 82:537–548
31. Shiina T, Tsunoyama Y, Nakahira Y, Khan MS (2005) Plastid RNA polymerases, promoters, and transcription regulators in higher plants. Int Rev Cytol 244:1–68
32. Schweer J, Turkeri H, Kolpack A, Link G (2010) Role and regulation of plastid sigma factors and their functional interactors during chloroplast transcription—recent lessons from *Arabidopsis thaliana*. Eur J Cell Biol 89:940–946

33. Morden CW, Wolfe KH, dePamphilis CW, Palmer JD (1991) Plastid translation and transcription genes in a non-photosynthetic plant: intact, missing and pseudo genes. EMBO J 10:3281–3288
34. Hajdukiewicz PT, Allison LA, Maliga P (1997) The two RNA polymerases encoded by the nuclear and the plastid compartments transcribe distinct groups of genes in tobacco plastids. EMBO J 16:4041–4048
35. Lerbs-Mache S (1993) The 110-kDa polypeptide of spinach plastid DNA-dependent RNA polymerase: single-subunit enzyme or catalytic core of multimeric enzyme complexes? Proc Natl Acad Sci U S A 90:5509–5513
36. Ikeda TM, Gray MW (1999) Identification and characterization of T3/T7 bacteriophage-like RNA polymerase sequences in wheat. Plant Mol Biol 40:567–578
37. Hedtke B, Borner T, Weihe A (1997) Mitochondrial and chloroplast phage-type RNA polymerases in *Arabidopsis*. Science 277:809–811
38. Hedtke B, Borner T, Weihe A (2000) One RNA polymerase serving two genomes. EMBO Rep 1:435–440
39. Liere K, Kaden D, Maliga P, Borner T (2004) Overexpression of phage-type RNA polymerase RpoTp in tobacco demonstrates its role in chloroplast transcription by recognizing a distinct promoter type. Nucleic Acids Res 32:1159–1165
40. Emanuel C, Weihe A, Graner A, Hess WR, Borner T (2004) Chloroplast development affects expression of phage-type RNA polymerases in barley leaves. Plant J 38:460–472
41. Silhavy D, Maliga P (1998) Mapping of promoters for the nucleus-encoded plastid RNA polymerase (NEP) in the *iojap* maize mutant. Curr Genet 33:340–344
42. Gray MW, Lang BF (1998) Transcription in chloroplasts and mitochondria: a tale of two polymerases. Trends Microbiol 6:1–3
43. Chang CC, Sheen J, Bligny M, Niwa Y, Lerbs-Mache S, Stern DB (1999) Functional analysis of two maize cDNAs encoding T7-like RNA polymerases. Plant Cell 11:911–926
44. Kusumi K, Yara A, Mitsui N, Tozawa Y, Iba K (2004) Characterization of a rice nuclear-encoded plastid RNA polymerase gene OsRpoTp. Plant Cell Physiol 45:1194–1201
45. Kobayashi Y, Dokiya Y, Sugiura M, Niwa Y, Sugita M (2001) Genomic organization and organ-specific expression of a nuclear gene encoding phage-type RNA polymerase in *Nicotiana sylvestris*. Gene 279:33–40
46. Kobayashi Y, Dokiya Y, Sugita M (2001) Dual targeting of phage-type RNA polymerase to both mitochondria and plastids is due to alternative translation initiation in single transcripts. Biochem Biophys Res Commun 289:1106–1113
47. Hedtke B, Legen J, Weihe A, Herrmann RG, Borner T (2002) Six active phage-type RNA polymerase genes in *Nicotiana tabacum*. Plant J 30:625–637
48. Bligny M, Courtois F, Thaminy S, Chang CC, Lagrange T, Baruah-Wolff J, Stern D, Lerbs-Mache S (2000) Regulation of plastid rDNA transcription by interaction of CDF2 with two different RNA polymerases. EMBO J 19:1851–1860
49. Kuhn K, Richter U, Meyer EH, Delannoy E, de Longevialle AF, O'Toole N, Borner T, Millar AH, Small ID, Whelan J (2009) Phage-type RNA polymerase RPOTmp performs gene-specific transcription in mitochondria of *Arabidopsis thaliana*. Plant Cell 21: 2762–2779
50. Courtois F, Merendino L, Demarsy E, Mache R, Lerbs-Mache S (2007) Phage-type RNA polymerase RPOTmp transcribes the rrn operon from the PC promoter at early developmental stages in *Arabidopsis*. Plant Physiol 145:712–721
51. Swiatecka-Hagenbruch M, Emanuel C, Hedtke B, Liere K, Borner T (2008) Impaired function of the phage-type RNA polymerase RpoTp in transcription of chloroplast genes is compensated by a second phage-type RNA polymerase. Nucleic Acids Res 36:785–792
52. Hricova A, Quesada V, Micol JL (2006) The SCABRA3 nuclear gene encodes the plastid RpoTp RNA polymerase, which is required for chloroplast biogenesis and mesophyll cell proliferation in *Arabidopsis*. Plant Physiol 141:942–956
53. Bottomley W, Smith HJ, Bogorad L (1971) RNA polymerases of maize: partial purification and properties of the chloroplast enzyme. Proc Natl Acad Sci U S A 68:2412–2416
54. McKown RL, Tewari KK (1984) Purification and properties of a pea chloroplast DNA polymerase. Proc Natl Acad Sci U S A 81:2354–2358
55. Lerbs S, Briat JF, Mache R (1983) Chloroplast RNA polymerase from spinach: purification and DNA-binding proteins. Plant Mol Biol 2:67–74
56. Igloi GL, Kössel H (1992) The transcriptional apparatus of chloroplasts. Crit Rev Plant Sci 10:525–558

57. Kaneko T, Sato S, Kotani H, Tanaka A, Asamizu E, Nakamura Y, Miyajima N, Hirosawa M, Sugiura M, Sasamoto S, Kimura T, Hosouchi T, Matsuno A, Muraki A, Nakazaki N, Naruo K, Okumura S, Shimpo S, Takeuchi C, Wada T, Watanabe A, Yamada M, Yasuda M, Tabata S (1996) Sequence analysis of the genome of the unicellular cyanobacterium *Synechocystis* sp. strain PCC6803. II. Sequence determination of the entire genome and assignment of potential protein-coding regions (supplement). DNA Res 3:185–209
58. Severinov K, Mustaev A, Kukarin A, Muzzin O, Bass I, Darst SA, Goldfarb A (1996) Structural modules of the large subunits of RNA polymerase. Introducing archaebacterial and chloroplast split sites in the beta and beta′ subunits of *Escherichia coli* RNA polymerase. J Biol Chem 271:27969–27974
59. Suzuki JY, Maliga P (2000) Engineering of the rpl23 gene cluster to replace the plastid RNA polymerase alpha subunit with the *Escherichia coli* homologue. Curr Genet 38:218–225
60. Shinozaki K, Sugiura M (1986) Organization of chloroplast genomes. Adv Biophys 21:57–78
61. Sugiura C, Kobayashi Y, Aoki S, Sugita C, Sugita M (2003) Complete chloroplast DNA sequence of the moss *Physcomitrella patens*: evidence for the loss and relocation of rpoA from the chloroplast to the nucleus. Nucleic Acids Res 31:5324–5331
62. Kabeya Y, Kobayashi Y, Suzuki H, Itoh J, Sugita M (2007) Transcription of plastid genes is modulated by two nuclear-encoded alpha subunits of plastid RNA polymerase in the moss *Physcomitrella patens*. Plant J 52:730–741
63. Smith AC, Purton S (2002) The principle apparatus of algal plastids. Eur J Phycol 37:301–311
64. Wilson RJ, Denny PW, Preiser PR, Rangachari K, Roberts K, Roy A, Whyte A, Strath M, Moore DJ, Moore PW, Williamson DH (1996) Complete gene map of the plastid-like DNA of the malaria parasite *Plasmodium falciparum*. J Mol Biol 261:155–172
65. Barbrook AC, Howe CJ, Kurniawan DP, Tarr SJ (2010) Organization and expression of organellar genomes. Philos Trans R Soc Lond B Biol Sci 365:785–797
66. Burgess RR, Travers AA, Dunn JJ, Bautz EKF (1969) Factor stimulating transcription by RNA polymerase. Nature 221:43–46
67. Darst SA, Polyakov A, Richter C, Zhang G (1998) Structural studies of *Escherichia coli* RNA polymerase. Cold Spring Harb Symp Quant Biol 63:269–276
68. Zhang G, Campbell EA, Minakhin L, Richter C, Severinov K, Darst SA (1999) Crystal structure of *Thermus aquaticus* core RNA polymerase at 3.3 A resolution. Cell 98: 811–824
69. Busby S, Ebright RH (1994) Promoter structure, promoter recognition, and transcription activation in prokaryotes. Cell 79:743–746
70. Lerbs S, Brautigam E, Parthier B (1985) Polypeptides of DNA-dependent RNA polymerase of spinach chloroplasts: characterization by antibody-linked polymerase assay and determination of sites of synthesis. EMBO J 4:1661–1666
71. Bulow S, Link G (1988) Sigma-like activity from mustard (*Sinapis alba* L) chloroplasts conferring DNA-binding and transcription specificity to *E. coli* core RNA polymerase. Plant Mol Biol 10:349–357
72. Liu B, Troxler RF (1996) Molecular characterization of a positively photoregulated nuclear gene for a chloroplast RNA polymerase sigma factor in *Cyanidium caldarium*. Proc Natl Acad Sci U S A 93: 3313–3318
73. Tanaka K, Oikawa K, Ohta N, Kuroiwa H, Kuroiwa T, Takahashi H (1996) Nuclear encoding of a chloroplast RNA polymerase sigma subunit in a red alga. Science 272: 1932–1935
74. Hallick RB, Lipper C, Richards OC, Rutter WJ (1976) Isolation of a transcriptionally active chromosome from chloroplasts of *Euglena gracilis*. Biochemistry 15:3039–3045
75. Pfalz J, Liere K, Kandlbinder A, Dietz KJ, Oelmuller R (2006) pTAC2, -6, and -12 are components of the transcriptionally active plastid chromosome that are required for plastid gene expression. Plant Cell 18:176–197
76. Link G (1984) DNA sequence requirements for the accurate transcription of a protein-coding plastid gene in a plastid in vitro system from mustard (*Sinapis alba* L.). EMBO J 3: 1697–1704
77. Gruissem W, Greenberg BM, Zurawski G, Prescott DM, Hallick RB (1983) Biosynthesis of chloroplast transfer RNA in a spinach chloroplast transcription system. Cell 35:815–828
78. Orozco EM Jr, Mullet JE, Chua NH (1985) An in vitro system for accurate transcription initiation of chloroplast protein genes. Nucleic Acids Res 13:1283–1302
79. Deng XW, Stern DB, Tonkyn JC, Gruissem W (1987) Plastid run-on transcription.

Application to determine the transcriptional regulation of spinach plastid genes. J Biol Chem 262:9641–9648
80. Mullet JE, Klein RR (1987) Transcription and RNA stability are important determinants of higher plant chloroplast RNA levels. EMBO J 6:1571–1579
81. Svab Z, Hajdukiewicz P, Maliga P (1990) Stable transformation of plastids in higher plants. Proc Natl Acad Sci U S A 87: 8526–8530
82. Pfannschmidt T, Link G (1994) Separation of two classes of plastid DNA-dependent RNA polymerases that are differentially expressed in mustard (*Sinapis alba* L.) seedlings. Plant Mol Biol 25:69–81
83. Pfannschmidt T, Ogrzewalla K, Baginsky S, Sickmann A, Meyer HE, Link G (2000) The multisubunit chloroplast RNA polymerase A from mustard (*Sinapis alba* L.). Integration of a prokaryotic core into a larger complex with organelle-specific functions. Eur J Biochem 267:253–261
84. Ogrzewalla K, Piotrowski M, Reinbothe S, Link G (2002) The plastid transcription kinase from mustard (*Sinapis alba* L.). A nuclear-encoded CK2-type chloroplast enzyme with redox-sensitive function. Eur J Biochem 269:3329–3337
85. Schroter Y, Steiner S, Matthai K, Pfannschmidt T (2010) Analysis of oligomeric protein complexes in the chloroplast sub-proteome of nucleic acid-binding proteins from mustard reveals potential redox regulators of plastid gene expression. Proteomics 10:2191–2204
86. Suzuki JY, Ytterberg AJ, Beardslee TA, Allison LA, Wijk KJ, Maliga P (2004) Affinity purification of the tobacco plastid RNA polymerase and in vitro reconstitution of the holoenzyme. Plant J 40:164–172
87. Satoh J, Baba K, Nakahira Y, Tsunoyama Y, Shiina T, Toyoshima Y (1999) Developmental stage-specific multi-subunit plastid RNA polymerases (PEP) in wheat. Plant J 18:407–415
88. Helmann JD, Chamberlin MJ (1988) Structure and function of bacterial sigma factors. Annu Rev Biochem 57:839–872
89. Wosten MM (1998) Eubacterial sigma-factors. FEMS Microbiol Rev 22:127–150
90. Gruber TM, Gross CA (2003) Multiple sigma subunits and the partitioning of the bacterial transcription space. Annu Rev Microbiol 57:441–456
91. Geszvain K, Landick R (2005) The structure of bacterial RNA polymerase. Website Edition (http://www. bact. wisc. edu/Landick) 1:1–10
92. Ishihama A (2000) Functional modulation of *Escherichia coli* RNA polymerase. Annu Rev Microbiol 54:499–518
93. Malhotra A, Severinova E, Darst SA (1996) Crystal structure of a sigma 70 subunit fragment from *E. coli* RNA polymerase. Cell 87: 127–136
94. Vassylyev DG, Sekine S, Laptenko O, Lee J, Vassylyeva MN, Borukhov S, Yokoyama S (2002) Crystal structure of a bacterial RNA polymerase holoenzyme at 2.6 A resolution. Nature 417:712–719
95. Campbell EA, Muzzin O, Chlenov M, Sun JL, Olson CA, Weinman O, Trester-Zedlitz ML, Darst SA (2002) Structure of the bacterial RNA polymerase promoter specificity sigma subunit. Mol Cell 9:527–539
96. Campbell EA, Tupy JL, Gruber TM, Wang S, Sharp MM, Gross CA, Darst SA (2003) Crystal structure of *Escherichia coli* sigma E with the cytoplasmic domain of its anti-sigma RseA. Mol Cell 11:1067–1078
97. Link G (1994) Plastid differentiation: organelle promoters and transcription factors. Results Probl Cell Differ 20:65–85
98. Tanaka K, Tozawa Y, Mochizuki N, Shinozaki K, Nagatani A, Wakasa K, Takahashi H (1997) Characterization of three cDNA species encoding plastid RNA polymerase sigma factors in *Arabidopsis thaliana*: evidence for the sigma factor heterogeneity in higher plant plastids. FEBS Lett 413:309–313
99. Isono K, Niwa Y, Satoh K, Kobayashi H (1997) Evidence for transcriptional regulation of plastid photosynthesis genes in *Arabidopsis thaliana* roots. Plant Physiol 114:623–630
100. Tozawa Y, Tanaka K, Takahashi H, Wakasa K (1998) Nuclear encoding of a plastid sigma factor in rice and its tissue- and light-dependent expression. Nucleic Acids Res 26:415–419
101. Kestermann M, Neukirchen S, Kloppstech K, Link G (1998) Sequence and expression characteristics of a nuclear-encoded chloroplast sigma factor from mustard (*Sinapis alba*). Nucleic Acids Res 26:2747–2753
102. Tan S, Troxler RF (1999) Characterization of two chloroplast RNA polymerase sigma factors from *Zea mays*: photoregulation and differential expression. Proc Natl Acad Sci U S A 96:5316–5321
103. Morikawa K, Ito S, Tsunoyama Y, Nakahira Y, Shiina T, Toyoshima Y (1999) Circadian-regulated expression of a nuclear-encoded plastid sigma factor gene (*sigA*) in wheat seedlings. FEBS Lett 451:275–278
104. Oikawa K, Fujiwara M, Nakazato E, Tanaka K, Takahashi H (2000) Characterization of two plastid sigma factors, *SigA1* and *SigA2*, that mainly function in matured chloroplasts in *Nicotiana tabacum*. Gene 261:221–228

105. Schweer J (2010) Plant sigma factors come of age: flexible transcription factor network for regulated plastid gene expression. Endocytobiosis Cell Res 20:1–20
106. Bohne AV, Irihimovitch V, Weihe A, Stern DB (2006) Chlamydomonas reinhardtii encodes a single sigma70-like factor which likely functions in chloroplast transcription. Curr Genet 49:333–340
107. Fujiwara M, Nagashima A, Kanamaru K, Tanaka K, Takahashi H (2000) Three new nuclear genes, *sigD*, *sigE* and *sigF*, encoding putative plastid RNA polymerase sigma factors in *Arabidopsis thaliana*. FEBS Lett 481:47–52
108. Kanamaru K, Fujiwara M, Seki M, Katagiri T, Nakamura M, Mochizuki N, Nagatani A, Shinozaki K, Tanaka K, Takahashi H (1999) Plastidic RNA polymerase sigma factors in *Arabidopsis*. Plant Cell Physiol 40:832–842
109. Lysenko EA (2007) Plant sigma factors and their role in plastid transcription. Plant Cell Rep 26:845–859
110. Lerbs-Mache S (2010) Function of plastid sigma factors in higher plants: regulation of gene expression or just preservation of constitutive transcription? Plant Mol Biol 76: 235–249
111. Ishizaki Y, Tsunoyama Y, Hatano K, Ando K, Kato K, Shinmyo A, Kobori M, Takeba G, Nakahira Y, Shiina T (2005) A nuclear-encoded sigma factor, *Arabidopsis* SIG6, recognizes sigma-70 type chloroplast promoters and regulates early chloroplast development in cotyledons. Plant J 42:133–144
112. Nagashima A, Hanaoka M, Motohashi R, Seki M, Shinozaki K, Kanamaru K, Takahashi H, Tanaka K (2004) DNA microarray analysis of plastid gene expression in an *Arabidopsis* mutant deficient in a plastid transcription factor sigma, SIG2. Biosci Biotechnol Biochem 68:694–704
113. Tsunoyama Y, Ishizaki Y, Morikawa K, Kobori M, Nakahira Y, Takeba G, Toyoshima Y, Shiina T (2004) Blue light-induced transcription of plastid-encoded *psbD* gene is mediated by a nuclear-encoded transcription initiation factor, AtSig5. Proc Natl Acad Sci U S A 101:3304–3309
114. Morikawa K, Shiina T, Murakami S, Toyoshima Y (2002) Novel nuclear-encoded proteins interacting with a plastid sigma factor, Sig1, in *Arabidopsis thaliana*. FEBS Lett 514:300–304
115. Xie YD, Li W, Guo D, Dong J, Zhang Q, Fu Y, Ren D, Peng M, Xia Y (2010) The *Arabidopsis* gene SIGMA FACTOR-BINDING PROTEIN 1 plays a role in the salicylate- and jasmonate-mediated defence responses. Plant Cell Environ 33:828–839
116. Kolpack A (2010) Regulatory protein-protein interactions in the plastid transcription. PhD thesis, Ruhr University, Bochum
117. Kanamaru K, Tanaka K (2004) Roles of chloroplast RNA polymerase sigma factors in chloroplast development and stress response in higher plants. Biosci Biotechnol Biochem 68:2215–2223
118. Privat I, Hakimi MA, Buhot L, Favory JJ, Mache-Lerbs S (2003) Characterization of *Arabidopsis* plastid sigma-like transcription factors SIG1, SIG2 and SIG3. Plant Mol Biol 51:385–399
119. Hanaoka M, Kanamaru K, Takahashi H, Tanaka K (2003) Molecular genetic analysis of chloroplast gene promoters dependent on SIG2, a nucleus-encoded sigma factor for the plastid-encoded RNA polymerase, in *Arabidopsis thaliana*. Nucleic Acids Res 31:7090–7098
120. Favory JJ, Kobayshi M, Tanaka K, Peltier G, Kreis M, Valay JG, Lerbs-Mache S (2005) Specific function of a plastid sigma factor for *ndhF* gene transcription. Nucleic Acids Res 33:5991–5999
121. Zghidi W, Merendino L, Cottet A, Mache R, Lerbs-Mache S (2007) Nucleus-encoded plastid sigma factor SIG3 transcribes specifically the *psbN* gene in plastids. Nucleic Acids Res 35:455–464
122. Yao J, Roy-Chowdhury S, Allison LA (2003) AtSig5 is an essential nucleus-encoded *Arabidopsis* sigma-like factor. Plant Physiol 132:739–747
123. Loschelder H, Schweer J, Link B, Link G (2006) Dual temporal role of plastid sigma factor 6 in *Arabidopsis* development. Plant Physiol 142:642–650
124. Schweer J, Loschelder H, Link G (2006) A promoter switch that can rescue a plant sigma factor mutant. FEBS Lett 580:6617–6622
125. Schweer J, Geimer S, Meurer J, Link G (2009) *Arabidopsis* mutants carrying chimeric sigma factor genes reveal regulatory determinants for plastid gene expression. Plant Cell Physiol 50:1382–1386
126. Chi W, Mao J, Li Q, Ji D, Zou M, Lu C, Zhang L (2010) Interaction of the pentatricopeptide-repeat protein DELAYED GREENING 1 with sigma factor SIG6 in the regulation of chloroplast gene expression in *Arabidopsis* cotyledons. Plant J 64:14–25
127. Lahiri SD, Yao J, McCumbers C, Allison LA (1999) Tissue-specific and light-dependent expression within a family of nuclear-encoded sigma-like factors from *Zea mays*. Mol Cell Biol Res Commun 1:14–20

128. Kubota Y, Miyao A, Hirochika H, Tozawa Y, Yasuda H, Tsunoyama Y, Niwa Y, Imamura S, Shirai M, Asayama M (2007) Two novel nuclear genes, OsSIG5 and OsSIG6, encoding potential plastid sigma factors of RNA polymerase in rice: tissue-specific and light-responsive gene expression. Plant Cell Physiol 48:186–192
129. Homann A, Link G (2003) DNA-binding and transcription characteristics of three cloned sigma factors from mustard (*Sinapis alba* L.) suggest overlapping and distinct roles in plastid gene expression. Eur J Biochem 270:1288–1300
130. Hara K, Morita M, Takahashi R, Sugita M, Kato S, Aoki S (2001) Characterization of two genes, *Sig1* and *Sig2*, encoding distinct plastid sigma factors(1) in the moss *Physcomitrella patens*: phylogenetic relationships to plastid sigma factors in higher plants. FEBS Lett 499:87–91
131. Ichikawa K, Sugita M, Imaizumi T, Wada M, Aoki S (2004) Differential expression on a daily basis of plastid sigma factor genes from the moss *Physcomitrella patens*. Regulatory interactions among *PpSig5*, the circadian clock, and blue light signaling mediated by cryptochromes. Plant Physiol 136:4285–4298
132. Browning DF, Busby SJ (2004) The regulation of bacterial transcription initiation. Nat Rev Microbiol 2:57–65
133. Tiller K, Link G (1995) Sigma-like plastid transcription factors. Methods Mol Biol 37:337–348
134. Tiller K, Link G (1993) Phosphorylation and dephosphorylation affect functional characteristics of chloroplast and etioplast transcription systems from mustard (*Sinapis alba* L.). EMBO J 12:1745–1753
135. Baginsky S, Tiller K, Pfannschmidt T, Link G (1999) PTK, the chloroplast RNA polymerase-associated protein kinase from mustard (*Sinapis alba*), mediates redox control of plastid in vitro transcription. Plant Mol Biol 39:1013–1023
136. Baginsky S, Tiller K, Link G (1997) Transcription factor phosphorylation by a protein kinase associated with chloroplast RNA polymerase from mustard (*Sinapis alba*). Plant Mol Biol 34:181–189
137. Meggio F, Pinna LA (2003) One-thousand-and-one substrates of protein kinase CK2? FASEB J 17:349–368
138. Johnson NL, Gardner AM, Diener KM, Lange-Carter CA, Gleavy J, Jarpe MB, Minden A, Karin M, Zon LI, Johnson GL (1996) Signal transduction pathways regulated by mitogen-activated/extracellular response kinase kinase kinase induce cell death. J Biol Chem 271:3229–3237
139. Schweer J, Turkeri H, Link B, Link G (2010) *AtSIG6*, a plastid sigma factor from *Arabidopsis*, reveals functional impact of cpCK2 phosphorylation. Plant J 62:192–202
140. Turkeri H, Schweer J, Link G (2012) Phylogenetic and functional features of the plastid transcription kinase cpCK2 from *Arabidopsis* signify a role of cysteinyl SH-groups in regulatory phosphorylation of plastid sigma factors. FEBS J 279:395–409
141. Shimizu M, Kato H, Ogawa T, Kurachi A, Nakagawa Y, Kobayashi H (2010) Sigma factor phosphorylation in the photosynthetic control of photosystem stoichiometry. Proc Natl Acad Sci U S A 107:10760–10764
142. Kornberg RD (2007) The molecular basis of eukaryotic transcription. Proc Natl Acad Sci U S A 104:12955–12961
143. Shiina T, Allison L, Maliga P (1998) rbcL transcript levels in tobacco plastids are independent of light: reduced dark transcription rate is compensated by increased mRNA stability. Plant Cell 10:1713–1722
144. Weihe A, Borner T (1999) Transcription and the architecture of promoters in chloroplasts. Trends Plant Sci 4:169–170
145. Liere K, Maliga P (2001) Plastid RNA polymerases. In: Aro E-M, Andersson B (eds) Regulation of photosynthesis. Kluwer Academic Publishers, Dordrecht, The Netherlands, pp 29–49
146. Liere K, Borner T (2007) Transcription and transcriptional regulation in plastids. In: Bock R (ed) Topics in current genetics: cell and molecular biology of plastids. Springer, Berlin, pp 121–174
147. Kapoor S, Sugiura M (1999) Identification of two essential sequence elements in the nonconsensus type II PatpB-290 plastid promoter by using plastid transcription extracts from cultured tobacco BY-2 cells. Plant Cell 11:1799–1810
148. Kapoor S, Suzuki JY, Sugiura M (1997) Identification and functional significance of a new class of non-consensus-type plastid promoters. Plant J 11:327–337
149. Sriraman P, Silhavy D, Maliga P (1998) Transcription from heterologous rRNA operon promoters in chloroplasts reveals requirement for specific activating factors. Plant Physiol 117:1495–1499
150. Sriraman P, Silhavy D, Maliga P (1998) The phage-type PclpP-53 plastid promoter comprises sequences downstream of the transcription initiation site. Nucleic Acids Res 26: 4874–4879

151. Eisermann A, Tiller K, Link G (1990) In vitro transcription and DNA binding characteristics of chloroplast and etioplast extracts from mustard (*Sinapis alba*) indicate differential usage of the *psbA* promoter. EMBO J 9:3981–3987
152. Christopher DA, Kim M, Mullet JE (1992) A novel light-regulated promoter is conserved in cereal and dicot chloroplasts. Plant Cell 4:785–798
153. To KY, Cheng MC, Suen DF, Mon DP, Chen LF, Chen SC (1996) Characterization of the light-responsive promoter of rice chloroplast *psbD-C* operon and the sequence-specific DNA binding factor. Plant Cell Physiol 37:660–666
154. Hoffer PH, Christopher DA (1997) Structure and blue-light-responsive transcription of a chloroplast *psbD* promoter from *Arabidopsis thaliana*. Plant Physiol 115:213–222
155. Cheng MC, Wu SP, Chen LF, Chen SC (1997) Identification and purification of a spinach chloroplast DNA-binding protein that interacts specifically with the plastid *psaA-psaB-rps14* promoter region. Planta 203:373–380
156. Suzuki JY, Sriraman P, Svab Z, Maliga P (2003) Unique architecture of the plastid ribosomal RNA operon promoter recognized by the multisubunit RNA polymerase in tobacco and other higher plants. Plant Cell 15:195–205
157. Pfannschmidt T, Link G (1997) The A and B forms of plastid DNA-dependent RNA polymerase from mustard (*Sinapis alba* L.) transcribe the same genes in a different developmental context. Mol Gen Genet 257:35–44
158. Iratni R, Diederich L, Harrak H, Bligny M, Lerbs-Mache S (1997) Organ-specific transcription of the *rrn* operon in spinach plastids. J Biol Chem 272:13676–13682
159. Hubschmann T, Borner T (1998) Characterisation of transcript initiation sites in ribosome-deficient barley plastids. Plant Mol Biol 36:493–496
160. Vera A, Sugiura M (1995) Chloroplast rRNA transcription from structurally different tandem promoters: an additional novel-type promoter. Curr Genet 27:280–284
161. Baeza L, Bertrand A, Mache R, Lerbs-Mache S (1991) Characterization of a protein binding sequence in the promoter region of the 16S rRNA gene of the spinach chloroplast genome. Nucleic Acids Res 19:3577–3581
162. Kim M, Thum KE, Morishige DT, Mullet JE (1999) Detailed architecture of the barley chloroplast *psbD-psbC* blue light-responsive promoter. J Biol Chem 274:4684–4692
163. Kim M, Mullet JE (1995) Identification of a sequence-specific DNA binding factor required for transcription of the barley chloroplast blue light-responsive *psbD-psbC* promoter. Plant Cell 7:1445–1457
164. Bogorad L, Bedbrook JR, Davidson JN, Hanson MR, Kolodner R (1977) Genes for plastid ribosomal proteins and RNAs. Brookhaven Symp Biol 1–15
165. Schmitz-Linneweber C, Small I (2008) Pentatricopeptide repeat proteins: a socket set for organelle gene expression. Trends Plant Sci 13:663–670
166. Mullet JE (1993) Dynamic regulation of chloroplast transcription. Plant Physiol 103: 309–313
167. Dillon SC, Dorman CJ (2010) Bacterial nucleoid-associated proteins, nucleoid structure and gene expression. Nat Rev Microbiol 8:185–195
168. Derelle E, Ferraz C, Rombauts S, Rouze P, Worden AZ, Robbens S, Partensky F, Degroeve S, Echeynie S, Cooke R, Saeys Y, Wuyts J, Jabbari K, Bowler C, Panaud O, Piegu B, Ball SG, Ral JP, Bouget FY, Piganeau G, De Baets B, Picard A, Delseny M, Demaille J, Van de Peer Y, Moreau H (2006) Genome analysis of the smallest free-living eukaryote *Ostreococcus tauri* unveils many unique features. Proc Natl Acad Sci U S A 103:11647–11652
169. Richter U, Kiessling J, Hedtke B, Decker E, Reski R, Borner T, Weihe A (2002) Two *RpoT* genes of *Physcomitrella patens* encode phage-type RNA polymerases with dual targeting to mitochondria and plastids. Gene 290:95–105
170. Kabeya Y, Hashimoto K, Sato N (2002) Identification and characterization of two phage-type RNA polymerase cDNAs in the moss *Physcomitrella patens*: implication of recent evolution of nuclear-encoded RNA polymerase of plastids in plants. Plant Cell Physiol 43:245–255
171. Yin C, Richter U, Borner T, Weihe A (2010) Evolution of plant phage-type RNA polymerases: the genome of the basal angiosperm *Nuphar advena* encodes two mitochondrial and one plastid phage-type RNA polymerases. BMC Evol Biol 10:379
172. Yagi Y, Ishizaki Y, Nakahira Y, Tozawa Y, Shiina T (2012) Eukaryotic-type plastid nucleoid protein pTAC3 is essential for transcription by the bacterial-type plastid RNA polymerase. Proc Natl Acad Sci U S A 109:7541–7546
173. Zhelyazkova P, Sharma CM, Forstner KU, Liere K, Vogel J, Borner T (2012) The primary transcriptome of barley chloroplasts: numerous noncoding RNAs and the dominating role of the plastid-encoded RNA polymerase. Plant Cell 24:123–136

Chapter 4

Plastid mRNA Translation

Masahiro Sugiura

Abstract

Overall translational machinery in plastids is similar to that of *E. coli*. Initiation is the crucial step for translation and this step in plastids is somewhat different from that of *E. coli*. Unlike the Shine-Dalgarno sequence in *E. coli*, *cis*-elements for translation initiation are not well conserved in plastid mRNAs. Specific *trans*-acting factors are generally required for translation initiation and its regulation in plastids. During translation elongation, ribosomes pause sometimes on photosynthesis-related mRNAs due probably to proper insertion of nascent polypeptides into membrane complexes. Codon usage of plastid mRNAs is different from that of *E. coli* and mammalian cells. Plastid mRNAs do not have the so-called rare codons. Translation efficiencies of several synonymous codons are not always correlated with codon usage in plastid mRNAs.

Key words Chloroplast, Codon, mRNA, Translation, 5′-UTR, Ribosome, tRNA, *Cis*-element, *Trans*-factor, In vitro translation system

1 Introduction

The plastid genome in flowering plants includes ca. 80 protein-coding genes [1, 2]. Many of these genes constitute operons and they are cotranscribed as polycistronic pre-mRNAs [3, 4]. These pre-mRNAs are cleaved into shorter mRNAs [5–8]. Ten to 12 genes contain introns, and these introns should be removed by RNA splicing [9, 10]. In addition, C-to-U RNA editing occurs over 30 sites in ca. 20 protein-coding sequences, which result in amino acid substitutions [11, 12]. Due to lose transcription termination, the 3′-untranslated region (3′-UTR) of pre-mRNAs is often cut and trimmed [13, 14]. A set of processing events results ultimately in the formation of mature mRNAs, mostly monocistronic mRNA but sometimes dicistronic and tricistronic mRNAs [7, 8, 15, 16]. Transcription and translation are generally coupled in bacteria, namely, translation can start during mRNA synthesis. Complex mRNA processing steps are likely to prevent such

Pal Maliga (ed.), *Chloroplast Biotechnology: Methods and Protocols*, Methods in Molecular Biology, vol. 1132,
DOI 10.1007/978-1-62703-995-6_4, © Springer Science+Business Media New York 2014

coupling in plastids. Actually, plastid gene expression is mainly controlled at posttranscriptional steps: mRNA processing, translation, and mRNA stability [14, 17].

The plastid was originated from an ancestral photosynthetic prokaryote related to cyanobacteria. Hence, the translation machinery and translation mechanism of plastids are generally called prokaryotic-type. Prokaryotes are highly diverse, and detailed translation mechanisms have been unveiled almost exclusively using *E. coli* and its plasmids and phages [18]. Therefore, "*E. coli*-type" is suitable in many cases rather than prokaryotic-type. As described below, the translation initiation and its regulation in plastids differ significantly from those in *E. coli*.

These are several ways to analyze translation in plastids: (1) Biochemical approaches. Translation rates of endogenous mRNAs can roughly be measured by incorporation of radioactive amino acids, and specific antibodies can assay synthesized proteins after gel separations. Translation efficiencies of individual mRNAs can be estimated by polysome profiling, sucrose gradient centrifugation of plastid, or cell lysates followed by hybridization. Ribosome-mRNA interactions in the presence of the initiator formylmethionyl-tRNA (fMet-tRNA) can be detected by toeprint analysis; a ribosome-bound mRNA is used as a template for reverse transcriptase, which terminates cDNA synthesis at the edge of the bound ribosome. (2) Plastid transformation (in vivo systems). Translation mechanisms can be analyzed in vivo by introducing mutant genes or modifying authentic genes. This method can provide solid information how translation proceeds in plastids. In the *E. coli* system, extensive progresses have been made by using its plasmids and phages. However, no such tools are at present available for plastids. In addition, manipulation of plastid genomes may affect plant viability. (3) Genetic approaches. Genes involved in translation can be identified, but mechanisms of the identified gene products are not easy to analyze. (4) Cell-free (in vitro) systems. Basic processes of translation can be analyzed rapidly using an active plastid extract with an exogenous mRNA species. However, it is not easy to analyze effects of environmental and developmental conditions. (5) *In organelle* systems. Translation of endogenous mRNAs can be studied using radioactive amino acids. Effects of light, ATP, and other cofactors can be monitored.

Though extensive studies on translation have been made using *Chlamydomonas* plastids, this review focuses on the translation machinery and translation mechanisms of plastids from flowering plants, especially those unique to flowering plants and different from *E. coli*. Other aspects of plastid translation, e.g., translation in *Chlamydomonas* plastids and mRNA stability, not covered here have been discussed in some recent reviews [19–27].

2 Translational Machinery

Overall components of the plastid translation machinery are similar to those in *E. coli*. The plastid ribosome consists of four ribosomal RNAs (rRNAs). The 4.5S rRNA [ca. 100 nucleotides (nt)] is unique to plastids and it is homologous to the 3′-end portion of the *E. coli* 23S rRNA [28, 29]. Many ribosomal proteins (r-proteins) were initially deduced based on the similarity to those of the *E. coli* ribosome [30, 31]. Protein analyses have then confirmed these proteins as genuine r-proteins in plastids [32–34]. In addition, six plastid-specific r-proteins (PSRP-1–6) were identified [32, 33, 35, 36]. The plastid ribosome in spinach and probably many other flowering plants comprises 59 r-proteins, 33 in the 50S subunit, 25 in the 30S subunit, and ribosome-recycling factor [37] in the 70S subunit. Plastid ribosome-recycling factor is present in approximately stoichiometric amount, specific to the 70S, so that this protein is included. Orthologues of the *E. coli* L25 and L30 are absent in plastids [33]. The *E. coli* ribosome is composed of 54 r-proteins [38], and most plastid r-proteins are longer than their *E. coli* counterparts [32, 33]. Hence, plastid ribosomes are somewhat larger than 70S. An early study indicated that the tobacco plastid ribosome (comprising 58–62 r-proteins) is larger than the *E. coli* ribosome [39].

The 30S subunit is responsible for translation initiation. Significant differences between plastid and *E. coli* 30S subunits exist not only in r-protein compositions but also in posttranslational modifications. r-Protein S1 exists in two forms, 43 kDa and 36~8 kDa [32]. The smaller S1 must be derived from the 43 kDa S1. The situation is similar to the cyanobacterium *Synechococcus* sp. PCC6301, in which two S1-like proteins were found but different genes encode these cyanobacterial proteins [40]. Recently, PSRP-1 was suggested to be not a bona fide r-protein but a functional homologue of the *E. coli* cold-shock protein pY [41]. Cryo-electron microscopic studies revealed that PSRP-1 binds in the decoding region of the 30S subunit [42]. PSRP-2 and -3 appear to structurally compensate for missing segments of the 16S rRNA within the 30S subunit, whereas PSRP-4 occupies a position buried within the head of the 30S subunit. PSRP-5 was tentatively assigned to locate near the tRNA-exit in the 50S subunit. On the other hand, tobacco L33 is not essential under a laboratory condition but is required in the cold [43]. Plastid ribosomes translate at most 80, whereas the *E. coli* counterparts translate over 4,000 cistrons. Although responsible for decoding only one-fiftieth of the number of cistrons, plastid ribosomes look more complex. It has been proposed that plastid ribosomes play a role in the light-dependent control of regulation, which is not the case in *E. coli*, and under many environmental stresses [32]. For example, the *psbA* mRNA level remains unchanged through dark and light conditions, while D1 protein (*psbA* product) synthesis increases dramatically upon illumination [44].

The plastid genome of most flowering plants contains 30 different tRNA genes [45]. In tobacco plastids, seven of them are located in the large inverted repeat, indicating that these seven genes are duplicated and the other 23 genes are single-copy genes. This suggests that the contents of individual tRNA species are not significantly different from each other in plastids. The minimum number of tRNA species required for translation of all 61 codons is 33 (including the start codon AUG) if normal wobble base pairing occurs in codon-anticodon recognition. The tRNA species that recognize leucine (CUU/C), proline (CCU/C), alanine (GCU/C), and arginine (CGC/A/G) codons (asterisks in Table 1) are not encoded by the plastid genome of

Table 1
Codon usage of 79 tobacco plastid mRNAs

	Codon	Fraction		Codon	Fraction		Codon	Fraction		Codon	Fraction
Phe	UUU	0.667	Ser	UCU	0.299	Tyr	UAU	0.804	Cys	UGU	0.752
Phe	UUC	0.333	Ser	UCC	0.152	Tyr	UAC	0.196	Cys	UGC	0.248
Leu	UUA	0.328	Ser	UCA	0.188	Stop	UAA	0.519	stop	UGA	0.228
Leu	UUG	0.200	Ser	UCG	0.059	Stop	UAG	0.253	Trp	UGG	1.000
Leu	CUU*	0.214	Pro	CCU*	0.393	His	CAU	0.770	Arg	CGU	0.219
Leu	CUC*	0.069	Pro	CCC*	0.187	His	CAC	0.230	Arg	CGC*	0.063
Leu	CUA	0.128	Pro	CCA	0.286	Gln	CAA	0.756	Arg	CGA*	0.251
Leu	CUG	0.062	Pro	CCG	0.135	Gln	CAG	0.244	Arg	CGG*	0.074
Ile	AUU	0.495	Thr	ACU	0.392	Asn	AAU	0.768	Ser	AGU	0.209
Ile	AUC	0.200	Thr	ACC	0.197	Asn	AAC	0.232	Ser	AGC	0.059
Ile	AUA	0.306	Thr	ACA	0.304	Lys	AAA	0.754	Arg	AGA	0.289
Met	AUG	0.853	Thr	ACG	0.107	Lys	AAG	0.246	Arg	AGG	0.104
fMet	AUG	0.141									
Val	GUU	0.372	Ala	GCU*	0.448	Asp	GAU	0.797	Gly	GGU	0.322
Val	GUC	0.120	Ala	GCC*	0.172	Asp	GAC	0.203	Gly	GGC	0.118
Val	GUA	0.380	Ala	GCA	0.280	Glu	GAA	0.757	Gly	GGA	0.389
Val	GUG	0.128	Ala	GCG	0.100	Glu	GAG	0.243	Gly	GGG	0.171
fMet	GUG	0.006									

Codon fraction is the ratio of each codon in the family of synonymous codons. Based on Table 1 of refs. 133 and 137. *Asterisks* indicate codons, for which no corresponding tRNA genes are present in the plastid genome flowering plants. The first position (A) of $tRNA^{Arg}$(ACG)s is generally modified to inosine (I), and $tRNA^{Arg}$(ICG) reads not only CGU but also CGC and CGA [46]. As no evidence has been reported for tRNA import into plastids, the above codons are probably read by the 30 tRNAs with modified nucleotides [45] and by the two-out-of-three (or the superwobble) mechanism [47]. The tRNA adenosine deaminase (A to I conversion) was identified in *Arabidopsis* plastids [48, 49].

Other components/factors in plastid translation, a set of aminoacyl-tRNA synthetases, initiation factors (IF-1–3), elongation factors (EF-Fu and EF-G), release factors (RF-1–2), and modification enzymes for RNA and protein components, were reported through their similarity with those of *E. coli*. Functions of these proteins are likely to be similar to those analyzed in *E. coli* and other bacteria. The existence of additional components/factors unique to plastids would be expected but these cannot be identified via similarity. Biochemical analyses will require to study action mechanisms of known components/factors and to find novel components/factors in plastids.

Among the components reported, two are unique to plastids (and mitochondria); glutamyl-tRNA formed by plastid glutamyl-tRNA synthetase is a precursor of chlorophyll [50], and the tRNA 5′-maturation enzyme RNase P lacks RNA components [51]. The proteinaceous RNase P is encoded in the nuclear genome and moves into both plastids and mitochondria [52].

3 Translation Initiation

3.1 Initiation Codons

A critical step of translation initiation is the selection of correct initiation (start) codons. The initiation codon of plastid mRNAs from flowering plants is AUG and rarely GUG. UUG has been assigned as a potential initiation codon for the IF-1 gene from *Chlorella* plastids [53], but there is no report in flowering plants. The *Chlorella* IF-1 mRNA was translated accurately from the UUG codon in a tobacco plastid extract [54]. The tobacco *psbC* mRNA whose GUG start codon was replaced with UUG could initiate translation [55]. Hence, UUG could also be an initiation codon, though low in efficiency, in flowering plants. Among the 79 protein-coding genes in tobacco plastids, only two mRNAs have GUG as initiation codons (*rps19* and *psbC*). The *ycf15* sequence has not been confirmed as a protein-coding gene and hence its start codon has not been defined (either AUG or GUG). The remaining mRNAs possess AUG codons in which two are created from ACG codons by RNA editing (*psbL* and *ndhD*) [15, 56, 57].

There are often multiple possible initiation codons in mRNAs and it is not easy to determine which is the real start codon. Determining the N-termini of translation products is straightforward, if the N-termini are not processed widely. Each of the potential initiation codons can be mutated by chloroplast transformation [58, 59], and phenotypic changes and protein accumulation are to be observed. This method is solid but laborious and time-consuming. Toeprint analysis is a general method to assign translation initiation regions (TIRs). Once TIRs are assigned, it is easy to define initiation codons if their reading frames are known (i.e., known genes) and multiple possible initiation codons are not closely located. In *E. coli*, the mRNA region from −20 to +13 (the A of AUG as +1) is protected by ribosomes during initiation so that toeprint signals are located around +13 [60]. The toeprint assay was used to locate TIRs of barley *rbcL* and *psbA* mRNAs, and strong signals were detected at +15 for both mRNAs [61]. Hence, their start codons could be defined as single AUGs within protected areas. The context of TIRs (a couple of nucleotides surrounding initiation codons) is likely to be important to start translation, but the context of plastid mRNAs has not been well characterized.

In contrast, in vitro translation assays allow us rapidly to define initiation codons. The tobacco *psbA* mRNA contains one possible start codon, and changing the AUG to ACG abolished translation completely in vitro [62]. The *ndhD* mRNA from tobacco contains potential start codons AUG and GUG, but the downstream ACG is edited to produce AUG [57], three possible start codons altogether. In vitro translation assays clearly indicated that only the edited AUG acts as the initiation codon [15]. The *ndhK* mRNA possesses four possible AUG codons in tobacco and many other dicot plants. In vitro analysis defined that the major initiation site of tobacco *ndhK* mRNAs is the third AUG codon [16]. The tobacco *psbC* mRNA has two possible initiation codons, AUG and GUG, and the GUG was determined to be the initiation site in vitro [55]. These observations indicate that exact protein-coding regions and then protein structures cannot always be predicted from their genomic sequences. The replacement of AUG by GUG decreased translation to about 10 % of the wild type in tobacco *atpE* mRNAs [63]. Change of the authentic *psbC* GUG initiation codon to AUG resulted in only twofold more in translation than the translation from the GUG [55], suggesting that the translation efficiency of initiation codons depend on mRNA species.

3.2 5′-Untranslated Regions

The most important step for translation is to select simultaneously both a right reading frame from three possible frames and a correct initiation codon among several to a dozen possible start codons on three reading frames around the initiation site. The *cis*-element in a 5′-untranslated region (5′-UTR) is a major determinant for correct initiation and regulation of plastid mRNA translation. Tobacco plastid transformation experiments have clearly shown in vivo that the 5′-UTR controls translation initiation of plastid

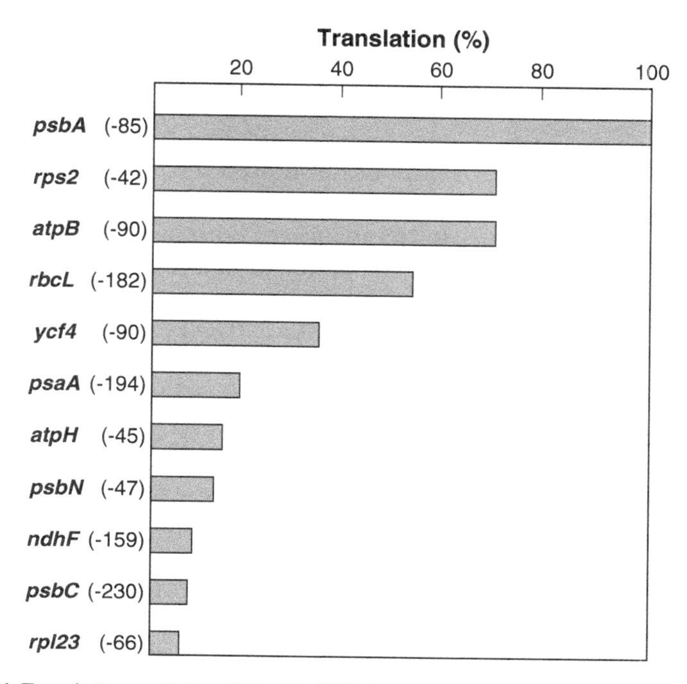

Fig. 1 Translation activity of the 5′-UTR from eleven tobacco plastid mRNAs. Numbers in parentheses represent 5′-ends relative to the first positions of start codons (A or G as +1). Bars indicate relative activity (*psbA* 5′-UTR as 100 %). (Redrawn based on ref. 68)

mRNA from *psbA*, *rbcL*, and several other genes [64, 65]. Furthermore, the *psbA* mRNA is present in all tissues and under light/dark conditions, and its translation is regulated by light via its 5′-UTR [66]. Transformation assays also indicated that the correct secondary structure of the tobacco *psbA* 5′-UTR is required for translation [67].

A highly active in vitro translation system has been developed from tobacco chloroplasts, which allows us to measure relative rates of translation [68]. Using this system, translation efficiencies of eleven 5′-UTRs from tobacco plastid mRNAs were determined. Highly efficient 5′-UTRs are those from *psbA*, *rps2*, and *atpB* mRNAs (Fig. 1).

The 5′-end of 5′-UTRs is either transcription initiation sites with triphosphates (pppN·····) or processed sites with monophosphates (pN······). Proper 5′-processing of mRNAs is sometimes required for efficient translation, for example, barley *rbcL* mRNAs [69]. In vitro experiments demonstrated that mRNAs carrying unprocessed and processed *atpH* and *rbcL* 5′-UTRs are translated at similar rates, whereas translation of mRNAs with processed *atpB* and *psbB* 5′-UTRs is more efficient than those with unprocessed 5′-UTRs [68].

Many plastid genes are transcribed as polycistronic mRNAs. Unlike *E. coli* polycistronic transcripts, plastid pre-mRNAs are generally cleaved into monocistronic mRNAs, suggesting that mRNA processing is required for translation of downstream cistrons. In vitro assays showed that tobacco *psaC-ndhD* dicistronic mRNAs is not functional and that the intercistronic processing is a prerequisite for translation of both *psaC* and *ndhD* cistrons [15]. Genetic experiments suggested that intercistronic processing is often required for translation of downstream cistrons, for example, maize and tobacco *petD* mRNAs [70, 71] and *Arabidopsis psbH* mRNAs [72] and *ndhB* mRNAs [73]. Knockout experiments suggested an important function of RNase E in intercistronic processing of polycistronic mRNAs [74]. However, this is not always the case. Each cistron of the tobacco *ndhC-K-J* tricistronic mRNA is efficiently translated [16], and the downstream cistron of an artificial dicistronic mRNA is efficiently translated in vivo [75, 76]. Therefore, effects of 5′-processing on translation differ from mRNA to mRNA.

In tobacco, the protein-coding regions (hereafter cistrons) of eight genes partially overlap; these are *atpB-atpE*, *psbD-psbC*, *ndhC-ndhK*, and *rpl22-rps3* [77]. This feature implies that translation of a downstream gene depends on that of its upstream gene, namely, translational coupling. Translational coupling is explained by the downstream cistron being translated exclusively by the ribosomes that completed translation of the upstream cistron. Hence, translation of the downstream cistron is usually low because the majority of the ribosomes are released at the stop codon. In vitro assays indicated that the *ndhC-ndhK* genes are translationally coupled [16]. Further in vitro studies demonstrated that free ribosomes enter at an internal AUG that is located in-frame in the middle of the upstream *ndhC* cistron and translate the 3′ half of the *ndhC* cistron and that some ribosomes resume *ndhK* translation in order to produce NdhC and NdhK in similar amounts [78]. However, translation of the *atpB-atpE* genes is not coupled [77], while translation of the *psbD-psbC* genes is partially coupled [79]. These downstream cistrons are fully or mostly independently translated. Hence, the upstream cistrons include sequences necessary for translation of the downstream cistrons. In addition, they have additional promoters within the upstream genes and produce monocistronic *atpE* and *psbC* mRNAs.

3.3 Cis-Elements in the 5′-UTR

Sequences (*cis*-elements) driving translation initiation reside mainly in upstream from start codons. In vitro assays are suitable to define *cis*-elements required for basic processes of translation initiation. This method uses a series of mutated 5′-UTRs in an mRNA of interest as template and in vitro translation reactions with active plastid extracts (S30 fractions). In vitro assays have shown that three *cis*-elements are required for translation of tobacco *psbA* mRNAs [62]. Two of them are complementary to the 3′-terminus of plastid 16S rRNA (termed RBS1 and RBS2) and the other is an AU-rich sequence (UAAAUAAA) located between RBS1 and RBS2 (termed the AU box). In vitro mRNA competition and gel mobility shift assays revealed the existence of a *trans*-acting factor, possibly interacting with the AU box. The 5′-UTR of tobacco *atpB* mRNAs is U-rich (the 20 nt upstream from the start codon consists of 15 Us and 5 As), and in vitro assays showed that the U-rich unstructured sequence is required for translation, possibly together with a 50 kDa protein factor [63].

Among the 79 protein-coding genes (including 6 *ycfs*) in tobacco plastids [80], 31 contain no Shine-Dalgarno (SD)-like sequences (*see* below) within 20 nt upstream from the start codon, and the remaining 48 have SD-like sequences but not in a conserved position [81]. Mutagenesis and in vitro translation assays showed that three SD-like sequences, GGA at positions −14 to −12 in *rps14* mRNAs [82], GGAG at −18 to −15 in *atpE* mRNAs, and GGAGG at −10 to −6 in *rbcL* mRNAs [83], are required for translational initiation. Assuming that SD-like sequences present between −18 and −6 are functional in tobacco chloroplasts, 38 of the 48 genes may have functional SD-like sequences. The distribution of SD-like sequences is similar to that from other plant species. Similarity is also observed in *Synechococcus* 6301 [81], again supporting the cyanobacterial origin of plastids. The tobacco *psbC* mRNA with the GUG start codon requires the extended SD-like sequence (GAGGAGGU) at −16 to −9 [55].

In *E. coli*, almost all mRNAs have the SD sequence (typically GGAGG), which is located several nt upstream from initiation codons and is complementary to the 3′-end of 16S rRNA (anti-SD, CCUCC) [84]. The 30S ribosomal subunit and fMet-tRNA bind to mRNA at two sites, namely, the SD sequence and the initiation codon. An optional spacing of 5 nt between the SD sequence and the initiation codon was determined experimentally [85]. The 3′-end (13 nt) of *E. coli* 16S rRNA is free and is not complexed with proteins or RNAs. Base pairing of the SD sequence and the anti-SD sequence has been proven experimentally [86, 87]. Hence, the proper spacing between two elements on mRNAs is critical for accurate and efficient initiation of translation. The *rbcS* mRNA from *Synechococcus* 6301 possesses two possible initiation codons (AUGs) separated by 3 nt. Using a homologous in vitro system, translation was found to start from the upstream AUG. Mutation analysis indicated that SD-like sequences in positions similar to those of *E. coli* mRNAs are not required for translation [88]. The tobacco *psbA* mRNA contains GGAG (−36 to −33) in the 5′-UTR. Mutation of this sequence showed little effect on translation, indicating that the GGAG is not a ribosome-binding site [62]. The previous toeprint assay indicated that *E. coli* ribosomes associate with the corresponding GGAG but not with the TIR of barley *psbA* mRNAs [61]. In this case, fMet-tRNA seemed to recognize

UUG or AUU (as start codon) several nt downstream from this SD-like sequence. The tobacco *rps2* mRNA possesses the SD-like element (GGAG) at a proper position, whereas mutations of the GGAG sequence resulted in a large increase in translation in vitro [89]. An additional in vitro analysis indicated the SD-like sequence in tobacco *atpE* mRNAs is a part of the *cis*-element for *atpE* translation [77]. Using the plastid transformation assay, translation of tobacco *atpI* mRNAs was shown to require element(s) upstream of the SD-like sequence [90]. Hence, results obtained by mutation of SD-like sequences should be interpreted carefully. In addition, big differences were observed translation efficiencies of identical mRNAs between transformed *E. coli* and transformed plastids [91]. However, the 5′-UTR from *E. coli* phage T7 gene 10 is highly efficient [92]. These observations suggest that a limited number of plastid mRNAs use SD-like sequences as a sole determinant of translation initiation.

3.4 Trans-Acting Factors

Many of the plastid mRNA levels remain relatively unchanged during light and dark conditions [93, 94], whereas translation rates increase dramatically upon illumination or by change of developmental and environmental cues, e.g., fruit ripening [95]. Such regulation of translation can most easily be understood if *trans*-acting factors are present. Genetic and biochemical approaches proposed the presence of such factors for many plastid mRNAs. For example, maize nuclear genes *crp1* and *atp-1* products are required for translation of *petA-D* mRNAs and *atpB-E* mRNAs, respectively [70, 96, 97], and the *Arabidopsis* HCF173 protein is involved in translation initiation of *psbA* mRNA [98]. Biochemical assays showed that spinach plastid proteins interact with the 5′-UTRs of *psbA* [99] *atpI*, *atpH*, and *atpA* [100–102] and of *atpB* and *atpE* [102]. Further results support the hypothesis that the spinach *atpI* 5′-UTR is divided into two *cis*-elements and each is bound by different *trans*-factor(s) [101]. A 43 kDa protein binds efficiently to a U-rich sequence preceding the AUG start codon in the spinach *psbA* mRNA [103]. Binding of this protein is detectable in the light and therefore it is a candidate for a light regulatory factor. The r-proteins S1 (43 and 47 kDa) were reported to bind the spinach *psbA* 5′-UTR [104]. Hence, the above 43 kDa protein may be S1 since S1 is likely to require for translation of a set of plastid mRNAs (ref. 105, our unpublished observation). A TAB2, the *Arabidopsis* orthologue of *Chlamydomonas* TAB1, was shown to be essential for *psaB* translation initiation [106].

In the thylakoid membrane complex, some plastid-encoded proteins are subjected to assembly-dependent translational regulation or control by epistasy of synthesis (CES) [107]. Unassembled components repress its own translation initiation through the 5′-UTR, and this autoregulation is a general feature in *Chlamydomonas* plastids. Tobacco *rbcL* translation is regulated in a similar manner [108]. However, this regulation is likely to be released when tobacco leaves are expanded [109]. *Trans*-acting factors (or candidates) are reported one after another. It will be necessary to investigate in more detail relationships between *cis-elements* and the corresponding *trans*-factors and how *trans*-factors work.

4 Translation Elongation and Termination

Basic processes of translation elongation and termination in plastids seem to be similar to those found in *E. coli*. Notable observations are that translation elongation of photosynthesis-related mRNAs is regulated by light [110], namely, mRNA-ribosome initiation complexes can be formed even in the dark and translation elongation starts/increases by light [111]. During translation of *psbA* mRNAs, the light-induced level of ATP is an essential determinant [112] and in addition other factor(s) caused by light is also necessary [113]. Translation initiation complexes for *rbcL* mRNAs are normally formed in the dark, but the elongation step might be inhibited. The release of such a translation block upon illumination may contribute to light-activated translation of *rbcL* mRNAs [114]. Ribosome pausing was reported during translation elongation of *rbcL and atp* mRNAs [115, 116]. Ribosomes pause at distinct sites during the elongation of *psbA* proteins (D1 proteins) probably to allow efficient assembly of PSII [117–119]. These analyses have been done successfully *in organelle*, isolated plastids, partially lysed plastids, crude thylakoid fractions, and in vitro translation systems.

Sequences downstream of the start codon were often important for efficient translation [120]. It has not been well investigated whether the downstream sequence is responsible for initiation or elongation.

5 mRNA Storage/Stabilization

The levels of many mRNAs are relatively unchanged under light and dark conditions though their transcription is generally low in the dark. Therefore, mRNA storage or stabilizing mechanisms should operate in plastids. It is thought that thylakoid proteins are synthesized by ribosomes attached on the thylakoid membrane and soluble proteins by ribosomes in stroma. This theory raises a possibility that the thylakoid membrane store/stabilize certain mRNAs. However, *psbA* and *rbcL* mRNAs are present in both stroma and thylakoids, and the *rbcL* mRNA associated with thylakoid membranes and the *psbA* mRNA in the stroma are not translated [121]. There is another report that the large subunit of

RuBisCO is also synthesized on thylakoid-associated ribosomes [122]. The stroma-localized *psbA* mRNA was not associated with ribosomes and was suggested to associate with protein(s) [123]. These observations suggest the existence of mRNA storage mechanisms in stroma.

Abundant RNA-binding proteins have been found in tobacco plastids [124]. These proteins (cp28, cp29A, cp29B, cp31, and cp33) contain two consensus sequence-type RNA-binding domains (or RRMs) of ca. 80 amino acids in size and have affinity for poly(G) and poly(U) in vitro [125]. Biochemical studies indicated that these cp proteins exist as complexes with ribosome-free mRNAs in the stroma [126]. An in vitro mRNA degradation assay showed that exogenous *psbA* mRNA is more rapidly degraded in cp protein-depleted extracts than in non-depleted extracts. Hence, these cp proteins are likely to act as storage/stabilizing factors for mRNAs in the stroma [127]. For example, cp31A is required for editing and stability of specific plastid mRNAs [128, 129]. Pentatricopeptide repeat (PPR) proteins were found to bind to the *atpH* 5′-UTR and stabilize the mRNA thereby to enhance *atpH* translation [130]. PPR proteins are also RNA-binding proteins and function in various posttranscriptional processes [131]. Hence, these PPR proteins are also likely to participate in mRNA stabilization. Individual PPR proteins are usually low in amount, and these proteins may not be enough to store large amounts of mRNAs.

6 Codon Usage

Codon usage in plastids differs from that in prokaryotes and in eukaryotic nuclei. The tobacco plastid possesses 79 protein-coding genes that consist of 22,976 codons. C-to-U RNA editing has been reported at 38 sites in the mRNA coding regions, and 37 of them cause amino acid conversion [11, 132]. As mentioned in Subheading 3.1, several genes contain multiple possible initiation codons, and the genuine start codon for these mRNAs has been determined experimentally. Based on these data, codon usage was calculated from tobacco plastid mRNAs [133] (Table 1). This codon usage table is based exclusively on single mature chloroplast mRNAs because codons function on mRNAs but not on genomes (DNAs). Codon usage tables available in databases have been usually constructed by simple summation of automatically collected gene sequences. One example for tobacco plastids was constructed from almost 1.5 times more than the existing genes (over one-third of the genes were added two to three times, whereas some plastid genes were missing and several nuclear genes were included). As can be seen in Table 1, the so-called rare codon is not present, and four codons each for leucine, proline, alanine, and arginine are decoded by the corresponding single tRNA species, which are not seen in eubacteria and mammals. Early observations in *E. coli* and yeast showed that codon usage is positively correlated with tRNA content, especially for highly expressed genes [134]. According to this correlation, codon optimization generally improved the expression of mammalian proteins in *E. coli* [135].

Based on an in vitro translation system from tobacco plastids, an in vitro assay to measure translation efficiencies of synonymous codons was devised [133, 136, 137]. Using the system, the translation efficiency of nine synonymous codon groups was measured (Table 2). Unexpectedly, the translation efficiency of phenylalanine, tyrosine, and arginine codons is opposite to their codon usage. Therefore, the translation efficiencies of synonymous codons are not always correlated with codon usage in tobacco plastids. These results raise a question concerning the usefulness of the so-called codon optimization.

Table 2
Codon usage and translation efficiency

	Codon	Fraction	Translation
Asn	AAU	0.768	+++
	AAC	0.232	+
Asp	GAU	0.797	+++
	GAC	0.203	+
Ala	GCU	0.448	+++
	GCC	0.172	++
	GCA	0.280	++
	GCG	0.100	+
Tyr	UAU	0.804	+
	UAC	0.196	+++
Phe	UUU	0.667	+
	UUC	0.333	+++
Arg	CGU	0.219	+
	CGA	0.251	±
	AGA	0.289	+++
Leu	UUA	0.328	+++
	CUA	0.128	+
His	CAU	0.770	+++
	CAC	0.230	+
Glu	GAA	0.757	+++
	GAG	0.243	+

Fraction, from Table 1. Translation: high (+++), medium (++), and low (+) efficiencies. Based on refs. 133 and 137

7 Problems and Prospects

E. coli is estimated to require ca. 130 genes for components of the translation machinery. Plastids possess mRNA-specific *trans*-acting factors. Hence, plastids are likely to need ca. 200 genes for translation. Many genes still remain unidentified. Plastids should reserve excess translation activity to respond quickly to environmental changes. Quantitative analysis will be necessary, for example, the amount of ribosomes, tRNAs and other translation factors per plastid and also the amount of functional mRNA molecules (not mRNA levels) per plastid. A large amount of some mRNA species, e.g., *psbA* mRNAs, is stored as an inactive form in the dark, and these mRNAs are promptly translated upon illumination. How standby mRNAs are stored is an important problem that remains to be elucidated. Finally, all translation systems should possess proofreading steps. In bacteria, tmRNA is one of the important components of proofreading but the plastid genomes of flowering plants encode no tmRNA. Plastids may have a unique means not found in other translation systems. Understanding the mechanism of plastid translation will widen our knowledge on translation and its regulation.

References

1. Sugiura M (1992) The chloroplast genome. Plant Mol Biol 19:149–168
2. Bock R (2007) Structure, function, and inheritance of plastid genomes. In: Bock R (ed) Cell and molecular biology of plastids. Springer, Potsdam-Golm, pp 29–63
3. Shinozaki K, Ohme M, Tanaka M et al (1986) The complete nucleotide sequence of the tobacco chloroplast genome: its gene organization and expression. EMBO J 5: 2043–2049
4. Sugita M, Sugiura M (1996) Regulation of gene expression in chloroplasts of higher plants. Plant Mol Biol 32:315–326
5. Tanaka M, Obokata J, Chunwongse J et al (1987) Rapid splicing and stepwise processing of a transcript from the *psbB* operon in tobacco chloroplasts. Mol Gen Genet 209:427–431
6. Matsubayashi T, Wakasugi T, Shinozaki K et al (1987) Six chloroplast genes (*ndhA-F*) homologous to human mitochondrial genes encoding components of the respiratory chain NADH dehydrogenase are actively expressed: determination of the splice sites in *ndhA* and *ndhB* pre-mRNAs. Mol Gen Genet 210:385–393
7. Barkan A (1988) Proteins encoded by a complex chloroplast transcription unit are each translated from both monocistronic and polycistronic mRNAs. EMBO J 7:2637–2644
8. Westhoff P, Herrmann RG (1988) Complex RNA maturation in chloroplasts. The psbB operon from spinach. Eur J Biochem 171: 551–564
9. Shinozaki K, Deno H, Sugita M et al (1986) Intron in the gene for the ribosomal protein Sl6 of tobacco chloroplast and its conserved boundary sequences. Mol Gen Genet 202:1–5
10. Stern DB, Goldschmidt-Clermont M, Hanson MR (2010) Chloroplast RNA metabolism. Annu Rev Plant Biol 61:125–155
11. Sugiura M (2008) RNA editing in chloroplasts. In: Goringer HU (ed) RNA editing. Springer, Berlin, pp 123–142
12. Chateigner-Boutin AL, Small I (2010) Plant RNA editing. RNA Biol 7:213–219
13. Stern DB, Gruissem W (1987) Control of plastid gene expression: 3′-inverted repeats act as mRNA processing and stabilizing elements, but do not terminate transcription. Cell 51:1145–1157

14. Schuster G, Lisitsky I, Klaff P (1999) Polyadenylation and degradation of mRNA in the chloroplast. Plant Physiol 120:937–944
15. Hirose T, Sugiura M (1997) Both RNA editing and RNA cleavage are required for translation of tobacco chloroplast *ndhD* mRNA: a possible regulatory mechanism for the expression of a chloroplast operon consisting of functionally unrelated genes. EMBO J 16:6804–6811
16. Yukawa M, Sugiura M (2008) Termination codon-dependent translation of partially overlapping *ndhC-ndhK* transcripts in chloroplasts. Proc Natl Acad Sci USA 105: 19550–19554
17. Gruissem W (1989) Chloroplast gene expression: how plants turn their plastids on. Cell 56:161–170
18. Romby P, Springer M (2007) Translational control in prokaryotes. In: Mathews MB, Sonenberg N, Hershey JWB (eds) Translational control in biology and medicine. Cold Spring Harbor Laboratory Press, Cold Spring Harbor, pp 803–827
19. Marín-Navarro J, Manuell AL, Wu J et al (2007) Chloroplast translation regulation. Photosynth Res 94:359–374
20. Peled-Zehavi H, Danon A (2007) Translation and translational regulation in chloroplasts. In: Bock R (ed) Cell and molecular biology of plastids. Springer, Potsdam-Golm, pp 249–281
21. Waters MT, Langdale JA (2009) The making of a chloroplast. EMBO J 28:2861–2873
22. Wobbe L, Schwarz C, Nickelsen J et al (2008) Translational control of photosynthetic gene expression in phototrophic eukaryotes. Physiol Plant 133:507–515
23. Rochaix J-D (2001) Posttranscriptional control of chloroplast gene expression. From RNA to photosynthetic complex. Plant Physiol 125:142–144
24. Choquet Y, Wollman F-A (2002) Translational regulations as specific traits of chloroplast gene expression. FEBS Lett 529:39–42
25. Zerges W (2000) Translation in chloroplasts. Biochimie 82:583–601
26. Manuell A, Beligni MV, Yamaguchi K et al (2004) Regulation of chloroplast translation: interactions of RNA elements, RNA-binding proteins and the plastid ribosome. Biochem Soc Trans 32:601–605
27. Schuster G, Stern D (2009) RNA polyadenylation and decay in mitochondria and chloroplasts. Prog Mol Biol Transl Sci 85:393–422
28. Whitfeld PR, Jeaver CJ, Bottomley W et al (1978) Low-molecular-weight (4.5S) ribonucleic acid in higher-plant chloroplast ribosomes. Biochem J 175:1103–1112
29. Bowman CM, Dyer TA (1979) 4.5S ribonucleic acid, a novel ribosome component in the chloroplasts of flowering plants. Biochem J 183:605–613
30. Gantt JS (1988) Nucleotide sequences of cDNAs encoding four complete nuclear-encoded plastid ribosomal proteins. Curr Genet 14:519–528
31. Sugiura M, Torazawa K, Wakasugi T et al (1991) Chloroplast genes coding for ribosomal proteins in land plants. In: Mache R, Stutz E, Subramanian AR (eds) The translational apparatus of photosynthetic organelles. Springer, Berlin, pp 59–69
32. Yamaguchi K, von Knoblauch K, Subramanian AR (2000) The plastid ribosomal proteins: identification of all the proteins in the 30S subunit of an organelle ribosome (chloroplast). J Biol Chem 275:28455–28465
33. Yamaguchi K, Subramanian AR (2000) The plastid ribosomal proteins: identification of all the proteins in the 50S subunit of an organelle ribosome (chloroplast). J Biol Chem 275:28466–28482
34. Maki Y, Tanaka A, Wada A (2000) Stoichiometric analysis of barley plastid ribosomal proteins. Plant Cell Physiol 41: 289–299
35. Yamaguchi K, Subramanian AR (2003) Proteomic identification of all plastid-specific ribosomal proteins in higher plant chloroplast 30S ribosomal subunit PSRP-2 (U1A-type domains), PSRP-3α/β (ycf65 homologue) and PSRP-4 (Thx homologue). Eur J Biochem 270:190–205
36. Carol P, Li YF, Mache R (1991) Conservation and evolution of the nucleus-encoded and chloroplast-specific ribosomal proteins in pea and spinach. Gene 103:139–145
37. Rolland N, Janosi L, Block MA et al (1999) Plant ribosome recycling factor homologue is a chloroplastic protein and is bactericidal in *Escherichia coli* carrying temperature-sensitive ribosome recycling factor. Proc Natl Acad Sci USA 96:5464–5469
38. Wittmann HG (1983) Architecture of prokaryotic ribosomes. Annu Rev Biochem 52:35–65
39. Capel MS, Bourque DP (1982) Characterization of *Nicotiana tabacum* chloroplast and cytoplasmic ribosomal proteins. J Biol Chem 257:7746–7755
40. Sugita C, Sugiura M, Sugita M (2000) A novel nucleic acid-binding protein in the cyanobacterium *Synechococcus* sp. PCC6301: a

soluble 33-kDa polypeptide with high sequence similarity to ribosomal protein S1. Mol Gen Genet 263:655–663

41. Sharma MR, Dönhöfer A, Barat C et al (2010) PSRP1 is not a ribosomal protein, but a ribosome-binding factor that is recycled by the ribosome-recycling factor (RRF) and elongation factor G (EF-G). J Biol Chem 285:4006–4014
42. Sharma MR, Wilson DN, Datta PP et al (2007) Cryo-EM study of the spinach chloroplast ribosome reveals the structural and functional roles of plastid-specific ribosomal proteins. Proc Natl Acad Sci USA 104: 19315–19320
43. Rogalski M, Schöttler MA, Thiele W et al (2008) Rpl33, a nonessential plastid-encoded ribosomal protein in tobacco, is required under cold stress conditions. Plant Cell 20:2221–2237
44. Fromm H, Devic M, Fluhr R et al (1985) Control of psbA gene expression; in mature *Spirodela* chloroplasts light regulation of 32-kd protein synthesis is independent of transcript level. EMBO J 4:291–295
45. Sugiura M (1987) Structure and function of the tobacco chloroplast genome. Bot Mag Tokyo 100:407–436
46. Curran JF (1995) Decoding with the A:I wobble pair is inefficient. Nucleic Acids Res 23:683–688
47. Rogalski M, Karcher D, Bock R (2008) Superwobbling facilitates translation with reduced tRNA sets. Nat Struct Mol Biol 15:192–198
48. Delannoy E, Le Ret M, Faivre-Nitschke E et al (2009) Arabidopsis tRNA adenosine deaminase arginine edits the wobble nucleotide of chloroplast $tRNA^{Arg}$(ACG) and is essential for efficient chloroplast translation. Plant Cell 21:2058–2071
49. Karcher D, Bock R (2009) Identification of the chloroplast adenosine-to-inosine tRNA editing enzyme. RNA 15:1251–1257
50. Jahn D, Verkamp E, Söll D (1992) Glutamyl-transfer RNA: a precursor of heme and chlorophyll biosynthesis. Trends Biochem Sci 17:215–218
51. Wang MJ, Davis NW, Gegenheimer PA (1988) Novel mechanisms for maturation of chloroplast transfer RNA precursors. EMBO J 7:1567–1574
52. Gobert A, Gutmann B, Taschner A et al (2010) A single *Arabidopsis* organellar protein has RNase P activity. Nat Struct Mol Biol 17:740–744
53. Wakasugi T, Nagai T, Kapoor M et al (1997) Complete sequencing of the chloroplast genome of the green alga *Chlorella vulgaris* reveals the existence of genes possibly involved in chloroplast division. Proc Natl Acad Sci USA 94:5967–5972
54. Hirose T, Ideue T, Wakasugi T et al (1999) The chloroplast *infA* gene with a functional UUA initiation codon. FEBS Lett 445: 169–172
55. Kuroda H, Suzuki H, Kusumegi T et al (2007) Translation of *psbC* mRNAs starts from the downstream GUG, not the upstream AUG, and requires the extended Shine-Dalgarno sequence in tobacco chloroplasts. Plant Cell Physiol 48:1374–1378
56. Kudla J, Igloi GL, Metzlaff M et al (1992) RNA editing in tobacco chloroplasts leads to the formation of a translatable *psbL* mRNA by a C to U substitution within the initiation codon. EMBO J 11:1099–1103
57. Neckermann K, Zeltz P, Igloi GL et al (1994) The role of RNA editing in conservation of start codons in chloroplast genomes. Gene 146:177–182
58. Svab Z, Maliga P (1993) High-frequency plastid transformation in tobacco by selection for a chimeric *aadA* gene. Proc Natl Acad Sci USA 90:913–917
59. Maliga P (2004) Plastid transformation in higher plants. Annu Rev Plant Biol 55: 289–313
60. Hartz D, McPheeters DS, Traut R et al (1988) Extension inhibition analysis of translation initiation complexes. Methods Enzymol 164:419–425
61. Kim J, Mullet JE (1994) Ribosome-binding sites on chloroplast *rbcL* and *psbA* mRNAs and light-induced initiation of D1 translation. Plant Mol Biol 25:437–448
62. Hirose T, Sugiura M (1996) *Cis*-acting elements and trans-acting factors for accurate translation of chloroplast psbA mRNAs: development of an *in vitro* translation system from tobacco chloroplasts. EMBO J 15:1687–1695
63. Hirose T, Sugiura M (2004) Multiple elements required for translation of plastid *atpB* mRNA lacking the Shine-Dalgarno sequence. Nucleic Acids Res 32:3503–3510
64. Staub JM, Maliga P (1993) Accumulation of D1 polypeptide in tobacco plastids is regulated via the untranslated region of the *psbA* mRNA. EMBO J 12:601–606
65. Eibl C, Zou Z, Beck A et al (1999) In vivo analysis of plastid *psbA*, *rbcL* and *rpl32* UTR elements by chloroplast transformation: tobacco plastid gene expression is controlled by modulation of transcript levels and translation efficiency. Plant J 19:333–345

66. Staub JM, Maliga P (1994) Translation of *psbA* mRNA is regulated by light via the 5′-untranslated region in tobacco plastids. Plant J 6:547–553
67. Zou Z, Eibl C, Koop HU (2003) The stem-loop region of the tobacco *psbA* 5′UTR is an important determinant of mRNA stability and translation efficiency. Mol Genet Genomics 269:340–349
68. Yukawa M, Kuroda H, Sugiura M (2007) A new *in vitro* translation system for non-radioactive assay from tobacco chloroplasts: effect of pre-mRNA processing on translation in vitro. Plant J 49:367–376
69. Reinbothe S, Reinbothe C, Heintzen C et al (1993) A methyl jasmonate-induced shift in the length of the 5′ untranslated region impairs translation of the plastid rbcL transcript in barley. EMBO J 12:1505–1512
70. Barkan A, Walker M, Nolasco M et al (1994) A nuclear mutation in maize blocks the processing and translation of several chloroplast mRNAs and provides evidence for the differential translation of alternative mRNA forms. EMBO J 13:3170–3181
71. Monde RA, Greene JC, Stern DB (2000) Disruption of the *petB-petD* intergenic region in tobacco chloroplasts affects *petD* RNA accumulation and translation. Mol Gen Genet 263:610–618
72. Felder S, Meierhoff K, Sane AP et al (2001) The nucleus-encoded HCF107 gene of *Arabidopsis* provides a link between intercistronic RNA processing and the accumulation of translation-competent *psbH* transcripts in chloroplasts. Plant Cell 13:2127–2141
73. Hashimoto M, Endo T, Peltier G et al (2003) A nucleus-encoded factor, CRR2, is essential for the expression of chloroplast *ndhB* in *Arabidopsis*. Plant J 36:541–549
74. Walter M, Piepenburg K, Schöttler MA et al (2010) Knockout of the plastid RNase E leads to defective RNA processing and chloroplast ribosome deficiency. Plant J 64:851–863
75. Staub JM, Maliga P (1995) Expression of a chimeric *uidA* gene indicates that polycistronic mRNAs are efficiently translated in tobacco plastids. Plant J 7:845–848
76. Quesada-Vargas T, Ruiz ON, Daniell H (2005) Characterization of heterologous multigene operons in transgenic chloroplasts: transcription, processing, and translation. Plant Physiol 138:1746–1762
77. Suzuki H, Kuroda H, Yukawa Y, Sugiura M (2011) The downstream *atpE* cistron is efficiently translated via its own *cis*-element in partially overlapping *atpB-atpE* dicistronic mRNAs in chloroplasts. Nucleic Acids Res 39:9405–9412
78. Yukawa M, Sugiura M (2013) Additional pathway to translate the downstream *ndhK* cistron in partially overlapping *ndhC-ndhK* mRNAs in chloroplasts. Proc Natl Acad Sci USA 110:5701–5706
79. Adachi Y, Kuroda H, Yukawa Y, Sugiura M (2012) Translation of partially overlapping *psbD-psbC* mRNAs in chloroplasts: the role of 5′-processing and translational coupling. Nucleic Acids Res 40:3152–3158
80. Yukawa M, Tsudzuki T, Sugiura M (2005) The 2005 version of the chloroplast DNA sequence from tobacco (*Nicotiana tabacum*). Plant Mol Biol Rep 23:359–365
81. Sugiura M, Hirose T, Sugita M (1998) Evolution and mechanism of translation in chloroplasts. Annu Rev Genet 32:437–459
82. Hirose T, Kusumegi T, Sugiura M (1998) Translation of tobacco chloroplast *rps14* mRNA depends on a Shine-Dalgarno-like sequence in the 5′-untranslated region but not on internal RNA editing in the coding region. FEBS Lett 430:257–260
83. Hirose T, Sugiura M (2004) Functional Shine-Dalgarno-like sequences for translational initiation of chloroplast mRNAs. Plant Cell Physiol 45:114–117
84. Shine J, Dalgarno L (1974) The 3′-terminal sequence of *Escherichia coli* 16S ribosomal RNA: complementarity to nonsense triplets and ribosome binding sites. Proc Natl Acad Sci USA 71:1342–1346
85. Chen H, Bjerknes M, Kumar R et al (1994) Determination of the optimal aligned spacing between the Shine-Dalgarno sequence and the translation initiation codon of *Escherichia coli* mRNAs. Nucleic Acids Res 22:4953–4957
86. Hui A, de Boer HA (1987) Specialized ribosome system: preferential translation of a single mRNA species by a subpopulation of mutated ribosomes in *Escherichia coli*. Proc Natl Acad Sci USA 84:4762–4766
87. Jacob WF, Santer M, Dahlberg AE (1987) A single base change in the Shine-Dalgarno region of 16S rRNA of *Escherichia coli* affects translation of many proteins. Proc Natl Acad Sci USA 84:4757–4761
88. Mutsuda M, Sugiura M (2006) Translation initiation of cyanobacterial *rbcS* mRNAs requires the 38-kDa ribosomal protein S1 but not the Shine-Dalgarno sequence: development of a cyanobacterial *in vitro* translation system. J Biol Chem 281:38314–38321
89. Plader W, Sugiura M (2003) The Shine-Dalgarno-like sequence is a negative regulatory

element for translation of tobacco chloroplast *rps2* mRNA: an additional mechanism for translational control in chloroplasts. Plant J 34:377–382

90. Baecker JJ, Sneddon JC, Hollingsworth MJ (2009) Efficient translation in chloroplasts requires element(s) upstream of the putative ribosome binding site from *atpI*. Am J Bot 96:627–636
91. Drechsel O, Bock R (2010) Selection of Shine-Dalgarno sequences in plastids. Nucleic Acids Res. doi:10.1093/nar/gkq978
92. Kuroda H, Maliga P (2001) Complementarity of the 16S rRNA penultimate stem with sequences downstream of the AUG destabilizes the plastid mRNAs. Nucleic Acids Res 29:970–975
93. Inamine G, Nash B, Weissbach H et al (1985) Light regulation of the synthesis of the large subunit of ribulose-1, 5-bisphosphate carboxylase in peas: evidence for translational control. Proc Natl Acad Sci USA 82:5690–5694
94. Shiina T, Allison L, Maliga P (1998) RbcL transcript levels in tobacco plastids are independent of light: reduced dark transcription rate is compensated by increased mRNA stability. Plant Cell 10:1713–1722
95. Kahlau S, Bock R (2008) Plastid transcriptomics and translatomics of tomato fruit development and chloroplast-to-chromoplast differentiation: chromoplast gene expression largely serves the production of a single protein. Plant Cell 20:856–874
96. McCormac DJ, Barkan A (1999) A nuclear gene in maize required for the translation of the chloroplast *atpB/E* mRNA. Plant Cell 11:1709–1716
97. Schmitz-Linneweber C, Williams-Carrier R, Barkan A (2005) RNA immunoprecipitation and microarray analysis show a chloroplast pentatricopeptide repeat protein to be associated with the 5′ region of mRNAs whose translation it activates. Plant Cell 17:2791–2804
98. Schult K, Meierhoff K, Paradies S et al (2007) The nuclear-encoded factor HCF173 is involved in the initiation of translation of the *psbA* mRNA in *Arabidopsis thaliana*. Plant Cell 19:1329–1346
99. Klaff P, Mundt SM, Steger G (1997) Complex formation of the spinach chloroplast *psbA* mRNA 5′ untranslated region with proteins is dependent on the RNA structure. RNA 3:1468–1479
100. Hotchkiss TL, Hollingsworth MJ (1999) ATP synthase 5′ untranslated regions are specifically bound by chloroplast polypeptides. Curr Genet 35:512–520
101. Merhige PM, Both-Kim D, Robida MD et al (2005) RNA-protein complexes that form in the spinach chloroplast atpI 5′ untranslated region can be divided into two subcomplexes, each comprised of unique *cis*-elements and *trans*-factors. Curr Genet 48:256–264
102. Robida MD, Merhige PM, Hollingsworth MJ (2002) Proteins are shared among RNA-protein complexes that form in the 5′untranslated regions of spinach chloroplast mRNAs. Curr Genet 41:53–62
103. Klaff P, Gruissem W (1995) A 43 kD light-regulated chloroplast RNA-binding protein interacts with the *psbA* 5′ non-translated leader RNA. Photosynth Res 46:235–248
104. Alexander C, Faber N, Klaff P (1998) Characterization of protein-binding to the spinach chloroplast *psbA* mRNA 5′untranslated region. Nucleic Acid Res 26:2265–2272
105. Shteiman-Kotler A, Schuster G (2000) RNA-binding characteristics of the chloroplast S1-like ribosomal protein CS1. Nucleic Acids Res 28:3310–3315
106. Barneche F, Winter V, Crevecoeur M et al (2006) ATAB2 is a novel factor in the signalling pathway of light-controlled synthesis of photosystem proteins. EMBO J 25: 5907–5918
107. Choquet Y, Vallon O (2000) Synthesis, assembly and degradation of thylakoid membrane proteins. Biochimie 82:615–634
108. Wostrikoff K, Stern D (2007) Rubisco large-subunit translation is autoregulated in response to its assembly state in tobacco chloroplasts. Proc Natl Acad Sci USA 104:6466–6471
109. Ichikawa K, Miyake C, Iwano M et al (2008) Ribulose 1,5-bisphosphate carboxylase/oxygenase large subunit translation is regulated in a small subunit-independent manner in the expanded leaves of tobacco. Plant Cell Physiol 49:214–225
110. Klein RR, Mason HS, Mullet JE (1988) Light-regulated translation of chloroplast proteins. I. Transcripts of *psaA-psaB*, *psbA*, and *rbcL* are associated with polysomes in dark-grown and illuminated barley seedlings. J Cell Biol 106:289–301
111. Edhofer I, Mühlbauer SK, Eichacker LA (1998) Light regulates the rate of translation elongation of chloroplast reaction center protein D1. Eur J Biochem 257:78–84
112. Kuroda H, Inagaki N, Satoh K (1992) The level of stromal ATP regulates translation of the D1 protein in isolated chloroplasts. Plant Cell Physiol 33:33–39
113. Taniguchi M, Kuroda H, Satoh K (1993) ATP-dependent protein synthesis in isolated pea chloroplasts: evidence for accumulation of a translation intermediate of the D1 protein. FEBS Lett 317:57–61

114. Kim J, Mullet JE (2003) A mechanism for light-induced translation of the *rbcL* mRNA encoding the large subunit of ribulose-1,5-bisphosphate carboxylase in barley chloroplasts. Plant Cell Physiol 44:491–499
115. Mullet JE, Klein RR, Grossman AR (1986) Optimization of protein synthesis in isolated higher plant chloroplasts. Identification of paused translation intermediates. Eur J Biochem 155:331–338
116. Stollar NE, Kim J-K, Hollingsworth MJ (1994) Ribosomes pause during the expression of the large ATP synthase gene cluster in spinach chloroplasts. Plant Physiol 105: 1167–1177
117. van Wijk KJ, Bingsmark S, Aro E-M et al (1995) In vitro synthesis and assembly of photosystem II core proteins. The D1 protein can be incorporated into photosystem II in isolated chloroplasts and thylakoids. J Biol Chem 270:25685–25695
118. Nilsson R, Brunner J, Hoffman NE et al (1999) Interactions of ribosome nascent chain complexes of the chloroplast-encoded D1 thylakoid membrane protein with cpSRP54. EMBO J 18:733–742
119. Zhang L, Paakkarinen V, van Wijk KJ et al (2000) Biogenesis of the chloroplast-encoded D1 protein: regulation of translation elongation, insertion, and assembly into photosystem II. Plant Cell 12:1769–1781
120. Kuroda H, Maliga P (2001) Sequences downstream of the translation initiation codon are important determinants of translation efficiency in chloroplasts. Plant Physiol 125:430–436
121. Minami E, Shinohara K, Kawakami N et al (1988) Localization and properties of transcripts of *psbA* and *rbcL* genes in the stroma of spinach chloroplast. Plant Cell Physiol 29:1303–1309
122. Mühlbauer SK, Eichacker LA (1999) The stromal protein large subunit of ribulose-1,5-bisphosphate carboxylase is translated by membrane-bound ribosomes. Eur J Biochem 261:784–788
123. Ruhlman T, Verma D, Samson N et al (2010) The role of heterologous chloroplast sequence elements in transgene integration and expression. Plant Physiol 152:2088–2104
124. Li Y, Sugiura M (1990) Three distinct ribonucleoproteins from tobacco chloroplasts: each contains a unique amino terminal acidic domain and two ribonucleoprotein consensus motifs. EMBO J 9:3059–3066
125. Li Y, Sugiura M (1991) Nucleic acid-binding specificities of tobacco chloroplast ribonucleoproteins. Nucleic Acids Res 19: 2893–2896
126. Nakamura T, Ohta M, Sugiura M et al (1999) Chloroplast ribonucleoproteins are associated with both mRNAs and intron-containing precursor tRNAs. FEBS Lett 460:437–441
127. Nakamura T, Ohta M, Sugiura M et al (2001) Chloroplast ribonucleoproteins function as a stabilizing factor of ribosome-free mRNAs in the stroma. J Biol Chem 276:147–152
128. Hirose T, Sugiura M (2001) Involvement of a site-specific *trans*-acting factor and a common RNA-binding protein in the editing of chloroplast mRNAs: development of a chloroplast *in vitro* RNA editing system. EMBO J 20:1144–1152
129. Tillich M, Hardel SL, Kupsch C et al (2009) Chloroplast ribonucleoprotein CP31A is required for editing and stability of specific chloroplast mRNAs. Proc Natl Acad Sci USA 106:6002–6007
130. Prikryl J, Rojas M, Schuster G et al (2010) Mechanism of RNA stabilization and translational activation by a pentatricopeptide repeat protein. Proc Natl Acad Sci USA 108: 1415–1420
131. Schmitz-Linneweber C, Small I (2008) Pentatricopeptide repeat proteins: a socket set for organelle gene expression. Trends Plant Sci 13:663–670
132. Sasaki T, Yukawa Y, Miyamoto T et al (2003) Identification of RNA editing sites in chloroplast transcripts from the maternal and paternal progenitors of tobacco (*Nicotiana tabacum*): comparative analysis shows the involvement of distinct *trans*-factors for *ndhB* editing. Mol Biol Evol 20: 1028–1035
133. Nakamura M, Sugiura M (2007) Translation efficiencies of synonymous codons are not always correlated with codon usage in tobacco chloroplasts. Plant J 49:128–134
134. Ikemura T (1985) Codon usage and tRNA content in unicellular and multicellular organisms. Mol Biol Evol 2:13–34
135. Gustafsson C, Govindarajan S, Minshull J (2004) Codon bias and heterologous protein expression. Trends Biotechnol 22:346–353
136. Nakamura M, Sugiura M (2009) Selection of synonymous codons for better expression of recombinant proteins in tobacco chloroplasts. Plant Biotechnol 26:53–56
137. Nakamura M, Sugiura M (2011) Translation efficiencies of synonymous codons for arginine differ dramatically and are not correlated with codon usage in chloroplasts. Gene 472:50–54

Chapter 5

Engineering Chloroplasts for High-Level Foreign Protein Expression

Ralph Bock

Abstract

Expression of transgenes from the plastid genome offers a number of attractions to biotechnologists, with the potential to attain very high protein accumulation levels arguably being the most attractive one. High-level transgene expression is of particular importance in resistance engineering (e.g., via expression of insecticidal proteins) and molecular farming. Over the past years, the production of many commercially valuable proteins in chloroplast-transgenic (transplastomic) plants has been attempted, including pharmaceutical proteins (such as subunit vaccines and protein antibiotics) and industrial enzymes. Although, in some cases, spectacularly high foreign protein accumulation levels have been obtained, expression levels were disappointingly poor in other cases. In this review, I summarize our current knowledge about the factors influencing the efficiency of plastid transgene expression and highlight possible optimization strategies to alleviate problems with poor expression levels.

Key words 3′-UTR, 5′-UTR, Chloroplast, Codon optimization, Downstream box, Integration site, Plastid, Plastid transformation, Operon expression, Promoter strength, Protein localization, Protein stability, Transcription, Translational regulation

1 Introduction

Expression of transgenes from the plastid (chloroplast) genome has recently received considerable attention [1–4]. In addition to the high level of transgene containment conferred by transgene accommodation in the plastid genome (due to maternal plastid inheritance in most crops leading to exclusion of plastid transgenes from pollen transmission [5, 6]), the chloroplast's enormous capacity to accumulate foreign proteins to very high levels represents a particularly big attraction of the technology [2, 3, 7, 8]. A number of studies have reported examples of foreign proteins that could be produced in transgenic chloroplasts to levels exceeding 10 % of the total soluble protein of the plant. However, by far not all attempts to express transgenes to high levels via transformation of the plastid genome have been successful, and it is not always clear why a given transgene works or does not work well in chloroplasts.

Pal Maliga (ed.), *Chloroplast Biotechnology: Methods and Protocols*, Methods in Molecular Biology, vol. 1132,
DOI 10.1007/978-1-62703-995-6_5,

Here, I review our current knowledge about the factors determining the efficiency of plastid transgene expression and discuss suitable strategies to maximize expression levels. As in most of the published work on high-level transgene expression in chloroplasts the model plant tobacco was used, this article focuses on plastid genome engineering in higher plants. It should be borne in mind that, in addition to the genetic factors discussed here, a number of other parameters are known to influence the yield of recombinant proteins expressed from the plant's plastid genome. These include the choice of the host plant [9], the growth conditions [10], the tissue type [11], and the developmental stage of the plant material harvested [10–12].

2 Choice of Expression Elements

In a prokaryotic system like the chloroplast, expression elements typically reside immediately upstream and downstream of the coding region (Fig. 1). They comprise the promoter, the 5′-untranslated region (5′-UTR), and the 3′-untranslated region (3′-UTR). The promoter determines the transcription rate, the 5′-UTR harbors the translation initiation signals (and therefore determines the translation rate of the mRNA), and the 3′-UTR is chiefly responsible for determining mRNA stability. All three types of expression signals have been demonstrated to influence plastid transgene expression, albeit to different magnitudes.

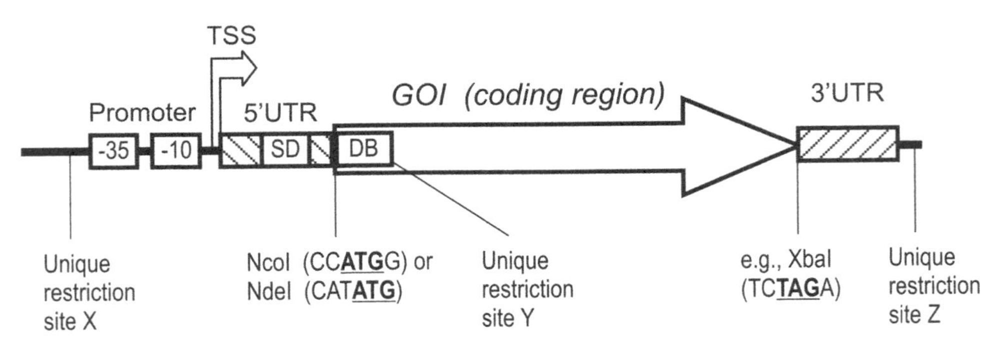

Fig. 1 Elements involved in the efficiency of plastid transgene expression and modular design of expression cassettes. The promoter is indicated by the two sequence elements characteristic of PEP promoters (−35 box and −10 box). The 5′ UTR harbors the Shine-Dalgarno sequence (SD) as ribosome-binding site. *DB* denotes a "downstream box" (N-terminal peptide fusion), *GOI* the coding region of the gene of interest. UTRs are indicated as *hatched boxes. See* text for details. The *restriction sites* below the schematic drawing indicate a possible modular design of the expression cassette. Restriction sites X and Z can be used to transfer the fully assembled cassette into a plastid targeting vector. *NcoI* or *NdeI* restriction sites can accommodate the ATG start codon (*underlined* and *in bold*); *XbaI* is an example of a site for a rare-cutting restriction enzyme that harbors a stop codon in its recognition sequence (*underlined* and *in bold*)

2.1 Promoter Strength

Higher plant plastids possess two distinct RNA polymerase activities. One of them is of the bacterial type and encoded by genes in the plastid genome (PEP: plastid-encoded RNA polymerase) that were retained from the cyanobacterial endosymbiont. The second RNA-polymerizing activity is provided by enzymes that are strikingly similar to bacteriophage-type polymerases. These so-called NEPs (nuclear-encoded RNA polymerases) are encoded by nuclear genes that presumably were acquired early in eukaryotic evolution through horizontal gene transfer [13, 14]. The two RNA polymerase activities recognize different types of promoters. While PEP promoters closely resemble bacterial σ^{70}-type promoters and comprise a −35 and a −10 (TATA) box, most NEP promoters have a core sequence motif (YRTA) that is similar to the consensus sequence of promoters in plant mitochondrial genomes.

In general, PEP promoters appear to be much stronger than NEP promoters [15], and highly expressed genes in the plastid genome (e.g., most photosynthesis genes) are usually transcribed from PEP promoters. For this reasons, PEP promoters have been predominantly used to drive the expression of plastid transgenes (Fig. 1).

The strongest promoter in plastids of higher plants is the promoter driving the ribosomal RNA operon, P*rrn*, and in most of the studies reporting high expression levels of plastid transgenes, the P*rrn* promoter was used (e.g., [11, 16–18]). As rRNAs are not translated, the P*rrn* promoter needs to be combined with appropriate translation initiation signals (*see* Subheading 2.2) to confer high-level foreign protein accumulation.

An alternative to high-level transgene transcription from the strong P*rrn* promoter relies on co-transcription with highly expressed plastid operons. The corresponding vector design was named "operon-extension" vector [10]. When the strongly expressed plastid *psbA* operon was extended by the reporter gene *uidA* (encoding β-glucuronidase), the attained expression level was even higher than that obtained with the P*rrn* promoter in a standard vector. However, it seems unlikely that this is solely due to high transcription rates. The chloroplast *psbA* transcript is highly translated and it is conceivable that dense ribosome coverage of the upstream *psbA* cistron exerts a stimulating effect on translation of the transgene(s) inserted downstream (*see* Subheading 2.2).

In addition to plastid promoters, some foreign promoters (from non-plastid source genomes) have been tested for their capacity to drive transgene expression in chloroplasts. These include highly active bacterial promoters and plant mitochondrial promoters [19, 20]. However, although these promoters were demonstrated to be recognized by the chloroplast transcriptional apparatus (and thus provide valuable tools for fine-tuning of plastid transgene expression), they turned out to be significantly weaker than strong chloroplast PEP promoters.

2.2 Choice of the 5′-UTR

Chloroplast protein biosynthesis occurs on bacterial-type 70S ribosome. Similar to bacteria, the efficiency of translation initiation in plastids largely depends on the sequence and structure of the 5′-UTR. Ribosome recruitment to many plastid mRNAs is mediated by bacterial-type ribosome-binding sites, which exhibit sequence complementarity to the 3′-end of the 16S ribosomal RNA and are also referred to as Shine-Dalgarno sequences (SD sequences; Fig. 1). However, not all 5′-UTRs of plastid mRNAs harbor canonical Shine-Dalgarno sequences in the conserved spacing of four to nine nucleotides upstream of the initiator codon, and it has been suggested that, in a process called Shine-Dalgarno-independent translation initiation, mRNA-specific translational activator proteins bind to the 5′-UTR and direct the ribosomal 30S subunit to the AUG start codon. Although systematic comparative studies are still largely lacking, extrapolating from bacterial translation, it is generally assumed that Shine-Dalgarno-dependent translation rates are overall stronger than Shine-Dalgarno-independent translation rates.

As plastid gene expression is predominantly controlled at the level of translation [21, 22], it is unsurprising that the choice of the 5′-UTR is one of the most important factors determining the expression level of a given plastid transgene. Many different 5′-UTRs have been tested in transgenic experiments, mainly by determining their efficiency in driving translation of reporter genes [10, 23–25]. In addition to 5′-UTRs taken from endogenous plastid genes, also a number of bacterial and synthetic sequences have been tested in chloroplasts. The 5′-UTR ("leader sequence") of *gene 10* from bacteriophage T7 harbors the strongest translation initiation signals identified so far [24, 26]. The T7 *gene 10* leader contains a perfect Shine-Dalgarno sequence and also drives extraordinarily high rates of mRNA translation in *E. coli*. When combined with the strong ribosomal RNA operon promoter (P*rrn*), it confers the highest foreign protein accumulation levels obtained to date [11, 18].

A recent study has explored the possibility of boosting translation of transgenic mRNAs by providing multiple Shine-Dalgarno sequences [27]. The data indicate that it is possible to enhance translation initiation rates in this way, suggesting that combination of several ribosome-binding sites may be another viable strategy to maximize transgene expression from the plastid genome. To optimize foreign protein accumulation in nongreen plastid types (e.g., amyloplasts in potato tubers or chromoplasts in tomato fruits), the design of chimeric expression elements (promoter and 5′-UTR combinations) has proven a successful strategy [28–30].

2.3 Choice of the 3′-UTR

The 3′-UTRs of plastid mRNAs do not function as efficient terminators of transcription. Instead, they fold into stable stem-loop-type RNA secondary structures that are believed to protect the mRNA from 3′ to 5′ exoribonucleolytic degradation [31]. A number of studies have compared the expression of reporter genes from the plastid genome that differ only in their 3′-UTR [23, 25, 32]. The picture that has emerged from this work is that, while RNA accumulation levels can be influence by the choice of the 3′-UTR, the impact of the 3′-UTR on expression levels is rather limited. For example, plants expressing *gfp* with the terminator from the *Escherichia coli* ribosomal RNA operon *rrnB* accumulate four times more *gfp* transcripts than plants expressing *gfp* with the *rbcL* or *rpoA* 3′-UTRs [32]. Nonetheless, increased mRNA accumulation did not result in higher protein accumulation indicating that GFP expression levels are not limited by transcript abundance. This is consistent with the finding that the expression of plastid genes is predominantly controlled at the translational level in that translational regulation can override even relatively large changes in transcript accumulation [21, 25]. However, it is conceivable that transcript accumulation can limit the expression of certain transgenes under certain conditions, and in these cases, testing different 3′-UTRs may be a suitable strategy for maximizing expression levels.

3 Coding Region

The sequence of the coding region can also influence the expression level of a given transgene. This can be an effect of codon usage (the differential use of synonymous codons), which affects the efficiency of translation elongation or, alternatively, may sometimes be related to sequences immediately downstream of the initiation codon ("downstream box") affecting the efficiency of translation initiation. The sequence of the coding region can also affect translational efficiency via RNA secondary structure formation. For example, in bacteria, partial sequence complementarity between the ribosome-binding site (Shine-Dalgarno sequence) and sequences elsewhere in the mRNA can make the Shine-Dalgarno sequence inaccessible to the ribosome by masking it in a double-stranded structure [33–35]. Recent work has shown that secondary structure formation at the translation initiation site can also block translation in plastids [36]. As RNA secondary structures are also involved in triggering various RNA processing events in chloroplasts [31, 37–39], it may be desirable to try to avoid formation of stable stem-loop-type secondary structures not only at the translation initiation site but also within the coding region. However, although highly stable short-range interactions between RNA sequences can be computed using appropriate software tools (e.g., Mfold; http://frontend.bioinfo.rpi.edu/applications/mfold/cgi-bin/rna-form1.cgi), reliable prediction of mRNA secondary structures in vivo remains notoriously difficult. Also, it is currently unknown whether aberrant RNA secondary structure formation is responsible for the (few) described cases in which lack of plastid transgene expression was found to be due to mRNA instability [40].

3.1 Codon Usage

Codon usage is often blamed for problems with the expression of heterologous transgenes, and this is also the case in plastid engineering. However, only few studies have directly compared codon-optimized and non-optimized transgenes to determine the influence of codon usage on foreign protein accumulation. The few studies that have been conducted [17, 24] confirm that codon usage does indeed influence the efficiency of plastid transgene expression. The plastid genome of higher plants is highly AT rich, and therefore, AT-rich transgenes should express better than GC-rich ones. This was first confirmed for the transplastomic expression of TetC, the nontoxic 47 kDa polypeptide fragment ("fragment C") of tetanus toxin. The TetC protein can be used as a subunit vaccine against tetanus. When a bacterial high-AT (72.3 % AT) gene version was compared with a synthetic gene with much lower AT content (52.5 % AT), TetC accumulation in tobacco chloroplasts reached 25 % of the total soluble protein for the high-AT gene version, but only 10 % for the low-AT gene version [17].

Some of the highest expression levels reported so far have been obtained with codon-optimized transgenes [11, 18, 41]. However, in these cases, only synthetic (codon usage-optimized) coding regions were used, and no side-by-side comparison with the unmodified coding regions was conducted. Therefore, the contribution of the codon usage adjustment to the attained extraordinarily high expression levels is unknown in all these cases.

3.2 Downstream Boxes

Coding sequences immediately downstream of the translation initiation codon have been shown to significantly influence expression levels of plastid transgenes and are sometimes referred to as "downstream boxes" [10, 42–44]. The downstream box was originally described as a translational enhancer of several bacterial and bacteriophage mRNAs. In plastids, it was found that certain N-terminal fusions can result in significant enhancement of recombinant protein accumulation [10, 42–44]. However, whether or not these N-terminal fusions really act as translational enhancers in chloroplasts is not known. Recently, the N-terminus of plastid proteins has been shown to harbor key determinants of protein stability [45]. It, therefore, seems conceivable that, rather than stimulating translation, the insertion of specific sequences downstream of the start codon can stabilize an otherwise unstable recombinant protein. Consequently, in the absence of compelling evidence for these sequences really acting by increasing translation rates, caution should be exercised with calling them "downstream boxes." For this reason, they are discussed here as "N-terminal peptide fusions" (*see* Subheading 5) and not as "downstream boxes."

4 Choice of the Integration Site in the Genome

Over the years, many different insertion sites in the plastid genome have been tested. Although relatively few studies have been conducted to compare identical transgenes in different genomic locations side by side, currently, there is no reason to assume that plastid transgene expression levels would strongly depend on the insertion site in the genome. This is consistent with the general belief that, due to the absence of a complex chromatin-like DNA organization, the expression of plastid genes is not influenced by position effects or other epigenetic mechanisms that modulate gene expression in eukaryotes.

Transgene targeting to the inverted repeat region (IR) of the plastid genome doubles the gene dosage, and at least in some cases, this may result in increased expression levels [10]. However, in view of the pivotal role of translational regulation in plastids (*see* Subheading 2.3), even if transcript levels can be doubled by placing the transgene into the IR, this cannot necessarily be expected to translate into an increase in recombinant protein accumulation.

5 Protein Stability and Folding

Growing evidence suggests that unsuccessful expression of recombinant proteins from the chloroplast genome is often due to protein instability [12, 46]. Unfortunately, there is no easy solution to protein instability, because the factors influencing the half-life of plastid proteins are largely unknown. Only recently, some determinants of protein stability in chloroplasts have been uncovered [45]. The data available so far suggest an important role of the N-terminus in determining the protein half-life, whereas the C-terminal sequence does not seem to contribute significantly to protein stability. Similar to proteins in both prokaryotes and eukaryotes, the identity of the N-terminal amino acid residue (i.e., the penultimate residue following the initiator methionine) influences protein stability, suggesting existence of an N-end rule-like pathway of protein degradation in plastids. However, additional sequence determinants residing in the N-terminal region may play an even more important role in determining the half-life of plastid proteins. Together, the currently available data indicate that the stability of plastid proteins is largely determined by three factors: (1) the action of methionine aminopeptidase, the enzyme that cleaves off the initiator methionine and thereby exposes the penultimate amino acid residue, (2) an N-end rule-like pathway of protein degradation, and (3) additional sequence motifs in the N-terminal region [45].

It seems highly unlikely that protein stability in plastids is exclusively determined by the primary sequence of the protein.

Instead, the three-dimensional structure is expected to play an important role in defining the accessibility of the protein to proteases [47, 48]. However, next to nothing is known about the nature of these structural determinants of protein (in)stability. Thus, while the N-terminus of the recombinant protein can be protected by addition of suitable stabilizing sequences (*see* below), the impact of internal sequences and structural motifs on protein stability remains unpredictable for the time being.

Consistent with the importance of N-terminal sequences for protein stability in plastids, several N-terminal peptide fusions have been described to strongly enhance recombinant protein accumulation, presumably by stabilizing otherwise relatively unstable proteins (*see* also Subheading 3.1). These stabilizing peptide sequences include foreign sequences (e.g., the N-terminus of GFP), synthetic peptide sequences, and sequences from the N-termini of endogenous plastid proteins [10, 24, 43–45]. Consequently, if poor expression of the recombinant protein of interest is suspected to be due to low protein stability, testing some of these N-terminal peptide fusions may provide an appropriate remedy.

Protein folding represents an important aspect of protein stability in that improperly folded proteins are often condemned to rapid proteolytic degradation. Biochemical work and bioinformatics analyses of genome sequences and EST data suggest that the protein folding machinery of the chloroplast resembles that of bacteria. However, whether or not folding problems contribute to poor accumulation of some chloroplast-expressed recombinant proteins is not known.

The activity and stability of some recombinant proteins depends on proper formation of intramolecular disulfide bonds. Several studies have documented that the chloroplast is capable of correctly forming disulfide bonds [49, 50], presumably due to the presence of plastid-localized protein disulfide isomerases. Interestingly, faithful disulfide bond formation was shown to occur both in the chloroplast stroma and the thylakoid lumen [50].

6 Protein Localization

Expression of a transgene from the chloroplast genome offers three choices of protein localization: (1) in the stroma, (2) in the inner membrane and/or the thylakoid membrane, or (3) in the thylakoid lumen.

In the vast majority of studies reports so far, the recombinant proteins expressed from the chloroplast genome accumulated as soluble proteins in the stroma. The protein accommodation capacity of the chloroplast stroma is known to be very high, with Rubisco (ribulose-1,5-bisphosphate carboxylase/oxygenase), the most abundant protein on our planet, accumulating to more than 50 % of the total soluble leaf protein in many plant species.

Several transplastomic studies have demonstrated that (soluble) recombinant proteins that accumulate in the stroma can be expressed to very high levels, which, in the most extreme cases, can even exceed the accumulation levels of Rubisco in the wild type [18].

Accumulating evidence suggests that insertion of recombinant proteins into the thylakoid membrane is not very well tolerated by the photosynthetic apparatus and can result in severe mutant (pigment-deficient) phenotypes [51, 52]. The reason may be that the thylakoid membrane is a highly protein-rich membrane, densely packed with the multiprotein complexes of the photosynthetic electron transport chain (photosystem II, cytochrome b_6f complex, photosystem I, ATP synthase). Given the large number of subunits and cofactors these complexes are comprised of, the highly coordinated assembly process required to produce them [53], and their intricate three-dimensional arrangement in the thylakoid membrane [54, 55], it is conceivable that insertion of foreign proteins can lead to severe disturbances in the structure and function of the photosynthetic apparatus. However, more work is needed to determine whether the chloroplast is generally not a suitable site for the high-level expression of hydrophobic proteins.

Only relatively few studies have attempted to target recombinant proteins to the thylakoid lumen [50, 56]. Although, to date, no systematic comparison between expression levels attainable in the thylakoid lumen versus the stroma has been conducted, it seems conceivable that the capacity of the lumen to accommodate foreign proteins is overall smaller than that of the stroma.

7 Multigene Expression from Operons

Gene organization in operons and their transcription as polycistronic RNAs is another shared feature of bacteria and plastids. However, whereas in bacteria, polycistronic transcripts are usually directly translated, they often undergo posttranscriptional cleavage into monocistronic or oligocistronic units in plastids. Moreover, intercistronic processing (sometimes also referred to as RNA cutting) can be a prerequisite for efficient translation [57]. Whether or not intercistronic processing is required to promote efficient plastid transgene expression from operons needs to be determined on a case-by-case basis and, unfortunately, is currently unpredictable. At least some transgenes can be efficiently expressed from unprocessed polycistronic transcripts [58, 59]. Transplastomic experiments have defined a minimum sequence element (dubbed "intercistronic expression element"; IEE) that is necessary and sufficient to trigger intercistronic processing and, in this way, facilitates the expression of stable translatable monocistronic mRNAs from operons [38, 60]. Inclusion of this element in synthetic operons reduces the probability that individual cistrons in polycistronic transcripts are poorly expressed.

8 Other Considerations

In addition to the factors discussed above, a number of other features of the gene/protein of interest may influence expression levels and/or activity of the recombinant protein to be expressed from the plastid genome. These include posttranslational modifications, potential deleterious effects of the recombinant protein on plastid gene expression, photosynthesis or other metabolic processes [52, 61, 62], and genome instability as occasionally induced by unwanted homologous recombination between duplicated expression signals. The use of homologous expression signals (promoters and UTRs taken from resident plastid genes) to drive transgene expression duplicates endogenous expression signals present elsewhere in the plastid genome. Depending on the relative orientation of the duplicated elements, recombination can either cause deletions in the genome or, alternatively, result in inversions by flip-flop recombination [63–67]. Use of heterologous expression signals from other species reduces the level of sequence homology and, in this way, the risk of unwanted recombination. However, such heterologous expression elements need to be carefully tested and compared to the corresponding homologous elements to ascertain that they work equally efficiently.

9 Summary and Guidelines for Maximizing Expression Levels

For achieving high-level expression of a single transgene in higher plant chloroplasts, a number of factors should be considered (Fig. 1). The following general recommendations can be made:

Promoter choice: The ribosomal RNA operon promoter (P*rrn*) gives the highest transcription rates.

Choice of 5′-UTR: The Shine-Dalgarno sequence derived from the *gene 10* leader sequence of phage T7 (and combined with the ribosomal RNA operon promoter; [24, 26]) confers the highest translation rates reported so far.

"*Downstream box*" (*N-terminal peptide fusions*): No general recommendation possible. Inclusion of short N-terminal fusion peptides can be beneficial at least in some cases [10, 24, 42–44, 46], presumably by stabilizing otherwise unstable recombinant proteins [45].

Choice of 3′-UTR: The 3′-UTR from the *Escherichia coli* ribosomal RNA operon *rrnB* seems to confer higher RNA stability than most plastid 3′-UTRs [32]. Overall, the influence of the 3′-UTR on expression levels appears to be lower than that of promoter and 5′-UTR.

Genomic integration site: Transgene integration into the IR region of the plastid genome doubles the gene dosage and, if foreign protein accumulation is limited by transcript amounts, may result in increased expression levels [10].

Codon usage: Codon optimization (http://www.kazusa.or.jp/codon/) can help to obtain higher expression levels, at least in certain cases [17, 24]. Whether or not it is generally useful remains to be investigated.

RNA secondary structure: RNA folding programs (e.g., Mfold; http://frontend.bioinfo.rpi.edu/applications/mfold/cgi-bin/rna-form1.cgi) can be used to assess the risk that the Shine-Dalgarno sequence is inaccessible to the ribosome.

Protein stability: Hardly predictable. Based on a preliminary N-end rule for plastids, Cys and His should be avoided as N-terminal (penultimate) amino acid residues [45]. N-terminal peptide fusions can improve expression levels by increasing protein stability (*see* above: "Downstream box").

Protein localization: For most soluble proteins, the stroma of the chloroplast appears to be the preferred location in that it probably has the highest protein accumulation capacity. Overexpression of membrane proteins and other hydrophobic proteins in plastids may be problematic.

Acknowledgements

Work on plastid transformation in the author's laboratory is supported by grants from the European Union (FP7), the Bundesministerium für Bildung und Forschung (BMBF), the Deutsche Forschungsgemeinschaft (DFG), and the Max Planck Society.

References

1. Maliga P (2004) Plastid transformation in higher plants. Annu Rev Plant Biol 55:289–313
2. Daniell H, Singh ND, Mason H, Streatfield SJ (2009) Plant-made vaccine antigens and biopharmaceuticals. Trends Plant Sci 14:669–679
3. Bock R (2007) Plastid biotechnology: prospects for herbicide and insect resistance, metabolic engineering and molecular farming. Curr Opin Biotechnol 18:100–106
4. Koop H-U, Herz S, Golds TJ, Nickelsen J (2007) The genetic transformation of plastids. Top Curr Genet 19:457–510
5. Ruf S, Karcher D, Bock R (2007) Determining the transgene containment level provided by chloroplast transformation. Proc Natl Acad Sci U S A 104:6998–7002
6. Svab Z, Maliga P (2007) Exceptional transmission of plastids and mitochondria from the transplastomic pollen parent and its impact on transgene containment. Proc Natl Acad Sci U S A 104:7003–7008
7. Daniell H (2006) Production of biopharmaceuticals and vaccines in plants via the chloroplast genome. Biotechnol J 1:1071–1079
8. Bock R, Warzecha H (2010) Solar-powered factories for new vaccines and antibiotics. Trends Biotechnol 28:246–252
9. McCabe MS, Klaas M, Gonzalez-Rabade N, Poage M, Badillo-Corona JA, Zhou F, Karcher D, Bock R, Gray JC, Dix PJ (2008) Plastid transformation of high-biomass tobacco variety Maryland Mammoth for production of human

immunodeficiency virus type 1 (HIV-1) p24 antigen. Plant Biotechnol J 6:914–929
10. Herz S, Füßl M, Steiger S, Koop H-U (2005) Development of novel types of plastid transformation vectors and evaluation of factors controlling expression. Transgenic Res 14:969–982
11. Zhou F, Badillo-Corona JA, Karcher D, Gonzalez-Rabade N, Piepenburg K, Borchers A-MI, Maloney AP, Kavanagh TA, Gray JC, Bock R (2008) High-level expression of HIV antigens from the tobacco and tomato plastid genomes. Plant Biotechnol J 6:897–913
12. Birch-Machin I, Newell CA, Hibberd JM, Gray JC (2004) Accumulation of rotavirus VP6 protein in chloroplasts of transplastomic tobacco is limited by protein stability. Plant Biotechnol J 2:261–270
13. Liere K, Börner T (2007) Transcription and transcriptional regulation in plastids. Top Curr Genet 19:121–174
14. Filée J, Forterre P (2005) Viral proteins functioning in organelles: a cryptic origin? Trends Microbiol 13:510–513
15. Hajdukiewicz PTJ, Allison LA, Maliga P (1997) The two RNA polymerases encoded by the nuclear and the plastid compartments transcribe distinct groups of genes in tobacco plastids. EMBO J 16:4041–4048
16. De Cosa B, Moar W, Lee S-B, Miller M, Daniell H (2001) Overexpression of the Bt cry2Aa2 operon in chloroplasts leads to formation of insecticidal crystals. Nat Biotechnol 19:71–74
17. Tregoning JS, Nixon P, Kuroda H, Svab Z, Clare S, Bowe F, Fairweather N, Ytterberg J, van Wijk KJ, Dougan G, Maliga P (2003) Expression of tetanus toxin fragment C in tobacco chloroplasts. Nucleic Acids Res 31:1174–1179
18. Oey M, Lohse M, Kreikemeyer B, Bock R (2009) Exhaustion of the chloroplast protein synthesis capacity by massive expression of a highly stable protein antibiotic. Plant J 57:436–445
19. Newell CA, Birch-Machin I, Hibberd JM, Gray JC (2003) Expression of green fluorescent protein from bacterial and plastid promoters in tobacco chloroplasts. Transgenic Res 12:631–634
20. Bohne A-V, Ruf S, Börner T, Bock R (2007) Faithful transcription initiation from a mitochondrial promoter in transgenic plastids. Nucleic Acids Res 35:7256–7266
21. Eberhard S, Drapier D, Wollman F-A (2002) Searching limiting steps in the expression of chloroplast-encoded proteins: relations between gene copy number, transcription, transcript abundance and translation rate in the chloroplast of Chlamydomonas reinhardtii. Plant J 31:149–160
22. Kahlau S, Bock R (2008) Plastid transcriptomics and translatomics of tomato fruit development and chloroplast-to-chromoplast differentiation: chromoplast gene expression largely serves the production of a single protein. Plant Cell 20:856–874
23. Staub JM, Maliga P (1994) Translation of the psbA mRNA is regulated by light via the 5′-untranslated region in tobacco plastids. Plant J 6:547–553
24. Ye G-N, Hajdukiewicz PTJ, Broyles D, Rodriguez D, Xu CW, Nehra N, Staub JM (2001) Plastid-expressed 5-enolpyruvylshikimate-3-phosphate synthase genes provide high level glyphosate tolerance in tobacco. Plant J 25:261–270
25. Eibl C, Zou Z, Beck A, Kim M, Mullet J, Koop H-U (1999) In vivo analysis of plastid psbA, rbcL and rpl32 UTR elements by chloroplast transformation: tobacco plastid gene expression is controlled by modulation of transcript levels and translation efficiency. Plant J 19:333–345
26. Kuroda H, Maliga P (2001) Complementarity of the 16S rRNA penultimate stem with sequences downstream of the AUG destabilizes the plastid mRNAs. Nucleic Acids Res 29:970–975
27. Drechsel O, Bock R (2011) Selection of Shine-Dalgarno sequences in plastids. Nucleic Acids Res 39:1427–1438
28. Valkov VT, Gargano D, Manna C, Formisano G, Dix PJ, Gray JC, Scotti N, Cardi T (2011) High efficiency plastid transformation in potato and regulation of transgene expression in leaves and tubers by alternative 5′ and 3′ regulatory sequences. Transgenic Res 20:137–151
29. Zhang J, Ruf S, Hasse C, Childs L, Scharff LB, Bock R (2012) Identification of cis-elements conferring high levels of gene expression in non-green plastids. Plant J 72:115–128
30. Caroca R, Howell KA, Hasse C, Ruf S, Bock R (2012) Design of chimeric expression elements that confer high-level gene activity in chromoplasts. Plant J 73:368–379
31. Stern DB, Goldschmidt-Clermont M, Hanson MR (2010) Chloroplast RNA metabolism. Annu Rev Plant Biol 61:125–155
32. Tangphatsornruang S, Birch-Machin I, Newell CA, Gray JC (2011) The effect of different 3′ untranslated regions on the accumulation and stability of transcripts of a gfp transgene in chloroplasts of transplastomic tobacco. Plant Mol Biol 76:385–396
33. Hall MN, Gabay J, Debarbouille M, Schwartz M (1982) A role for mRNA secondary structure in the control of translation initiation. Nature 295:616–618
34. Neupert J, Karcher D, Bock R (2008) Design of simple synthetic RNA thermometers for

temperature-controlled gene expression in Escherichia coli. Nucleic Acids Res 36:e124

35. Kudla G, Murray AW, Tollervey D, Plotkin JB (2009) Coding-sequence determinants of gene expression in Escherichia coli. Science 324:255–258
36. Verhounig A, Karcher D, Bock R (2010) Inducible gene expression from the plastid genome by a synthetic riboswitch. Proc Natl Acad Sci U S A 107:6204–6209
37. Hayes R, Kudla J, Schuster G, Gabay L, Maliga P, Gruissem W (1996) Chloroplast mRNA 3′-end processing by a high molecular weight protein complex is regulated by nuclear encoded RNA binding proteins. EMBO J 15:1132–1141
38. Zhou F, Karcher D, Bock R (2007) Identification of a plastid Intercistronic Expression Element (IEE) facilitating the expression of stable translatable monocistronic mRNAs from operons. Plant J 52:961–972
39. Walter M, Piepenburg K, Schöttler MA, Petersen K, Kahlau S, Tiller N, Drechsel O, Weingartner M, Kudla J, Bock R (2010) Knockout of the plastid RNase E leads to defective RNA processing and chloroplast ribosome deficiency. Plant J 64:851–863
40. Wurbs D, Ruf S, Bock R (2007) Contained metabolic engineering in tomatoes by expression of carotenoid biosynthesis genes from the plastid genome. Plant J 49:276–288
41. Oey M, Lohse M, Scharff LB, Kreikemeyer B, Bock R (2009) Plastid production of protein antibiotics against pneumonia via a new strategy for high-level expression of antimicrobial proteins. Proc Natl Acad Sci U S A 106:6579–6584
42. Kuroda H, Maliga P (2001) Sequences downstream of the translation initiation codon are important determinants of translation efficiency in chloroplasts. Plant Physiol 125:430–436
43. Lenzi P, Scotti N, Alagna F, Tornesello ML, Pompa A, Vitale A, De Stradis A, Monti L, Grillo S, Buonaguro FM, Maliga P, Cardi T (2008) Translational fusion of chloroplast-expressed human papillomavirus type 16L1 capsid protein enhances antigen accumulation in transplastomic tobacco. Transgenic Res 17:1091–1102
44. Gray BN, Ahner BA, Hanson MR (2009) High-level bacterial cellulase accumulation in chloroplast-transformed tobacco mediated by downstream box fusions. Biotechnol Bioeng 102:1045–1054
45. Apel W, Schulze WX, Bock R (2010) Identification of protein stability determinants in chloroplasts. Plant J 63:636–650
46. Bellucci M, de Marchis F, Mannucci R, Bock R, Arcioni S (2005) Cytoplasm and chloroplasts are not suitable subcellular locations for ß-zein accumulation in transgenic plants. J Exp Bot 56:1205–1212
47. Adam Z (2007) Protein stability and degradation in plastids. Top Curr Genet 19:315–338
48. Sakamoto W (2006) Protein degradation machineries in plastids. Annu Rev Plant Biol 57:599–621
49. Staub JM, Garcia B, Graves J, Hajdukiewicz PTJ, Hunter P, Nehra N, Paradkar V, Schlittler M, Carroll JA, Spatola L, Ward D, Ye G, Russell DA (2000) High yield production of a human therapeutic protein in tobacco chloroplasts. Nat Biotechnol 18:333–338
50. Bally J, Paget E, Droux M, Job C, Job D, Dubald M (2008) Both the stroma and thylakoid lumen of tobacco chloroplasts are competent for the formation of disulphide bonds in recombinant proteins. Plant Biotechnol J 6:46–61
51. Glenz K, Bouchon B, Stehle T, Wallich R, Simon MM, Warzecha H (2006) Production of a recombinant bacterial lipoprotein in higher plant chloroplasts. Nat Biotechnol 24:76–77
52. Hennig A, Bonfig K, Roitsch T, Warzecha H (2007) Expression of the recombinant bacterial outer surface protein A in tobacco chloroplasts leads to thylakoid localization and loss of photosynthesis. FEBS J 274:5749–5758
53. Albus C, Ruf S, Schöttler MA, Lein W, Kehr J, Bock R (2010) Y3IP1, a nucleus-encoded thylakoid protein, co-operates with the plastid-encoded Ycf3 protein in photosystem I assembly. Plant Cell 22:2838–2855
54. Kirchhoff H, Haferkamp S, Allen JF, Epstein DBA, Mullineaux CW (2008) Protein diffusion and macromolecular crowding in thylakoid membranes. Plant Physiol 146:1571–1578
55. Kirchhoff H (2008) Molecular crowding and order in photosynthetic membranes. Trends Plant Sci 13:201–207
56. Tissot G, Canard H, Nadai M, Martone A, Botterman J, Dubald M (2008) Translocation of aprotinin, a therapeutic protease inhibitor, into the thylakoid lumen of genetically engineered tobacco chloroplasts. Plant Biotechnol J 6:309–320
57. Hirose T, Sugiura M (1997) Both RNA editing and RNA cleavage are required for translation of tobacco chloroplast ndhD mRNA: a possible regulatory mechanism for the expression of a chloroplast operon consisting of functionally unrelated genes. EMBO J 16:6804–6811
58. Staub JM, Maliga P (1995) Expression of a chimeric uidA gene indicates that polycistronic mRNAs are efficiently translated in tobacco plastids. Plant J 7:845–848
59. Quesada-Vargas T, Ruiz ON, Daniell H (2005) Characterization of heterologous multigene operons in transgenic chloroplasts. Transcription, processing, and translation. Plant Physiol 138:1746–1762
60. Lu Y, Rijzaani H, Karcher D, Ruf S, Bock R (2013) Efficient metabolic pathway engineering

in transgenic tobacco and tomato plastids with synthetic multigene operons. Proc Natl Acad Sci U S A 110:E623–E632

61. Magee AM, Coyne S, Murphy D, Horvath EM, Medgyesy P, Kavanagh TA (2004) T7 RNA polymerase-directed expression of an antibody fragment transgene in plastids causes a semi-lethal pale-green seedling phenotype. Transgenic Res 13:325–337
62. Petersen K, Bock R (2011) High-level expression of a suite of thermostable cell wall-degrading enzymes from the chloroplast genome. Plant Mol Biol 76:311–321
63. Iamtham S, Day A (2000) Removal of antibiotic resistance genes from transgenic tobacco plastids. Nat Biotechnol 18:1172–1176
64. Rogalski M, Ruf S, Bock R (2006) Tobacco plastid ribosomal protein S18 is essential for cell survival. Nucleic Acids Res 34:4537–4545
65. Rogalski M, Karcher D, Bock R (2008) Superwobbling facilitates translation with reduced tRNA sets. Nat Struct Mol Biol 15:192–198
66. Rogalski M, Schöttler MA, Thiele W, Schulze WX, Bock R (2008) Rpl33, a nonessential plastid-encoded ribosomal protein in tobacco, is required under cold stress conditions. Plant Cell 20:2221–2237
67. Gray BN, Ahner BA, Hanson MR (2009) Extensive homologous recombination between introduced and native regulatory plastid DNA elements in transplastomic plants. Transgenic Res 18:559–572

Chapter 6

Excision of Plastid Marker Genes Using Directly Repeated DNA Sequences

Elisabeth A. Mudd, Panagiotis Madesis, Elena Martin Avila, and Anil Day

Abstract

Excision of marker genes using DNA direct repeats makes use of the predominant homologous recombination pathways present in the plastids of algae and plants. The method is simple, efficient, and widely applicable to plants and microalgae. Marker excision frequency is dependent on the length and number of directly repeated sequences. When two repeats are used a repeat size of greater than 600 bp promotes efficient excision of the marker gene. A wide variety of sequences can be used to make the direct repeats. Only a single round of transformation is required, and there is no requirement to introduce site-specific recombinases by retransformation or sexual crosses. Selection is used to maintain the marker and ensure homoplasmy of transgenic plastid genomes. Release of selection allows the accumulation of marker-free plastid genomes generated by marker excision, which is spontaneous, random, and a unidirectional process. Positive selection is provided by linking marker excision to restoration of the coding region of an herbicide resistance gene from two overlapping but incomplete coding regions. Cytoplasmic sorting allows the segregation of cells with marker-free transgenic plastids. The marker-free shoots resulting from direct repeat-mediated excision of marker genes have been isolated by vegetative propagation of shoots in the T_0 generation. Alternatively, accumulation of marker-free plastid genomes during growth, development and flowering of T_0 plants allows the collection of seeds that give rise to a high proportion of marker-free T_1 seedlings. The simplicity and convenience of direct repeat excision facilitates its widespread use to isolate marker-free crops.

Key words Chloroplast transformation, DNA direct repeats, Herbicide tolerant plants, Homologous recombination, Homoplasmic, Heteroplasmic, Marker gene excision

1 Introduction

Marker genes combined with positive selection are required for the selective multiplication of the small number of transgenic plastids generated by plastid transformation. Following transformation, the first resistant shoots that are isolated are usually heteroplasmic and continued selection for marker genes present in plastids is required to remove untransformed plastids. Stable plastid transformation is achieved once transplastomic plants with a homoplasmic population of transgenic plastids have been isolated. Marker genes,

Pal Maliga (ed.), *Chloroplast Biotechnology: Methods and Protocols*, Methods in Molecular Biology, vol. 1132,
DOI 10.1007/978-1-62703-995-6_6, © Springer Science+Business Media New York 2014

combined with prolonged positive selection to attain homoplasmy, are essential for isolating stable transplastomic plants. Once stable plastid transformants have been isolated, marker genes have normally served their purpose and their continued presence can be undesirable.

In flowering plants, the first markers used were based on antibiotic resistance mutations in plastid 16S ribosomal RNA and protein (*rps12*) genes [1–3]. They were relatively inefficient and are not in routine use today. The *aadA* gene is the most widely used and successful marker gene in the plastid transformation field [4] allowing antibiotic selection (spectinomycin and streptomycin) of transplastomic microalgae [5] and flowering plants [6]. Other antibiotic–marker gene combinations used to select plastid transformants include kanamycin–*npt*II [7], kanamycin–*aphA6* [8], and chloramphenicol–*cat* [9]. Removal of marker genes conferring antibiotic resistance is particularly important for gaining the regulatory approval needed for commercialisation of transplastomic crops. Within the European Union, Directive 2001/18/EC of the European Parliament and Council covers the deliberate release of transgenic organisms into the environment. The directive requires the removal of marker genes against antibiotics in medical or veterinary use, which may have harmful consequences on human health and the environment.

2 The Arguments in Favor of Removing Marker Genes from Plastids

The high copy number of plastid genomes and the functionality of plastid marker genes in bacteria increase the risk of possible transmission of marker genes from plants to bacteria in the environment [10–12]. Further bacteria–bacteria spread and possible acquisition of *aadA* by pathogenic bacteria would make them resistant to spectinomycin and streptomycin. Both antibiotics are used occasionally in the clinic; streptomycin is used to treat tuberculosis [13, 14], whilst spectinomycin can be used as a second line of treatment against gonorrhea if beta lactam antibiotics cannot be used due to unavailability, patient allergenicity, or bacterial resistance [13, 15]. Removal of marker genes from edible transplastomic crops addresses the additional and remote risk of an enzyme conferring antibiotic resistance being ingested and inactivating a therapeutic dose of its target antibiotic. Toxicity and allergenicity tests on proteins encoded by marker proteins would also be required to ensure their safety [16]. These arguments would also apply to the *nptII* and *aphA6* genes conferring resistance to kanamycin and related amino glycoside antibiotics, and the *cat* gene conferring resistance to chloramphenicol. Whilst the risk of transmission of plastid-localized antibiotic-resistance genes to pathogenic bacteria, and the current clinical importance of their target antibiotics, are subject to scrutiny and debate [13], lengthy risk assessments on the hazards of these genes are circumvented by excising these controversial genes from plastids. Excision of antibiotic resistance genes from plastids is the most direct and simplest approach to comply with Directive 2001/18/EC and shift the focus of attention away from marker genes to the all-important plastid-localized trait genes that add value to a crop.

The use of alternative markers for plastid transformation avoids the unnecessary propagation of antibiotic resistance genes in transplastomic crops. Marker genes comprised of native plastid photosynthesis genes combined with photosynthesis-deficient plastome mutants allow for selection based on restoration of photosynthesis. Because such schemes involve the repair of defective plastid genes with wild type plastid genes, no foreign marker genes are involved and regulatory issues are less of a concern. Plastid transformation based on rescue of photosynthesis works well in Chlamydomonas [17] but in flowering plants it cannot be used for direct selection and has been used to assist selection based on antibiotics [18, 19]. The recent development of a new plastid marker based on a feedback-insensitive anthranilate synthase α-subunit gene (ASA2 marker) from tobacco and non-antibiotic 4-methylindole or 7-methyl-DL-tryptophan selection [20] is a positive step to address concerns on potential hazards associated with antibiotic resistance genes in transplastomic crops. Widespread use of non-antibiotic selection regimes and associated marker genes will ultimately depend on their overall efficiencies for isolating stable transplastomic plants compared to the *aadA* marker gene. Alternative markers would still be subject to risk assessment and regulatory approval, which can be avoided by their removal. Marker excision has additional advantages beyond curtailing the unnecessary dissemination of antibiotic resistance genes and their products into the environment. Elimination of marker genes from a transplastomic crop removes the metabolic load associated with their expression and can improve plant fitness. A further advantage of marker excision is that it allows reuse of a plastid marker for additional rounds of transformation, which is particularly relevant to plastid transformation where few efficient marker genes are available [4, 6].

3 Marker Gene Excision from Plastids

The arguments in favor of marker gene excision from plastids include the relative ease with which these genes can be removed from plastids. The excision methods can be divided into two groups. The first group of methods exploits native properties of the plastid genetic system including the predominance of homologous DNA recombination and sorting out of marker-free genomes from a heteroplasmic mixture of plastid genomes. The second group of methods involves the introduction of foreign site-specific recombinases and their target sites into plastids (*see* Chapter 12). Methods exploiting the natural DNA maintenance mechanisms in plastids include the following: (1) Marker excision using directly repeated DNA sequences (DNA direct repeats), (2) Transient co-integration of a marker gene followed by its excision [21], (3) Co-transformation and integration of marker and herbicide resistance genes into different plastid genomes followed by segregation of marker-free herbicide resistant genomes from wild type and other transgenic plastid genomes [22]. Marker excision by direct repeats and transient co-integration rely on the efficient homologous recombination pathway present in plastids.

4 Native DNA Recombination Pathways in Plastids

Homologous recombination is responsible for integration of transgenes into the plastid genome (*see* Chapter 1) and is the predominant recombination pathway operating in plastids [23]. The Rec A protein is central to homologous recombination and DNA repair in bacteria such as *Escherichia coli* [24]. It promotes DNA strand transfer and heteroduplex formation. Homologues of bacterial Rec A proteins are present in plastids and are encoded by nuclear genes [25, 26]. T-DNA knockouts in Arabidopsis genes indicate that a plastid-targeted RecA protein (RecA1) and a dual mitochondria-and-plastid targeted Rec A protein (RecA2) are probably essential [27] in plants. Partial loss of function of the *RecA1* gene appears to alter the copy number and topology of chloroplast DNA molecules [28]. Little is known on the other plastid enzymes involved in the homologous DNA recombination pathway. The predominance of homologous recombination in plastids has been suggested to result from suppression of illegitimate recombination by plastid-localized members of the whirly family of single-stranded DNA binding proteins [29]. The observation that a variety of plastid DNA sequences can act as targeting arms to insert foreign genes by homologous recombination into different regions of the plastid genome in a range of species suggests recombination is not restricted to specific DNA sequences. However, the influence of DNA sequences on rates of homologous recombination in plastids has not been studied in a systematic series of experiments. Our current knowledge of the recombination substrates in plastids does not rule out the possibility of short specific sequences acting as "hot spots" for initiating homologous recombination between target and donor sequences. Sequences that appear to enhance recombination have been identified in the large inverted region of the *Chlamydomonas reinhardtii* chloroplast genome [30]. In *E. coli*, eight base chi sequences recognized by the *recBCD* complex [24] stimulate homologous recombination.

The absence, or limited number, of chi sequences in plastid genomes and apparent lack of genes encoding homologues of *recBCD* in cyanobacteria and flowering plants, suggests that a *recBCD* pathway is unlikely to operate in plastids [23].

5 Excision of Marker Genes by Recombination Between Direct DNA Repeats

A recombination event between directly repeated DNA sequences (DNA direct repeats) excises the sequences lying between the direct repeats plus one copy of the repeat. This is illustrated in Fig. 1. The process is dependent on two components: the direct repeats and the native chloroplast proteins that act on these DNA sequences. In many cases the direct repeats are comprised of duplicated 5′ or 3′ regulatory elements that flank the coding sequence to be excised in plastids (Table 1). The excision process gives rise to two products: a marker-free plastid genome with one copy of the repeat and an excised small DNA circle (the excision cassette) containing the marker gene plus the second copy of the repeat. Both DNA products can be detected in transplastomic plants early in the excision process. The small DNA circle is unstable and is lost

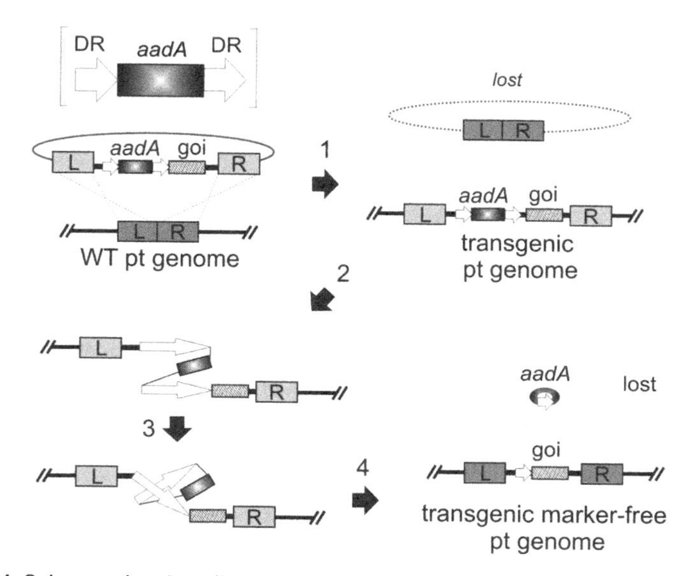

Fig. 1 Scheme showing direct repeat-mediated excision of the *aadA* marker gene. *1*. Integration of transgenes into the WT plastid (pt) genome by homologous recombination. The gene of interest (goi), *aadA* gene flanked by DNA direct repeats (DR), and left (L) and right (R) plastid targeting arms are shown. The unselected vector lacking *aadA* cannot replicate in plastids and is lost. *2*. Pairing of DNA direct repeats. *3*. Cross-over event leading to loop-out and excision of *aadA* and one copy of the direct repeat. *4*. The *aadA* circle cannot replicate and is lost leading to the isolation of transgenic marker-free plastids

Table 1
Recombination events detected between direct DNA repeats in transgenic plastids

Species	Repeat length (bp)	Sequence	Reference
Tobacco	84[d]	P*rrn*	[34]
	117[d]	P*rrn*	[45]
	174[b]	P*rrn* + *rbcL* 5′UTR	[38]
	210[d]	*rbcL* 3′UTR	[35]
	403[b]	*hppd* gene	[35]
	418[a]	*psbA* 3′UTR	[38]
	649[c]	P*atpB* & 5′ UTR	[19]
Lettuce	~150[d]	P*rrn* & RBS	[36]
Soybean	367[b]	*bar* gene	[46]
Chlamydomonas	216[a]	*chlL* gene	[39]
	483[a]	pACYC184	[40]

Shown are events that led to: [a]marker-free algae or plants, [b]*aadA*-free herbicide resistant plants, [c]precise *rbcL* deletion mutant lacking foreign genes, [d]plants lacking the trait gene or both trait and *aadA* genes resulting from undesirable excision events. For[a,b] in tobacco and soybean, multiple repeats or positive selection were used to enhance desirable excision events between repeats of less than 600 bp. Undesirable recombination events[d] between 84 bp and 117 bp repeats were stimulated by the introduction of foreign site specific-recombinases into plastids [34, 35]

during the following divisions of plastids and cells. The marker-free plastid genome is maintained by the replication and segregation mechanisms present in plastids. The complexity of this simple scheme is increased when applied to the plastid genetic system which is comprised of multiple copies of the unit plastid genome per plastid (up to ~100 genomes) and multiple plastids per leaf mesophyll cell (up to ~150 plastids).

The recombination process giving rise to marker excision is spontaneous and takes place in a fraction of plastid genomes. This gives rise to a heteroplasmic population of plastid genomes comprised of marker-containing and marker-free genomes together with the excised small DNA circles containing the marker gene. In the absence of selection, spontaneous marker-excision and the instability of the resulting small DNA circles will lead to an accumulation of marker-free plastid genomes. Excision and loss of the marker gene means the recombination event cannot be reversed and the process is unidirectional. The process of isolating homoplasmic marker-free shoots will be influenced by the DNA-replication-repair-recombination pathways present in plastids that normally ensure the uniformity of plastid DNA within a plant including

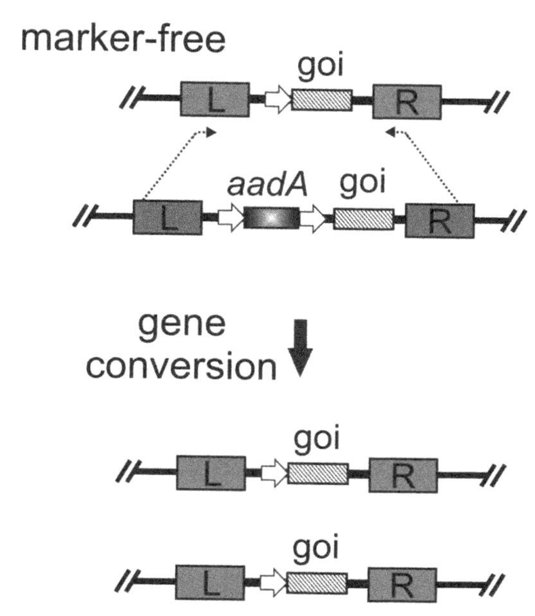

Fig. 2 Marker-free plastid genomes can act as templates for gene conversion. In the scheme the region between the left (L) and right (R) arms is copied using the marker-free genome containing the gene of interest (goi) as the template. DNA strands shown as *dotted lines* from the *aadA* plastid genome replicate the region between the L and R arms from the marker-free genome. This leads to loss of *aadA*

gene conversion [31, 32]. Gene conversion pathways include marker-free plastid genomes acting as templates to convert marker-containing genomes to marker-free plastid genomes (Fig. 2). Gene conversion events can act both ways to accelerate the production of homoplasmic cells with marker-free plastids or homoplasmic cells with only marker-containing plastids. However, whereas homoplasmic cells with marker-free plastids are the stable end point of the direct repeat excision event, marker containing plastids are unstable and will continue to excise the marker gene.

Isolation of marker-free cells is facilitated by cytoplasmic sorting of plastids, which is an inherent property of plastid genetics. In cytoplasmic sorting, a heteroplasmic population of plastids segregates into pure populations of each plastid type during the cell divisions that accompany the growth and development of plants; this is also known as vegetative segregation [33]. The combined results of marker excision, gene conversion and cytoplasmic sorting lead to the isolation of marker-free cells and shoots. Bottle necks that reduce the number of plastids and plastid genomes per cell [33] will enhance the segregation of homoplasmic cells only containing marker-free transgenic plastids. One such bottle-neck may be present during the formation of egg cells. Another bottle neck that promotes segregation of plastid genomes has been found in regenerating shoots [34]. In species that exhibit maternal inheritance of chloroplasts, segregation of marker-free plastid genomes into individual egg cells will give rise to marker-free seeds following their fertilization. The process of marker excision mediated by direct repeats is controlled by selection for the marker gene. In the presence of selection, the plastids containing marker genes will be resistant and will divide whereas the proliferation of marker-free plastids will be restricted. In the absence of selection, marker-free plastids will survive and proliferate allowing their accumulation and the eventual isolation of marker-free cells resulting from the combined actions of marker excision, gene conversion and cytoplasmic sorting.

6 Factors Influencing Marker Excision by Direct Repeats

6.1 Cis-Acting DNA Components Present in the Excision Cassette

6.1.1 Length of DNA Direct Repeats

Recombination between relatively short direct repeats of ~200 bp leading to excision of foreign genes has been observed in transplastomic plants (Table 1). For example, two direct repeats of 210 bp comprised of the 3′ UTR of the *rbcL* gene led to the excision of the 4-hydroxyphenylpuruvate dioxygenase (*hppd*) gene conferring resistance to the herbicide diketonitrile in tobacco [35]. In lettuce, excision of a green fluorescent protein (GFP) gene in two out of four transplastomic lines tested resulted from recombination between two ~150 bp direct repeats comprised of the tobacco *rrn* promoter and a synthetic 5′ UTR [36]. On the other hand there are many reports in the literature of the stable maintenance of ~150 bp to ~400 bp duplications of regulatory elements in transgenic plastid genomes [37, 38]. Similarly, in Chlamydomonas, recombination between 216 bp direct repeats of the *chlL* gene could be detected using an assay based on restoration of pigmentation [39] but larger repeats of 483 bp from the pACYC184 plasmid were required for efficient excision of the *aadA* marker gene [40]. Whilst it is clear that recombination can take place between relatively short DNA repeats of ~100–200 bp, this length of sequence gives rise to low excision frequencies and is not suitable for efficient marker removal. As mentioned above, the influence of DNA sequence composition on excision frequency is not known and it remains possible that some direct repeats are more effective substrates for the plastid recombination machinery.

In the absence of selection for the excision event, we have found that two direct repeats of 649 bp flanking the marker gene was effective in promoting spontaneous excision of marker genes from tobacco plastids [19]. Whilst further increases in repeat length above this size might be expected to increase excision frequency, the relationship between repeat length and recombination rate is not known. The point at which an increase in recombination rate plateaus against repeat size remains to be determined. A balance needs to be found between the maintenance and excision of the marker

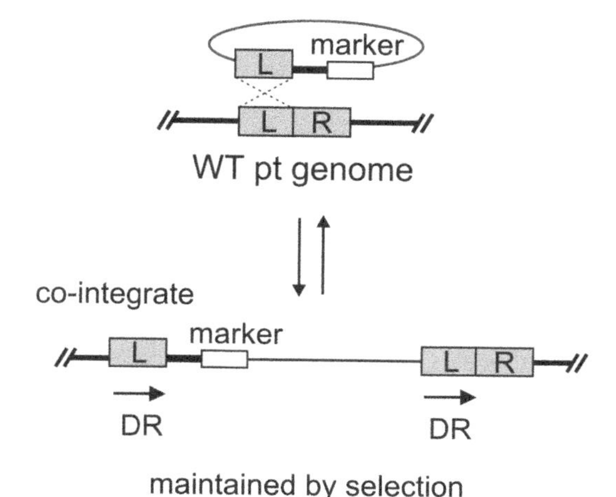

Fig. 3 Large DNA direct repeats (DR) maintained by selection. Integration of a vector with a single left (L) targeting arm leads to a co-integrate plastid genome with large direct repeats comprised of the left targeting arm. The co-integrate genome is maintained by selecting the marker gene [41, 42]. Recombination between the directly repeated left arms excises the vector. The vector lacks a right targeting arm (R)

gene used to select plastid transformants. In the examples cited above [19, 38], marker genes can be maintained in plastomes by selection, even when high rates of marker excision are observed. In Chlamydomonas [41], vectors containing a single 4.9 kbp targeting arm create a large duplication on insertion of the entire vector into the plastid genome following a single recombination event (Fig. 3). The resulting co-integrate plastid genome is highly unstable due to duplication of the targeting arm but can be maintained by selection. When selection is released the integrated vector is lost as a result of recombination between duplicated 4.9 kbp DNA sequences. Co-integrate plastid genomes can also be maintained in tobacco plastids [42]. The fact that selection can be used to isolate and maintain marker genes flanked by large sequence duplications in plastids may indicate that even higher rates of marker excision mediated by long direct repeats are achievable.

6.1.2 Number of DNA Direct Repeats

An alternative method for enhancing recombination rates is to increase the number of direct repeats in a plastid transformation construct [34]. We found that including three 418 bp direct repeats of the *psbA* 3′ UTR and two 174 bp direct repeats of the *rrn* promoter and the *rbcL* 5′ UTR led to high excision rates of the intervening genes (Fig. 4). The construct was designed to promote multiple excision events. The presence of multiple DNA repeats created a local zone of enhanced recombination which was stabilized once all repeats were removed by excision events to leave either a solo 418 bp *psbA* repeat or solo 174 bp *rrn* repeat. DNA blot

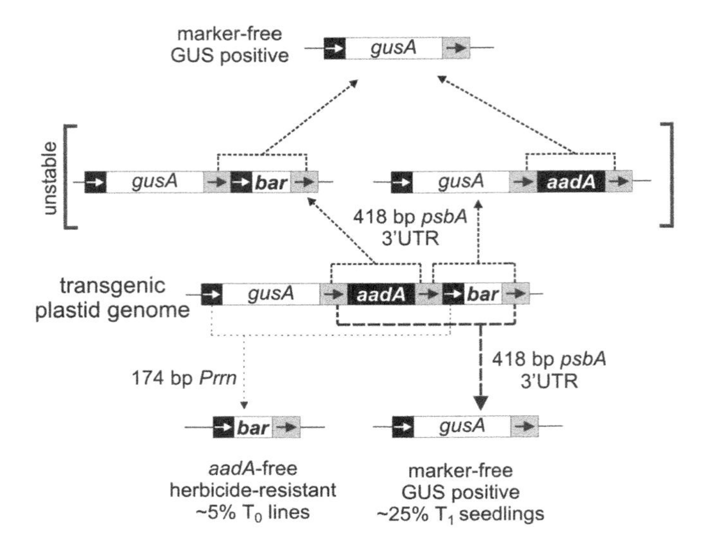

Fig. 4 Multiple direct repeats increase the frequency of gene excision from plastids [38]. Transgenic plastids contain three foreign genes (*gusA, aadA, bar*). The *rrn* promoter (P*rrn*) is repeated twice and the *psbA* 3′UTR is repeated three times. Two excision products were isolated: *aadA*-free plants resistant to the herbicide phosphinothricin and marker-free plants containing the gene-of-interest β-glucuronidase (GUS). Predicted intermediates containing two genes were not detected indicating that once recombination is activated all local DNA direct repeats are removed

analysis showed that excision events between the outermost *psbA* repeats were the most frequent and gave rise to marker-free genomes containing the *gusA* gene. Excision of a single gene giving rise to intermediates containing direct repeats was not detected (Fig. 4). This indicates that once recombination was activated the reaction goes to completion until all direct repeats have been removed. Release of phosphinothricin (herbicide) selection for the *bar* gene allowed accumulation of marker-free genomes containing *gusA* during plant growth and development. The next generation contained a high proportion (~25 %) of homoplasmic marker-free seedlings. Under herbicide selection, we were able to isolate *aadA-free* transplastomic plants containing the bar gene resulting from recombination between the 174 bp *rrn* repeats in the T_0 generation albeit at a low frequency (~5 %). This indicates that within the zone of enhanced recombination the 418 bp repeats were more effective substrates for the plastid recombination apparatus than the 174 bp repeats.

6.1.3 Influence of Genes Within the Direct Repeats on Excision Frequency

The distance between direct repeats may influence the excision process. Although the relationship between distance and recombination rates is not known, increasing the distance between repeats might be expected to reduce the recombination frequency.

High rates of recombination were observed between two direct repeats of 649 bp comprised of the *atpB* promoter and 5′ UTR separated by 5,500 bp of intervening sequence [19]. Because most marker genes used in plastid transformation are less than 2,000 bp, the distance between direct repeats flanking a marker is unlikely to play a major role in limiting marker excision.

Negative selection against sequences present in an excision cassette might be expected to promote cassette loss. Selection could act against the expression of foreign genes that increase metabolic load and reduce plastid fitness. Metabolic load could result from very high levels of foreign gene expression thereby competing with the expression of native genes. Alternatively, a recombinant protein may have a negative impact on plastid functions. Negative selection might explain the excision of GFP [36] and *hppd* [35] genes resulting from recombination between relatively short direct repeats: ~150 bp (GFP) and 210 bp (*hppd*). Spontaneous marker excision using direct repeats of 649 bp was efficient and was not reliant on negative selection [17]. However, negative selection as an add-on feature rather than a strict requirement enhances the tools available to influence the excision process. A negative selectable marker that is functional in plastids has been described [43] and could be included in the excision cassette to influence the timing of marker removal.

6.2 Trans-Acting Recombination Proteins Acting on Direct Repeats

A homologue of the RecA protein has been implicated in homologous recombination in plastids. Manipulating plastid RecA levels provides an additional approach to influence marker gene excision. Over-expression of the *E. coli* RecA protein in Chlamydomonas chloroplasts increased recombination rates between two 216 bp direct repeats by over 15-fold [39]. This suggests that recombination between direct repeats is limited by Rec A activity in WT Chlamydomonas chloroplasts. Analysis of the influence of proteins on recombination between direct repeats in flowering plants requires a method to monitor the excision process. A visual scheme to monitor recombination between direct repeats is provided by an excision cassette (Fig. 5) that includes the native *rbc L* gene together with the *aadA* and *gusA* genes [19]. Loss of the *rbc L* gene results in a pigment-deficient phenotype. The appearance of pigment-deficient cells is therefore indicative of the loss of the excision cassette which includes the *aadA* marker gene. The combined processes of marker excision and cytoplasmic sorting give rise to pale-green sectors. This allows the number and spatial distribution of marker-free cells to be visualized as pigment-deficient sectors. The variable pattern of pale-green sectors in different leaves is indicative of a random process (Fig. 6). The large number of sectors demonstrates the efficiency of the excision process. This visual assay can be used to measure recombination rates between direct repeats in plastids. We have used the assay to demonstrate a role for

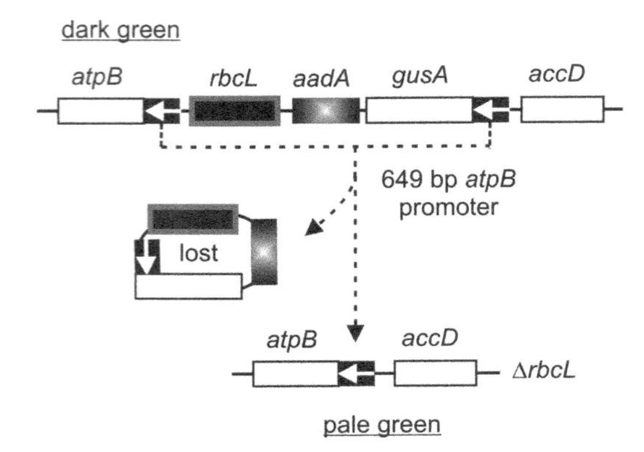

Fig. 5 Excision of a native plastid gene using DNA direct repeats. A recombination event between 5′ *atp B* regulatory sequences driving expression of a *gus A* transgene and the native *atp B* gene leads to loss of *aadA* and *gusA* transgenes, and the plastid *rbc L* gene. Loss of *rbc L* leads to a pale green phenotype. Because the loss of *aadA* and *rbc L* are mediated by the same recombination event it allows the marker-excision process to be visualized

Fig. 6 Visualization of marker excision by recombination between DNA direct repeats in a tobacco leaf. *Pale areas* are indicative of the products of marker excision. The recombination assay is shown in Figure 5. The varying sizes and distribution of *pale green sectors* is indicative of a random process

plastid Rec A in direct repeat-mediated excision (Madesis and Day, unpublished results). Such studies involving functional assays provide fundamental information on the enzymes involved in marker excision and provide a route towards identifying the key enzymes that regulate the process.

The frequency of recombination between short direct repeats of ~120 bp (Table 1) appears to be stimulated by the presence of Cre site-specific recombinase in plastids [44, 45]. Similarly, recombination between *attP* and *attB* sites in transgenic plastid DNA by phiC31 site-specific recombinase stimulated recombination between 84 bp direct repeats of the *rrn* promoter [34].

The transient double-strand DNA breaks in transgenic plastid DNA induced by Cre recombinase at *loxP* sites or by phiC31 recombinase at *attB/P* sites appear to stimulate the homologous recombination pathway in plastids.

7 Isolation of aadA-Free Herbicide-Resistant Transplastomic Plants

Transformation of plastids using herbicides as primary selection agents has not been successful. Herbicides can be used as secondary selection agents soon after the isolation of antibiotic resistant cells using the *aadA* marker. Switching to herbicide selection allows removal of the *aadA* marker gene by direct repeat-mediated excision [38] or segregation [22], whilst maintaining the herbicide resistance gene in plastids. In these examples, loss of the *aadA* gene is not coupled to expression of an herbicide resistance gene, which is active in transgenic plastids right from the start of the transformation.

Restoration of an *aadA*-disrupted herbicide resistance gene using direct repeat-mediated recombination couples excision of *aadA* with the gain of herbicide resistance [35]. The method is illustrated in Fig. 7 and allows positive herbicide selection for the

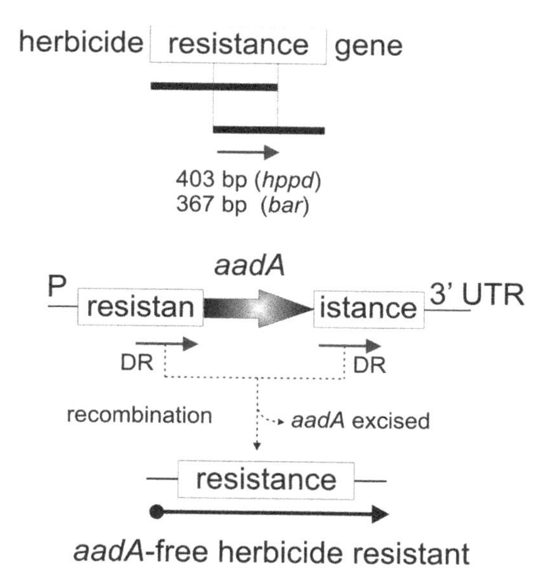

Fig. 7 Restoration of the coding sequence of an herbicide resistance gene by recombination between direct DNA repeats. An *aadA* marker gene disrupts the coding region of an herbicide resistance gene. Overlapping sequences of the herbicide resistance gene provide direct DNA repeats (DR). Restoration of a functional herbicide resistance gene by recombination confers resistance to the relevant herbicide, facilitating the isolation of *aadA*-free plants. The procedure has been exemplified using the *hppd* [35] and *bar* [46] genes in tobacco and soybean, respectively

product of the excision event. This drives the excision event by selection for herbicide resistant cells and plastids. The method was first demonstrated using 403 bp direct repeats from the *hppd* gene and diketonitrile selection [35]. The precision of homologous recombination allows the intact reading frame to be restored without any insertions or deletions within the 403 bp recombination region. In this procedure, the accumulation of plastomes with the restored *hppd* gene is required before herbicide selection is applied. This means that once plastid transformants have been isolated using *aadA*-based antibiotic selection, a period of plant growth in the absence of antibiotics is required before herbicide selection is applied. The method combines spontaneous direct repeat-mediated excision of *aadA* with positive selection and is efficient. The method is suitable for direct repeats comprised of coding regions conferring positive selection such as herbicide resistance genes. Restoration of the *bar* gene conferring resistance to the herbicide phosphinothricin has been achieved using direct repeats of 367 bp in soybean plastids [46]. A similar scheme involving restoration of the *chlL* gene by direct repeat-mediated excision of *aadA* in Chlamydomonas resulted in a change from a yellow-in-the-dark to a green-in-the-dark phenotype [39].

8 Conclusions

DNA direct repeat-mediated excision of marker genes exploits the native and efficient homologous recombination pathway present in diverse plastids ranging from those present in microalgae to flowering plant plastids. The homologous recombination pathway in plastids is not restricted to specific DNA targets. This adds flexibility in the design of constructs and allows a variety of DNA sequences to be used as direct repeats to excise marker genes. Duplicated regulatory elements flanking the marker gene provide a convenient source of direct repeats. The method is efficient and simple. Excision rates are influenced by the length and number of direct repeats. Because native plastid enzymes are responsible for marker excision, only a single round of plant transformation is required. There is no requirement to introduce site-specific recombinases by retransformation or sexual crosses. The procedure should be applicable to species exhibiting biparental inheritance of plastids. We have shown that direct repeat excision can give rise to ~25 % of transplastomic seedlings that are marker-free [38]. For a biparental species, where ~25 % of eggs and sperm cells are marker-free, transplastomic seedlings free of markers will arise at a frequency of ~6 % ($0.25 \times 0.25 \times 100$ %). Increasing the sizes of repeats (~600 bp) and the number of repeats increases excision frequency. Direct repeat-mediated excision and reuse of *aadA* to retransform transgenic plastids demonstrates the complete removal of *aadA* from the first round of transformation [19]. Direct repeat excision leaves one copy of the repeat in the genome. This means that DNA repeats with different sequences are required if the same marker is used to introduce genes to a variety of locations within the plastid genome via multiple rounds of plastid transformation.

Selection is used to maintain the marker and ensure homoplasmy of transgenic plastid genomes. Release of selection allows the accumulation of marker-free plastid genomes generated by marker excision, which is largely a unidirectional process. Cytoplasmic sorting allows the segregation of cells with marker-free transgenic plastids. The marker-free shoots resulting from direct repeat-mediated excision of marker genes have been isolated by vegetative propagation of shoots in the T_0 generation. Alternatively, accumulation of marker-free plastid genomes during growth, development and flowering of T_0 plants allows the collection of seeds that give rise to a high proportion of marker-free T_1 seedlings. The efficacy of direct repeat excision is illustrated by its use to remove native photosynthesis related genes resulting in the production of marker-free photosynthetic mutants. Direct repeats drive excision of the *rbcL* gene despite the deleterious consequences of the process on chloroplast function. However, direct repeat excision is not applicable to the deletion of essential plastid genes such as *clpP* and *accD,* because simultaneous excision of marker genes from all plastids within a cell is not possible. Excision of essential genes is possible with site-specific recombinases [47].

Whilst direct repeat-mediated excision of markers is an effective method for isolating marker-free transplastomic plants, the timing of excision is spontaneous and takes place at all times during the growth and development of plants. The limitations of lack of control over the timing of appearance of marker-free cells need to be balanced against the spontaneous accumulation of marker-free cells which has the advantages of incurring very limited staff time or experimental costs. A degree of control can be achieved by coupling loss of *aadA* to restoration of an herbicide resistance gene allowing positive selection for the excision event [35, 46]. The simplicity and convenience of direct repeat excision facilities its widespread use in transplastomic technologies.

Acknowledgements

Work in the authors' laboratory was supported by research grants BB/E020445 and BB/I011552 from the Biotechnology and Biological Sciences Research Council (UK).

References

1. Svab Z, Hajdukiewicz P, Maliga P (1990) Stable transformation of plastids in higher plants. Proc Natl Acad Sci U S A 87:8526–8530
2. O'Neill C, Horvath GV, Horvath E, Dix PJ, Medgyesy P (1993) Chloroplast transformation in plants: polyethylene glycol (PEG) treatment of protoplasts is an alternative to biolistic delivery systems. Plant J 3:729–738
3. Staub JM, Maliga P (1992) Long regions of homologous DNA are incorporated into the tobacco plastid genome by transformation. Plant Cell 4:39–45
4. Day A, Goldschmidt-Clermont M (2011) The chloroplast transformation toolbox: selectable markers and marker removal. Plant Biotechnol J 9:540–553
5. Goldschmidt-Clermont M (1991) Transgenic expression of aminoglycoside adenine transferase in the chloroplast: a selectable marker of site-directed transformation of chlamydomonas. Nucleic Acids Res 19:4083–4089
6. Svab Z, Maliga P (1993) High-frequency plastid transformation in tobacco by selection for a chimeric *aadA gene*. Proc Natl Acad Sci U S A 90:913–917
7. Carrer H, Hockenberry TN, Svab Z, Maliga P (1993) Kanamycin resistance as a selectable marker for plastid transformation in tobacco. Mol Gen Genet 241:49–56
8. Huang FC, Klaus SM, Herz S, Zou Z, Koop HU, Golds TJ (2002) Efficient plastid transformation in tobacco using the *aphA-6* gene and kanamycin selection. Mol Genet Genomics 268:19–27
9. Li W, Ruf S, Bock R (2011) Chloramphenicol acetyltransferase as selectable marker for plastid transformation. Plant Mol Biol 76:443–451
10. Kay E, Vogel TM, Bertolla F, Nalin R, Simonet P (2002) In situ transfer of antibiotic resistance genes from transgenic (transplastomic) tobacco plants to bacteria. Appl Environ Microbiol 68:3345–3351
11. Ceccherini MT, Pote J, Kay E, Van VT, Marechal J, Pietramellara G, Nannipieri P, Vogel TM, Simonet P (2003) Degradation and transformability of DNA from transgenic leaves. Appl Environ Microbiol 69:673–678
12. Pontiroli A, Rizzi A, Simonet P, Daffonchio D, Vogel TM, Monier JM (2009) Visual evidence of horizontal gene transfer between plants and bacteria in the phytosphere of transplastomic tobacco. Appl Environ Microbiol 75: 3314–3322
13. Goldstein DA, Tinland B, Gilbertson LA, Staub JM, Bannon GA, Goodman RE, McCoy RL, Silvanovich A (2005) Human safety and genetically modified plants: a review of antibiotic resistance markers and future transformation selection technologies. J Appl Microbiol 99:7–23
14. Sacchettini JC, Rubin EJ, Freundlich JS (2008) Drugs versus bugs: in pursuit of the persistent predator *Mycobacterium tuberculosis*. Nat Rev Microbiol 6:41–52
15. Kubanova A, Frigo N, Kubanov A, Sidorenko S, Lesnaya I, Polevshikova S, Solomka V, Bukanov N, Domeika M, Unemo M (2010) The Russian gonococcal antimicrobial susceptibility programme (RU-GASP) - national resistance prevalence in 2007 and 2008, and trends during 2005–2008. Euro Surveill 15: 10–14
16. Fuchs RL, Ream JE, Hammond BG, Naylor MW, Leimgruber RM, Berberich SA (1993) Safety assessment of the neomycin phosphotransferase II (NPTII) protein. Bio/Technology 11:1543–1547
17. Boynton JE, Gillham NW, Harris EH, Hosler JP, Johnson AM, Jones AR, Randolph-Anderson BL, Robertson D, Klein TM, Shark KB et al (1988) Chloroplast transformation in *Chlamydomonas* with high velocity microprojectiles. Science 240:1534–1538
18. Klaus SM, Huang FC, Eibl C, Koop HU, Golds TJ (2003) Rapid and proven production of transplastomic tobacco plants by restoration of pigmentation and photosynthesis. Plant J 35:811–821
19. Kode V, Mudd EA, Iamtham S, Day A (2006) Isolation of precise plastid deletion mutants by homology-based excision: a resource for site-directed mutagenesis, multi-gene changes and high-throughput plastid transformation. Plant J 46:901–909
20. Barone P, Zhang XH, Widholm JM (2009) Tobacco plastid transformation using the feedback-insensitive anthranilate synthase [alpha]-subunit of tobacco (ASA2) as a new selectable marker. J Exp Bot 60:3195–3202
21. Klaus SM, Huang FC, Golds TJ, Koop HU (2004) Generation of marker-free plastid transformants using a transiently cointegrated selection gene. Nat Biotechnol 22:225–229
22. Ye GN, Colburn SM, Xu CW, Hajdukiewicz PT, Staub JM (2003) Persistence of unselected transgenic DNA during a plastid transformation and segregation approach to herbicide resistance. Plant Physiol 133:402–410
23. Day A, Madesis P (2007) DNA replication, recombination, and repair in plastids. In: Bock R (ed) Cell and molecular biology of plastids. Springer, Berlin, pp 65–119

24. Kowalczykowski SC (2000) Initiation of genetic recombination and recombination-dependent replication. Trends Biochem Sci 25:156–165
25. Cerutti H, Osman M, Grandoni P, Jagendorf AT (1992) A homolog of *Escherichia coli* RecA protein in plastids of higher plants. Proc Natl Acad Sci U S A 89:8068–8072
26. Nakazato E, Fukuzawa H, Tabata S, Takahashi H, Tanaka K (2003) Identification and expression analysis of cDNA encoding a chloroplast recombination protein REC1, the chloroplast RecA homologue in *Chlamydomonas reinhardtii*. Biosci Biotechnol Biochem 67:2608–2613
27. Shedge V, Arrieta-Montiel M, Christensen AC, Mackenzie SA (2007) Plant mitochondrial recombination surveillance requires unusual *RecA* and *MutS* homologs. Plant Cell 19:1251–1264
28. Rowan BA, Oldenburg DJ, Bendich AJ (2010) RecA maintains the integrity of chloroplast DNA molecules in Arabidopsis. J Exp Bot 61:2575–2588
29. Marechal A, Parent JS, Veronneau-Lafortune F, Joyeux A, Lang BF, Brisson N (2009) Whirly proteins maintain plastid genome stability in Arabidopsis. Proc Natl Acad Sci U S A 106:14693–14698
30. Newman SM, Harris EH, Johnson AM, Boynton JE, Gillham NW (1992) Nonrandom distribution of chloroplast recombination events in *Chlamydomonas reinhardtii*—evidence for a hotspot and an adjacent cold region. Genetics 132:413–429
31. Staub JM, Maliga P (1995) Marker rescue from the Nicotiana tabacum genome using a plastid *Escherichia coli* shuttle vector. Mol Gen Genet 249:37–42
32. Khakhlova O, Bock R (2006) Elimination of deleterious mutations in plastid genomes by gene conversion. Plant J 46:85–94
33. Birky CW (2001) The inheritance of genes in mitochondria and chloroplasts: laws, mechanisms, and models. Annu Rev Genet 35: 125–148
34. Lutz KA, Maliga P (2008) Plastid genomes in a regenerating tobacco shoot derive from a small number of copies selected through a stochastic process. Plant J 56:975–983
35. Dufourmantel N, Dubald M, Matringe M, Canard H, Garcon F, Job C, Kay E, Wisniewski JP, Ferullo JM, Pelissier B, Sailland A, Tissot G (2007) Generation and characterization of soybean and marker-free tobacco plastid transformants over-expressing a bacterial 4-hydroxyphenylpyruvate dioxygenase which provides strong herbicide tolerance. Plant Biotechnol J 5:118–133
36. Lelivelt CLC, McCabe MS, Newell CA, deSnoo CB, van Dun KMP, Birch-Machin I, Gray JC, Mills KHG, Nugent JM (2005) Stable plastid transformation in lettuce (*Lactuca sativa* L.). Plant Mol Biol 58: 763–774
37. Zoubenko OV, Allison LA, Svab Z, Maliga P (1994) Efficient targeting of foreign genes into the tobacco plastid genome. Nucleic Acids Res 22:3819–3824
38. Iamtham S, Day A (2000) Removal of antibiotic resistance genes from transgenic tobacco plastids. Nat Biotechnol 18:1172–1176
39. Cerutti H, Johnson AM, Boynton JE, Gillham NW (1995) Inhibition of chloroplast DNA recombination and repair by dominant negative mutants of *Escherichia coli* RecA. Mol Cell Biol 15:3003–3011
40. Fischer N, Stampacchia O, Redding K, Rochaix JD (1996) Selectable marker recycling in the chloroplast. Mol Gen Genet 251:373–380
41. Kindle KL, Richards KL, Stern DB (1991) Engineering the chloroplast genome: techniques and capabilities for chloroplast transformation in *Chlamydomonas reinhardtii*. Proc Natl Acad Sci U S A 88:1721–1725
42. Ahlert D, Ruf S, Bock R (2003) Plastid protein synthesis is required for plant development in tobacco. Proc Natl Acad Sci U S A 100: 15730–15735
43. Serino G, Maliga P (1997) A negative selection scheme based on the expression of cytosine deaminase in plastids. Plant J 12:697–701
44. Corneille S, Lutz K, Svab Z, Maliga P (2001) Efficient elimination of selectable marker genes from the plastid genome by the CRE-lox site-specific recombination system. Plant J 27: 171–178
45. Hajdukiewicz PTJ, Gilbertson L, Staub JM (2001) Multiple pathways for Cre/lox-mediated recombination in plastids. Plant J 27:161–170
46. Lestrade C, Pelissier B, Roland A, Dubald M (2009) Construct for obtaining transplastomic plant or plant cell comprises at least chimeric gene encoding selectable marker and chimeric color gene, or chimeric gene encoding luminescent protein, or chimeric gene encoding a negative marker Bayer Cropscience Ag. Patent Application WO2010079117
47. Kuroda H, Maliga P (2003) The plastid *clpP1* protease gene is essential for plant development. Nature 425:86–89

Chapter 7

Fluorescent Labeling and Confocal Microscopic Imaging of Chloroplasts and Non-green Plastids

Maureen R. Hanson and Amirali Sattarzadeh

Abstract

While chlorophyll has served as an excellent label for plastids in green tissue, the development of fluorescent proteins has allowed their ready visualization in all tissues of the plants, revealing new features of their morphology and motility. Gene regulatory sequences in plastid transgenes can be optimized through the use of fluorescent protein reporters. Fluorescent labeling of plastids simultaneously with other subcellular locations reveals dynamic interactions and mutant phenotypes. Transient expression of fluorescent protein fusions is particularly valuable to determine whether or not a protein of unknown function is targeted to the plastid. Particle bombardment and agroinfiltration methods described here are convenient for imaging fluorescent proteins in plant organelles. With proper selection of fluorophores for labeling the components of the plant cell, confocal microscopy can produce extremely informative images at high resolution at depths not feasible by standard epifluorescence microscopy.

Key words Agroinfiltration, Biolistic DNA delivery, Chloroplast, Confocal microscopy, dsRed, Fluorescent protein, GFP, mCherry, Particle bombardment, RFP, Transient expression, YFP

1 Introduction

Fluorescent labeling of plastids followed by microscopic imaging is a valuable research tool in chloroplast biotechnology. Fluorescent proteins (FPs) can be used as reporters of gene expression from both nuclear and chloroplast transgenes for assessment of gene regulatory elements and for screening for transgenic plants among putative transformants. The chloroplast is naturally labeled by chlorophyll autofluorescence; altered autofluorescence can reveal mutant or abnormal phenotypes [1] and the position and morphology of chloroplasts can be readily monitored [2]. FPs can label non-green plastids to observe their dynamics, morphology, and association with other fluorescently labeled subcellular organelles in living cells [3, 4]. Interactions between proteins can be assessed by imaging transfer of energy from one molecule to a nearby fluorophore or by causing a split fluorescent protein to associate [5, 6].

Pal Maliga (ed.), *Chloroplast Biotechnology: Methods and Protocols*, Methods in Molecular Biology, vol. 1132,
DOI 10.1007/978-1-62703-995-6_7,

Transgenes carrying a localization signal [7] or all of a coding region can be fused to the coding region of FPs and expressed from a stably integrated transgene or transiently through particle bombardment [8], DNA uptake into protoplasts [9], or by Agrobacterium-mediated transfer, which is most commonly performed by infiltrating leaves [10, 11].

Traditional epifluorescence microscopy is suitable for imaging chloroplasts in thin sections or in surface layers, but confocal laser scanning microscopy (CSLM) has allowed great improvement in image quality and depth of imaging. CSLM, which allows noninvasive optical sectioning of living tissue due to the invention of an iris diaphragm (pinhole) that eliminates out-of-focus fluorescence [12] has often been applied to image chloroplasts and other intracellular organelles. When multiphoton microscopes, which avoid out-of-focus fluorescence by exciting the fluorophore only in a focal plane, become more widely available, undoubtedly they will also find important applications in chloroplast biology [13, 14].

1.1 Fluorescent Labels

The major sources of autofluorescence in chloroplasts are chlorophyll a and b, which absorb blue and red photons. The natural autofluorescence of chlorophyll can be exploited to observe altered morphology or light responses of chloroplasts in mutant or transgenic lines. Chloroplasts in wild-type plants move to position themselves to acquire light or to avoid light, depending on light intensity [15]. The location of chloroplasts in wild-type plants under different growth conditions can readily be observed by fluorescence microscopy (Fig. 1).

Chlorophyll autofluorescence is particularly useful to locate chloroplasts with respect to other subcellular organelles in green tissue. Fluorescent proteins can be deliberately targeted to particular organelles to observe them simultaneously with autofluorescent chloroplasts in wild-type and mutants in different environmental conditions (Fig. 2). ORFS of unknown location can be labeled with GFP or other FPs and infiltrated into wild-type leaves in order to determine whether they co-localize with the chlorophyll autofluorescence (Fig. 3).

Plastids in both green and non-green tissues can be imaged by expressing fluorescent proteins from nuclear or plastid transgenes. Higher levels of GFP expression are usually achieved in transplastomic plants in comparison to nuclear transgenic plants. Several codon-altered GFP coding regions have been reported for use in plastids [16, 17]. There is no evidence that fluorescent proteins are toxic to plants even at high expression levels [18]. With a suitable nuclear promoter or chloroplast gene regulatory regions, expression of the fluorescent protein will occur in most tissues of a regenerated transgenic plant. For nuclear transgenes, an appropriate N-terminal transit sequence must be selected that will be recognized by the chloroplast import apparatus in the tissues of interest.

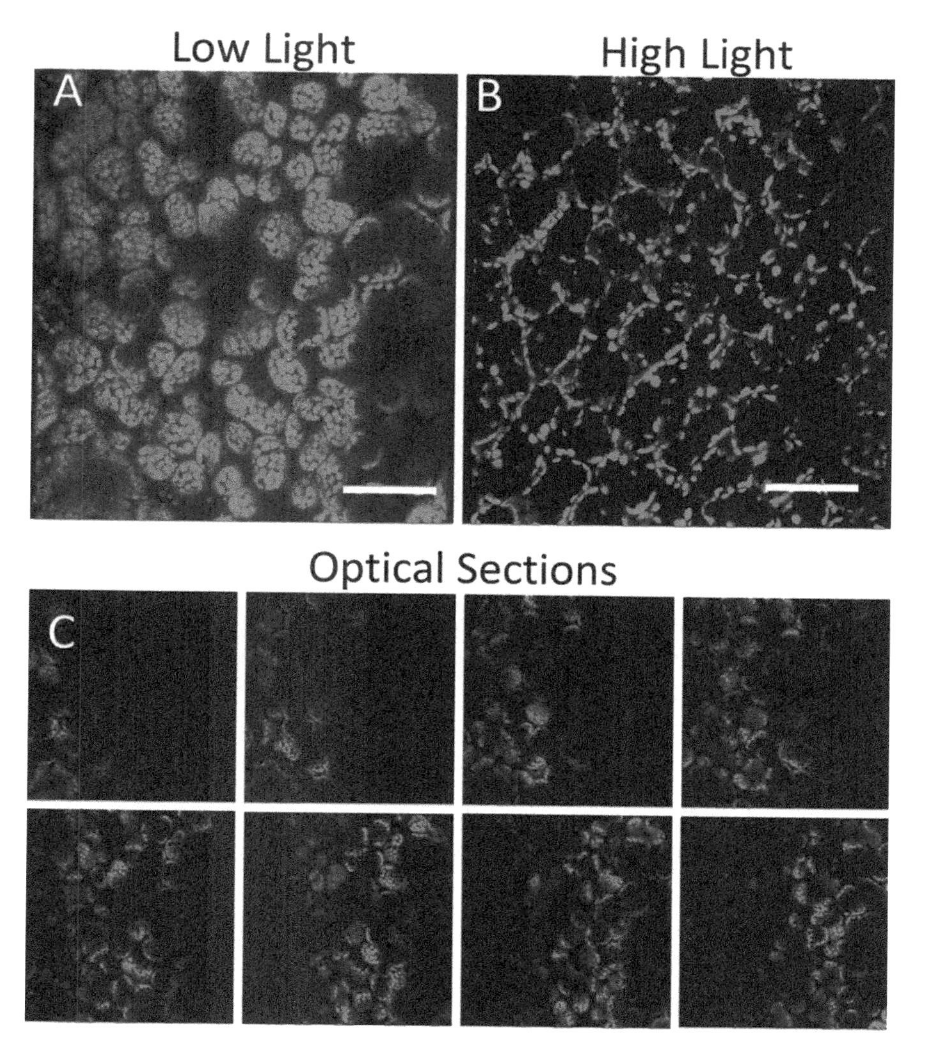

Fig. 1 Chloroplast distribution in *N. benthamiana* leaf cells after different light treatments. Leaves were dark adapted and then exposed to (**a**) low light for 2 h (**b**) or high light for 30 min. The images (**a** and **b**) are displayed as maximum projections generated from optical slices taken along the *z*-axis. Scale bars = 50 μm. (**c**) Eight optical slices covering total distance of 33 μm for image **a** are shown. The slices are separated from one another by 4.7 μm

The transit sequences as well as the 5′ and 3′ gene regulatory sequences can affect the intensity of the fluorescent protein signal [19]. Because chloroplast targeting information is usually located at the N-terminus of nuclear-encoded proteins, the coding region for the fluorescent protein is engineered onto the C-terminus. In stably transformed plants, the diverse morphology and number of plastids in different tissues can be readily monitored by fluorescence microscopy (Fig. 4). Images of the same cells can be acquired by phase, bright-field, or differential interference contrast (DIC) and merged with the fluorescent signals in order to locate the fluorescence

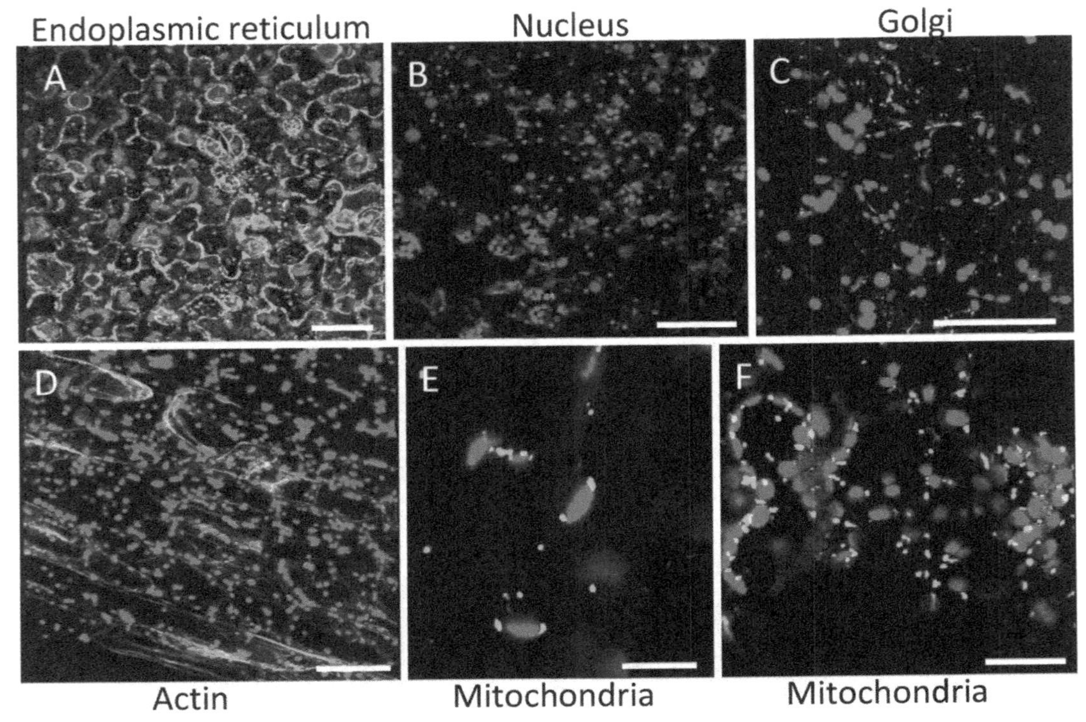

Fig. 2 Use of chloroplast autofluorescence to examine association of plastids with other subcellular organelles. GFP expression in stable transgenic lines is shown in **a**, **b**, **d**, and **f** and transient expression is shown in **c** and **e**. GFP is targeted to (**a**) Endoplasmic reticulum [94, 95], (**b**) nucleus [4], (**c**) Golgi [96], (**d**) actin [97], (**e** and **f**) mitochondria [24]; (**a**, **b**, and **f**) tobacco, (**d**) Arabidopsis, (**c** and **e**) tomato leaf epidermal cells. Bar (**e**) 10 μm, all others, 50 μm

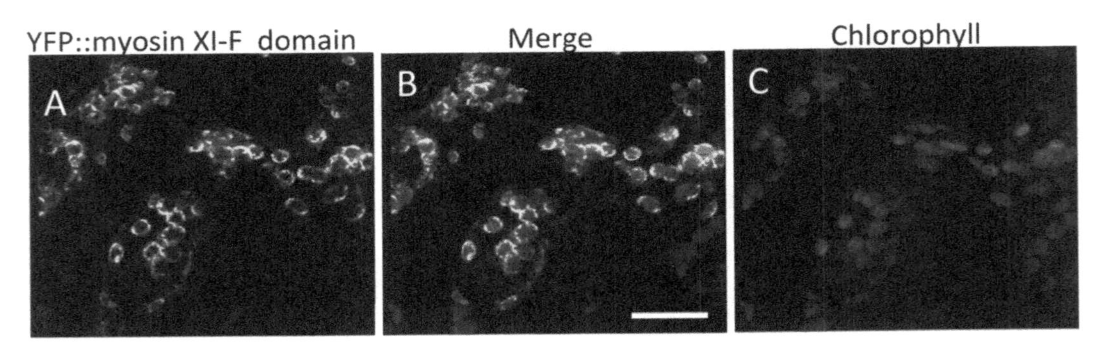

Fig. 3 Use of chlorophyll autofluorescence and transient expression for localization of an ORF. (**a**) YFP::myosin XI-F subdomain from *N. benthamiana* agroinfiltrated into *N. benthamiana* leaves [81]. YFP (*yellow*). (**b**) Merged images of YFP and chlorophyll autofluorescence (*red*). (**c**) Chlorophyll autofluorescence (*red*). Images are maximum projections of 10 confocal images taken along the *z*-axis. Scale bar = 20 μm. Leaves were not co-infiltrated with a strain expressing a silencing inhibitor

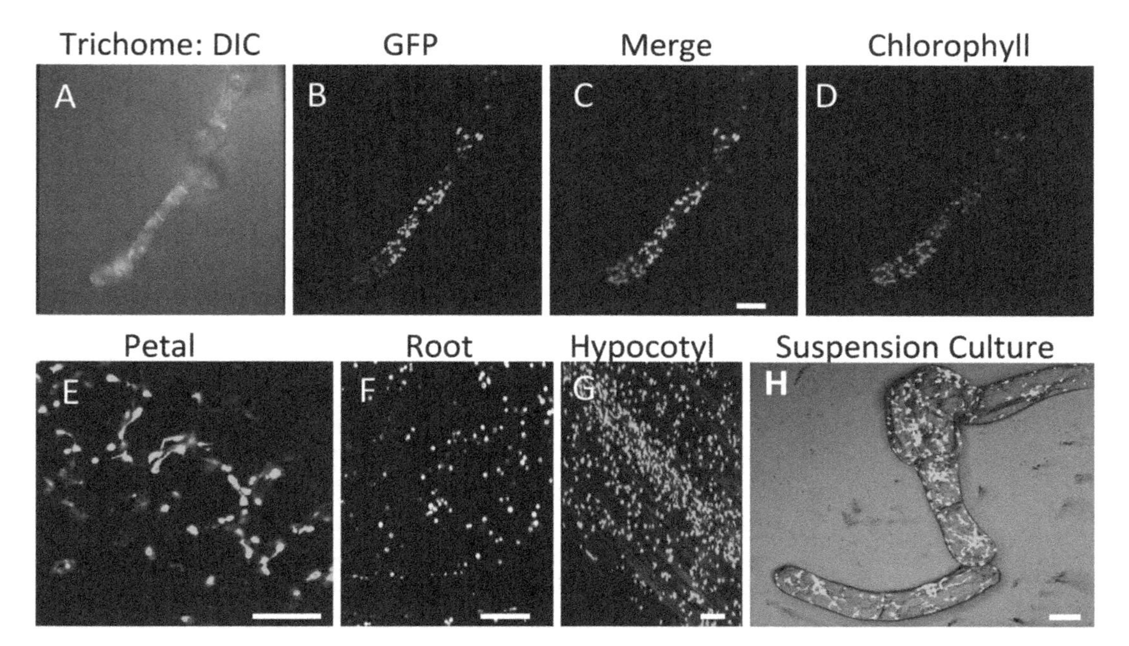

Fig. 4 Plastid morphology in various tissues of transgenic plants carrying a nuclear transgene encoding plastid-targeted FPs [21, 98–100]. (**a**) Differential interference contrast (DIC) image of trichome. The other images are maximum projections of several optical sections taken along the *z*-axis. Images (**b**, **e**, **g**) show GFP fluorescence as *green*, (**f**) YFP as *yellow*. Image (**c**) shows merged images of the GFP fluorescence (**b**) and *red* chlorophyll autofluorescence (**d**). Image (**h**) shows merged images of DIC, GFP (*green*) and propidium iodide (*red*) staining of tobacco cell culture carrying a nuclear transgene encoding plastid-targeted GFP. Bars = 25 μm. (**a**–**d**) tobacco trichome, (**e**) tobacco epidermal cell in the limb of a petal, (**f**) maize root cell (**g**) Arabidopsis hypocotyls and (**h**) tobacco cell culture. The fluorescent protein constructs used for tobacco transformation are described in Kohler et al. [24], for maize in Sattarzadeh et al. [101] and for Arabidopsis in Hanson and Sattarzadeh [21]

within a cell or tissue (Fig. 4). Fluorescent protein labeling reveals unexpected morphology of non-green plastids, especially striking in cultured cells [20], Fig. 4). The first reported transgenic plants with GFP located in the stroma confirmed the existence of long-forgotten plastid tubular extensions, now termed stromules [21–24]. The presence of tubular structures on small fluorescent organelles in non-green tissues can aid in their identification as plastids.

1.2 Selection of the Fluorescent Protein for Labeling

One factor to consider in selecting a particular fluorescent protein is the natural autofluorescence of the tissue of interest, as overlap of autofluorescent signal and the fluorescent protein is best avoided. Autofluorescent molecules are often found in the plant cell wall, vacuoles, and chloroplasts [25]. While GFP has been used most frequently to label chloroplasts, yellow fluorescent proteins (YFP) and red fluorescent proteins (e.g., dsRed, RFP, mCherry) are also suitable, as their fluorescence can be distinguished from that of chlorophyll. Chlorophyll is not the only autofluorescent compound in plants, however. Examining a wild-type tissue of interest prior to transformation for autofluorescence can be a valuable guide for choosing a fluorescent protein [26, 27]. Review articles and microscope and fluorophore vendor Web sites list the excitation and emission maxima of the various fluorescent proteins that have been expressed in plants [28–31]. A Web tool is available to compare fluorescence excitation and emission curves of several fluorescent proteins in comparison to various dyes that are commercially available to label subcellular locations (http://www.invitrogen.com/site/us/en/home/support/Research-Tools/Fluorescence-SpectraViewer.html). Fluorescent dyes should not be neglected as an effective means to label particular subcellular locations. For example, the Mitotracker dyes are effective for monitoring plant mitochondria [24], propidium iodide is convenient for locating the plant cell wall (Fig. 4h), and DAPI (4′6-diamidino-2-phenylindole) reacts with DNA, including nucleoids within the chloroplast [32].

Other properties of different fluorescent proteins should also be considered. Mutations have been introduced into genes encoding fluorescent proteins in order to create FPs proteins with different absorbance spectra. Deliberate mutation as well as cloning of genes for fluorescent proteins from additional organisms has resulted in shifting of excitation and emission optima to a variety of wavelengths, reduced the rate of photodestruction, created photoactivatable and photoconvertible proteins, and increased brightness [33–39]. For imaging the plastid envelope, proplastids, or thin stromules, a bright fluorescent protein as well as high-level expression is often essential. Although all fluorophores are susceptible to destruction during illumination, variants have been selected that are more resistant to photobleaching than others [35, 40]. For example, citrine YFP is relatively less affected by illumination, pH and anions than many other FPs, thus making it suitable for expression in acidic compartments within a cell or tissue, and therefore was chosen to make a collection of maize transgenic plants labeled in specific intracellular locations [41]. Another factor for FP selection is temperature sensitivity; some fluorescent proteins have been optimized for folding at 37 °C for bacterial and mammalian cell expression rather than the temperature optima of most plants.

Certain fluorescent proteins can be used to assess protein–protein interactions by bioluminescence resonance energy transfer (BRET), fluorescence resonance energy transfer (FRET) [42], and fluorescence lifetime imaging microscopy (FLIM) [31, 43, 44]. All these methods are based on transfer of energy from one molecule to another, with transfer only occurring when the molecules are in close proximity. In BRET, energy transfer occurs from luciferase to a fluorescent protein, usually YFP, and thus requires a substrate but not irradiation of the donor [45–48]. To date fewer studies with plant cells have used BRET than FRET. The most common FRET combinations are CFP (Cyan FP) and YFP; FRET has been monitored in plant cells transformed both stably or transiently [28, 49–54]. CFP and YFP can also transfer energy to mCherry. A two-step FRET process involving CFP, YFP, and mCherry was used to image a chloroplast enzyme complex [55]. The use of FRET and FRET-FLIM for studying protein–protein interactions in plant cells has been described in Subheading 3 and review articles [5, 56, 57].

FPs have been altered for detection of protein–protein interactions. In bimolecular fluorescence complementation (BiFC), the fluorescent protein is divided into two parts that are not fluorescent unless brought into close proximity. By attaching sequences encoding one of the partners to a gene encoding one protein, and the other partner to a second protein, whether or not the two proteins closely interact can be monitored by examining cells transformed with both constructs for fluorescence. BiFC has already been exploited to study whether chloroplast proteins are present in complexes. Split YFP fused to Arabidopsis MinD and chloroplast-targeted *E. coli* MinC, respectively, resulted in punctuate YFP signals in protoplasts [58]. BiFC could be observed when the chloroplast ChrD protein, known to dimerize, was fused to split YFP [59]. Furthermore, interactions between a sigma factor and a pentatricopeptide repeat protein [60] and between a chloroplast-localized heat-shock protein and a viral movement protein [61] were demonstrated by BiFC.

Fluorescent sensors have been created by incorporating a sensing domain between two FPs (typically CFP and YFP) that will undergo FRET only if the FPs are in close proximity. Binding of the ligand of interest to the sensing domain will affect whether or not FRET occurs. Single-FP sensors have also been created in which a particular FP's fluorescence exhibits dynamic changes due to intrinsic sensitivity to some factor or due to altered conformation of a protein fused to the single FP [62–64]. The potential for use of these biosensors in plastids is indicated by the successful targeting of a redox sensor to mitochondria [65, 66].

1.3 Introduction of Fluorescent Protein-Encoding Genes into Plants

Stably transformed plants with fluorescent proteins located in chloroplasts have been reported in a number of species. Both nuclear transgenes and plastid transgenes have been used to express FPs in plastids. Transiently expressed transgenes have become increasingly important with the development of convenient vectors for Agrobacterium-mediated introduction of transgenes through agroinfiltration [67–70]. Transient expression is suitable not only for determination of protein location and labeling of non-green plastids, but also for BRET, FRET, FLIM, and BiFC. Because protoplasts are readily obtained from leaves of many species, including those recalcitrant to regeneration from cell cultures, they are valuable for transient expression of transgenes introduced by disruption of membranes by electronic or chemical means, for which a

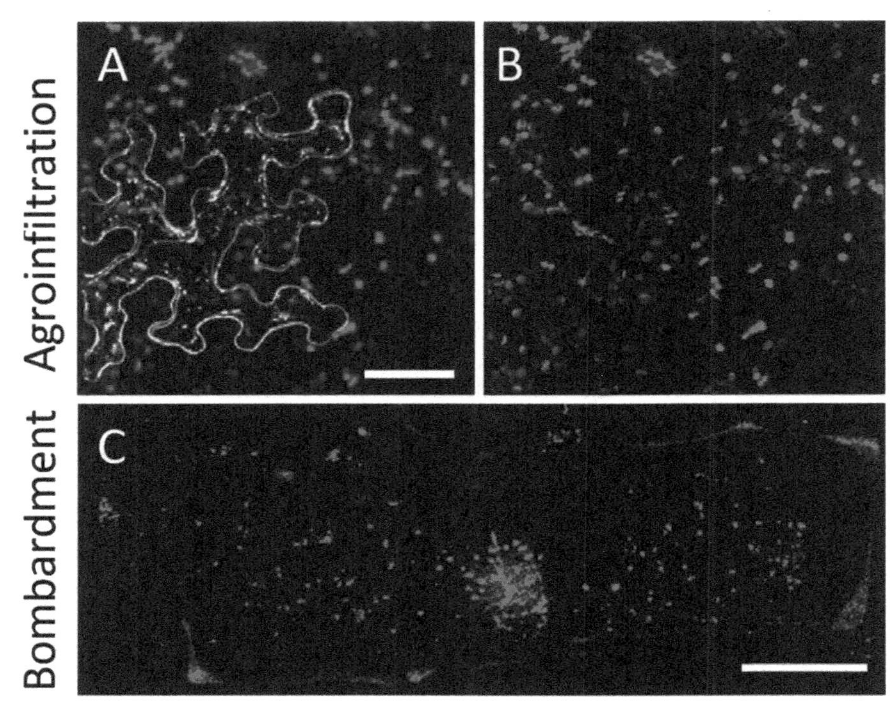

Fig. 5 Use of co-infiltration for localization of an ORF and/or labeling of plastids. Transient expression of plastid-targeted mCherry in epidermal cells of *N. benthamiana* and onion. (**a**) YFP::myosin subdomain from *A. thaliana* [81] co-infiltrated into *N. benthamiana* leaves with vector PT-RK, which targets mCherry to plastids [90], and with a vector expressing the tomato bushy stunt P19 silencing inhibitor [78]. (**b**) Vector PT-RK co-infiltrated with the silencing inhibitor into *N. benthamiana* leaves. (**c**) Transient expression of vector PT-RK in onion epidermal cells following biolistic bombardment. Images are maximum projections of 10 confocal images taken along the *z*-axis. Scale bar = 50 μm

number of protocols are available [71–74]. While agroinfiltration and protoplast isolation is most often practiced on leaf material, biolistic bombardment can be used to express fluorescent proteins transiently in a wide variety of tissues. A convenient method has been developed to express transgenic fluorescent protein in cotyledons of both dicots and monocots following incubation of seedling roots with Agrobacterium and a surfactant [75]. Agroinoculation with viral vectors can be used to label tissues in addition to leaves due to systemic spreading of the engineered virus [67].

All transient expression experiments potentially suffer from the possibility of posttranscriptional gene silencing [76]. To mitigate this problem, Agrobacterium strains that can express viral silencing suppressors are sometimes co-infiltrated with the strains expressing the genes of interest, often resulting in marked improvement in level and duration of expression [67, 77, 78] (Fig. 5). A variety of suppressors from different viruses are available for use in co-infiltration [79]. Co-bombardment with DNA encoding a silencing suppressor has also been shown to enhance transient GFP expression in bombarded bean cotyledons [80].

2 Materials

2.1 Plants, Plasmids, and Bacterial Strains

1. For biolistic transient assay, seedlings (e.g., *Nicotiana benthamiana* or *N. tabacum*) are grown in culture on MS agar petri dishes or in Magenta boxes (PlantMedia, Dublin, Ohio) until leaves are 1–2 cm diameter. A large white onion may be purchased fresh from a market.
2. For agroinfiltration, *Nicotiana benthamiana* or *N. tabacum* seedlings are grown in Metromix for 3–4 weeks under 14 h light/10 h dark cycle. Tomato seedlings were grown under the same light conditions until two to four leaves were present.
3. *Agrobacterium tumefaciens* strain (GV3101 *pMP90*) carrying a plasmid encoding a YFP::Myosin XI-F subdomain fusion protein gene [81]. For information on additional FP fusion genes and transgenic plants *see* **Note 1**.

2.2 Disposables

1. Disposable supplies (macrocarriers, 1.0–1.1 μ gold or tungsten microcarriers, rupture disks) for the Bio-Rad PDS/1000 system.
2. 95 % ethanol for sterilizing scalpels and forceps for plant tissue culture.
3. One milliliter needle-less plastic syringes (Becton-Dickinson, New Jersey).

2.3 Buffers, Media, and Solutions

1. MS agar: basal salts for Murashige and Skoog medium (Sigma-Aldrich) are combined with 0.8 % Phytagar (GIBCO) and 2–3 % sucrose *see* **Note 2**.
2. Luria–Bertani (LB) medium: 10 g/L Bacto-tryptone, 10 g/L NaCL, and 5 g/L yeast extract, pH 7.0. Solid medium includes 1.5 % Bacto-agar (Difco, Detroit, MI).
3. Selection agents: antibiotics rifampicin, kanamycin, spectinomycin (Sigma-Aldrich, St. Louis, Missouri) and the herbicide BASTA (ammonium glufosinate, Sigma-Aldrich, St. Louis, Missouri).
4. Stock solutions: Rifampicin at 50 μg/ml. 50 mg/ml stock solution in methanol can be stored at −20 °C. Kanamycin at 50 μg/ml. 50 mg/ml stock solution in distilled water can be stored at −20 °C. Spectinomycin at 50 μg/ml. 50 mg/ml stock solution in distilled water can be stored at −20 °C. Acetosyringone (Roth, Germany) at 150 μM. 100 mM stock solution in DMSO can be stored at −20 °C.
5. Infiltration medium: 10 mM $MgSO_4$, 10 mM MES, pH to 5.6. Autoclave, then add acetosyringone to150 μM.

3 Methods

3.1 Bombardment of Aseptic Leaves and Onion Explants

1. Because onion epidermal cells are large and colorless, they are particularly useful for observing intracellular location of fluorescent proteins introduced by bombardment [82]. Unless damaged or invaded by a plant pathogen, the internal tissues of the onion are aseptic. After flaming a scalpel, slice out a cube 1 cm × 1 cm and a few millimeters deep and discard. Then slice a cube 1–2 cm × 1–2 cm and a few millimeters deep and place on a MS agar plate, internal side upward. Continue cutting slices until sufficient slices are available to cover a circle in the middle of the plate of approximate diameter 4–5 cm (Fig. 6). Bombardment is usually most efficient in this region.
2. Remove leaves from the aseptic seedling cultures and place them with lower epidermis facing upward on the MS agar plate (Fig. 6; *see* **Note 3**). Cluster the leaves in the middle of the plates as described above.
3. Place the plate with the explants in the platform at the third position from the bottom.
4. Coat 1.1 μ tungsten microcarriers (particles) with DNA encoding fluorescent proteins as described as the Bio-Rad Web site or in other chapters of this book. DNA from *E. coli* plasmids or Agrobacterium binary vectors may be used as well as PCR products [41, 83]. While gold particles are less toxic to cells than tungsten and can lead to higher transformation frequencies, tungsten is usually sufficient for transient assay and is considerably less expensive.
5. Pipet 5–10 μl of DNA-coated particle solution onto the macrocarriers. Load biolistic device and operate according to manufacturer's instructions. The same tissue can be bombarded twice, resulting in an increased number of transformed cells *see* **Notes 4** and **5**.

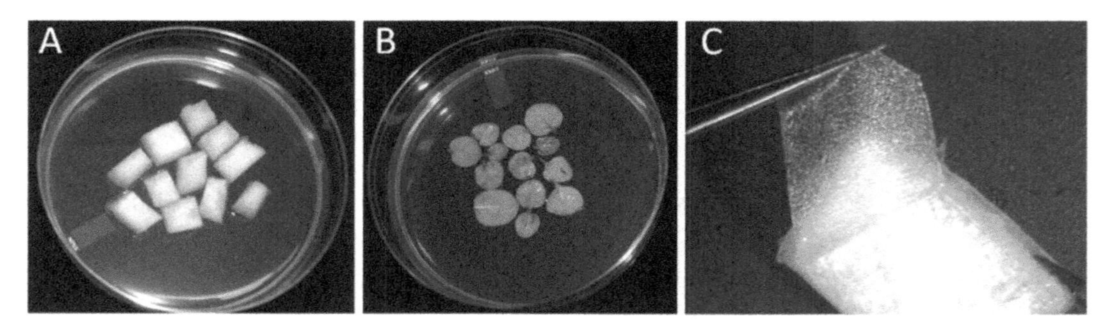

Fig. 6 Explants for biolistic transformation and microscopic analysis. (**a**) Market onion (**b**) tobacco leaf from plants grown on MS agar (**c**) onion epidermis stripped from a bombarded explant

6. Wrap the bombarded plates with parafilm and incubate for 2 days before examining the tissue for expression. Peel the epidermis from the onion explants (Fig. 6) and place onto a microscope slide in a drop of water under a coverslip. The bombarded leaves should be sliced into small sections that will lie flat in water under the coverslip.

3.2 Agroinfiltration of Leaves

1. Agrobacterium strains carrying FP expression vectors are streaked on LB medium agar plates with appropriate antibiotic and incubated at 30 °C for 2 days. Agrobacterium vectors typically contain a selective marker for maintenance in bacteria and another for selection in plants. Agrobacterium cells from agar plates can be taken directly from the plates and put into the Infiltration medium instead of following **steps 2** and **3**.
2. Inoculate the strain into 5 ml LB broth with appropriate antibiotic and grow overnight at 30 °C on a shaker (150 rpm).
3. Collect the cells by centrifugation for 10 min at 3,500 × *g*.
4. Resuspend the culture in Infiltration Medium to an OD_{600} of 0.5 and incubate for 2–5 h at room temperature in the dark. For tomato agroinfiltration, use an OD_{600} of 0.1.
5. Two or more Agrobacterium strains carrying fluorescent protein markers can be mixed together in order to express two or more markers simultaneously (Fig. 5). Level and duration of expression can be improved by coinfiltration with a virus-encoded suppressor of gene silencing. A commonly used suppressor is the p19 protein of tomato bushy stunt virus (TBSV), which prevents posttranscriptional gene silencing (PTGS) in the infiltrated tissues [78]. By co-expressing the p19 protein, maximal protein expression is usually seen on the 5th day after infiltration. Expression is often detectable without the use of a silencing inhibitor (Fig. 2c, e).
6. The Agrobacterium strain(s) is infiltrated into the intracellular spaces from the abaxial side into young leaves of 3–4 week-old Nicotiana plants with a needleless syringe. Press the tip of the syringe onto the underside of the leaf while pressing on the leaf with a finger of the other hand on the opposite side and slowly introduce liquid. Successful infiltration will result in visible darkening of the leaf surface *see* **Note 6**.

3.3 Monitoring of Expression

1. Sufficient transgene expression in bombarded tissue should have occurred by 2 days to merit examination. With the use of a silencing suppressor, a typical optimum day for observation is 5 days after infiltration.
2. Before selecting a particular fluorescent protein to use in transient or stable transformation, it is advisable to determine whether the available fluorescence microscopes are equipped

with the appropriate band-pass excitation filter or laser, which will allow light of appropriate wavelength to excite the molecule. Because of the strong output of light from lasers, fluorescent proteins and chlorophyll can be imaged even if the wavelength available is not the optimal excitation wavelength. Likewise, the microscope must be equipped with a suitable emission filter to eliminate the exciting light from the detector or light from other fluorophores.

3. Confocal microscopes are usually available in common facilities, where training on a particular manufacturer's microscope can be obtained. The microscopic images provided in this chapter are from a Leica TCS SP2 CLSM equipped with 458, 478, 488, 514, 543, and 638 nm laser lines for excitation. This microscope has four fluorescence detectors allowing simultaneous capture of up to 4 fluorescence channels plus bright field and is also equipped with DIC optics. The major fluorescence microscope vendors have Web sites with valuable tutorials about fluorescence imaging, fluorophore selection, and optimal type of microscope to use for different applications. Excellent introductions to fluorescence can be viewed at http://www.invitrogen.com/site/us/en/home/support/Tutorials.html
4. Confocal microscopy provides enhanced resolution compared to standard epifluorescence microscopy. Common alternatives often available in the laboratory or an institution's facilities are epifluorescence microscopes, often equipped with deconvolution software (Metamorph, Deltavision) to remove out-of-focus fluorescence. Peeled onion or leaf epidermis explants are sufficiently thin that quality images can often be obtained in a standard epifluorescence microscope. Spinning disk confocal microscopes are particular valuable for imaging rapidly moving organelles such as Golgi, secretory vesicles or mitochondria because only a short time interval is needed for image acquisition. Plastids move relatively slower than other organelles, and thus their movement can be followed easily in a typical CSLM. Multiphoton microscopes can reach greater depths within the tissue and produce less photodamage.
5. CSLMs allow the users to obtain optical sections at different depths within a tissue. A single optical section is usually much less informative than a number of optical sections that can be "stacked" together by the CSLM's operating software (*see* Fig. 1c for an example of 8 individual optical sections vs. the stacked image in Fig. 1a). Typical CLSM software allows assembly of individual optical sections to construct of a 3-D image that can be rotated to observe the spatial relationships between different fluorescent signals. Another common image processing performed on the CSLM is production of a movie

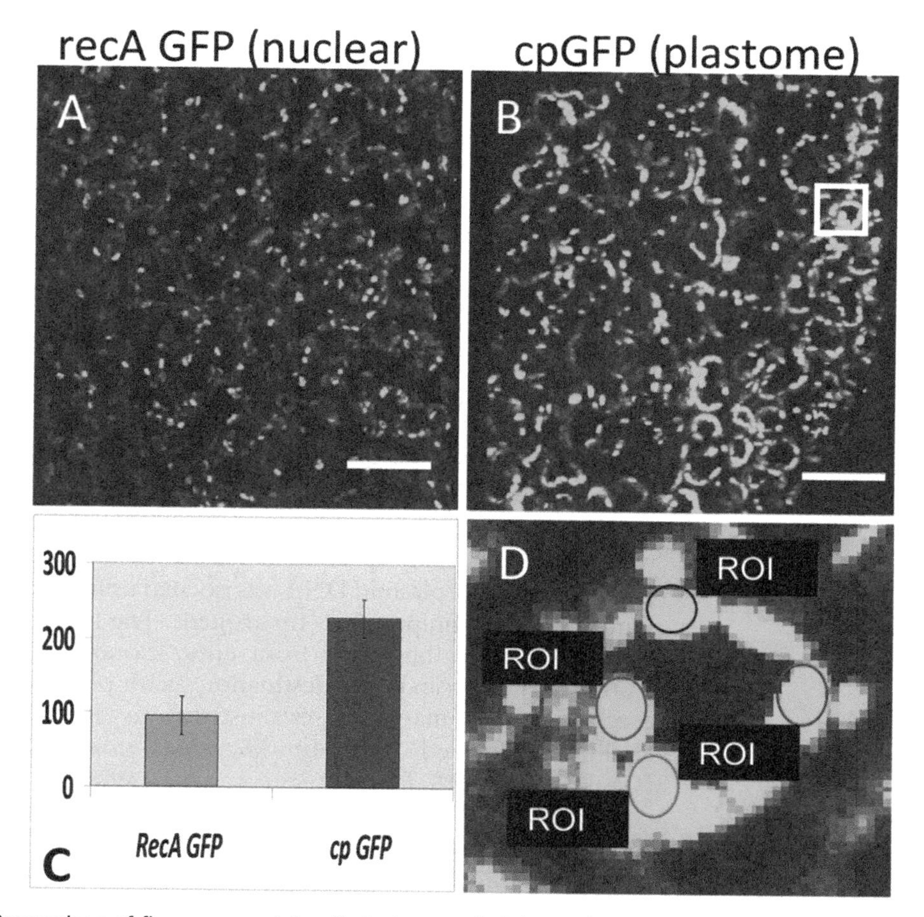

Fig. 7 Comparison of fluorescence intensity in transgenic tobacco leaf cells containing carrying (**a**) nuclear-encoded, plastid-targeted GFP [24] and (**b**) plastome-encoded GFP [17]. (**c**) The average of frequency of fluorescence intensity calculated via Leica quantification tools for 35 selected ROIs in image **a** and **b**. (**d**) Four selected ROI (s) from image **b**. (**d**). Scale bars = 50 μm

from a time series. Images can be acquired at defined intervals and then played sequentially. The quality of the movie will depend in part on how often the images are acquired relative to the movement rate of the objects being imaged.

6. Relative fluorescence of different fluorophores can be quantified using Adobe Photoshop or Image J, an open source, public domain Java image processing program (http://rsbweb.nih.gov/ij/). Most confocal systems include software for quantifying fluorescent signals. For example, the use of the quantification tools from Leica Microsystems (LCS, Wetzlar, Germany) is illustrated in Fig. 7. LCS has three quantification tools: *Line Profile*, *Histogram* and *Stack Profile* which are categorized under the Quantify tab. *Profile* can be used for measuring the intensities along a line segment within a single image

and *Histogram* tool measure the frequency of intensity values in a given the region of interest in a single image. Stacks of the images can be analyzed with the *Stack* Profile tool. After selecting the region of interest (ROI) a statistics window will show the data for each selected area *see* **Note 7**.

4 Notes

1. A standard TOPO and Gateway cloning system (Invitrogen, Carlsbad, CA) can be used to generate FP fusions to the gene of interest [84] using a variety of Gateway-compatible vectors [85–87]. In the example shown in Fig. 3, cDNA sequences encoding YFP and myosin fusion were amplified from pENSG-YFP:: Myosin XI-F subdomain [81] using *Accu prime* Taq DNA polymerase (Invitrogen) to generate PCR products and were cloned using pCR8/GW/TOPO TA Cloning Kit (Invitrogen). The plasmid DNA was isolated using the PureLink Quick Plasmid Miniprep Kit (Invitrogen). The LR reaction was used to transfer the insert from entry clones pCR8/GW/TOPO-YFP::myosin to the destination vector pEarleyGate 202 according to the manufacturer's instructions (Invitrogen) and finally transformed into *Agrobacterium tumefaciens* strain (GV3101 *pMP90*) for transient assay. A suite of Gateway-compatible binary vectors known as pEarleyGate vectors [86] allow fusion of open reading frames C-terminally to CFP, GFP, or YFP. Among the pEarleyGate vectors are ones that incorporate N- or C-terminal epitope tags onto the protein. pEarley Gate vectors can be obtained from the Arabidopsis Biological Resource Center (ABRC) (http://www.arabidopsis.org/).

   A set of TMV-based vectors that can be modified with Gateway technology and allow fusion to fluorescent proteins to express proteins at high levels by agroinoculation [88]. A modified strategy for Gateway cloning allows addition of a fluorescent tag to proteins without introducing extra linker sequences [89].

   A useful set of binary plasmids and Arabidopsis lines carrying four different fluorescent proteins (CFP, GFP, YFP, mCherry) fused to the transit sequence of the small subunit of Rubisco was produced by Nelson et al [90]. Additional binary plasmids that label mitochondria, peroxisomes, ER, Golgi, plasma membrane, and tonoplast are also available, and can be used in coinfiltration experiments to label multiple locations transiently. The constructs can be obtained from the ABRC and transformed into *Agrobacterium tumefaciens* strain GV3101 pMP90 [91] for transient or stable transformation. Arabidopsis lines stably transformed with these organelle markers [90] are also available from the ABRC. Methods for introducing constructs into Agrobacterium and handling strains have been described by Wise et al. [51, 92].

2. Before pouring agar medium in the petri dishes, the medium can be degassed by placing it under vacuum until gas bubbles no longer escape. This optional step will prevent gases escaping from the solid medium in the biolistic chamber vacuum from disrupting samples during the bombardment process.
3. Because the onion and leaf explants used for transient expression will not be cultured for long after bombardment, the medium onto which they are placed is often not critical. An alternative to agar medium is a moist paper towel. An explant of onion tissue can be made that is sufficiently thick so that the bombarded layer remains healthy due to the presence of the layers below, as long as humidity is kept high.
4. Many types of explants can be bombarded for transient expression. Conditions will need to be optimized for explants with thick cuticles or protective layers. The force of bombardment can be altered by changing the pressure at which rupture disks break (disks of different strengths are sold by Bio-Rad) or by moving the plate to a shelf closer or further away from the firing chamber.
5. Leaves and other explants can be surface-sterilized (typically 10 % bleach, 0.1 % detergent for 20 min) before bombardment, to avoid spraying microorganisms into the biolistic device chamber during bombardment. Surface sterilization also allows explants to be cultured on sucrose-containing medium for a few days, which is valuable for delicate tissues or ones that are easily dehydrated.
6. Agroinfiltration into *N. benthamiana* usually produces higher expression levels than *N. tabacum*, tomato or Arabidopsis. Although Arabidopsis agroinfiltration is more difficult, procedures are available [93]. Protoplast DNA uptake or bombardment have more frequently been the choices for transient expression in Arabidopsis.
7. In comparing the relative fluorescence of two different samples, it is absolutely necessary to have the same setting for both samples. Recent software allows saving of settings so that samples imaged at different times can be performed with the same values.

Acknowledgments

This work was supported by grants from the Chemical Sciences, Geosciences, and Biosciences Division, Office of Basic Energy Sciences, Office of Science, US Department of Energy to M.R.H., including DE-89-ER14030 and DE–FG02–09ER16070.

References

1. Maxwell K, Johnson GN (2000) Chlorophyll fluorescence—a practical guide. J Exp Bot 51: 659–668
2. Wada M (2013) Chloroplast movement. Plant Sci 210:177–182
3. Kwok EY, Hanson MR (2004) In vivo analysis of interactions between GFP-labeled microfilaments and plastid stromules. BMC Plant Biol 4:2
4. Kwok EY, Hanson MR (2004) Plastids and stromules interact with the nucleus and cell membrane in vascular plants. Plant Cell Rep 23:188–195
5. Bhat RA, Lahaye T, Panstruga R (2006) The visible touch: *in planta* visualization of protein-protein interactions by fluorophore-based methods. Plant Methods 2:12
6. Wong KA, O'Bryan JP (2011) Bimolecular fluorescence complementation. J Vis Exp 50: 2643
7. Vothknecht UC, Soll J (2000) Protein import: the hitchhikers guide into chloroplasts. Biol Chem 381:887–897
8. Seki M, Iida A, Morikawa H (1998) Transient expression of foreign genes in tissues of *Arabidopsis thaliana* by bombardment-mediated transformation. Methods Mol Biol 82:219–225
9. Lee DW, Hwang I (2011) Transient expression and analysis of chloroplast proteins in Arabidopsis protoplasts. Methods Mol Biol 774:59–71
10. Fischer R, Vaquero-Martin C, Sack M, Drossard J, Emans N, Commandeur U (1999) Towards molecular farming in the future: transient protein expression in plants. Biotechnol Appl Biochem 30:113–116
11. Kapila J, De Rycke R, Van Montagu M, Angenon G (1997) An Agrobacterium-mediated transient gene expression system for intact leaves. Plant Sci 122:101–108
12. Pawley JB (ed) (1995) Handbook of biological confocal microscopy. Plenum Press, New York
13. Benediktyova Z, Nedbal L (2009) Imaging of multi-color fluorescence emission from leaf tissues. Photosynth Res 102:169–175
14. Feijo JA, Moreno N (2004) Imaging plant cells by two-photon excitation. Protoplasma 223:1–32
15. Wada M, Kagawa T, Sato Y (2003) Chloroplast movement. Annu Rev Plant Biol 54:455–468
16. Franklin S, Ngo B, Efuet E, Mayfield SP (2002) Development of a GFP reporter gene for *Chlamydomonas reinhardtii* chloroplast. Plant J 30:733–744
17. Reed ML, Wilson SK, Sutton CA, Hanson MR (2001) High-level expression of a synthetic red-shifted GFP coding region incorporated into transgenic chloroplasts. Plant J 27:257–265
18. Millwood RJ, Moon HS, Stewart CNJ (2008) Fluorescent proteins in transgenic plants. In: Geddes CD (ed) Reviews in fluorescence 2008. Springer, New York, pp 387–403
19. Primavesi LF, Wu H, Mudd EA, Day A, Jones HD (2008) Visualisation of plastids in endosperm, pollen and roots of transgenic wheat expressing modified GFP fused to transit peptides from wheat SSU RubisCO, rice FtsZ and maize ferredoxin III proteins. Transgenic Res 17:529–543
20. Pyke KA (2013) Divide and shape: an endosymbiont in action. Planta 237:381–387
21. Hanson MR, Sattarzadeh A (2008) Dynamic morphology of plastids and stromules in angiosperm plants. Plant Cell Environ 31:646–657
22. Kwok EY, Hanson MR (2004) Stromules and the dynamic nature of plastid morphology. J Microsc 214:124–137
23. Natesan SK, Sullivan JA, Gray JC (2005) Stromules: a characteristic cell-specific feature of plastid morphology. J Exp Bot 56:787–797
24. Kohler RH, Zipfel WR, Webb WW, Hanson MR (1997) The green fluorescent protein as a marker to visualize plant mitochondria in vivo. Plant J 11:613–621
25. Fang Y, Spector DL (2010) Live cell imaging of plants. Cold Spring Harb Protoc 2012, pdb top68
26. Groover A, Jackson D (2007) Live-cell imaging of GFP in plants. CSH protocols 2007, pdb ip31
27. Ross FWD (1995) Fluorescence microscopy, vol 2. Cambridge University Press, Cambridge, England
28. Berg RH, Beachy RN (2008) Fluorescent protein applications in plants. Methods Cell Biol 85:153–177
29. Geddes CD (ed) (2008/2010) Reviews in fluorescence 2008/2010, vol 2008/2010. Springer, New York
30. Mathur J (2007) The illuminated plant cell. Trends Plant Sci 12:506–513
31. Shaw SL, Ehrhardt DW (2013) Smaller, faster, brighter: advances in optical imaging of living plant cells. Annu Rev Plant Biol 64:351–375
32. Coleman AW (1979) Use of the fluorochrome 4′6-diamidino-2-phenylindole in genetic and developmental studies of chloroplast DNA. J Cell Biol 82:299–305

33. Lippincott-Schwartz J, Patterson GH (2003) Development and use of fluorescent protein markers in living cells. Science 300:87–91
34. Nienhaus GU, Nienhaus K, Holzle A, Ivanchenko S, Renzi F, Oswald F, Wolff M, Schmitt F, Rocker C, Vallone B, Weidemann W, Heilker R, Nar H, Wiedenmann J (2006) Photoconvertible fluorescent protein EosFP: biophysical properties and cell biology applications. Photochem Photobiol 82:351–358
35. Shaner NC, Patterson GH, Davidson MW (2007) Advances in fluorescent protein technology. J Cell Sci 120:4247–4260
36. Shaner NC, Steinbach PA, Tsien RY (2005) A guide to choosing fluorescent proteins. Nat Methods 2:905–909
37. Dixit R, Cyr R, Gilroy S (2006) Using intrinsically fluorescent proteins for plant cell imaging. Plant J 45:599–615
38. Lippincott-Schwartz J, Patterson GH (2008) Fluorescent proteins for photoactivation experiments. Methods Cell Biol 85:45–61
39. Lippincott-Schwartz J, Patterson GH (2009) Photoactivatable fluorescent proteins for diffraction-limited and super-resolution imaging. Trends Cell Biol 19:555–565
40. Shaner NC, Lin MZ, McKeown MR, Steinbach PA, Hazelwood KL, Davidson MW, Tsien RY (2008) Improving the photostability of bright monomeric orange and red fluorescent proteins. Nat Methods 5:545–551
41. Mohanty A, Yang Y, Luo A, Sylvester AW, Jackson D (2009) Methods for generation and analysis of fluorescent protein-tagged maize lines. Methods Mol Biol 526:71–89
42. Tsien RY (2009) Indicators based on fluorescence resonance energy transfer (FRET). Cold Spring Harb Protoc 2009, pdb top57
43. Ishikawa-Ankerhold HC, Ankerhold R, Drummen GP (2012) Advanced fluorescence microscopy techniques—FRAP, FLIP, FLAP, FRET and FLIM. Molecules 17:4047–4132
44. Padilla-Parra S, Tramier M (2012) FRET microscopy in the living cell: different approaches, strengths and weaknesses. BioEssays 34:369–376
45. Robitaille M, Heroux I, Baragli A, Hebert TE (2009) Novel tools for use in bioluminescence resonance energy transfer (BRET) assays. Methods Mol Biol 574:215–234
46. Subramanian C, Woo J, Cai X, Xu X, Servick S, Johnson CH, Nebenfuhr A, von Arnim AG (2006) A suite of tools and application notes for in vivo protein interaction assays using bioluminescence resonance energy transfer (BRET). Plant J 48:138–152
47. Subramanian C, Xu Y, Johnson CH, von Arnim AG (2004) In vivo detection of protein-protein interaction in plant cells using BRET. Methods Mol Biol 284:271–286
48. Xu X, Soutto M, Xie Q, Servick S, Subramanian C, von Arnim AG, Johnson CH (2007) Imaging protein interactions with bioluminescence resonance energy transfer (BRET) in plant and mammalian cells and tissues. Proc Natl Acad Sci U S A 104: 10264–10269
49. Gremillon L, Kiessling J, Hause B, Decker EL, Reski R, Sarnighausen E (2007) Filamentous temperature-sensitive Z (FtsZ) isoforms specifically interact in the chloroplasts and in the cytosol of *Physcomitrella patens*. New Phytol 176:299–310
50. Seidel T, Kluge C, Hanitzsch M, Ross J, Sauer M, Dietz KJ, Golldack D (2004) Colocalization and FRET-analysis of subunits c and a of the vacuolar H+-ATPase in living plant cells. J Biotechnol 112:165–175
51. Wise AA, Liu Z, Binns AN (2006) Three methods for the introduction of foreign DNA into Agrobacterium. Methods Mol Biol 343: 43–53
52. Behera S, Kudla J (2013) High-resolution imaging of cytoplasmic Ca^{2+} dynamics in Arabidopsis roots. Cold Spring Harb. Protoc 2013(7). pii: pdb.prot073023. doi: 10.1101/pdb.prot073023.
53. Swanson SJ, Gilroy S (2013) Imaging changes in cytoplasmic calcium using the Yellow Cameleon 3.6 biosensor and confocal microscopy. Methods Mol Biol 1009:291–302
54. Wanke D, Hohenstatt ML, Dynowski M, Bloss U, Hecker A, Elgass K, Hummel S, Hahn A, Caesar K, Schleifenbaum F, Harter K, Berendzen KW (2011) Alanine zipper-like coiled-coil domains are necessary for homotypic dimerization of plant GAGA-factors in the nucleus and nucleolus. PloS One 6:e16070
55. Seidel T, Seefeldt B, Sauer M, Dietz KJ (2010) In vivo analysis of the 2-Cys peroxiredoxin oligomeric state by two-step FRET. J Biotechnol 149:272–279
56. Bucherl C, Aker J, de Vries S, Borst JW (2010) Probing protein-protein Interactions with FRET-FLIM. Methods Mol Biol 655:389–399
57. Wolf H, Barisas BG, Dietz KJ, Seidel T (2013) Kaede for detection of protein oligomerization. Mol Plant. doi:10.1093/mp/sst039
58. Zhang M, Hu Y, Jia J, Gao H, He Y (2009) A plant MinD homologue rescues *Escherichia coli* HL1 mutant (DeltaMinDE) in the absence of MinE. BMC Microbiol 9:101
59. Citovsky V, Lee LY, Vyas S, Glick E, Chen MH, Vainstein A, Gafni Y, Gelvin SB, Tzfira

T (2006) Subcellular localization of interacting proteins by bimolecular fluorescence complementation *in planta*. J Mol Biol 362: 1120–1131
60. Citovsky V, Gafni Y, Tzfira T (2008) Localizing protein-protein interactions by bimolecular fluorescence complementation *in planta*. Methods 45:196–206
61. Krenz B, Windeisen V, Wege C, Jeske H, Kleinow T (2010) A plastid-targeted heat shock cognate 70 kDa protein interacts with the Abutilon mosaic virus movement protein. Virology 401:6–17
62. Frommer WB, Davidson MW, Campbell RE (2009) Genetically encoded biosensors based on engineered fluorescent proteins. Chem Soc Rev 38:2833–2841
63. Jones AM, Grossmann G, Danielson JA, Sosso D, Chen LQ, Ho CH, Frommer WB (2013) In vivo biochemistry: applications for small molecule biosensors in plant biology. Curr Opin Plant Biol 16:389–395
64. Lalonde S, Ehrhardt DW, Frommer WB (2005) Shining light on signaling and metabolic networks by genetically encoded biosensors. Curr Opin Plant Biol 8:574–581
65. Hanson GT, Aggeler R, Oglesbee D, Cannon M, Capaldi RA, Tsien RY, Remington SJ (2004) Investigating mitochondrial redox potential with redox-sensitive green fluorescent protein indicators. J Biol Chem 279: 13044–13053
66. Jiang K, Schwarzer C, Lally E, Zhang S, Ruzin S, Machen T, Remington SJ, Feldman L (2006) Expression and characterization of a redox-sensing green fluorescent protein (reduction-oxidation-sensitive green fluorescent protein) in Arabidopsis. Plant Physiol 141:397–403
67. Lindbo JA (2007) High-efficiency protein expression in plants from agroinfection-compatible tobacco mosaic virus expression vectors. BMC Biotechnol 7:52
68. Lindbo JA (2007) TRBO: a high-efficiency tobacco mosaic virus RNA-based overexpression vector. Plant Physiol 145:1232–1240
69. Sainsbury F, Thuenemann EC, Lomonossoff GP (2009) pEAQ: versatile expression vectors for easy and quick transient expression of heterologous proteins in plants. Plant Biotechnol J 7:682–693
70. Wroblewski T, Tomczak A, Michelmore R (2005) Optimization of Agrobacterium-mediated transient assays of gene expression in lettuce, tomato and Arabidopsis. Plant Biotechnol J 3:259–273
71. Miao Y, Jiang L (2007) Transient expression of fluorescent fusion proteins in protoplasts of suspension cultured cells. Nat Protoc 2:2348–2353
72. Wu FH, Shen SC, Lee LY, Lee SH, Chan MT, Lin CS (2009) Tape-Arabidopsis Sandwich—a simpler Arabidopsis protoplast isolation method. Plant Methods 5:16
73. Yoo SD, Cho YH, Sheen J (2007) Arabidopsis mesophyll protoplasts: a versatile cell system for transient gene expression analysis. Nat Protoc 2:1565–1572
74. Hellwege EM, Raap M, Gritscher D, Willmitzer L, Heyer AG (1998) Differences in chain length distribution of inulin from *Cynara scolymus* and *Helianthus tuberosus* are reflected in a transient plant expression system using the respective 1-FFT cDNAs. FEBS Lett 427:25–28
75. Li JF, Park E, von Arnim AG, Nebenfuhr A (2009) The FAST technique: a simplified Agrobacterium-based transformation method for transient gene expression analysis in seedlings of Arabidopsis and other plant species. Plant Methods 5:6
76. Johansen LK, Carrington JC (2001) Silencing on the spot. Induction and suppression of RNA silencing in the Agrobacterium-mediated transient expression system. Plant Physiol 126:930–938
77. Lombardi R, Circelli P, Villani ME, Buriani G, Nardi L, Coppola V, Bianco L, Benvenuto E, Donini M, Marusic C (2009) High-level HIV-1 Nef transient expression in *Nicotiana benthamiana* using the P19 gene silencing suppressor protein of Artichoke Mottled Crinckle Virus. BMC Biotechnol 9:96
78. Voinnet O, Rivas S, Mestre P, Baulcombe D (2003) An enhanced transient expression system in plants based on suppression of gene silencing by the p19 protein of tomato bushy stunt virus. Plant J 33:949–956
79. Dhillon T, Chiera JM, Lindbo JA, Finer JJ (2009) Quantitative evaluation of six different viral suppressors of silencing using image analysis of transient GFP expression. Plant Cell Rep 28:639–647
80. Chiera JM, Lindbo JA, Finer JJ (2008) Quantification and extension of transient GFP expression by the co-introduction of a suppressor of silencing. Transgenic Res 17:1143–1154
81. Sattarzadeh A, Krahmer J, Germain AD, Hanson MR (2009) A myosin XI tail domain homologous to the yeast myosin vacuole-binding domain interacts with plastids and stromules in *Nicotiana benthamiana*. Mol Plant 2:1351–1358
82. Scott A, Wyatt S, Tsou PL, Robertson D, Allen NS (1999) Model system for plant cell biology: GFP imaging in living onion epidermal cells. Biotechniques 26(1125):1128–1132
83. Xiao YL, Redman JC, Monaghan EL, Zhuang J, Underwood BA, Moskal WA, Wang W, Wu HC, Town CD (2010) High throughput generation of promoter reporter (GFP)

transgenic lines of low expressing genes in Arabidopsis and analysis of their expression patterns. Plant Methods 6:18

84. Xu R, Li QQ (2008) Protocol: streamline cloning of genes into binary vectors in agrobacterium via the gateway(R) TOPO vector system. Plant Methods 4:4
85. Curtis MD, Grossniklaus U (2003) A gateway cloning vector set for high-throughput functional analysis of genes *in planta*. Plant Physiol 133:462–469
86. Earley KW, Haag JR, Pontes O, Opper K, Juehne T, Song K, Pikaard CS (2006) Gateway-compatible vectors for plant functional genomics and proteomics. Plant J 45:616–629
87. Tzfira T, Tian GW, Lacroix B, Vyas S, Li J, Leitner-Dagan Y, Krichevsky A, Taylor T, Vainstein A, Citovsky V (2005) pSAT vectors: a modular series of plasmids for autofluorescent protein tagging and expression of multiple genes in plants. Plant Mol Biol 57:503–516
88. Kagale S, Uzuhashi S, Wigness M, Bender T, Yang W, Borhan MH, Rozwadowski K (2012) TMV-Gate vectors: gateway compatible tobacco mosaic virus based expression vectors for functional analysis of proteins. Sci Rep 2:874
89. Dubin MJ, Bowler C, Benvenuto G (2010) Overexpressing tagged proteins in plants using a modified gateway cloning strategy. Cold Spring Harb Protoc 2010, pdb prot5401. doi: 10.1101/pdb.prot5401.
90. Nelson BK, Cai X, Nebenfuhr A (2007) A multicolored set of in vivo organelle markers for co-localization studies in Arabidopsis and other plants. Plant J 51:1126–1136
91. Koncz C, Schel l J (1986) The promoter of the TL-DNA gene 5 controls the tissue-specific expression of chimeric genes carried by a novel type of Agrobacterium binary vector. Mol Gen Genet 204:383–396
92. Wise AA, Liu Z, Binns AN (2006) Culture and maintenance of agrobacterium strains. Methods Mol Biol 343:3–13
93. Lee MW, Yang Y (2006) Transient expression assay by agroinfiltration of leaves. Methods Mol Biol 323:225–229
94. Haseloff J (1999) GFP variants for multispectral imaging of living cells. Methods Cell Biol 58:139–151
95. Haseloff J, Siemering KR, Prasher DC, Hodge S (1997) Removal of a cryptic intron and subcellular localization of green fluorescent protein are required to mark transgenic Arabidopsis plants brightly. Proc Natl Acad Sci U S A 94:2122–2127
96. Hawes C, Brandizzi F, Batoko H, Moore I (2001) Organelle motility in plant cells: imaging golgi and ER dynamics with GFP. Curr Protoc Cell Biol 13:13.3.1–13.3.10. doi:10.1002/0471143030.cb0107s19
97. Sheahan MB, Staiger CJ, Rose RJ, McCurdy DW (2004) A green fluorescent protein fusion to actin-binding domain 2 of Arabidopsis fimbrin highlights new features of a dynamic actin cytoskeleton in live plant cells. Plant Physiol 136:3968–3978
98. Hanson MR, Sattarzadeh A (2011) Stromules: recent insights into a long neglected feature of plastid morphology and function. Plant Physiol 155:1486–1492
99. Holzinger A, Buchner O, Lutz C, Hanson MR (2007) Temperature-sensitive formation of chloroplast protrusions and stromules in mesophyll cells of Arabidopsis thaliana. Protoplasma 230:23–30
100. Kohler RH, Cao J, Zipfel WR, Webb WW, Hanson MR (1997) Exchange of protein molecules through connections between higher plant plastids. Science 276: 2039–2042
101. Sattarzadeh A, Fuller J, Moguel S, Wostrikoff K, Sato S, Covshoff S, Clemente T, Hanson M, Stern DB (2010) Transgenic maize lines with cell-type specific expression of fluorescent proteins in plastids. Plant Biotechnol J 8:112–125

Part II

Nicotiana tabacum the Model Species of Chloroplast Biotechnology

Chapter 8

Plastid Transformation in *Nicotiana tabacum* and *Nicotiana sylvestris* by Biolistic DNA Delivery to Leaves

Pal Maliga and Tarinee Tungsuchat-Huang

Abstract

The protocol we report here is based on biolistic delivery of the transforming DNA to tobacco leaves, selection of transplastomic clones by spectinomycin resistance and regeneration of plants with uniformly transformed plastid genomes. Because the plastid genome of *Nicotiana tabacum* derives from *Nicotiana sylvestris*, and the two genomes are highly conserved, vectors developed for *N. tabacum* can be used in *N. sylvestris*. Also, the tissue culture responses of *N. tabacum* cv. Petit Havana and *N. sylvestris* accession TW137 are similar, allowing plastid engineering protocols developed for *N. tabacum* to be directly applied to *N. sylvestris*. However, the tissue culture protocol is applicable only in a subset of *N. tabacum* cultivars. Here we highlight differences between the protocols for the two species. We describe updated vectors targeting insertions in the unique and repeated regions of the plastid genome as well as systems for marker excision. The simpler genetics of the diploid *N. sylvestris*, as opposed to the allotetraploid *N. tabacum*, make it an attractive model for plastid transformation.

Key words AAD, Aminoglycoside-3″-adenylyltransferase, Chloroplast transformation, Kanamycin selection, Neomycin phosphotransferase, *Nicotiana tabacum*, *Nicotiana sylvestris*, NPTII, Plastid transformation, ptDNA, Spectinomycin selection, Tobacco

1 Introduction

The plastid genome (ptDNA) of *Nicotiana tabacum* is relatively small (155 kb), encodes 112 genes [1], and is present in up to 10,000 copies per cell [2]. The protocol we use for the transformation of the tobacco plastid genome is based on biolistic delivery of the transforming DNA into leaf cells on the surface of gold particles, integration of the transforming DNA into the plastid genome by homologous recombination and selective enrichment of the transformed genome copies during shoot regeneration on a selective medium. Spectinomycin or kanamycin are selective agents inhibiting cell proliferation, chlorophyll biosynthesis and shoot regeneration from leaf sections when they are cultured on RMOP shoot regeneration medium. Transplastomic clones are identifiable by the absence of the detrimental effects of the antibiotics: green

Pal Maliga (ed.), *Chloroplast Biotechnology: Methods and Protocols*, Methods in Molecular Biology, vol. 1132,
DOI 10.1007/978-1-62703-995-6_8, © Springer Science+Business Media New York 2014

callus and shoots on the regeneration medium. Because shoot regeneration occurs early in clonal history, the shoots regenerating from leaf cells are heteroplastomic, often having wild-type and transplastomic cells in the different leaf cell layers [3].

Plastid transformation in tobacco was first carried out in this laboratory in *Nicotiana tabacum* cv. Petit Havana [4]. Because of its relatively short life cycle (3–4months from seed to seed), and the speed of shoot regeneration from leaf explants (2–3 weeks), this variety is widely used as the model plant of plastid engineering. *N. tabacum* has a drawback: it is an allotetraploid species, with most genes present in two copies derived from the parental species *Nicotiana sylvestris* and *Nicotiana tomentosiformis*. We use *N. sylvestris* line T137 as the diploid alternative to *N. tabacum*. *N. sylvestris* TW137 is a long-day plant forming rosettes until flowering is induced whereas *N. tabacum* cv. Petit Havana is day neutral and the internodes elongate as the plants grow. The seed-to-seed generation time of *N. sylvestris* is about the same as that of N. *tabacum* cv Petit Havana, but it is more difficult to grow, because it is more sensitive to diseases and pests. With the protocol described here it takes 4–6 months to obtain a genetically stable homo-transplastomic plant in *N. tabacum*. The same process in *N. sylvestris* takes about 2 months longer. The longer time requirement in *N. sylvestris* is due to slower shoot regeneration and rooting, processes that may be optimized in the future.

Plastid genome manipulations include insertion of the marker gene (and linked genes-of-interest) into the plastid genome, post-transformation excision of the marker gene, replacement of wild-type genes with mutant forms, gene knockouts by replacement of plastid genes with selectable markers genes, and co-transformation based on transformation with two plasmids and recovering a non-selectable gene encoded in one of the plasmids by selection for the marker gene encoded in the second plasmid. In this chapter we review only our currently recommended insertion vectors and describe a protocol for plastid transformation by selection for spectinomycin resistance. This article is an update of earlier protocols on plastid transformation and CRE-mediated marker excision in *N. tabacum* [5], transformation of the plastid genome to study RNA editing [6] and plastid transformation in *N. sylvestris* [7]. For additional information on the technology and its applications see recent general reviews [8–12], specialized reviews on marker gene excision [13, 14], and applications in biotechnology [15–17].

1.1 Marker Genes

Our first generation vectors conferred spectinomycin resistance by replacing the spectinomycin sensitive 16S rRNA (*rrn16*) gene in the plastid genome with a resistant allele carried by the vector [4]. Transformation efficiency was significantly increased when the recessive *rrn16* resistance gene was replaced with the dominant *aadA* gene, which encodes the aminoglycoside-3″-adenylyltransferase

(AAD) inactivating enzyme [18]. A variant spectinomycin resistance gene encodes an AAD-GFP fusion protein so that transplastomic clones can be selected by spectinomycin resistance in culture and plastid transformation confirmed by GFP accumulation in chloroplasts [19]. Another variant of the *aadA* gene, *aadA*^{au} confers spectinomycin resistance to cells in culture and a golden pigment phenotype to plants [20]. Spectinomycin selection yields about one transplastomic clone per bombarded sample using the standard PDS-1000/He biolistic gun, and 4–5 transplastomic clones per bombardment with the Hepta Adaptor. Low levels of AAD (<1 % of total soluble cellular protein) are sufficient to recover transplastomic clones [21].

In contrast, recovery of transplastomic clones by selection for kanamycin resistance requires high-level expression of the marker gene. Kanamycin resistance is based on inactivation of the antibiotic by neomycin phosphotransferase (NPTII). Our first generation kanamycin resistance gene was inefficient (~1 % NPTII), yielding only one transplastomic clone in 50 bombarded samples [22]. Bombardment with highly expressed kanamycin resistance genes (NPTII is ≥5 % of total soluble cellular protein) significantly improved plastid transformation efficiency with vectors carrying a kanamycin resistance as marker gene [23] now approaching the efficiency of spectinomycin selection.

1.2 Vectors Targeting Insertions at Alternative Sites

Plastid transformation is based on homologous recombination between ptDNA flanking the marker gene and the plastid genome. Depending on the site of recombination between the vector and plastid genome sequences, the sequence in the plastid genome may derive from the vector or be the native sequence [21]. Figure 1 shows a vector set targeting insertions in the repeated region of plastid genome between the *trnV* and 3′-rps12-rps7 operon. Figure 2 depicts vectors targeting insertions in the *trnI-trnA* and *rbcL-accD* intergenic regions. The unifying theme of vectors in Fig. 3 is the option for post-transformation removal of marker genes by the P1 phage Cre/*loxP* and PhiC31 phage INT/*attP-attB* site specific recombination system. Table 1 lists the vectors which are suitable for the expression of the Cre or Int site-specific recombinase, for the excision of *loxP* or *attP/attB* flanked marker genes, respectively.

Most of the vectors carry a selectable spectinomycin resistance (*aadA*) gene. A black bar at the *aadA* C-terminal end symbolizes a c-myc tag facilitating detection of AAD, the *aadA* gene product on protein gel blots. Please note that early vectors, such as pPRV111A, carry the unintended Spc1 spectinomycin resistance mutation in the *rrn16* gene, encoding the spectinomycin resistant 16S rRNA [24]. We became aware of this when excision of *aadA* gene did not yield spectinomycin sensitive plants. The Spc1 mutation has been removed from the targeting region of excision

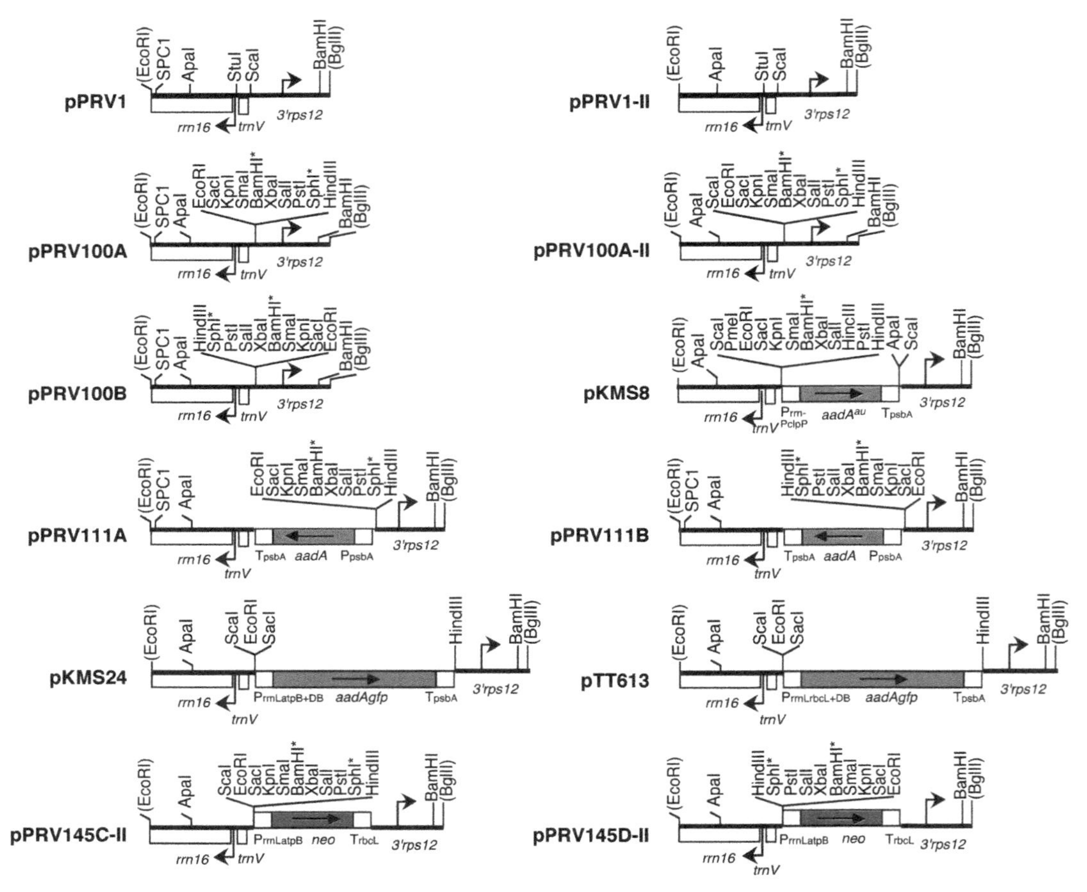

Fig. 1 Plastid transformation vectors targeting insertions in the *trnV* and *3′-rps12* intergenic region located in the inverted repeat region of ptDNA. Shown here are plasmid pPRV1 (GenBank Accession No. U12809); plasmid pPRV100A (GenBank Accession No. U12810); plasmid pPRV100B (GenBank Accession No. U12811); plasmid pKMS8 (GenBank Accession No. HQ023426); plasmid pPRV111A (GenBank Accession No. U12812); plasmid pPRV111B (GenBank Accession No. U12813); plasmid pPRV145C-II (GenBank Accession No. EU224422); plasmid pPRV145D-II (GenBank Accession No. EU224423). Plasmids pKMS24 and pTT613 are similar to plasmids pMSK56 and pKMS57 [19], respectively, and differ only by a few restriction sites

vectors and the modified vectors are identified by the –II extension after the original plasmid name, such as the pPRV123Blox-II vector shown in Fig. 3. The flanking sequences typically integrate, with about equal frequency, at sites adjacent or distal to the marker gene [21]. If *aadA* integration involves recombination adjacent to *aadA* (Fig. 1), the Spc1 mutation will be absent in the transplastome, an event that occurs about 50 % of the time.

Because the *N. tabacum* and *N. sylvestris* plastid genomes are highly similar (there are only seven ptDNA polymorphic sites between the two species) [25] plastid transformation vectors carrying *N. tabacum* plastid genome targeting sequences can also be used for plastid transformation in *N. sylvestris.*

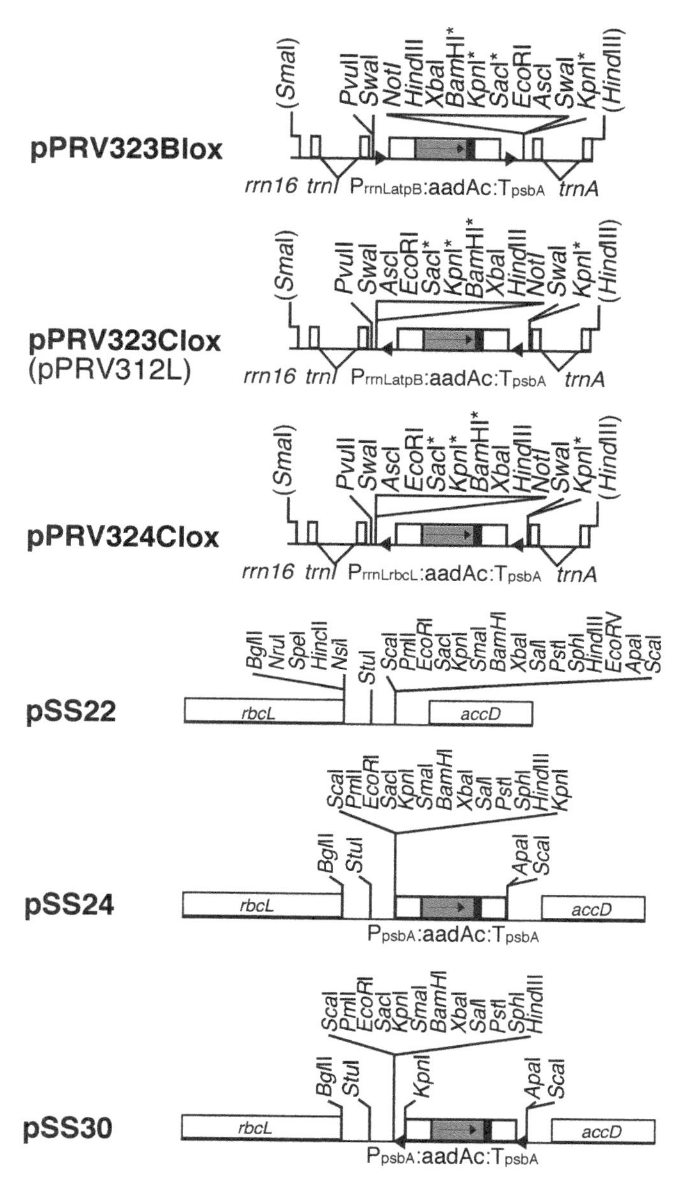

Fig. 2 Plastid transformation vectors targeting insertions in the *trnI-trnA* (inverted repeat region) and *rbcL-accD* (large unique region) of the plastid genome. Shown here are plasmid pPRV323Blox (GenBank Accession No. EU224429); plasmid pPRV323Clox (pPRV312L) (GenBank Accession No. DQ489715); plasmid pPRV-324Clox (GenBank Accession No. EU224430); plasmid pSS22 (GenBank Accession No. FJ416604); plasmid pSS24 (GenBank Accession No. FJ416605); plasmid pSS30 (GenBank Accession No. FJ416606). *Triangles* and *boxed* B,B′ and P,P′ mark *loxP* and *attP*/*attB* sites for post-transformation excision of the flanked *aadA* marker gene by the Cre and PhiC31 recombinases, respectively

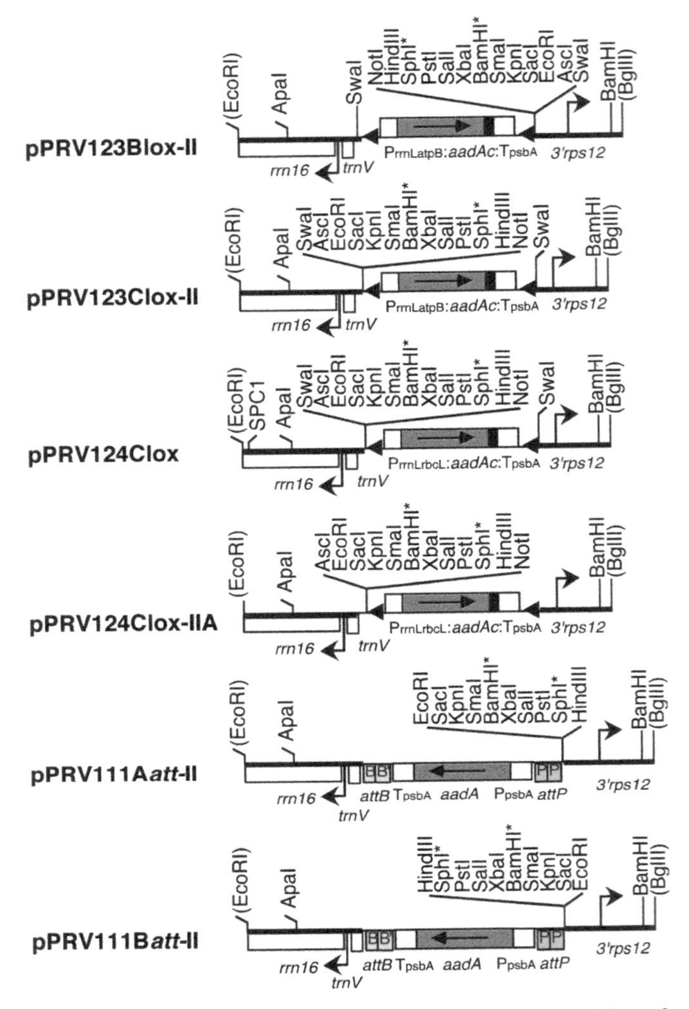

Fig. 3 Plastid transformation vectors for post-transformation excision of marker genes. Shown here are the maps of plasmid pPRV123Blox-II (GenBank Accession No. EU224424); plasmid pPRV123Clox-II (GenBank Accession No. EU224425); plasmid pPRV124Clox (GenBank Accession No. EU224426); plasmid pPRV111A*att*-II (GenBank Accession No. EF416277); plasmid pPRV111B*att*-II (GenBank Accession No. EF416276). *Triangles* flanking *aadA* in plasmids pPRV323Blox, pPRV323Clox and pPRV324Clox mark *loxP* sites for post-transformation excision of the marker gene

1.3 Need for Cultivar-Specific Protocols

The tissue culture response of *N. tabacum* cv. Petit Havana and *N. sylvestris* TW137 [7] is very similar, thus the protocols of tissue culture selection and plant regeneration are applicable to both of these genetic lines. The cv. Petit Havana plants are relatively small. To obtain plants with a larger biomass, the tissue culture protocols have been modified to enable plastid transformation in cultivars Wisconsin 38 [26], Xanthi, Burley [27], Samsun, K327 (22X-1; [28]

Table 1
Agrobacterium binary vectors for Cre or Int expression in chloroplasts

Plasmid	Agro vector	Plant marker gene	Plant selection	Recombinase[a]				Reference
				P	TP + aa	Rec	3′-UTR	
pKO27	pPZP212	*neo*	Kanamycin 100 mg/L	P2′	TP + 22aa	*Cre*	Tnos	[36]
pKO28	pPZP212	*neo*	Kanamycin 100 mg/L	P2′	TP + 5aa	*Cre*	Tnos	[36]
pKO30	pPZP222	*aacCI*	Gentamycin 100 mg/L	P2′	TP + 22aa	*Cre*	Tnos	[37]
pKO31	pPZP222	*aacCI*	Gentamycin 100 mg/L	P2′	TP + 5aa	*Cre*	Tnos	[37]
pKO117	pPZP222	*aacCI*	Gentamycin 100 mg/L	P2′	TP + 22aa	*Int*	Tnos	[23]

[a]Abbreviations: *P* promoter, *TP+aa* the Rubisco small subunit transit peptide (TP) and the number of rubisco N-terminal amino acids fused with the recombinase, *3′-UTR* 3′ untranslated region

and Maryland Mammoth [29]. Plastid transformation has also been reported in *Nicotiana plumbaginifolia* [30] and *Nicotiana benthamiana* [31].

2 Materials

2.1 Plant Propagation

1. Desiccator jar.
2. Commercial Bleach.
3. Concentrated HCl.
4. Eppendorf tubes.
5. Deep (100 × 20 mm) petri dishes containing RM plant maintenance medium. The RM plant maintenance medium (MS medium, ref. [32]) per liter contains: 100 ml 10× macronutrients, 10 ml 100× micronutrients, 5 ml 1 % Fe-EDTA, 30 g sucrose, pH 5.6–5.8 with 1 M KOH; 7 g agar). RM Medium 10× Macronutrient solution per liter contains: 19 g KNO_3, 3.7 g $MgSO_4 \cdot 7H_2O$, 4.4 g $CaCl_2 \cdot 2H_2O$, 1.7 g KH_2PO_4, 16.5 g $(NH_4)NO_3$. RM Medium 100× Micronutrient solution per liter contains: 169 mg $MnSO_4 \cdot H_2O$, 62 mg H_3BO_3, 86 mg $ZnSO_4 \cdot 7H_2O$, 8.3 mg KI, 2.5 ml $Na_2MoO_4 \cdot 2H_2O$ (1 mg/ml), 2.5 ml $CuSO_4 \cdot 5H_2O$ (1 mg/ml), 0. 25 ml $CoCl_2 \cdot 6H_2O$ (1 mg/ml). Agar, plant tissue culture tested (Sigma, St Louis, MO; Catalog No. A7921). Add 50 ml RM plant maintenance medium to deep (100 × 20 mm) petri dishes and 60 ml medium to Magenta boxes.
6. Magenta boxes containing 60 ml RM plant maintenance medium. For medium composition *see* Subheading 2.1, **item 5**.
7. Sterile scalpels in container.
8. Sterile forceps in metal box.

2.2 Preparation of DNA-Coated Gold Particles

1. 0.6 μm gold microcarriers available from Bio-Rad (Catalog No. 165-2262).
2. Plasmid DNA prepared using the QIAGEN Maxi Kit (Qiagen Inc., Valencia, CA).
3. Ice-cold 70 % ethanol.
4. Ice-cold 100 % ethanol.
5. Ice-cold sterile distilled water.
6. 50 % glycerol, sterilized by autoclaving.
7. 2.5 M $CaCl_2$ solution.
8. 0.1 M spermidine free base (Sigma, St Louis, MO; Catalog No. S4139).

2.3 DNA Delivery by the Biolistic Process

1. Bio-Rad PDS1000/He biolistic gun (Bio-Rad Laboratories, Hercules, CA; Catalog No. 165-2257).
2. Hepta Adaptor for PDS-1000/He system (Bio-Rad Laboratories, Hercules, CA; Catalog No. 165-2225).
3. Macrocarrier disk for PDS1000/He biolistic gun (Bio-Rad Catalog No. 165-2335).
4. Rupture disk (1,100 psi) for PDS1000/He biolistic gun (Bio-Rad Catalog No. 165-2329).
5. Stopping screen for standard PDS1000/He biolistic gun (Bio-Rad Catalog No. 165-2336).
6. Stopping screen for PDS1000/He biolistic gun with Hepta Adaptor (Bio-Rad Catalog No. 165-2226).
7. ThermoSavant VLP285 vacuum pump for the gun (ThermoSavant, Holbrook, NY).
8. Pressurized helium in tank, 99.999 % pure, moisture free.
9. Laminar flow hood for bombardment and tissue culture.
10. Whatman No. 4 filter paper disks sterilized by autoclaving in glass petri dish.
11. Sterile *N. tabacum* cv. Petit Havana and *N. sylvestris* TW137 plants in Magenta boxes grown on RM plant maintenance medium obtained as described in Subheading 3.1.
12. Petri plate (100 × 15 mm) containing solid RMOP medium (20 ml) for leaf bombardment. The RMOP shoot regeneration medium [33] per liter contains: 100 ml RM medium 10× macronutrients, 10 ml RM medium 100× micronutrients, 5 ml 1 % Fe-EDTA, 1 ml thiamine (1 mg/ml), 0.1 ml alpha-naphthaleneacetic acid (NAA 1 mg/ml in 0.1MNaOH), 1 ml 6-benzylaminopurine (BAP 1 mg/ml in 0.1 M HCl), 0.1 g myo-inositol, 30 g sucrose, pH 5.8 with 5 M KOH, 7 g agar.
13. Deep (100 × 20 mm) petri plates with 50 ml selective RMOP medium. The selective RMOP medium contains filter sterilized

500 mg/l spectinomycin dihydrochloride (Sigma, St Louis, MO, Catalog No. S4014), or 500 mg/l streptomycin sulfate (Sigma, St Louis, MO, Catalog No. S9137), or both. For the composition of RMOP medium *see* Subheading 2.3, **item 12**.

14. 70 % ethanol.
15. Plastic wrap (strips) that is permeable to gas exchange, for example Glad ClingWrap.

2.4 Identification of Transplastomic Events

1. Deep (100×20 mm) petri dishes containing RM plant maintenance medium (*see* Subheading 2.1, **item 5**).
2. Deep (100×20 mm) petri dishes containing a selective spectinomycin medium (500 mg/l; Sigma, St Louis, MO; Catalog No. S9007) RMOP plant regeneration medium (*see* Subheading 2.3, **item 13**).
3. Deep (100×20 mm) petri dishes containing 500 mg/l each of spectinomycin (Sigma, St Louis, MO; Catalog No. S9007) and streptomycin (Sigma, St Louis, MO; Catalog No. S9137) in the RMOP plant regeneration medium (*see* Subheading 2.3, **item 13**).
4. Magenta boxes containing selective spectinomycin (500 mg/l; Sigma, St Louis, MO; Catalog No. S9007) RM plant maintenance medium (*see* Subheading 2.3, **item 12**).
5. Deep (100×20 mm) petri dishes containing RM plant maintenance medium (*see* Subheading 2.1, **item 5**).

2.5 Transfer of Plants to Greenhouse

1. Captan (Bonide Products Inc., Oriskany, NY), a fungicide; active ingredient *N*-trichloromethylthio-4-cyclohexane-1, 2-dicarboximide.
2. Sand and garden soil.
3. Plastic trays (10″×20″) with holes on bottom and transparent plastic domes.
4. Pots.

3 Methods

3.1 Plant Propagation

1. Sterilize seed in open Eppendorf tubes placed in a rack inside a desiccator jar by exposure to chlorine vapor generated by adding 3 ml concentrated HCl to 100 ml bleach in a 250 ml beaker. Vapor-phase sterilization should be carried out in a fume hood for 3–12 h. Seal tubes when opening the desiccator. Sterile seed may be stored, but chlorine vapor fume should be removed by opening the tube in sterile hood (*see* **Note 1**).
2. Germinate about 100 seeds in deep (100×20 mm) petri dishes containing RM plant maintenance medium. Sprinkle dry seed

on medium and incubate the dishes in the culture room at 26 °C illuminated for 16 h per day. When seedlings reach 1 cm in size, cut off shoots with sterile scalpels holding stem with sterile forceps. Root shoots in fresh plates by inserting them in fresh medium in deep dishes (20 seedlings per plate). Leaf size increases faster if root system is removed.

3. Transfer shoots to Magenta boxes containing RM plant maintenance medium (one shoot per box) and let them grow into plants. When shoot tip riches lid, prepare one-node cuttings with scalpel (cut stem using sterile petri dishes as platform in laminar flow hood) and grow new plants by inserting cuttings with forceps in fresh RM plant maintenance medium.
4. Repeat process of propagation by cuttings to maintain a stock of sterile plants. The internodes of *N. tabacum* will be 1–2 cm long. *N. sylvestris* grows as a rosette plant with short internodes. *N. tabacum* cuttings will grow into plants in 3–4 week; *N. sylvestris* in 5–6 weeks.
5. Conservatively, you will need one plant (four to five leaves) for each bombardment with the Hepta Adaptor. You may harvest the leaves for bombardment and use the stem with axillary bud for plant propagation. Use buds only from the lower half of the stem to keep newly developing shoots in the vegetative phase. You may cut back the shoot at the time of leaf harvest and let a new shoot develop from a bud at the base for a second harvest of leaves.
6. The stock plants can be maintained by propagation for 12–18 months. Loss of vigorous growth is the sign indicating that it is time to initiate a fresh stock. If buds are used for propagation from the upper half of the plants, the shoot will enter the flowering phase producing small narrow leaves, shortening the useful life of the stock.

3.2 Preparation of DNA-Coated Gold Particles

This protocol is based on Bio-Rad Bulletin 9075.

1. First prepare gold particles. Weigh out 30 mg 0.6 μm gold microcarrier in a 1.5 ml Eppendorf tube and add 1 ml ice-cold 70 % ethanol. Shake tube vigorously in a Vortex microtube holder for 5 min, and then let the particles settle in the tube for 15 min at room temperature (20–25 °C).
2. Spin the tube in microcentrifuge at 3,000 rpm (600 × *g*) for 1 min to sediment and compact gold. Remove the ethanol with a pipette. Add 1 ml ice-cold sterile distilled water, suspend the particles by vortex, and then allow the gold particles to settle at room temperature for 10 min.
3. Compact the gold particles by spinning in a microcentrifuge at 3,000 rpm (600 × *g*) for 1 min. Remove the water with a pipette and add 1 ml ice-cold sterile distilled water. Suspend the gold

particles by vortex, and then allow them to settle at room temperature for 10 min.

4. Repeat washing the gold particles with water by repeating **step 3**.
5. Spin the tube in a microcentrifuge at 5,000 rpm (1,700 × *g*) for 15 s, then remove water completely. Add 500 μl 50 % glycerol and vortex for 1 min to resuspended particles. Gold concentration will be 60 mg/ml. The clean gold can be stored for 2 weeks at room temperature.
6. You may use freshly prepared gold particles, or stored gold for DNA coating. If using stored gold, vortex tube for 5 min before coating with DNA. To start DNA coating, place Eppendorf tube containing clean gold in a Vortex microtube holder and shake at setting 3. While tube is shaking, remove 50 μl aliquots of gold and pipette into ten 1.5 ml Eppendorf tubes in a rack. (*See* **Note 2** to calculate the number of bombardments needed.) While tubes are shaking, add 5 μl plasmid DNA (1 μg/μl), 50 μl 2.5 M $CaCl_2$ and 20 μl 0.1 M spermidine free base. Make sure to add the components in this order and that the contents are thoroughly mixed before adding the next component.
7. Shake tubes on Vortex for 5 min at setting 3.
8. Sediment gold by spinning in microcentrifuge at 3,000 rpm (600 × *g*) for 1 min. Remove the supernatant with a pipette and add 140 μl 70 % ethanol. Tap tubes lightly until the pellet just starts to come into solution to make sure pellet is not tightly packed. If gold does not go into solution by gently tapping the tube, break up pellet by pipetting up and down.
9. Sediment gold by spinning in a microcentrifuge at 3,000 rpm for 1 min. Remove supernatant and add 140 μl ice-cold 100 % ethanol to each tube. Lightly tap tube until the pellet just starts to come into solution.
10. Sediment gold by spinning in a microcentrifuge at 5,000 rpm (1,700 × *g*) for 15 s. Resuspend coated gold pellet in 50 μl 100 % ethanol by gently tapping tube. Pellet should easily enter solution. Shake tubes at setting 3 while waiting to use them for bombardment. If tubes are sitting for a long period of time before bombardment, replace ethanol in tube with fresh 100 % ethanol.

3.3 DNA Delivery by the Biolistic Process

1. Prepare the leaves for bombardment. Put two sterile Whatman No. 4 filter papers on top of solid RMOP medium (20 ml) in a 100 × 15 mm petri plate, then place sterile leaves abaxial side up (flat upper surface is laying on the medium) on them. Use one leaf if bombarding with the standard PDF1000/He gun and cover the entire plate surface if using the Hepta Adaptor.

Leaves of greenhouse-grown plants are also suitable for plastid transformation (*see* **Note 3**). Leaves may be stored on top of the filter paper in a "ready-to-shoot" state for at least a day.

2. Set up the biolistic gun in a sterile laminar flow hood. Sterilize the main chamber, rupture disk retaining cap, microcarrier launch assembly (Hepta Adaptor, if used) and the target shelf by wiping with a tissue soaked in 70 % ethanol.
3. Sterilize rupture disks (1,100 psi), macrocarriers, macrocarrier holders, and stopping screens by soaking in 100 % ethanol (5 min) then air-dry them in the laminar flow hood in an open petri dish.
4. Turn on helium tank and set pressure in regulator (distal to tank) for 1,300 psi (200–300 psi above rupture disk value).
5. Turn on vacuum pump and gene gun. Set the vacuum rate on the gene gun to 7 and the vent rate to 2.
6. Pipette 10 μl of DNA-coated gold onto the macrocarrier or "flying disk" in holder and let the disk air-dry for 5 min. 5 samples may be made up at one time if bombardment is carried out with one macrocarrier (or seven, if Hepta Adaptor is used).
7. Place rupture disk into retaining cap and screw in place tightly.
8. Put stopping screen and flying disk (face down) in microcarrier launch assembly and place in chamber just below rupture disk. For description see Bio-Rad Bulletin 9075.
9. Place leaf on thin RMOP plate into chamber 9 cm (fourth shelf from top) below the microcarrier launch assembly and close the door.
10. Press vacuum button to open valve. When vacuum reaches 28 in. Hg hold down fire button until the pop from the gas breaking the rupture disk is heard (*see* **Note 4**).
11. Release vacuum immediately and remove the leaf sample.
12. Repeat **steps 6–11** until all leaf samples are bombarded. When finished, turn off helium tank, and release pressure by holding down the fire button while vacuum is on. Turn off vacuum pump.
13. Place clear plastic sleeves over plates containing bombarded leaf samples and incubate in culture room. Incubation allows time for marker gene expression before selection is started.
14. After 2 days, cut bombarded leaves into small (1 cm square) pieces and place seven pieces per plate abaxial side up on selective RMOP medium (500 mg/l spectinomycin) in deep plates (100 × 20 mm) (*see* **Note 5**).
15. Seal each plate with plastic wrap that is permeable to gas exchange and incubate plates in culture room at 26 °C, illuminated for 16 h.

Fig. 4 The phenotype of transplastomic clones in bombarded leaf cultures. Note single shoot in *N. tabacum* (**a**) and bushy callus with many shoots in *N. sylvestris* (**b**) plates. In sterile culture *N. tabacum* plants (**c**) have longer internodes than *N. sylvestris* (**d**), which tend to grow as rosettes

3.4 Identification of Transplastomic Events

1. Incubate plates in culture room at 26 °C, with a 16 h/8 h light-dark cycle. Sensitive cells of the bombarded leaves bleach and from scanty callus that is yellow or brown in color on the selective RMOP plant regeneration medium containing 500 mg/L spectinomycin HCl. Cells carrying transformed spectinomycin resistant plastids turn green and overcome the inhibition of shoot regeneration by spectinomycin. Putative transplastomic clones appear as green shoot only, green callus only, or green callus with shoots 4–12 weeks after bombardment (Fig. 4). For designation of clones (*see* **Note 6**). However, mutations in the 16S rRNA also confer a similar spectinomycin resistance phenotype, as does expression of the *aadA* gene in the nucleus [18]. Positive identification of transplastomic clones is the next task.
2. Transfer small (1 cm^2) leaf sections of the regenerated shoots or callus pieces onto (a) selective streptomycin–spectinomycin RMOP medium and (b) spectinomycin RMOP medium (*see* **Note 7**). Transgenic clones carrying an *aadA* gene are resistant to both spectinomycin and streptomycin, whereas

spontaneous spectinomycin resistant mutants are resistant only to spectinomycin [18]. Resistance is manifested as formation of green calli with regenerating shoots; sensitivity is indicated by formation of scanty white callus in the leaves. However, integration of *aadA* derived from the plastid transformation vector in the nucleus will also confer spectinomycin and streptomycin resistance. Positively confirm integration of transgenes in the plastid genome by PCR and DNA gel blot analyses (*see* Chapter 12). Evaluate plants from three to four independently transformed clones because ~10 % of plants regenerated in tissue culture are sterile due to somaclonal variation.

3. Repeat plant regeneration on the selective RMOP medium and verify uniform transformation of ptDNA by gel blot analyses. Plants regenerated twice on selective spectinomycin medium are normally homoplastomic.
4. Root shoots on a plant maintenance (MS) medium in sterile culture.

3.5 Transferring Plants to Greenhouse and Testing Maternal Inheritance

1. To transfer rooted transplastomic *N. tabacum* shoots to greenhouse gently break up the agar, wash roots in running tap water to remove the agar-solidified RM medium and plant in soil. Cover plants with a transparent dome to preserve moisture, and grow plants in shade. After a week remove plastic dome and expose the plants to full sunlight. When planting *N. sylvestris*, drench the soil after planting with the fungicide Captan to fend off fungal infection. If the *N. sylvestris* plants have no roots, dip the cut stem in commercial rooting hormone, plant the shoots in fine sand, drench the sand with Captan and cover the pots with a transparent plastic dome. Grow the plants in shade until new leaves develop, then treat them as rooted shoots.
2. Collect mature seedpods from selfing and crosses with wild type.
3. Germinate surface-sterilized seeds (*see* Subheading 3.1, **step 1**) on selective RM plant-maintenance medium containing (500 mg/l) spectinomycin. Seedlings carrying transformed ptDNA will be dark green, whereas sensitive seedlings will be white. One hundred percent green seedlings from the cross with transplastomic line as maternal parent confirms uniform population of transformed ptDNA in the transplastomic plants.

4 Notes

1. Vapor sterilization is preferable, when handling many samples. The disadvantages of vapor sterilization are its relatively long duration (hours) and potential for seed damage if residual moisture is present in the seed. Small seed samples can be

sterilized by a rapid, wet procedure in an Eppendorf tube. For wet sterilization place seed in tube, wet the seed with one drop of 70 % ethanol, immediately add 1 ml of 10-times diluted (0.6 %) Clorox bleach. After 3 min remove bleach with pipette and rinse seed five times with sterile distilled water to remove bleach. Spread seed on the surface of plate in last rinse, then tilt plate and remove water with pipette. A small seed sample of *N. tabacum* cv. Petit Havana would be 100 μl, containing about 500 seeds. The volume of about 500 *N. sylvestris* seed is 50 μl. A well-developed *N. tabacum* capsule contains 500–600 μl seed, an *N. sylvestris* capsule about 400 μl seed.

2. Each tube (30 mg) of gold is sufficient to prepare 50 macrocarriers. The standard PDS1000/He device uses one marocarrier per bombardment; the Hepta Adaptor accommodates seven. One standard bombardment, on average, yields one transplastomic clone whereas one bombardment with the Hepta Adaptor yields 4–5 transplastomic clones when selection is carried out for spectinomycin resistance. Note, however, that we bombard a single leaf with the standard device and a full plate if using Hepta Adaptor (entire plate surface covered by leaves, usually three or four of them).
3. For bombardment, select the first and second fully expanded leaves of rapidly growing plants that are not yet flowering. Surface-sterilize the leaves by rinsing in tap water containing a drop of surfactant (Tween 80 or liquid soap) to remove dirt; dipping the leaves in 70 % ethanol; and placing them in diluted commercial bleach (10-fold diluted Clorox; final sodium hypochlorite concentration is ~0.6 %) for 3 min. The bleach should be thoroughly removed by five times rinsing the leaves in sterile distilled water.
4. Vacuum is critical to achieve plastid transformation. If the gun is fired at a lower vacuum pressure no transplastomic lines will be obtained, because the DNA-coated particles will lack the momentum to penetrate the cells. We recommend testing the success of particle coating and DNA delivery in a transient expression assay, by histochemical staining for β-glucuronidase activity after bombardment with a nuclear *uidA* gene [34, 35].
5. You cannot make a mistake by placing too few pieces per plate, only by placing too many. Sections of one bombarded leaf are normally selected in three to five plates. If the leaf sections are too large, there will be insufficient nutrient in the medium to support growth for up to 12 weeks, the time frame within which transplastomic clones appear. Overcrowding may be also caused by less than the desired 50 ml culture medium in a deep plate. Diagnostic sign of overcrowding is absence of spontaneous spectinomycin resistant mutants, of which you should find up to five in a sample of 30 leaves. If spontaneous mutants are

found but no transplastomic clone is obtained, the problem is with the DNA coating.

6. Each shoot or callus at a distinct location derives from an independent event; therefore we treat them as independently derived clones. We identify independent clones by the initials of the species, the plasmid name and a serial number. Plants are typically regenerated 2× on selective spectinomycin medium to obtain homoplastomic plants. Letters are assigned to regenerated shoots (plants) in each cycle to individually track their history. Plant designated Ns-pCK2-6AB is *N. sylvestris* transformed with plasmid pCK2, is serial number (or event) 6 that was regenerated twice on selective medium.
7. Resistance to streptomycin and spectinomycin indicates the presence of *aadA* gene. However, the two antibiotics together inhibit shoot formation. Therefore, we score the presence of *aadA* gene by resistance to 500 mg/l streptomycin and 500 mg/l spectinomycin, but pick shoots for further studies from cultures on 500 mg/l spectinomycin.

Acknowledgements

This work was supported by grants from the USDA Biotechnology Risk Assessment Research Grant Program Award No. 2005-33120-16524, 2008-03012 and 2010-2716.

References

1. Shinozaki K, Ohme M, Tanaka M, Wakasugi T, Hayashida N, Matsubayashi T, Zaita N, Chunwongse J, Obokata J, Yamaguchi-Shinozaki K, Ohto C, Torazawa K, Meng B-Y, Sugita M, Deno H, Kamogashira T, Yamada K, Kusuda J, Takaiwa F, Kato A, Tohdoh N, Shimada H, Sugiura M (1986) The complete nucleotide sequence of the tobacco chloroplast genome: its gene organization and expression. EMBO J 5:2043–2049
2. Shaver JM, Oldenburg DJ, Bendich AJ (2006) Changes in chloroplast DNA during development in tobacco, *Medicago truncatula*, pea, and maize. Planta 224:72–82
3. Lutz KA, Maliga P (2008) Plastid genomes in a regenerating tobacco shoot derive from a small number of copies selected through a stochastic process. Plant J 56:975–983
4. Svab Z, Hajdukiewicz P, Maliga P (1990) Stable transformation of plastids in higher plants. Proc Natl Acad Sci U S A 87:8526–8530
5. Lutz KA, Svab Z, Maliga P (2006) Construction of marker-free transplastomic tobacco using the Cre-*loxP* site-specific recombination system. Nat Protoc 1:900–910
6. Lutz KA, Maliga P (2007) Transformation of the plastid genome to study RNA editing. Methods Enzymol 424:501–518
7. Maliga P, Svab Z (2011) Engineering the plastid genome of *Nicotiana sylvestris*, a diploid model species for plastid genetics. In: Birchler JJ (ed) Plant chromosome engineering: methods and protocols. Springer Science + Business Media, LLC, New York, pp 37–50
8. Maliga P (2012) Plastid transformation in flowering plants. In: Bock R, Knoop V (eds) Genomics of chloroplasts and mitochondria. Springer, New York, pp 393–414
9. Maliga P (2004) Plastid transformation in higher plants. Annu Rev Plant Biol 55: 289–313
10. Koop HU, Herz S, Golds TJ, Nickelsen J (2007) The genetic transformation of plastids. In: Bock R (ed) Cell and Molecular Biology of Plastids. Springer Verlag, Berlin, Heidelberg, pp 457–510

11. Day A (2012) Reverse genetics in flowering plant plastids. In: Bock R, Knoop V (eds) Genomics of chloroplasts and mitochondria. Springer Science + Business Media, LLC, New York, pp 415–441
12. Bock R (2001) Transgenic plastids in basic research and plant biotechnology. J Mol Biol 312:425–438
13. Day A, Goldschmidt-Clermont M (2011) The chloroplast transformation toolbox: selectable markers and marker removal. Plant Biotechnol J 9:540–553
14. Lutz KA, Maliga P (2007) Construction of marker-free transplastomic plants. Curr Opin Biotechnol 18:107–114
15. Maliga P, Bock R (2011) Plastid biotechnology: food, fuel and medicine for the 21st century. Plant Physiol 155:1501–1510
16. Bock R (2007) Plastid biotechnology: prospects for herbicide and insect resistance, metabolic engineering and molecular farming. Curr Opin Biotechnol 18:100–106
17. Daniell H, Chebolu S, Kumar S, Singleton M, Falconer R (2005) Chloroplast-derived vaccine antigens and other therapeutic proteins. Vaccine 23:1779–1783
18. Svab Z, Maliga P (1993) High-frequency plastid transformation in tobacco by selection for a chimeric *aadA* gene. Proc Natl Acad Sci U S A 90:913–917
19. Khan MS, Maliga P (1999) Fluorescent antibiotic resistance marker to track plastid transformation in higher plants. Nat Biotechnol 17:910–915
20. Tungsuchat-Huang T, Slivinski KM, Sinagawa-Garcia SR, Maliga P (2011) Visual spectinomycin resistance gene for facile identification of transplastomic sectors in tobacco leaves. Plant Mol Biol 76:453–461
21. Sinagawa-Garcia SR, Tungsuchat-Huang T, Paredes-Lopez O, Maliga P (2009) Next generation synthetic vectors for transformation of the plastid genome of higher plants. Plant Mol Biol 70:487–498
22. Carrer H, Hockenberry TN, Svab Z, Maliga P (1993) Kanamycin resistance as a selectable marker for plastid transformation in tobacco. Mol Gen Genet 241:49–56
23. Lutz K, Corneille S, Azhagiri AK, Svab Z, Maliga P (2004) A novel approach to plastid transformation utilizes the phiC31 phage integrase. Plant J 37:906–913
24. Svab Z, Maliga P (1991) Mutation proximal to the tRNA binding region of the *Nicotiana* plastid 16S rRNA confers resistance to spectinomycin. Mol Gen Genet 228:316–319
25. Yukawa M, Tsudzuki T, Sugiura M (2006) The chloroplast genome of *Nicotiana sylvestris* and *Nicotiana tomentosiformis*: complete sequencing confirms that the *Nicotiana sylvestris* progenitor is the maternal genome donor of *Nicotiana tabacum*. Mol Genet Genomics 275:367–373
26. Iamtham S, Day A (2000) Removal of antibiotic resistance genes from transgenic tobacco plastids. Nat Biotechnol 18:1172–1176
27. Lee SB, Kwon HB, Kwon SJ, Park SC, Jeong MJ, Han SE, Byun MO, Daniell H (2003) Accumulation of trehalose within transgenic chloroplasts confers draught tolerance. Mol Breeding 11:1–13
28. Yu LX, Gray BN, Rutzke CJ, Walsker LP, Wilson DB, Hanson MR (2007) Expression of thermostable microbial cellulases in the chloroplast of nicotine-free tobacco. J Biotechnol 131:362–369
29. McCabe MS, Klaas M, Gonzalez-Rabade N, Poage M, Badillo-Corona JA, Zhou F, Karcher D, Bock R, Gray JC, Dix PJ (2008) Plastid transformation of high-biomass tobacco variety Maryland Mammoth for production of human immunodeficiency virus type 1 (HIV-1) p24 antigen. Plant Biotechnol J 6:914–929
30. O'Neill C, Horvath GV, Horvath E, Dix PJ, Medgyesy P (1993) Chloroplast transformation in plants: polyethylene glycol (PEG) treatment of protoplasts is an alternative to biolistic delivery systems. Plant J 3:729–738
31. Davarpanah SJ, Jung SH, Kim YJ, Park YI, Min SR, Liu JR, Jeong WJ (2009) Stable Plastid Transformation in *Nicotiana benthamiana*. J Plant Biol 52:244–250
32. Murashige T, Skoog F (1962) A revised medium for the growth and bioassay with tobacco tissue culture. Physiol Plant 15:473–497
33. Sidorov V, Menczel L, Maliga P (1981) Isoleucine-requiring *Nicotian* plant deficient in threonine deaminase. Nature 294:87–88
34. Jefferson RA, Kavanagh TA, Bevan MW (1987) GUS fusions: beta-glucuronidase as a sensitive and versatile gene fusion marker in higher plants. EMBO J 6:3901–3907
35. Gallagher SR (ed) (1992) GUS protocols: using the GUS gene as a reporter of gene expression. Academic, San Diego
36. Corneille S, Lutz K, Svab Z, Maliga P (2001) Efficient elimination of selectable marker genes from the plastid genome by the CRE-*lox* site-specific recombination system. Plant J 72: 171–178
37. Corneille S, Lutz KA, Azhagiri AK, Maliga P (2003) Identification of functional *lox* sites in the plastid genome. Plant J 35:753–762

Chapter 9

Nicotiana tabacum: PEG-Mediated Plastid Transformation

Areli Herrera Díaz and Hans-Ulrich Koop

Abstract

Stable plastid transformation in *Nicotiana tabacum* has been achieved by using two different methods, the biolistic method, using a particle gun, and the polyethylene glycol (PEG)-mediated transformation. PEG-mediated plastid transformation involves the treatment of isolated protoplasts (plant cells without cell wall) with PEG in the presence of DNA. We have previously shown that in *Nicotiana tabacum* both methods are equally efficient. The PEG-mediated transformation efficiencies range between 20 and 50 plastid transformants per experiment (10^6 viable treated protoplasts). One advantage of the PEG method is that no expensive equipment such as a particle gun is required. The only crucial points are the handling and the cultivation of protoplasts. Furthermore, markers for the selection of transformed chloroplasts are required. One of the most often used selection markers is the *aadA* gene which encodes for spectinomycin and streptomycin resistance. Here we describe a simplified and inexpensive protocol for the transformation of chloroplasts in *Nicotiana tabacum* using an optimized protoplast culture protocol.

Key words Thin-alginate-layar cultur, Chloroplast transformation, PEG-mediated transformation, Polyethylene glycol, PEG, Protoplasts, *Nicotiana tabacum*

1 Introduction

Polyethylene glycol (PEG)-mediated plastid transformation is far less expensive than the biolistic procedure. Stable plastid transformation with the help of the PEG method in *Nicotiana tabacum* was reported for the first time by our group in 1993 [1] and shortly afterwards by O'Neill et al. in *Nicotiana plumbaginifolia* [2]. Although more than 15 years have passed, the exact mechanism of how DNA is transported through the double membrane of chloroplasts is still unknown [3].

To allow PEG-mediated transformation in plants, the cell wall must be removed by enzymatic treatment. Once protoplasts are isolated, they can be mixed with PEG in the presence of DNA. Because stable genetic transformation of plastids is based on homologous recombination, the transgene has to be flanked by plastome sequences ensuring the integration of the transgene into a specific region of the plastome [4]. PEG-mediated plastid transformation

Pal Maliga (ed.), *Chloroplast Biotechnology: Methods and Protocols*, Methods in Molecular Biology, vol. 1132,
DOI 10.1007/978-1-62703-995-6_9,

has been achieved not only in *Nicotiana tabacum* but also in *Nicotiana plumbaginifolia* [2], *Solanum lycopersicum* (tomato) [5], *Brassica oleracea* var. *botrytis* (cauliflower) [6], and *Lactuca sativa* (lettuce) [7]. In all cases, success of the PEG-mediated plastid transformation method strongly depends on a good protoplast regeneration protocol.

Detailed protocols and at least one review of plastid transformation using the PEG method in tobacco have been described [8, 9]. Usually, the expected transformation efficiency ranges between 20 and 50 plastid transformants per 10^6 viable treated protoplasts [9]. Overall, both methods, the PEG-mediated transformation and the particle gun method, are equally efficient in terms of number of transformed plants per experiment. Here we present a simplified protocol for PEG-mediated plastid transformation in tobacco. In this protocol the thin-alginate-layer technique developed in our laboratory [10] is combined with an improved PEG treatment protocol.

2 Materials

2.1 Culture of Plant Material

1. Dimanin C (Bayer Vital, Leverkusen, Germany): 5 % (w/v) in water (*see* **Note 1**).
2. "B5 modified" medium: supplement B5 medium (Duchefa, Haarlem, Netherlands) with 0.983 g of $MgSO_4 \cdot 7H_2O$ (Roth, Karlsruhe, Germany), 20 g of sucrose (Duchefa), and 8 g of plant agar (Duchefa); fill up to 1 l with autoclaved water; adjust to pH 5.8, with 1 N KOH (Merck, Darmstadt, Germany). 120 ml of medium are autoclaved in 720 ml glass jars, which are supplied with foam plugs (ceapren, 36 mm, Greiner, Germany) inserted into a hole of 25 mm diameter punched into the lid of the jar.

2.2 Protoplast Isolation

1. Preparation of cellulase R10 and macerozyme R10 (Duchefa) stock solutions (10 %): Dissolve 1 g of enzyme and 1.37 g sucrose (Duchefa) in 10 ml autoclaved water. Enzyme stock solutions should be centrifuged 10 min at 150 × *g* before sterilization through a 0.2 μm sterilization filter. Store aliquots at −20 °C.
2. MS micro salts (100×) for protoplast isolation medium [11]: 4 g of NaEDTA (Roth), 0.083 g KI (Merck), 0.620 g H_3BO_3 (USB, Cleveland, USA), 2.23 g $MnSO_4 \cdot H_2O$ (Merck), 0.860 g $ZnSO_4 \cdot 7H_2O$ (Sigma), 0.025 g $Na_2MoO_4 \cdot 2H_2O$ (Merck), 0.0025 g $CoCl_2$ (Merck), and 0.0025 g $CuSO_4 \cdot 5H_2O$ (Merck); fill up to 1 l with autoclaved water. Store in aliquots at −20 °C.
3. 2 M ammonium succinate for protoplast isolation medium [10]: Prepare 0.1 l, by dissolving 23.6 g of succinic acid (Sigma-Aldrich, Taufkirchen, Germany) and 10.6 g of NH_4Cl (Merck) in 60 ml autoclaved water. Adjust pH to 5.8 with

Table 1
Culture media

Components/l	F-PIN [10]	F-PCN [10]	RMOP [4]
MS basal medium (Duchefa)			4.4 g
MS macro mod:			
KNO_3 (Merck)	1.012 g	1.012 g	
$CaCl_2{\cdot}2H_2O$ (Merck)	0.440 g	0.440 g	
$MgSO_4{\cdot}7H_2O$ (Roth)	0.370 g	0.370 g	
KH_2PO_4 (Merck)	0.170 g	0.170 g	
Micro MS 100× (Subheading 2.2)	10 ml	10 ml	
MES	1.952 g	1.952 g	
2 M ammonium succinate (Subheading 2.2)	10 ml	10 ml	
PC vitamins 100× (Subheading 2.2)	10 ml	10 ml	
NT Vitamins 100× (Subheading 2.3)			10 ml
Benzylaminopurine (BAP, Sigma) (1 mg/ml)	1 ml	1 ml	1 ml
Naphthalene acetic acid (NAA, Sigma) (1 mg/ml)	100 µl	100 µl	100 µl
Sucrose (Duchefa)	130 g	20 g	30 g
Glucose(Duchefa)		65 g	
Plant agar (Duchefa)			8 g

All media are adjusted to pH 5.8 with 1 N KOH. The first two media are adjusted to 550 mOsm and filter sterilized. RMOP medium is autoclaved

approximately 22.4 g KOH (Merck), fill up to 0.1 l with autoclaved water, and sterilize by filtration through a 0.2 µm filter (*see* **Note 2**).

4. PC vitamins and salts (100×) for protoplast isolation medium [12]: 20 g of myoinositol (Serva, Heidelberg, Germany), 0.200 g nicotinic acid (Merck), 0.200 g pyridoxine HCl (Duchefa), 0.100 g thiamine HCl (Merck), 0.002 g d(+)-biotin (Duchefa), 20 g $CaCl_2{\cdot}2H_2O$ (Merck), and 0.200 g calcium pantothenate (Merck); fill up to 1 l with autoclaved water. Store in aliquots at −20 °C.
5. Preparation of F-PIN and F-PCN media for protoplast isolation is described in Table 1.

2.3 PEG-Mediated Plastid Transformation

1. PEG_{1500} solution: Dissolve 0.413 g of $Ca(NO_3)_2{\cdot}4H_2O$ (Roth) and 1.275 g mannitol (Duchefa) in 17.5 ml autoclaved water. Once these are dissolved, add 10 g of PEG_{1500} (Merck), adjust to pH 9.75 with 1 N KOH (Merck), and fill up to 26 ml with autoclaved water. Sterilize by filtration through a 0.2 µm filter. Single-use aliquots (125 µl) are stored at −20 °C (*see* **Note 3**).

2. Transformation medium: Prepare 0.2 l by dissolving 0.609 g of $MgCl_2 \cdot 6H_2O$ (AppliChem Darmstadt, Germany), 0.200 g MES (Sigma), and approximately 18 g of mannitol (Duchefa); fill up to 0.2 l with autoclaved water; and adjust to 550 mOsm with mannitol and to pH 5.8 with 1 N KOH (Merck). Sterilize by autoclaving.
3. TE pH 5.6: 10 mM Tris–HCl, pH 5.6 with HCl, 1 mM EDTA. Sterilize by autoclaving and store in 1 ml aliquots at −20 °C.

2.4 Protoplast Culture and Regeneration

1. F-alginate (2.4 %): Prepare 0.1 l by dissolving 0.137 g of MES (Sigma), 0.250 g $MgSO_4 \cdot 7H_2O$ (Roth), 0.204 g $MgCl_2 \cdot 6H_2O$ (AppliChem), and approximately 7.7 g mannitol (Duchefa); fill up to 0.1 l with autoclaved water; and adjust to 550 mOsm with mannitol and to pH 5.8 with 1 N KOH (Merck). Finally, add 2.4 g of alginic acid (Sigma). Sterilize by autoclaving (*see* **Note 4**).
2. Calcium-agar: Add 2.94 g of $CaCl_2 \cdot 2H_2O$ (Merck), 1.952 g MES (Sigma), and approximately 82 g of mannitol; fill up to 1 l with autoclaved water; and adjust to 550 mOsm with mannitol and to pH 5.8 with 1 N KOH. 10 g of plant agar (Duchefa) are added and aliquots of 200 ml are sterilized by autoclaving.
3. NT vitamins (100×) [4]: Add 10 g of myoinositol (Serva) and 0.100 g thiamine HCl (Merck); fill up to 1 l autoclaved water. Aliquots of 100 ml are stored at −20 °C.
4. Preparation of RMOP medium [4] is described in Table 1.
5. Spectinomycin dihydrochloride (Duchefa) at 1 mg/ml in autoclaved water is sterilized by filtration through a 0.2 μm filter. Aliquots of 10 ml are stored at −20 °C.

3 Methods

Protoplast isolation and regeneration highly depends on the quality of the donor plant material. Success in stable PEG-mediated plastid transformation in *Nicotiana tabacum* is influenced by the correct protoplast isolation, the careful treatment of protoplasts, and the use of appropriate vectors [8]. In this case, vector pKAA derived from vector pKCZ [13] is used. This vector has an *aadA* gene as selection marker and the AmCyan gene as a visual marker. To obtain protoplasts, healthy leaf strips are incubated with cellulase R10 and macerozyme R10. The digestion of the cell wall requires usually 14 h darkness, 25 °C, without shaking (*see* Subheading 3.2). Interestingly, this long enzymatic digestion is a prerequisite for highly efficient protoplast regeneration in *Nicotiana tabacum*. Because protoplasts have no cell wall, they are extremely sensitive and need the use of media carefully adjusted to the optimal osmotic value (550 mOsm). Successful isolation of intact mesophyll protoplasts can be achieved through filtration, floatation, and sedimentation. Once protoplasts have been isolated, the density of the protoplasts is determined by the use of a hemocytometer. Usually 1×10^6 protoplasts in 100 μl are used for one PEG-mediated plastid transformation experiment. After plastid transformation, protoplasts are alginate embedded in polypropylene grids. After PEG-mediated plastid transformation, grids containing embedded protoplasts are cultured in liquid medium for 1 week. Once protoplast-derived colonies are obtained, grids are transferred to solid RMOP medium containing appropriate antibiotics for selection and regeneration.

3.1 Germination of *Nicotiana tabacum*

1. Seeds of *Nicotiana tabacum* cv. Petite Havana are sterilized with 70 % ethanol (v/v) for 1 min, remove ethanol, and add Dimanin C (5 %) for 10 min. Dimanin C is removed and seeds are washed three times for 10 min with autoclaved water. Before use, seeds are air dried in a laminar flow, or they can be stored dry in sealed Petri dishes at 4 °C for several months.
2. Around 15 seeds are germinated in 120 ml "B5 modified" medium in 720 ml jars. Culture conditions comprise a photoperiod of 16 h light (60 μm/m^2/s) at 25 °C followed by 8 h dark (*see* **Note 5**).
3. Once seeds are germinated (usually after 1 week), seedlings are transferred individually to fresh 720 ml jars containing the same "B5 modified" medium. Plants are grown under the growth conditions described above for 3 weeks.

3.2 Protoplast Isolation from *Nicotiana tabacum* Leaves

1. Fully expanded leaves from 2 or 3 weeks old tobacco plants are harvested (Fig. 1). Leaf strips of circa 1 mm are cut perpendicular to the main vein using a fresh sharp scalpel. The main vein is discarded and strips from four to five leaves with the abaxial side down are incubated for preplasmolysis on 10 ml of F-PIN medium (Table 1) at room temperature in the dark for 1 h (*see* **Note 6**).
2. After 1 h incubation, F-PIN medium is discarded and exchanged by 10 ml F-PIN medium containing 0.25 ml of cellulase R10 and 0.25 ml macerozyme R10 (10 %) stock solutions. Strips are further incubated without shaking at 25 °C in the dark for 12–16 h.
3. After a careful swirling motion, the mixture containing the leaf strips should turn green. The protoplast suspension is passed through a 100 μm stainless steel sieve into a 12 ml sterile centrifuge tube.
4. This flow-through is overlayed with 2 ml of F-PCN medium (Table 1) and centrifuged at $70 \times g$ for 10 min (*see* **Note 7**).

Fig. 1 *Nicotiana tabacum* cv. petit Havana plant after 3 weeks of growth in a 720 ml jar on "B5 modified" medium

5. After centrifugation three bands are formed; the middle band contains the intact protoplasts (Fig. 2). Intact protoplasts are collected with a Pasteur pipette and the intact protoplasts are carefully transferred to a new sterile tube.
6. Finally, the protoplast suspension is filled up to 10 ml with transformation medium. Mix carefully by inverting the tube once and determine the cell density by the use of a hemocytometer (*see* **Note 8**).

3.3 PEG-Mediated Plastid Transformation

1. Once protoplasts have been quantified, the suspension of protoplasts is centrifuged at $50 \times g$ for 10 min.
2. The supernatant is discarded and protoplasts are adjusted to a cell density of 1×10^6 protoplasts per 100 μl of transformation medium.

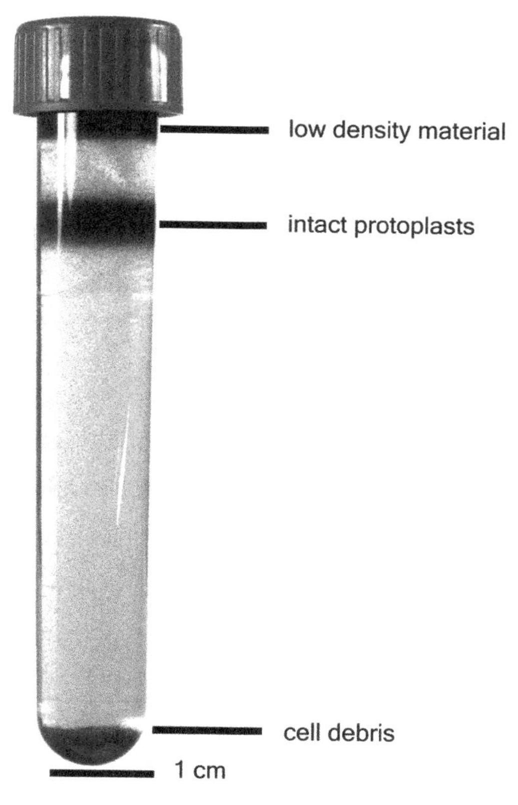

Fig. 2 Step-gradient purification of protoplasts after enzymatic cell wall degradation

3. Transfer 100 μl containing 1×10^6 protoplasts to a 3 cm Petri dish.
4. 50 μg of DNA (vector pKAA), dissolved in 18 μl TE pH 5.6, are added dropwise, and the protoplast suspension is carefully mixed by repeated tilting of the Petri dish.
5. 7 μl of F-PCN medium are added (*see* **Note 9**).
6. 125 μl of PEG_{1500} stock solution (40 %) are added dropwise and suspension is incubated for 7.5 min.
7. Add 125 μl of F-PCN medium dropwise and incubate for 2 min.
8. Finally, 2.6 ml of F-PCN medium are added gradually and mixed gently (*see* **Note 10**).

3.4 Protoplast Culture and Regeneration

1. Melt Ca^{2+} agar by boiling, and prepare Ca^{2+} agar plates by using approximately 25 ml per Petri dish (*see* **Note 11**).
2. Polypropylene grids (10×10 meshes, 2×2 mm mesh size; Sefar GmbH, Edling, Germany) are sterilized in water by autoclaving. Before use, polypropylene grids should be dried in a laminar flow bench.

Fig. 3 Leaf protoplasts of *Nicotiana tabacum* cv. petit Havana directly after embedding in alginate (**a**) and protoplast-derived colonies after 7 days of culture (**b**)

3. Once polypropylene grids and Ca^{2+} agar are prepared, the protoplast suspension is gently mixed with an equal volume of F-alginate (2.4 %).
4. Pipette 625 μl of the protoplast-alginate suspension to a Ca^{2+} agar plate using 1 ml tips with 1–2 mm removed from their ends (Eppendorf, Germany). Immediately afterwards, add one sterile grid. Usually, three grids per Ca^{2+} agar plate can be embedded. Gelling of the protoplast-alginate mixture can take between 30 min and 1 h.
5. Pipette 10 ml F-PCN medium into 6 cm Petri dishes.
6. After gelling on Ca^{2+} agar plates, grids are carefully removed and placed upside down in the 10 ml F-PCN medium in the 6 cm Petri dishes (*see* **Note 12**).
7. After 30 min of equilibration, grids are transferred in the same orientation to a new Petri dish containing 2 ml of F-PCN (*see* **Note 13**) (Figs. 3a and 4a). Culture conditions remain the same as in plant growing protocol with a photoperiod of 16 h light (60 μm/m^2/s) at 25 °C followed by 8 h dark for 7 days.
8. Colony formation takes between 7 and 8 days (Fig. 4b), and after this time, grids can be transferred to solid RMOP containing appropriate antibiotics (*see* **Note 14**). A grid containing colonies, 2 weeks of age, is shown in Fig. 3b.
9. Grids should be transferred to fresh medium every 3 weeks. Regular screening for green colonies begins at the third week, and green resistant calluses can be transferred to separate Petri dishes containing RMOP medium with appropriate antibiotics (*see* **Note 15**).

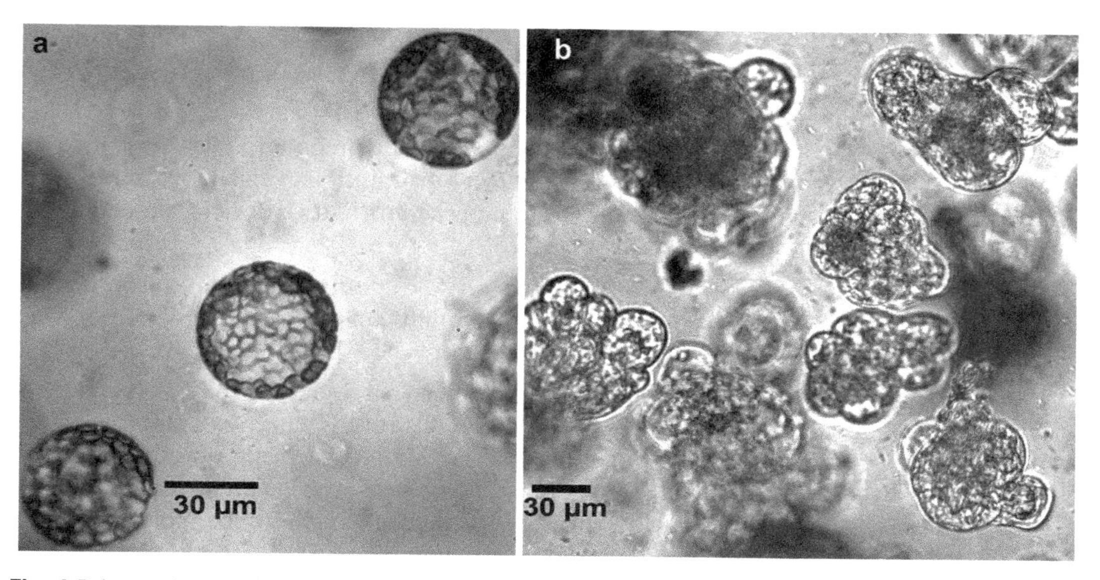

Fig. 4 Polypropylene grids containing alginate-embedded protoplasts immediately after embedding (**a**) and after 2 weeks of culture (**b**)

10. Once shoots from green colonies are obtained (cycle 0), small leaf explants 2 × 2 mm are subjected to additional rounds of regeneration in Petri dishes containing RMOP medium with the appropriate antibiotics. These additional rounds of regeneration amplify the transformed plastomes and thus eliminate the wild-type plastomes [14].
11. To allow rooting of tranplastomic shoots, shoots with a size between 2 and 4 cm are transferred to 720 ml jars containing "B5 modified" medium. After 2–3 weeks, plants can be transferred to the greenhouse.

4 Notes

1. All solutions should be prepared with deionized water with a resistivity of at least 18,2 MegaOhm per cm, and total organic content should be less than 1 part per billion. We use an Ultra Clear UV plus system from SG Wasseraufbereitung und Regenerierstation GmbH, Germany. This is the standard condition for the water used in this protocol.
2. Ammonium succinate is only soluble after the addition of KOH. However, to avoid boiling of the stock solution, KOH should be added slowly to ammonium succinate stock solution. Once the ammonium succinate has reached room temperature, adjust pH to 5.8 with KOH.

3. PEG_{1500} stock solution is prepared at room temperature to avoid changes of the pH. In our laboratory best results are obtained with PEG_{1500} (Merck).
4. Alginic acid, 2.4 %, needs to be prepared 2 weeks prior to use. Usually, it is necessary to wait until particulate matter has sedimented. The clear supernatant is used for the embedding of protoplasts.
5. Growth of seedlings on "B5 modified" medium is important, because this medium contains an increased Mg^{2+} content which reduces the appearance of yellow patches on leaves.
6. Usually, expanded leaves are harvested from the tobacco plants (Fig. 1). To avoid loss of leaf turgor, cutting of strips and transferring to F-PIN medium should be made quickly and gently. Dipping and submerging of leaf strips is important to obtain a high yield of protoplasts. Preplasmolysis can be prolonged to 3 h.
7. Carefully overlay protoplasts in F-PIN medium with F-PCN medium, avoiding any turbulence at the interface. This is important for the formation of a narrow band containing intact protoplasts.
8. Determination of cell density with hemocytometer should be carried out five times to avoid mistakes.
9. The quality of the DNA is also important for good results in PEG-mediated plastid transformation. In our experiments, best results are obtained with the use of Qiagen maxi prep kits (Qiagen, Hilden, Germany).
10. Instead of direct embedding of protoplasts, incubation at 25 °C for 12–14 h in the dark can be performed. This allows for an additional floatation purification before embedding to remove protoplasts which do not survive the PEG treatment.
11. Alginate gelling works best on freshly prepared Ca^{2+} agar plates. This simplifies the transfer of the embedded protoplasts to the new Petri dishes containing F-PCN medium.
12. Removing the grids from the Ca^{2+} agar plates can be difficult; adding 1 or 2 ml of F-PCN medium at the edges of the grids can facilitate this step. Grids are transferred to the fresh F-PCN medium with the help of flat forceps.
13. Protoplasts tend to sediment during initial stages of gelling. Inverting the grids exposes embedded protoplasts to the surface.
14. Alginate-embedded protoplasts should be cultured for 7 days in liquid F-PCN medium to allow them to reach the 32–64 cell stage. After colonies have reached this stage, they can be transferred to RMOP medium which has a lower osmolarity than the F-PCN medium.
15. In cases, in which the transformation vector contains an *aadA* cassette, 500 mg/l of spectinomycin and or streptomycin should be used.

Acknowledgments

The authors wish to express their gratitude to Stefan Kirchner for his valuable technical assistance and the preparation of the pictures for this work. Areli Herrera Díaz was supported by a research scholarship from the DAAD (Deutscher Akademischer Austausch Dienst).

References

1. Golds T, Maliga P, Koop H (1993) Stable plastid transformation in PEG-treated protoplasts of *Nicotiana tabacum*. Bio/Technology 11:95–97
2. O'Neill C, Horvath GV, Horvath E, Dix PJ, Medgyesy P (1993) Chloroplast transformation in plants: polyethylene glycol (PEG) treatment of protoplasts is an alternative to biolistic delivery systems. Plant J 3:729–738
3. Spörlein B, Streubel M, Dahlfeld G, Westhoff P, Koop H (1991) PEG-mediated plastid transformation: a new system for transient gene expression assays in chloroplasts. Theor Appl Genet 82:717–722
4. Svab Z, Hajdukiewicz P, Maliga P (1990) Stable transformation of plastids in higher plants. Proc Natl Acad Sci U S A 87:8526–8530
5. Nugent GD, Ten Have M, van der Gulik A, Dix PJ, Uijtewaal BA, Mordhorst AP (2005) Plastid transformants of tomato selected using mutations affecting ribosome structure. Plant Cell Rep 24:341–349
6. Nugent GD, Coyne S, Nguyen TH, Kavanagh TA, Dix PJ (2005) Nuclear and plastid transformation of *Brassica oleracea* var. botrytis (cauliflower) using PEG-mediated uptake of DNA into protoplasts. Plant Sci 170:135–142
7. Lelivelt CL, McCabe MS, Newell CA, Desnoo CB, van Dun KM, Birch-Machin I, Gray JC, Mills KH, Nugent JM (2005) Stable plastid transformation in lettuce (*Lactuca sativa* L.). Plant Mol Biol 58:763–774
8. Koop HU, Steinmuller K, Wagner H, Rossler C, Eibl C, Sacher L (1996) Integration of foreign sequences into the tobacco plastome *via* polyethylene glycol-mediated protoplast transformation. Planta 199:193–201
9. Kofer W, Eibl C, Steinmüller K, Koop H-U (1998) PEG-mediated plastid transformation in higher plants. In Vitro Cell Dev Biol Plant 34:303–309
10. Dovzhenko A, Bergen U, Koop HU (1998) Thin-alginate-layer technique for protoplast culture of tobacco leaf protoplasts: shoot formation in less than two weeks. Protoplasma 204:114–118
11. Murashige T, Skoog F (1962) A revised medium for rapid growth and bioassays with tobacco tissue cultures. Physiol Plant 15: 473–497
12. Koop HU, Schweiger HG (1985) Regeneration of plants from individually cultivated protoplasts using and improved microculture system. J Plant Physiol 121:245–257
13. Zou Z, Eibl C, Koop HU (2003) The stem-loop region of the tobacco *psbA* 5′-UTR is an important determinant of mRNA stability and translation efficiency. Mol Genet Genomics 269:340–349
14. Svab Z, Maliga P (1993) High-frequency plastid transformation in tobacco by selection for a chimeric *aadA* gene. Proc Natl Acad Sci U S A 90:913–917

Chapter 10

Plastid Transformation of Tobacco Suspension Cell Cultures

Jeffrey M. Staub

Abstract

Chloroplast transformation has been extremely valuable for the study of plastid biology and gene expression, but the tissue culture methodology involved can be laborious, and it can take several months to obtain homoplasmic regenerated plants useful for molecular or physiological studies. In contrast, transformation of tobacco suspension cell plastids provides an easy and efficient system to rapidly evaluate the efficacy of multiple constructs prior to plant regeneration. Suspension cell cultures can be initiated from many cell types, and once established, can be maintained by subculture for more than a year with no loss of transformation efficiency. Using antibiotic selection, homoplasmy is readily achieved in uniform cell colonies useful for comparative gene expression analyses, with the added flexibility to subsequently regenerate plants for *in planta* studies. Plastids from suspension cells grown in the dark are similar in size and cellular morphology to those in embryogenic culture systems of monocot species, thus providing a useful model for understanding the steps leading to plastid transformation in those recalcitrant species.

Key words Callus, Homoplasmic, Particle bombardment, Plastid transformation, Suspension cells, Tissue culture, Tobacco

1 Introduction

Transformation of tobacco leaf chloroplasts was first reported more than 20 years ago [1]. The technology for stable transformation has since been extended to numerous dicot species [2], where leaf chloroplast transformation is the most often used approach. As shown in Fig. 1a, the large size and peripheral localization of leaf chloroplasts make them an efficient target for transformation. In addition, homoplasmy of plastid transformants can be readily achieved under selection pressure in the light after multiple rounds of plant regeneration from leaf tissue [3]. This transformation system has been used extensively for studies of plastid gene expression [4] and introduction of agronomically important traits [5, 6]. Further, since the first report of overproduction of a human therapeutic protein in tobacco chloroplasts [7], a vast array of potential

Pal Maliga (ed.), *Chloroplast Biotechnology: Methods and Protocols*, Methods in Molecular Biology, vol. 1132,
DOI 10.1007/978-1-62703-995-6_10, © Springer Science+Business Media New York 2014

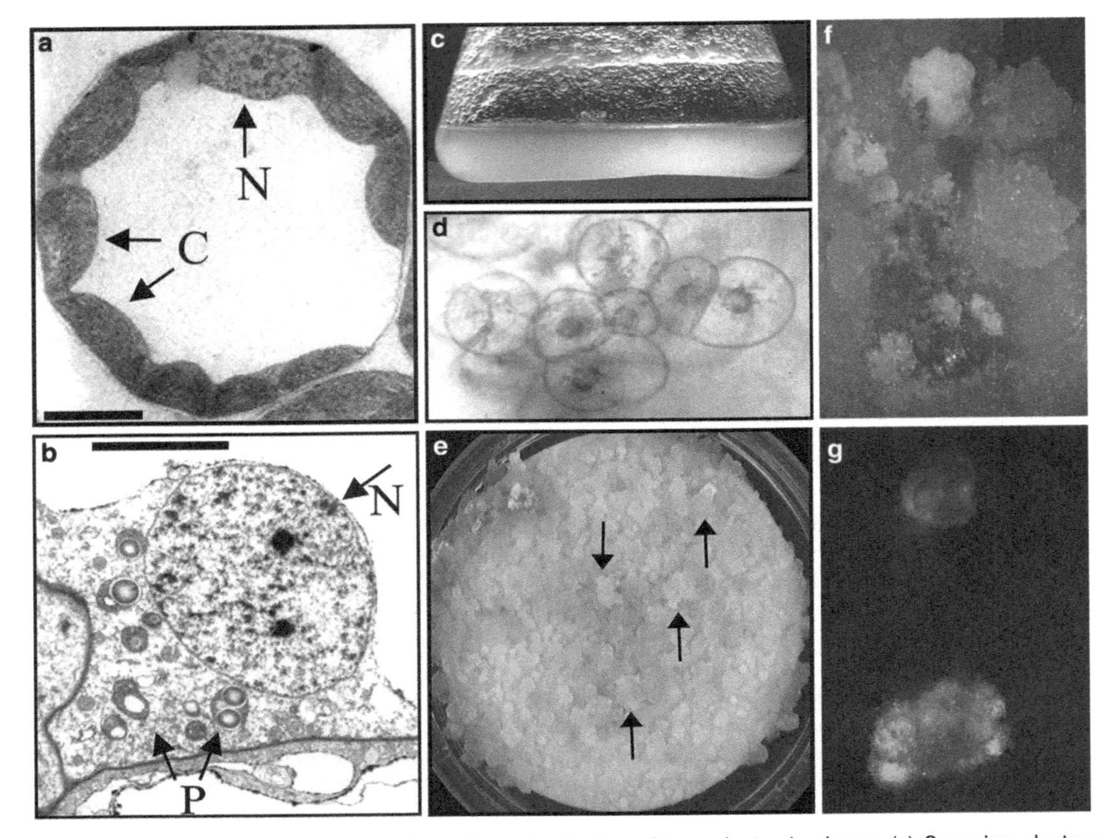

Fig. 1 Morphology of tobacco suspension cells and selection of transplastomic clones. (**a**) Scanning electron micrograph of in vitro grown leaf cell. Note that the chloroplasts (*C*) are similar in size as the nucleus (*N*). (**b**) Scanning electron micrograph of plastids in suspension cells. Note that plastids (*P*) are significantly smaller than the nucleus, and randomly distributed throughout the cell. (**c**) Dark-grown tobacco suspension cells at 3 days post-subculture to fresh medium. (**d**) Light microscope image of a suspension cell cluster. Note that the nucleus is visible in each cell. (**e**) Colonies from plastid transformation after 9 weeks on selective spectinomycin-containing medium. Note that greening is not apparent at this time point. (**f**) Close-up view of colonies selected during plastid transformation in visible light and (**g**) in fluorescent light for detection of GFP expression

therapeutics and high-value proteins have been overproduced in chloroplasts [8–10].

In contrast to the chloroplast transformation systems in dicot species, many monocot transformation and regeneration systems use dark-grown embryogenic calli or suspension cell cultures that carry undeveloped plastids. Furthermore, plant regeneration from leaf tissue is not possible in monocots, so amplification and purification of transformed plastids to the homoplasmic state has not been achieved. Thus, reports in rice [11, 12] and corn [13] have so far only produced heteroplasmic transformants.

Tobacco suspension cell cultures, such as BY2 cells [14], have been used for comparative studies of nuclear gene expression constructs. Although these cultures are readily propagated, this cell line has long ago lost its ability to regenerate plants.

On the other hand, primary suspension cell cultures derived from leaf cells have the advantages of rapid and uniform growth, high transformation efficiency and retain the ability to readily regenerate plants under appropriate tissue culture conditions. When grown under dark conditions, the plastids of this cell type remain small and undifferentiated and resemble embryogenic cultures in their morphology (Fig. 1b). We have used these unique properties of tobacco suspension cells to develop a plastid transformation system [15] that can be used for rapid evaluation of plastid transformation and gene expression constructs, and as a model to study the events leading to transformation of undeveloped plastids of embryogenic cultures systems. Moreover, cell suspensions are amenable to growth in bioreactors useful for potential commercial-scale protein overproduction [16]. Recently, a cell suspension system has also been developed for plastid transformation in the liverwort, *Marchantia polymorpha* [17].

2 Materials

2.1 Tissue Culture Media

1. Callus Medium: MS salts [18], B-5 vitamins [19], 4 mg/L *p*-chlorophenoxyacetic acid, 5 μg/L kinetin, 3 % sucrose, and 2.5 g/L Schweizer Hall gelling agent.
2. Suspension Medium: MS salts [18], MS vitamins, 4 mg/L *p*-chlorophenoxyacetic acid, 5 μg/L kinetin, 0.2 g/L myo-inositol, 0.15 g/L L-asparagine, and 3 % w/v sucrose.
3. Bombardment Medium: MS salts [18], B-5 vitamins [19], 18.2 g/L mannitol, 18.2 g/L sorbitol, 0.1 mg/L 1-naphthaleneacetic acid, 1 mg/L 6-benzylaminopurine, 3 % w/v sucrose, and 8 g/L TC agar.
4. Selection Medium: MS salts [18], B-5 vitamins [19], 3 % w/v sucrose, 0.1 mg/L NAA, 1 mg/L BAP, 750 mg/L spectinomycin, and 8 g/L TC agar.
5. Plant regeneration Medium: MS salts [18], B-5 vitamins [19], 3 % w/v sucrose, 0.1 mg/L NAA, 1 mg/L BAP, and 8 g/L TC agar.

2.2 Particle Bombardment

1. All materials for preparation of microprojectiles and operation of the PDS-1000 Helium Biolistic Delivery Device (stopping screens, rupture disks, macrocarriers) can be purchased from Bio-Rad.
2. M5 tungsten microparticles (average size 0.4 μM) can be purchased from Bio-Rad.
3. Gold nanoparticles can be purchased from Inbiogold (www.inbiogold.com).

4. 2.5 M $CaCl_2$ (filter sterilized).
5. 0.1 M Spermidine (free base, tissue culture grade; filter sterilized).
6. 70 % Ethanol.
7. 100 % Ethanol.
8. 70 % Isopropanol.

3 Methods

Suspension cell cultures may be initiated from any tissue type from which callus can be generated. Leaf tissue is chosen because it readily produces callus on appropriate culture medium and leaf-derived suspension cultures can retain plant regeneration capability for a long time, up to 2 years in our experiments [15]. When grown in the dark and subcultured at weekly intervals, suspensions grow mostly as small clumps of cells and single cells, with centrally located nucleus, large amount of cytoplasmic material and no obvious vacuole (Fig. 1c, d). Confocal scanning microscopy was used to show that the undeveloped plastids in these cultures have an average diameter and volume that is 4-fold and 50-fold less, respectively, than that of leaf chloroplasts [15].

Selection of plastid transformants is most efficient using the antibiotic spectinomycin in the light, where transformed cells are distinguished by greening under selective conditions. Resistance is conferred by a bacterial *aadA* gene driven by plastid gene expression signals [3]. An antibiotic kill-curve should be performed to determine the optimal concentration of suspension cells and spectinomycin that will allow continued growth of wild-type cells but inhibit their greening in the light. A suitable plastid transformation vector will carry the selectable *aadA* gene and the gene(s) of interest for testing [20].

After particle bombardment and selection for transformed plastids on spectinomycin-containing media, cell colonies can be used directly for gene expression studies or other assays. The cell colonies may continue to be propagated in the light as green colonies, or plants may be regenerated in the absence of antibiotic to be used for subsequent *in planta* studies.

3.1 Initiation and Maintenance of Tobacco Suspension Cell Cultures

1. Wild-type tobacco plants (cultivar Petit Havana) (*see* **Note 1**) grown in sterile tissue culture are used as starting material for callus formation. Young, mature leaves (~3–4 weeks after plant subculture by cuttings) are cut into ~1 cm square pieces and placed onto Callus Medium on agar plates. The plates are incubated in continuous dark until fluffy white callus is induced along the leaf edges.

2. At ~3 week intervals, callus tissue is excised from the leaf tissue and transferred to new Callus Medium plates for continued growth in the dark. At ~2–3 week intervals, plates should be observed for callus growth. If callus pieces have grown to ~1 cm square pieces, these should be subcultured further by cutting into three to four smaller pieces.
3. Callus that has been amplified for 3–5 months is collected and used for initiation of liquid suspension cell cultures by placing ~20 g of callus into 40 mL of Suspension Medium in a 250 mL Erlenmeyer flask (*see* **Note 2**). Suspensions are cultured in the dark at 25 °C with continuous shaking at 140–160 RPM.
4. The cell suspension should be subcultured twice weekly at a 3:4 suspended cell to medium ratio until a fine suspension is obtained. Typically, 30 mL of suspension cells are transferred into 40 mL of fresh medium at every subculture, and shaking in the dark is continued (*see* **Note 3**). A fine cell suspension suitable for transformation is obtained in ~5–8 weeks.

3.2 Preparation of Suspension Cells for Bombardment

1. Suspension cells are used for bombardment at 1 day after subculturing (*see* **Note 4**). For each plate to be bombarded, the amount of liquid suspension that contains 0.5 mL settled cell volume of suspension cells (usually ~5–8 mL) is collected on a 70 mm Whatman #1 filter paper (*see* **Note 5**). Settled cell volume is determined by allowing 10 mL of the cell suspension to settle for 20 min in a 15 mL conical tube.
2. The filter paper carrying the cells is placed on Bombardment Medium for 4 h to overnight prior to bombardment (*see* **Note 6**).

3.3 Preparation of Microparticles for Particle Bombardment

1. Gold particles are chosen because the commercially available bead preparations are very uniform in size as compared to tungsten preparations. As reported for plastid transformation in *Chlamydomonas* [21], smaller microparticles (0.4 μM) were found to be more efficient for plastid transformation than larger sized particles [15].
2. Carefully weigh 30 mg of microparticles and transfer to a 1.5 mL microcentrifuge tube with 1 mL of 70 % ethanol (v/v).
3. Vortex vigorously for 3–5 min and place on ice for 15 min.
4. Pellet the microparticles for 5 s in a microcentrifuge. Discard the supernatant.
5. Wash microparticles by vortexing for 1 min in 1 mL of sterile water. Allow the particles to settle for 1 min followed by a rapid pulse in the microcentrifuge. Repeat three times.
6. After the third wash, add 500 μL of sterile 50 % glycerol to bring the microparticle concentration to 60 mg/mL (*see* **Note 7**).

3.4 Coating Microcarriers with DNA

1. Vortex the microcarriers for 5 min to resuspend and pipette 50 μL (3 mg) of microparticles to a 1.5 mL microcentrifuge tube (*see* **Note 8**).
2. Add the following to the microparticles, in the order listed, with vortex mixing of each ingredient:
   (a) 5 μL of plasmid DNA at a concentration of 1 μg/μL (*see* **Note 9**).
   (b) 50 μL of 2.5 M $CaCl_2$.
   (c) 20 μL of 0.1 M spermidine.
3. Allow the microparticles to settle for 1 min and then microcentrifuge briefly to pellet. Remove the liquid and discard.
4. Add 200 μL of 70 % ethanol. Microcentrifuge briefly to pellet. Remove the liquid and discard.
5. Add 200 μL of 100 % ethanol. Microcentrifuge briefly to pellet. Remove the liquid and discard. Allow remaining pellet to air-dry briefly to remove any visible liquid.
6. Add 48 μL of 100 % ethanol. Resuspend the pellet by vortexing for 2–3 s.

3.5 Loading DNA-Coated Microparticles onto the Macrocarrier

1. Note that the macrocarrier disks and holders should be sterilized by autoclaving or with a 100 % ethanol wash prior to assembly. The rupture disk retaining cap should be similarly sterilized. All work should be performed in a laminar flow hood that has been sterilized with 70 % ethanol.
2. Assemble the macrocarrier disks onto their metal holders using sterile forceps.
3. For each bombardment, pipette one 6 μL drop (approximately 500 μg) of the microparticle solution into the center of the rupture disk. Allow the ethanol to evaporate. The loaded macrocarriers should be used within 2 h.
4. Using sterile forceps, quickly rinse a rupture disk in 70 % isopropanol and place it immediately into the rupture disk holder (*see* **Note 10**).
5. Ensure that the helium tank pressure is set at 200 dpi above the intended rupture pressure (1,100 dpi for 900 dpi rupture disks). The rupture disk pressure required for optimal transformation efficiency should be empirically determined, found to be 1,100 dpi for the cell suspensions described [15].
6. The PDS1000-Helium Biolistic Delivery Device requires an applied of vacuum of 28 in. of mercury. The agar plate containing the filter paper and suspension cells should be placed on the target shelf at 9 cm from the stopping screen.

3.6 Selection of Plastid-Transformed Colonies and Plants

1. On the day after bombardment, the filter paper carrying the cells should be moved from the Bombardment Media plates to Selection Media agar plates. Selection is typically carried out in the light using spectinomycin at 750 μg/mL (*see* **Note 11**).
2. Bombarded plates should be incubated in a controlled environment at 25 °C with 16 h light/8 h dark or continuous light. Greening cell colonies should become visible within 6–9 weeks (Fig. 1e, f).
3. Candidate plastid-transformed cell colonies should be transferred to fresh Selection Medium agar plates to amplify resistant cells (*see* **Note 12**). As an example, transformed colonies may also be identified at an early stage using a *gfp* reporter gene (Fig. 1g).
4. Resistant colonies can be tested for homoplasmy for the selectable marker and gene of interest using standard Southern blot or PCR procedures.
5. Once colonies with homoplasmic plastids are identified, cells can be stimulated to regenerate plants by placing cut pieces onto Plant Regeneration Medium.

4 Notes

1. Petit Havana cultivar was originally used for plastid transformation [1], but any cultivar of tobacco that has high transformation and regeneration potential may be used.
2. Suspension Medium should be filter-sterilized.
3. Suspension cultures should be allowed to settle for 10 min prior to subculture. Move only small floating cell clumps to new medium using a 20 mL pipette. Continued subculture of only the smallest cell clumps will result in a fine suspension cell culture.
4. Subculturing twice per week with bombardment at 1 day after subculture was found to have substantially higher plastid transformation frequency than less frequent subculture routines. We speculate that the rapid burst of cell and plastid division following subculture [22] may influence plastid transformation efficiency.
5. Using a Corning 500 mL flask with 0.45 μm vacuum filter system, small holes were cut into the 0.45 μm Whatman filter paper using sterile forceps to allow for more rapid adherence of the cells to the paper.
6. The influence of osmoticum treatment on plant plastid transformation has not been extensively studied. However, our unpublished data suggests that this treatment slightly increases transformation frequency.

7. Tungsten microparticles should be stored at −20 °C to prevent oxidation. Gold microparticles may be stored at 4 °C or at room temperature. Store dry tungsten and gold microparticles in a dry, nonoxidizing environment to minimize agglomeration.
8. The indicated volumes are sufficient for six bombardments.
9. The plasmid will carry the *aadA* gene suitable for selection of plastid transformants plus any additional plastid transgenes of interest. Plasmid DNA should be purified to high quality.
10. Isopropanol will evaporate on its own prior to reaching the required vacuum pressure.
11. Growth of dark-grown suspension cells was only moderately inhibited at even the highest spectinomycin concentrations tested, indicating that selection in the dark would be very difficult using this antibiotic. In the light, spectinomycin levels of 750 μg/mL were required to cause bleaching of wild-type cells with limited inhibition of cell growth.
12. Note that in most cases, the resistant cell colonies are heteroplasmic, containing both transformed and nontransformed plastids, until the second subculturing step.

References

1. Svab Z, Hajdukiewicz P, Maliga P (1990) Stable transformation of plastids in higher plants. Proc Natl Acad Sci U S A 87:8526–8530
2. Meyers B, Zaltsman A, Lacroix B, Kozlovsky SV, Krichevsky A (2010) Nuclear and plastid genetic engineering of plants: comparison of opportunities and challenges. Biotechnol Adv 28:747–756
3. Maliga P, Bock R (2011) Plastid biotechnology: food, fuel and medicine for the 21st century. Plant Physiol 155:1501–1510
4. Staub JM (2002) Expression of recombinant proteins *via* the plastid genome. In: Vinci VA, Parekh SR (eds) Handbook of industrial cell culture: mammalian, microbial and plant cells. Humana, Totowa, NJ, pp 259–278
5. Ye GN, Hajdukiewicz PT, Broyles D, Rodriguez D, Xu CW, Nehra N, Staub JM (2001) Plastid-expressed 5-enolpyruvylshikimate-3-phosphate synthase genes provide high level glyphosate tolerance in tobacco. Plant J 25:261–270
6. Bock R (2007) Plastid biotechnology: prospects for herbicide and insect resistance, metabolic engineering and molecular farming. Curr Opin Biotechnol 18:100–106
7. Staub JM, Garcia B, Graves J, Hajdukiewicz PT, Hunter P, Nehra N, Paradkar V, Schlittler M, Carroll JA, Spatola L, Ward D, Ye G, Russell DA (2000) High-yield production of a human therapeutic protein in tobacco chloroplasts. Nat Biotechnol 18:333–338
8. Cardi T, Lenzi P, Maliga P (2010) Chloroplasts as expression platforms for plant-produced vaccines. Expert Rev Vaccines 9:893–911
9. Daniell H, Singh ND, Mason H, Streatfield SJ (2009) Plant-made vaccine antigens and biopharmaceuticals. Trends Plant Sci 12:669–679
10. Bock R, Warzecha H (2010) Solar-powered factories for new vaccines and antibiotics. Trends Biotechnol 5:246–252
11. Khan MS, Maliga P (1999) Fluorescent antibiotic resistance marker for tracking plastid transformation in higher plants. Nat Biotechnol 17:910–915
12. Lee SM, Kang K, Chung H, Yoo SH, Xu XM, Lee SB, Cheong JJ, Daniell H, Kim M (2006) Plastid transformation in the monocotyledonous cereal crop, rice (*Oryza sativa*) and transmission of transgenes to their progeny. Mol Cells 21:401–410
13. Sidorov V, Staub JM, Wan Y, Ye G (2008) Plastid transformation of maize. US Patent Application 20080289063A1, 2008
14. Nagata T, Kumagai F (1999) Plant cell biology through the window of the highly synchronized tobacco BY-2 cell line. Methods Cell Sci 21:123–127

15. Langbecker CL, Ye GN, Broyles DL, Duggan LL, Xu CW, Hajdukiewicz PT, Armstrong CL, Staub JM (2004) High-frequency transformation of undeveloped plastids in tobacco suspension cells. Plant Physiol 135:39–46
16. Michoux F, Ahmad N, McCarthy J, Nixon PJ (2010) Contained and high-level production of recombinant protein in plant chloroplasts using a temporary immersion bioreactor. Plant Biotechnol J 9:575–584
17. Chiyoda S, Linley PJ, Yamato KT, Fukuzawa H, Yokota A, Kohchi T (2007) Simple and efficient plastid transformation system for the liverwort *Marchantia polymorpha* L. suspension-culture cells. Transgenic Res 16:41–49
18. Murashige T, Skoog F (1962) A revised medium for the growth and bioassay with tobacco tissue culture. Physiol Plant 15: 473–497
19. Gamborg OL, Miller RA, Ojima K (1968) Nutrient requirements of suspension cultures of soybean root cells. Exp Cell Res 50:151–158
20. Zoubenko OV, Allison LA, Svab Z, Maliga P (1994) Efficient targeting of foreign genes into the tobacco plastid genome. Nucleic Acids Res 19:3819–3824
21. Randolph-Anderson B, Boynton JE, Dawson J, Dunder E, Eskes R, Gillham NW, Johnson A, Perlman PS, Suttie J, Heiser WC (1995) Submicron gold particles are superior to larger particles for efficient biolistic transformation of organelles and some cell types. Bio-Rad Technical Bulletin 2015
22. Suzuki T, Kawano S, Sakai A, Fujie M, Kuroiwa H, Nakamura H, Kuroiwa T (1992) Preferential mitochondrial and plastid DNA synthesis before multiple cell divisions in *Nicotiana tabacum*. J Cell Sci 130:831–837

Chapter 11

Tryptophan and Indole Analog Mediated Plastid Transformation

Pierluigi Barone, Xing-Hai Zhang, and Jack M. Widholm

Abstract

A nonantibiotic/herbicide-resistance selection system for plastid transformation is described here in technical detail. This system is based on the feedback-insensitive anthranilate synthase (AS) α-subunit gene of tobacco (ASA2) as a selective marker and tryptophan (Trp) or indole analogs as selection agents. AS catalyzes the first reaction in the Trp biosynthetic pathway, naturally compartmentalized in the plastids, by converting chorismate to anthranilate and is subjected to feedback inhibition by Trp. In addition to Trp, various Trp analogs and indole compounds that can be converted to Trp analogs can also inhibit AS activity and therefore are toxic to cells. When cells are made to express the feedback-insensitive ASA2, they acquire resistance to these analogs and can be selected for during transformation process. We have demonstrated the feasibility of this selection system in tobacco (*Nicotiana tabacum* L. cv. *Petit Havana*). ASA2-expressing transplastomic plants were obtained on medium supplemented with either 7-methyl-DL-tryptophan (7-MT) or 4-methylindole (4-MI). These plants show normal phenotype and fertility and transmit the resistance to the selection agents strictly maternally.

Key words Anthranilate synthase (AS), 7-Methyl-DL-tryptophan (7-MT), 4-Methylindole (4-MI), Nonantibiotic selection, Plastid transformation, Selectable marker

1 Introduction

Most transplastomic lines have been selected using the antibiotics spectinomycin or streptomycin and the marker genes aminoglycoside 3″-adenylyltransferase (*aadA*) or the mutant plastid 16S rRNA (*rrn16*) gene [1]. These antibiotic selection agents bleach the tissues so that resistant clones are usually visualized by their green color to aid the resistance selection. Some tissue cultures such as *Zea mays L. ssp. mays* (maize) do not generally turn green in culture until shoot, and leaf structures are formed. Leaf tissue cannot be used since the leaf cells cannot be induced to divide. Thus, it would be desirable to have a selection system for plastid transformation that does not require greening. Toxic analogs of certain amino acids including tryptophan (Trp), lysine (Lys), and methionine (Met)

Pal Maliga (ed.), *Chloroplast Biotechnology: Methods and Protocols*, Methods in Molecular Biology, vol. 1132,
DOI 10.1007/978-1-62703-995-6_11, © Springer Science+Business Media New York 2014

might be possible selection agents since mutant cell lines have been selected with resistance to specific analogs, and a Trp analog has been used successfully to select transplastomic tobacco plants [2]. Cells selected as resistant to Trp, Lys, and Met analogs usually express a feedback-resistant form or higher levels of a control enzyme in the respective biosynthetic pathway [3–6]. These changes result in reversal of the toxicity of the analogs and in higher levels of the target free amino acid. Thus, one would expect that transforming with the genes of the enzymes that were altered should provide resistance that could be selected for. These amino acids, Trp, Lys, and Met, are essential amino acids, all of which are synthesized inside the plastids. The biosynthetic enzymes are encoded in the nucleus, but the enzymes are transported into the plastids. Thus, expressing the selectable marker genes in the plastid should be advantageous for the selection of transplastomic cells. These markers should also be dominant traits so some level of expression should enable selection. We have successfully used the tobacco anthranilate synthase (AS) α-subunit gene (ASA2) and the toxic Trp analog 7-methyltryptophan (7-MT) and the indole analog 4-methylindole (4-MI) that is converted to the toxic Trp analog 4-methyltryptophan to select transplastomic tobacco plants [2].

The AS enzyme converts chorismic acid to anthranilate, as the first step in the Trp biosynthesis branch, and is inhibited by Trp [7, 8]. The enzyme in plants is a heterotetramer consisting of two α-subunits and two β-subunits. The α-subunits contain the allosteric site that binds the feedback inhibitor Trp and false feedback inhibitors, Trp analogs. ASA2 is a naturally occurring feedback-insensitive form of the α-subunit, and high expression of this form can confer resistance to toxic Trp analogs and anthranilate and indole analogs that are converted by plant enzymes to toxic Trp analogs [4]. ASA2 expression also results in higher free Trp levels. In our selection of transplastomic tobacco plants using ASA2 and 7-MT and 4-MI, the selection efficiency was lower than with aadA and spectinomycin, but it may be possible to improve this by using both the α- and β-subunits of tobacco AS together as we have found with nuclear transformation of tobacco (F.-Y. Tsai, X.-H. Zhang and J.M. Widholm, unpublished). This might be especially important when ASA2 is used in other species since in this case the tobacco α-subunit must bind to the native β-subunit to form an active enzyme. However, there is evidence with soybean, Arabidopsis, and potato, among others, that the α-subunit from tobacco or rice, via nuclear transformation, can interact with the native β-subunit to form a feedback-insensitive enzyme [9–11].

To our knowledge only ASA2 of all the possible amino acid analog-related selectable marker genes has been used successfully to select for transplastomic plants, but it seems reasonable to believe that others could be used and that ASA2 might be used successfully with species other than tobacco.

2 Materials

2.1 Tissue Culture and Particle Bombardment

2.1.1 Plant Materials

1. Seeds of tobacco (*Nicotiana tabacum* L. cv. *Petit Havana*).
2. In vitro plants of *Nicotiana tabacum* L. cv. *Petit Havana*.

2.1.2 Media

1. RMOP: RMOP medium is used for selection of plastid transformants and for leaf callus/shoot formation test. The medium is prepared by dissolving in MilliQ grade water 4.33 g/l of Murashige and Skoog (MS) basal salt mixture (PhytoTechnology Laboratories®) and 30 g/l of sucrose supplemented with 0.1 mg/l naphthalene acetic acid (NAA), 1 mg/l 6-benzyl-aminopurine (BAP), 1 mg/l thiamine-HCl, and 100 mg/l myo-inositol. The pH is adjusted to 5.8 with 1 N KOH and the volume is brought to 1,000 ml with MilliQ grade water. 2 g of Gelzan™ (PhytoTechnology Laboratories®) is added to the medium that is autoclaved for 20 min at 121 psi in a 2 l bottle and allowed to cool to 60 °C before any analog is added. The medium is then poured into deep Petri dishes (100 mm×25 mm).
2. MS: MS medium is used to germinate seeds for segregation test and seedling growth inhibition test, to regenerate plant, and to induce root formation. MS medium is prepared by mixing 4.3 g of Murashige and Skoog (MS) basal salt mixture (PhytoTechnology Laboratories®) and 30 g of sucrose in a 2 l bottle and by adjusting the pH to 5.8 with 1 N KOH. Finally, the volume is made up to 1,000 ml and 2 g of Gelzan™ (PhytoTechnology Laboratories®) is added. The solution is autoclaved for 20 min at 121 psi and allowed to cool to 60 °C before any analog is added. The growth medium is poured into deep Petri dishes (100-mm×25-mm) or Magenta boxes.

2.1.3 Analog Stock Solutions

1. 7-Methyl-DL-tryptophan (7-MT): Weigh out 109.125 mg of 7-MT (Sigma-Aldrich) into a 15 ml polypropylene (PP) centrifuge tube (Thermo Scientific), add 2 ml of 1 M HCl, and sonicate it for 15 s. Bring up to final volume of 10 ml with water at a final concentration of 50 mM. The solution is filter-sterilized under a laminar flow hood and should be used immediately. *See* **Note 1**.
2. 4-Methylindole (4-MI): Weight out 65.59 mg of 4-MI (Acros Organics) into a 15 ml polypropylene (PP) centrifuge tube (Thermo Scientific), add 2 ml of 100 % ethanol (molecular biology grade), and vortex at high setting for 30 s. Bring up to final volume of 10 ml with water at a final concentration of 50 mM. The solution is filter-sterilized under a laminar flow hood and should be used immediately.

2.1.4 Particle Bombardment

1. Single-Channel Pipettes and autoclaved wide bore pipette tips.
2. Eppendorf DNA LoBind Tubes.
3. Particle Gun PDS1000/He (Bio-Rad).
4. Helium gas.
5. Rupture disks 1,100 psi; rupture disk retaining caps; stopping screens; macrocarrier holders; macrocarriers (Bio-Rad).
6. 0.6 μm gold microcarriers (Bio-Rad).
7. Benchtop microcentrifuge.
8. Ethanol (100 and 70 %).
9. 2.5 M calcium chloride: Weigh 184 g of $CaCl_2$ and dissolve in 50 ml of water; filter sterilize. May be stored at 4 °C for short periods; do not freeze.
10. 0.1 M spermidine-freebase: Since spermidine is highly hygroscopic, take a 1 g unopened bottle stored at 4 °C and add 6.8 ml of sterile water and filter sterilize; store a 500 μl aliquots at −20 °C. Do not use aliquots older than 3 months.
11. Tryptophan (Trp) and indole analogs stock solutions (*see* Subheading 2.1.3).

2.2 Plant Crosses

1. Seeds of *Nicotiana tabacum* L. cv. *Petit Havana*.
2. Seeds of transplastomic *Nicotiana tabacum* L. cv. *Petit Havana*.
3. Peat Jiffy Pellets.
4. 2 gallon pots.
5. Sunshine Mix #1 / LC1 soil.
6. Tweezers.
7. Dissecting needle.
8. Corn ear shoot bags (resized according to the need).

2.3 AS Enzyme Activity Assay

1. Leaf tissue from both wild-type (WT) tobacco and plastid transformants.
2. Autoclaved porcelain mortars and pestles.
3. 15 ml polypropylene (PP) centrifuge tubes (Thermo Scientific).
4. Extraction buffer: (a) 50 mM Tris–HCl at pH 8.0; (b) 5 mM $MgCl_2$; (c) 2.5 mM DTT; (d) 100 mM NH_4Cl; (e) 20 % glycerol; (f) 1 % PVPP.
5. Refrigerated ultracentrifuge.
6. Econo-Pac 10 DG column (Bio-Rad).
7. Bio-Rad Protein Assay Kit (Bio-Rad).
8. Assay buffer: (a) 50 mM Tris–HCl at pH 7.8; (b) 5 mM $MgCl_2$; (c) 1 mM EDTA; (d) 20 % glycerol.

9. Fluorescence spectrophotometer (excitation at 340 nm and emission at 400 nm).
10. 100 mM NH_4Cl.
11. 10 mM glutamine (Gln).
12. 1 mM chorismate.
13. 0.1, 1, 10, 100, 1,000 μM Trp.
14. 1 N HCl.
15. Ethyl acetate.

2.4 Trp Measurement

1. Leaf tissue from both wild-type (WT) tobacco and plastid transformants.
2. Autoclaved porcelain mortars and pestles.
3. Liquid nitrogen.
4. 2 ml screw-cap microcentrifuge tubes (Eppendorf).
5. 0.1 N HCl.
6. Fast Prep FP120 Cell Disrupter (Savant Instruments).
7. Refrigerated microcentrifuge.
8. Ultra Free-MC filter unit (Millipore).
9. HPLC.
   (a) Adsorbosil C18 250×4.6 mm, 5 U column (Alltech Associates).
   (b) Isocratic buffer system (85 % [v/v]: 140 mm sodium acetate, 17 mm triethylamine, adjusted to pH 5.05 using phosphoric acid, and 15 %: 60 % [v/v] acetonitrile in water at 1 ml min^1).
   (c) Kratos FS 970 fluorescence detector (excitation 215 nm; emission band pass filter >375 nm).

3 Methods

3.1 Dose-Response Test

A species-specific dose-response test is required to determine the selection level at which each analog is effective in inhibiting the growth of non-transformed plant material. A range of analogs and concentrations are listed in Table 1. The concentrations listed in Table 1 are for illustrative purposes and should not be construed as a definitive range. In the case of *Nicotiana tabacum* L. cv. *Petit Havana*, a leaf callus/shoot formation test and a seedling growth inhibition test of non-transformed plant material were performed. As a result the two analogs 7-MT (300 and 75 μM) and 4-MI (300 and 75 μM) exhibited effective selection properties (*see* Fig. 1).

Table 1
List of the Trp and indole analogs and relative concentrations used for the seedling growth inhibition and leaf callus/shoot formation inhibition tests

TRP analogs	Concentrations
5-Methyltrypthophan (5MT)	10, 30, 100, and 300 μm
α-Methyltrypthophan (αMT)	1, 3, 10, 30, 100, and 300 μm
4-Methyltrypthophan (4MT)	10, 30, 100, and 300 μm
5-Hydroxyl-L-tryptophan (5HT)	10, 30, 100, and 300 μm
7-Methyl-DL-tryptophan (7MT)	10, 30, 100, 300, 400, and 500 μm
6-Methyl-DL-tryptophan (6MT)	10, 30, 100, 300, 600, and 1,000 μm
DL-4-Fluorotryptophan (4FT)	10, 30, 100, and 300 μm
DL-5-Fluorotryptophan (5FT)	10, 30, 100, and 300 μm
DL-6-fluorotryptophan (6FT)	10, 30, 100, and 300 μm
Indole analogs	**Concentrations**
6-Fluoroindole (6FI)	10, 30, 50,70, 100, and 300 μm
4-Fluoroindole (4FI)	10, 20, 30, 100, and 300 μm
5-Fluoroindole (5FI)	10, 30, 100, 150, 200, and 300 μm
5-Methoxyindole (5MI)	10, 30, 100, and 300 μm
7-Methoxyindole (7MI)	10, 30, 100, and 300 μm
6-Aminoindole (6AI)	10, 30, 100, and 300 μm
4-Methylindole (4MI)	10, 30, 100, 300, and 600 μm

3.2 Preparation of Leaf Plant Tissue

The leaf material for both the leaf callus formation test and particle bombardment is obtained from tobacco plants maintained as in vitro stock cultures in Magenta boxes with MS medium by propagation of nodal sections. Plants are initially generated aseptically from sterile seeds (*see* Subheading 3.4), and after 6 weeks stem segments with two leaves are cut out from grown plantlets and transplanted to fresh MS medium. The growth conditions are 16 h light-8 h dark photoperiod, 28 °C, and fluorescent lamps providing a photosynthetic photon flux density of 150 μmol/m^2/s at the shelf surface. The leaf material to be used as target for the biolistic delivery needs to be fully expanded, dark-green, and healthy with an approximate size of 2.5 cm × 2.5 cm (*see* **Note 2**).

3.3 Leaf Callus/ Shoot Formation Test

Using a sterile cork borer with a 6 mm inner diameter, cut out 30 leaf disks from fully expanded, dark-green leaves obtained as described in Subheading 3.2 and place them on RMOP medium supplemented with various concentrations of different Trp and indole analogs (*see* Table 1). Callus formation and shoot production

Fig. 1 Bombarded tobacco leaf material on RMOP/7-MT 300 μM (**a**) or RMOP/4-MI 300 μM (**b**) with green resistant shoots. Seed germination of WT (*Nicotiana tabacum* L. cv. *Petit Havana*) on either 4-MI (**c**) or 7-MT (**d**) 75 μM. Seedlings from seeds of self-pollinated transplastomic plant 27 on 4MI (**e**) and plant C on 7MT (**f**) 75 μM. Germination of seeds on 7 MT 75 μM from either WT plant pollinated with pollen from transplastomic plant C (**g**) or from transplastomic plant C pollinated with pollen from WT (**h**). Reproduced from ref. [2] with permission from Oxford University Press

are visually evaluated after 2, 4, and 6 weeks (*see* Fig. 1). The growth condition is 16 h photoperiod with a temperature of 28 °C under fluorescent lamps providing a photosynthetic photon flux density of 150 μmol/m^2/s at the shelf surface.

3.4 Preparation of Seeds

In a laminar flow hood, wash tobacco seeds with 70 % ethanol for 30 s in a 1.5 ml Eppendorf tube. Rinse off ethanol with sterile distilled water, three times. Surface sterilize with a solution of distilled water–Clorox (80:20; 1.25 % Na hypochlorite) with 0.1 % (vol/vol) of Tween-20™ for 20 min. Wash three times with sterile distilled water.

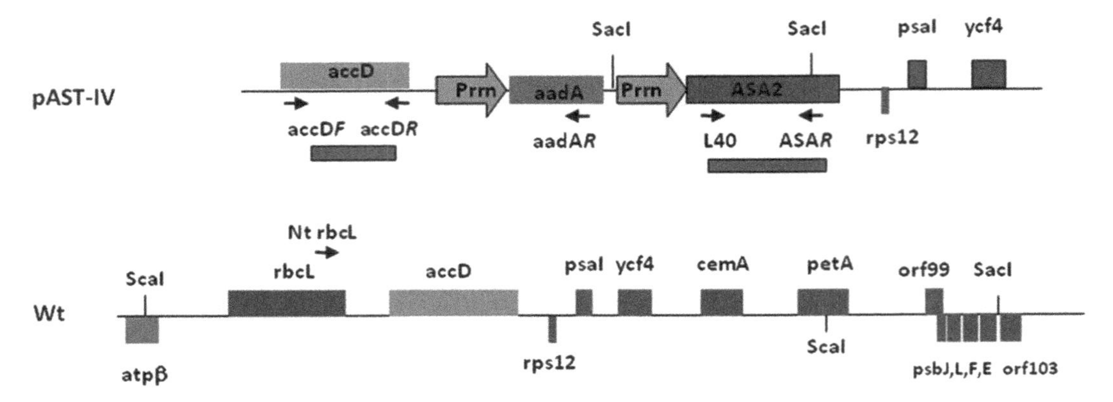

Fig. 2 Structure of the vector pASTIV used for tobacco transformation and its corresponding region of the wild-type tobacco plastid genome based on GenBank Z00044 and NC_001879. Relevant plastid genes and their orientation are presented. Prrn, plastid 16S rRNA operon promoter; *aad*A, spectinomycin resistance gene with the *Chlamydomonas* rbcL 3′-untranslated region (500 bp); ASA2, modified version of ASA2 (1,671 bp) including its termination sequence (204 bp). The locations of *Sac*I and *Sca*I used for Southern hybridization are shown. Location and orientation of PCR primers are indicated by *arrows*. accD probe (accD F–accD R) and ASA2 probe (L40–ASAR) are shown. The figure size is not to scale. Reproduced from ref. [2] with permission from Oxford University Press

3.5 Seedling Growth Inhibition Test

Place a total of 50 surface-sterilized seeds on MS medium plates supplemented with various concentrations of different Trp and indole analogs (*see* Table 1). Do not seal the Petri dishes as this will reduce gas exchange and negatively affect the germination. Seed germination and root/leaf formation are visually evaluated after 2 and 4 weeks (*see* Fig. 1). The growth condition is 16 h photoperiod with a temperature of 28 °C under fluorescent lamps providing a photosynthetic photon flux density of 150 μmol/m^2/s at the shelf surface.

3.6 Construction of Tobacco Chloroplast Expression Vector

The chloroplast transformation vector pASTIV was constructed by our lab as described previously [2] and is shown in Fig. 2. The vector contains a 1,671 bp long version of the feedback-insensitive anthranilate synthase (AS) α-subunit gene of tobacco (ASA2) isolated from the 5MT (5-methyltryptophan)-resistant tobacco cell line AB15–12-1 [3]. The putative transit peptide present in the original gene (GenBank gi: 3348123) has been deleted, and the 3′-noncoding region (204 bp) is used as termination sequence. The plasmid also has the *aadA* gene with the *Chlamydomonas rbcL* 3′-untranslated region. Both genes are driven by a chloroplast 16S rRNA promoter Prrn [12, 13]. The site of insertion in the plastid genome is the region between *accD* and *ycf4* genes at nucleotide 59029 and nucleotide 63410 (GenBank Z00044 or NC_001879), respectively (*see* Fig. 2).

3.7 Biolistic Plant Transformation of Tobacco Leaves

3.7.1 Preparation of Gold Microcarriers

1. Gold particle washing: Weigh out 9 mg of 0.6 μm gold particles into a 1.5 ml DNA LoBind Tubes (Eppendorf), add 500 μl of cold 100 % ethanol (molecular biology grade), and sonicate it for 15 s. Centrifuge the suspension for 60 s at 3,000 × *g* in a benchtop centrifuge and remove the supernatant. Add 1 ml of ice-cold, sterile distilled water, slightly finger vortex the tube to dislodge the gold pellet, and centrifuge for 60 s at 3,000 × *g*. Repeat three times to rinse off ethanol. After removing the final rinse, add 250 μl of cold sterile water to the gold pellet and sonicate it again for 15 s. Make 50 μl aliquots of the gold suspension in 1.5 ml DNA LoBind Tubes (Eppendorf). Store the gold suspension aliquots (1.8 mg of gold in 50 μl of water) at −20 °C and use within 1 month.
2. DNA coating of the gold particles: Thaw completely one of the gold suspension aliquots stored at −20 °C and sonicate it for 15 s. Add 10 μl of pASTIV vector (concentration of 1 μg/μl) and finger vortex for 5 s. Add 50 μl of 2.5 M $CaCl_2$ and vortex for 5 s and then add 20 μl of spermidine (0.1 M). After 5–10 min of continuous vortexing (low setting), allow the gold suspension to settle and centrifuge the sample at 5,000 × *g* for 15 s in a benchtop centrifuge. Discard the supernatant and resuspend the pellet completely in 250 ml of cold 100 % ethanol (molecular biology grade) by gently tapping the tube. Centrifuge the mixture for 15 s at 5,000 × *g*, discard the supernatant, and resuspend the pellet in 36 μl of cold 100 % ethanol. Keep the tube with the coated gold particles on ice until ready to use. Make sure to use the DNA-coated gold particles within 2 h.

3.7.2 Leaf Particle Bombardment

1. Sterilize the particle gun by spraying the chamber and its components with 70 % ethanol and let it air-dry.
2. Place one fully expanded, dark-green leaf of tobacco plants grown in sterile culture abaxial side up on each RMOP medium plate.
3. Resuspend the coated gold particle mixture completely by vortexing to eliminate any clumps. Pipette 6 μl of the mixture onto the center of each macrocarrier and spread it out evenly. Between each loading of the DNA-coated gold particles onto the macrocarrier, make sure to vortex the mixture for at least 10 s and pipette it immediately.
4. Place the microcarrier launch assembly in the top slot (L0) inside the bombardment chamber.
5. Select the following bombardment parameters:
   (a) Rupture disk rating 1,100 psi.
   (b) Gap distance = 6 mm (distance from rupture disk to macrocarrier).

(c) Target shelf for tissue is L3 = 9 cm (3rd shelf from the bottom).

(d) Vacuum pressure of 28 in Hg.

(e) Bombard the sample following the manufacturer's directions.

3.8 Selection and Regeneration of Chloroplast Transgenic Plants

1. Incubate the plates with the bombarded leaves in the dark at 28 °C for 2 days.
2. Cut out leaf pieces (6 mm × 6 mm) and transfer them to RMOP medium containing the analog (e.g., 300 μM of 4-MI or 7-MT). Place the bombarded (abaxial) side of the leaf in direct contact with the selection medium.
3. Incubate the plates with the leaf pieces at 16 h photoperiod with a temperature of 28 °C under fluorescent lamps providing a photosynthetic photon flux density of 150 μmol/m^2/s at the shelf surface.
4. Every other week transfer the plant material onto new fresh analog-containing RMOP medium and further chop into small pieces when needed in order to ensure a continuous contact between the leaf section and the selection medium.
5. 6–12 weeks after bombardment, the transgenic clones appear as green shoots (1st round of selection).
6. A first PCR screening (*see* Subheading 3.9.1) of the plastid transformants can be performed at this stage of the selection/regeneration protocol to discriminate between the true chloroplast transgenic plants and escapes or nuclear transgenic plants.
7. Cut out small sections from leaves of the primary transformants obtained from the 1st round of selection and place them again on RMOP selection medium (2nd round of selection).
8. Repeat **steps 3–5** to obtain secondary transformants.
9. Cut out small sections from leaves of the secondary plastid transformants and place them one more time on RMOP selection medium (3rd round of selection) to obtain green shoots (*see* Fig. 1).
10. Detach the green shoots from the leaf tissue and transfer them to MS medium supplemented with a lower concentration of the analog (e.g., 75 μM of 4-MI or 7-MT) for root formation.
11. After 4–6 weeks when roots are well established, gently wash off the agar-solidified rooting medium from rooted plantlets and move them to soil in the green house.
12. A second PCR screening is performed at this stage to confirm the chloroplast transgenic nature of the plants (*see* Subheading 3.9.1).
13. Collect seeds from T0 generation transgenic plants.

3.9 Molecular Analyses of Transformed Plants

3.9.1 PCR Analysis

1. Extract total cellular DNA using the DNeasy Plant Mini Kit (Qiagen) from both WT and plastid transformants generated after the 3rd round of selection. Alternatively, total cellular DNA can be extracted using the CTAB extraction method [14].
2. Quantify the DNA concentration using UV spectrophotometer readings at 260 and 280 nm and check its integrity on a 1 % or 2 % agarose gel.
3. Mix in a 0.2 ml PCR tube:
   (a) 100 ng total cellular DNA.
   (b) 5 μl of 10× PCR reaction buffer (New England Biolabs Inc.).
   (c) 1 μl of 10 μM Primer 1 (this primer anneals within the transgene cassette, e.g., aadAR 5′-ACCTTAGTGATCTCGCCTTTCACG-3′; *see* Fig. 2).
   (d) 1 μl of 10 μM Primer 2 (this primer anneals to the native chloroplast genome, e.g., NtrbcL 5′-TTTGCAGCAGTGGACGTTTTGGATAA-3′; *see* Fig. 2).
   (e) 200 μM each of dNTPs final concentration (Invitrogen, Carlsbad, CA).
   (f) 0.5 μl (2.5 U) Taq DNA polymerase (New England Biolabs Inc.).
   (g) Bring to final volume (50 μl) with sterile distilled water.
4. Perform PCR analysis under following reaction conditions:
   (a) Denaturation at 94 °C for 5 min.
   (b) 35 cycles of denaturation at 94 °C for 30 s, annealing at 58 °C for 30 s, and extension at 72 °C for 1.5 min.
   (c) Extension at 72 °C for 5 min.
5. Resolve the PCR products on 0.8 % agarose gel stained with ethidium bromide.
6. Plastid transgenic plants will show the PCR product confirming the site-specific integration of the expression cassettes in the plastid genome (ptDNA), while WT escapes and nuclear transgenic plants are not expected to produce a PCR product (*see* Fig. 3).

3.9.2 Southern Analysis

1. Isolate total cellular DNA from the chloroplast transgenic and WT plants, using a DNeasy Plant Maxi Kit (Qiagen). Alternatively, total cellular DNA can be extracted using the CTAB extraction method.
2. Quantify the DNA concentration using UV spectrophotometer readings at 260 and 280 nm and check its integrity on a 1 % or 2 % agarose gel.

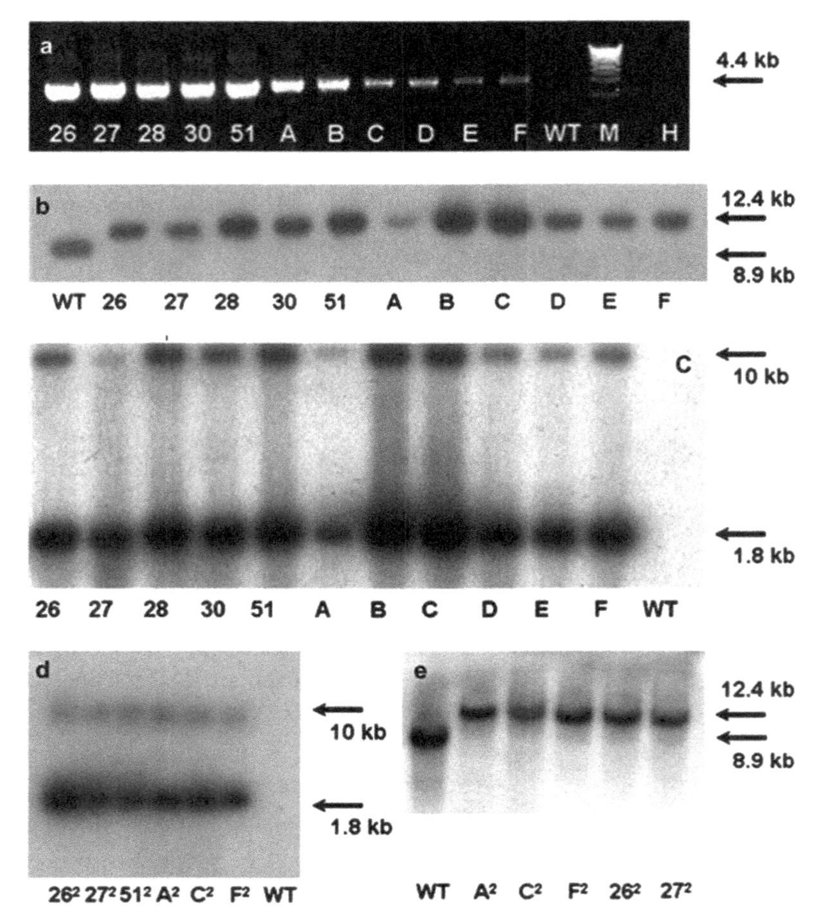

Fig. 3 Plastid genomic analysis of site-specific integration of the *ASA2*gene: (**a**) PCR amplification with primer aadAR (transgene cassette specific) and primer Nt*rbc*L (plastid site specific; *see* Fig. 2); for Southern blot DNA was digested with either *Sca*I and hybridized to the plastid site-specific probe *accD* (**b, e**) or *Sac*I and hybridized to an ASA2 probe (**c, d**). Lines 26–51 selected on 300 μM 7MT and lines A–F selected on 300 μM 4MI. WT, untransformed control; H, blank control/water; M, DNA ladder. Lines 26^2–51^2F_1 crosses (male non-transgenic × female transgenic) germinated on 75 μM 7MT and lines A^2–F^2F_1 crosses (male non-transgenic × female transgenic) germinated on 75 μM 4 MI. WT, untransformed control; H, blank control/water. Reproduced from ref. [2] with permission from Oxford University Press

3. Digest 5 μg DNA per sample using 5 U/μg of restriction enzyme(s) in the appropriate buffer following the manufacturer's directions. The chosen enzyme(s) should cut the DNA creating fragment(s) that consent to clearly distinguish between WT, heteroplastic, and homoplastic lines.
4. Separate the digested DNA by 0.8 % agarose gel electrophoresis in 1× TAE buffer, blot it onto a nylon membrane

(Hybond-N+, GE Healthcare Limited) using capillary transfer, and cross-link it to the membrane by UV light.

5. Probe the Southern blot with a radiolabeled (or non-radioisotope-labeled) sequence according to the standard protocols.
6. Due to the site-specific integration of transgene(s) into the chloroplast genome, the DNA fragment of a homoplastic plant will show on the autoradiography film as a higher molecular weight band as compared to the one from a WT plant (*see* Fig. 3). Southern blot analysis confirms not only the homoplastic condition of the transformants but also their degree of homoplastomy since a heteroplastomic event will show the smaller WT fragment along with the larger transgenic one.

3.10 Maternal Inheritance

1. Soak Peat Jiffy Pellets in water for 20 min and plant 10 seeds (1 seed/pellet) from both the produced transplastomic and the WT plants. Put the pellets in a tray covered with a plastic dome to maintain humidity and keep it in growth chamber at 28 °C and 16 h photoperiod.
2. Once the seedlings are big enough, transfer them to the 2 gallon pots filled with Sunshine Mix #1 / LC1 soil in the green house.
3. At the beginning of flowering (*see* **Note 3**), start to remove the anthers with tweezers prior to their dehiscence and cover the emasculated flower(s) with a corn ear shoot bags resized to fit the tobacco flower.
4. At the beginning of color formation of the petals of the emasculated flowers (1–2 days after emasculation), collect with a dissecting needle paternal pollen (from WT or from the transplastomic plants) and resuspend it in a few drops of sterile distilled water.
5. Transfer the pollen suspension onto the top of the stigma of each emasculated flower and bag the flower again. Perform the following crosses:
   (a) Female non-transgenic (WT) × male transgenic (homoplastic).
   (b) Female transgenic (homoplastic) × male non-transgenic (WT).
   (c) Female transgenic (homoplastic) × male transgenic (homoplastic).
6. If pollination and subsequent fertilization do not occur, the flowers will abort and abscise (*see* **Note 4**).
7. Harvest the seeds after 3–5 weeks.
8. Sterilize the seeds as described in Subheading 3.4 and proceed with the seedling growth inhibition test as described in Subheading 3.5.

9. Seeds from cross (a) will not be able to germinate on selection medium, while seeds from cross (b) and (c) will germinate on selection medium confirming integration of the transgene(s) in the ptDNA and its expected maternal inheritance (*see* Fig. 1).

3.11 AS Enzyme Activity Assay

1. Collect 2 g of leaves from both plastid transformants and WT plants and homogenize with 1.5 volume of ice-cold extraction buffer using a sterile porcelain mortar and pestle.
2. Transfer the homogenate into a 15 ml PP tube and centrifuge at 30,000 × *g* at 4 °C for 10 min.
3. Desalt the supernatant using an Econo-Pac 10 DG column according to the manufacturer's directions.
4. Determine protein concentration using the Bio-Rad Protein Assay Kit following the manufacturer's instructions.
5. Mix 1.5 ml DNA LoBind Tubes (Eppendorf):
   (a) 100 μl of desalted extract.
   (b) 400 μl of assay buffer.
   (c) 1 mM chorismate (*see* **Note 5**).
   (d) 100 mM NH_4Cl or 10 mM Gln (for α-subunit or holoenzyme activity, respectively) (*see* **Note 5**).
   (e) Trp at different concentrations (0.1, 1, 10, 100, 1,000 μM).
6. Incubate at 30 °C for 30 min.
7. Stop the reaction by adding 100 μl of 1 N HCl.
8. Add 1 ml of ethyl acetate and centrifuge at 16,000 × *g* for 3 min in a microcentrifuge at room temperature.
9. Quantify the anthranilate produced in the upper phase with a fluorescence spectrophotometer (excitation at 340 nm and emission at 400 nm).
10. AS (α-subunit or holoenzyme) activity is measured as the conversion rate of chorismate to anthranilate and, in the transgenic plants, is less sensitive to Trp inhibition compared to the one from the untransformed tobacco (*see* Fig. 4).

3.12 Trp Measurement

1. Precool a sterile mortar/pestle with liquid nitrogen.
2. Grind with liquid nitrogen into a coarse powder a young fully expanded leaf of either a 35-day-old WT plant or a homoplastic plant.
3. Transfer immediately after grinding approximately 100 mg of tissue into a 2 ml screw-cap microcentrifuge tube.
4. Add 200 μl of 0.1 N HCl and homogenized the sample using a Fast Prep FP120 Cell Disrupter at 4.0 m/s for 20 s.

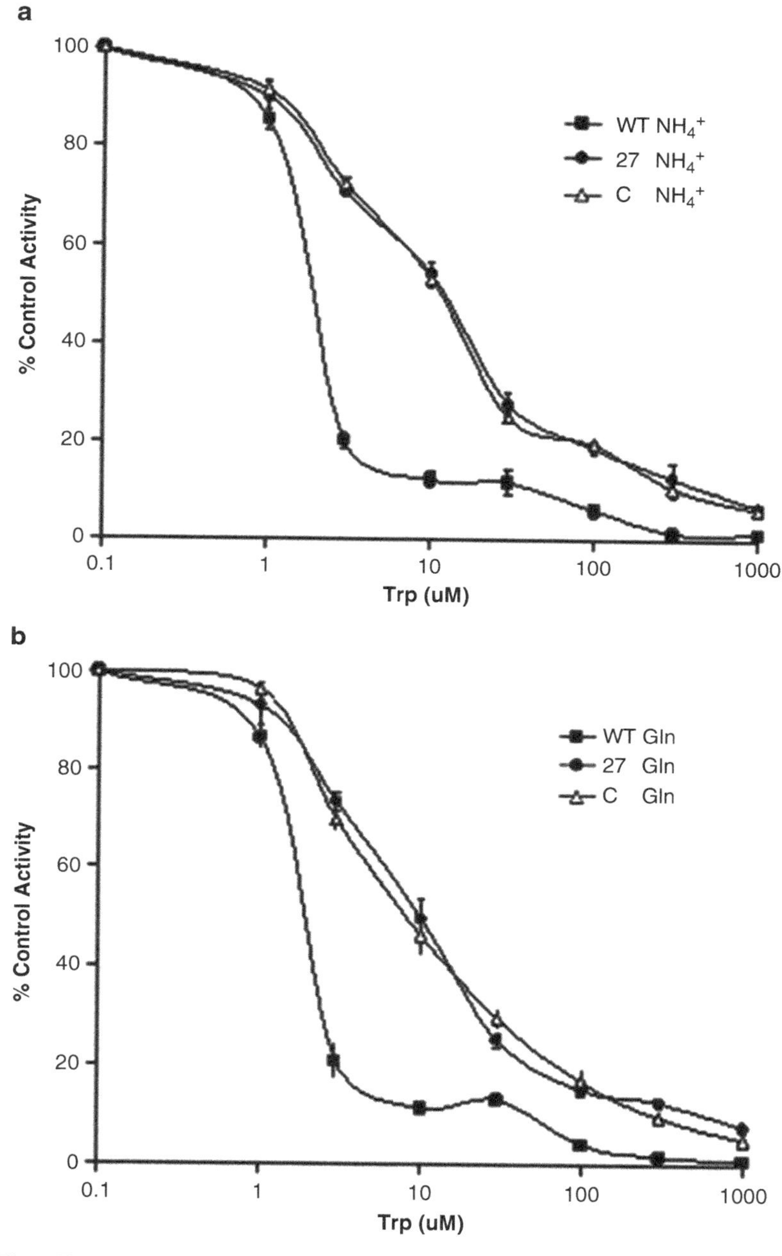

Fig. 4 Trp inhibition of AS enzyme activity in total leaf extracts from wild-type (WT) (filled squares) and transplastomic plants 27 selected on 7-MT (*filled circles*) and C selected on 4-MI (*open triangles*), with 100 mM NH4Cl (**a**) (α-subunit activity) or 10 mM Gln (**b**) (holoenzyme activity) as substrate. The values are means ± SD of three replicates. Reproduced from ref. [2] with permission from Oxford University Press

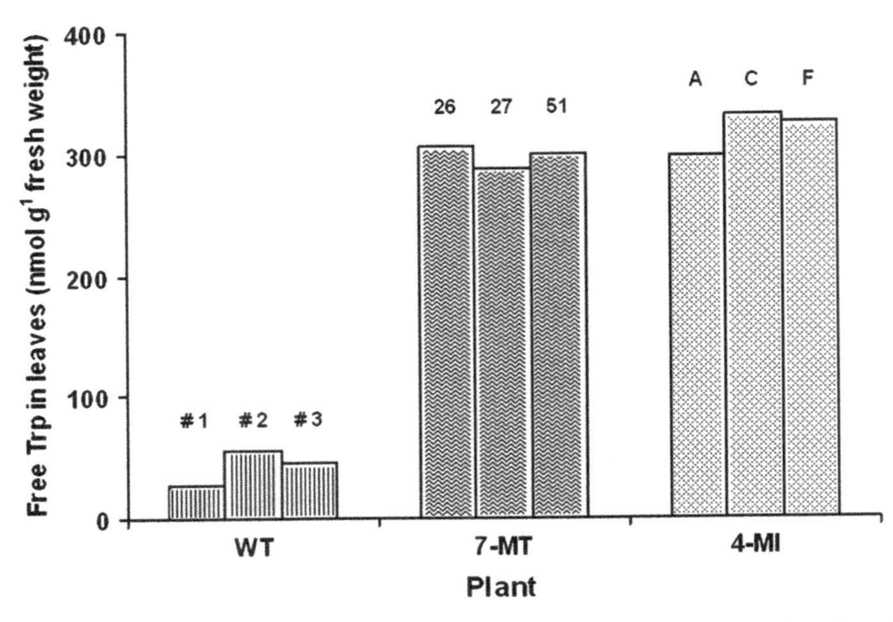

Fig. 5 Free Trp content in leaves of three WT plants and three transplastomic plants selected on either 7MT (26, 27, 51) or 4MI (A, C, F). Reproduced from ref. [2] with permission from Oxford University Press

5. Centrifuge at 4 °C and 13,000 × *g* for 10 min.
6. Transfer the supernatant to a new 2 ml screw-cap tube, freeze the sample in liquid nitrogen, thaw it, and centrifuge again as in **step 5** to completely sediment debris.
7. Filter the supernatant using an Ultra Free-MC (10,000) filter unit according to the manufacturer's directions.
8. Dilute the filtrate 1: 10 with 0.1 N HCl.
9. Use 10 μl of the diluted sample for the HPLC analyses.
10. Since the ASA2 enzyme is insensitive to feedback inhibition by Trp, the transplastomic lines overexpressing ASA2 show a higher Trp level than the WT (*see* Fig. 5).

4 Notes

1. When the 7-MT stock solution (50 mM in 0.2 N HCl) is added to the medium after autoclaving, make sure to readjust the pH to the optimal value of 5.8 by adding the same volume of filter-sterilized 0.2 N KOH.
2. Tissue obtained from plants that are flowering is not recommended since the age of explants negatively affects the transformation efficiency. The use of leaves harvested from greenhouse grown plants is not suggested due to the difficulty of sterilizing the material.

3. Avoid flowers that have begun to develop substantial coloration in the unopened corolla, as these may have shed some pollen internally.
4. Damage to the pistil during emasculation, low pollen fertility, extreme temperatures, etc. can decrease the rate of success of pollination, so always make extra crosses to be sure that at least a few will produce seeds.
5. As explained in "Introduction," plant AS is a heterotetramer of two α- and two β-subunits: The α-subunit binds the substrate chorismate and carries out its aromatization, while the β-subunit transfers an amino group from the other substrate glutamine. This glutamine-dependent AS reaction requires both α- and β-subunits. The α-subunit alone can synthesize anthranilate from chorismate using ammonia as the amino donor rather than glutamine, if ammonia is present in high concentrations.

References

1. Maliga P (2004) Plastid transformation in higher plants. Ann Rev Plant Biol 55:289–313
2. Barone P, Zhang XH, Widholm JM (2009) Tobacco plastid transformation using the feedback-insensitive anthranilate synthase alpha-subunit of tobacco (ASA2) as a new selectable marker. J Exp Bot 60:3195–3202
3. Song HS, Brotherton JE, Gonzales RA, Widholm JM (1998) Tissue culture-specific expression of a naturally occurring tobacco feedback-insensitive anthranilate synthase. Plant Physiol 117:533–543
4. Widholm JM (1981) Utilization of indole analogs by carrot and tobacco cell tryptophan synthase in vivo and in vitro. Plant Physiol 67:1101–1104
5. Negrutiu I, Cattoirreynearts A, Verbruggen I, Jacobs M (1984) Lysine overproducer mutants with an altered dihydrodipicolinate synthase from protoplast culture of *Nicotiana sylvestris* (Spegazzini and Comes). Theor Appl Genet 68:11–20
6. Gonzales RA, Das PK, Widholm JM (1984) Characterization of cultured tobacco cell-lines resistant to ethionine, a methionine analog. Plant Physiol 74:640–644
7. Bohlmann J, Lins T, Martin W, Eilert U (1996) Anthranilate synthase from Ruta graveolens—duplicated AS alpha genes encode tryptophan-sensitive and tryptophan-insensitive isoenzymes specific to amino acid and alkaloid biosynthesis. Plant Physiol 111:507–514
8. Li J, Last RL (1996) The Arabidopsis thaliana trp5 mutant has a feedback-resistant anthranilate synthase and elevated soluble tryptophan. Plant Physiol 110:51–59
9. Inaba Y, Brotherton JE, Ulanov A, Widholm JM (2007) Expression of a feedback insensitive anthranilate synthase gene from tobacco increases free tryptophan in soybean plants. Plant Cell Rep 26:1763–1771
10. Ishihara A, Asada Y, Takahashi Y, Yabe N, Komeda Y, Nishioka T, Miyagawa H, Wakasa K (2006) Metabolic changes in Arabidopsis thaliana expressing the feedback-resistant anthranilate synthase alpha subunit gene OASA1D. Phytochemistry 67:2349–2362
11. Yamada T, Tozawa Y, Hasegawa H, Terakawa T, Ohkawa Y, Wakasa K (2004) Use of a feedback-insensitive at subunit of anthranilate synthase as a selectable marker for transformation of rice and potato. Mol Breeding 14:363–373
12. Goldschmidtclermont M (1991) Transgenic expression of aminoglycoside adenine transferase in the chloroplast—a selectable marker for site-directed transformation of chlamydomonas. Nucleic Acids Res 19:4083–4089
13. Eibl C, Zou ZR, Beck A, Kim M, Mullet J, Koop HU (1999) In vivo analysis of plastid psbA, rbcL and rpl32 UTR elements by chloroplast transformation: tobacco plastid gene expression is controlled by modulation of transcript levels and translation efficiency. Plant J 19:333–345
14. Murray MG, Thompson WF (1980) Rapid isolation of high molecular-weight plant DNA. Nucleic Acids Res 8:4321–4325

Chapter 12

Plastid Marker Gene Excision in Greenhouse-Grown Tobacco by Agrobacterium-Delivered Cre Recombinase

Tarinee Tungsuchat-Huang and Pal Maliga

Abstract

Uniform transformation of the thousands of plastid genome (ptDNA) copies in a cell is driven by selection for plastid markers. When each of the plastid genome copies is uniformly altered, the marker gene is no longer needed. Plastid markers have been efficiently excised by site-specific recombinases expressed from nuclear genes either by transforming tissue culture cells or introducing the genes by pollination. Here we describe a protocol for the excision of plastid marker genes directly in tobacco (*Nicotiana tabacum*) plants by the Cre recombinase. Agrobacterium encoding the recombinase on its T-DNA is injected at an axillary bud site of a decapitated plant, forcing shoot regeneration at the injection site. The excised plastid marker, the *bar*^{au} gene, confers a visual aurea leaf phenotype; thus marker excision via the flanking recombinase target sites is recognized by the restoration of normal green color of the leaves. The *bar*^{au} marker-free plastids are transmitted through seed to the progeny. PCR and DNA gel blot (Southern) protocols to confirm transgene integration and plastid marker excision are also provided herein.

Key words Agrobacterium transformation, Bud injection, Cre site-specific recombinase, DNA gel blot analyses, Marker-free, *Nicotiana tabacum*, PCR protocol, Plastid transformation, Plastid marker excision, Tobacco

1 Introduction

The 155-kb tobacco plastid genome (ptDNA) is highly polyploid. Given the thousands of plastid genome copies per cell, selective enrichment for marker genes is essential to obtaining a uniformly transformed plastid genome population. For protocols on plastid transformation in *Nicotiana tabacum* (tobacco), *see* Chapters 8, 9, and 10 in this volume. When a uniformly transformed plastid genome population is obtained, the marker gene becomes unnecessary, as the transplastome remains stable in the absence of selection. Removal of the marker gene is highly desirable because it removes the metabolic burden imposed by its high-level expression and because it allows us to circumvent any regulatory implications and public perception issues associated with an antibiotic resistance phenotype.

Pal Maliga (ed.), *Chloroplast Biotechnology: Methods and Protocols*, Methods in Molecular Biology, vol. 1132,
DOI 10.1007/978-1-62703-995-6_12,

There are currently four methods for marker gene removal: direct-repeat-mediated marker excision [1, 2], transient cointegration [3, 4], cotransformation and segregation [5], and recombinase-mediated marker excision. The last approach requires flanking the marker gene with recombinase target sites in the transformation vector. The plastid genomes transformed with these vectors are stable in the absence of recombinases. However, upon introduction of the recombinase into plastids, the target site flanked marker genes are efficiently excised. The enzymes tested for plastid marker excision are the P1 phage Cre and PhiC31 phage Int site-specific recombinases [6–8]. Methods for plastid marker excision rely on the expression of a plastid-targeted recombinase from a nuclear gene. The gene may be introduced by Agrobacterium-mediated transformation of tissue culture cells [6–8], by pollen of a recombinase-expressing plant [6], or by agroinfiltration resulting in transient expression of a recombinase gene encoded in the T-DNA [9]. For further information on recombinase-mediated plastid marker gene excision, see review articles [10–13] and protocols [14].

Here we describe a protocol for the excision of plastid marker genes directly in tobacco (*Nicotiana tabacum*) plants. A visual marker that enables tracking marker excision by a change in leaf color enables successful manipulation of the plastid genome in plants. The marker system in Nt-pSS42 plants used in the protocol requires two genes: the *bar*^{au} gene that confers a yellow leaf phenotype and a spectinomycin resistance (*aadA*) gene necessary for the introduction of the *bar*^{au} gene into the plastid genome. The *bar* gene confers resistance to phosphinothricin when most ptDNA copies carry the gene. However, the *bar* gene is not suitable for selective enrichment when it is present in only a few ptDNA copies [5, 15]. Recently, we developed a novel *aadA* gene that fulfills both functions: it is a conventional selectable spectinomycin resistance (*aadA*) gene in culture and allows detection of transplastomic sectors in the greenhouse by leaf color. Conversion of the *aadA* gene into an aurea version was achieved by incorporating a segment of the plastid *clpP1* gene leader in the *aadA* 5′-untranslated region, an approach that is applicable to any marker gene [16].

The protocol that we describe here employs Agrobacterium-mediated transformation of meristematic cells in the shoot apex with a nuclear gene encoding a plastid-targeted Cre recombinase in the pKO31 plasmid T-DNA [17] (Fig. 1b). Agrobacterium cells are delivered by injection at the axillary bud site, destroying the shoot meristem (Fig. 2a). Differentiation of a new shoot apex is forced by decapitating the plants above the injection site. Detection of *in planta* plastid marker excision is enabled by the excision of a visual plastid marker, the aurea *bar* (*bar*^{au}) gene, via the *loxP*-flanked target sites in the *N. tabacum* Nt-pSS42 plastid genome (Fig. 1a). Excision of the *bar*^{au} gene restores the green pigmentation in young leaves (Fig. 2b, c). Note that the mature leaves of the

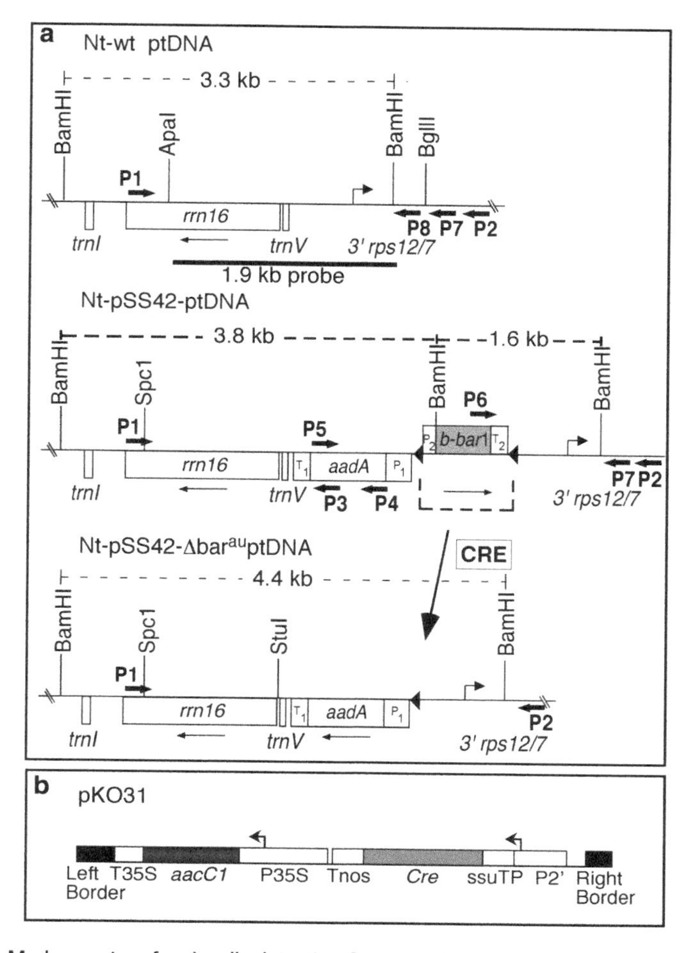

Fig. 1 Marker system for visually detecting Cre recombinase-mediated excision of the aurea *bar*^{au} gene in tobacco leaves [17]. (**a**) Shown is the Nt-pSS42 transplastome carrying the *bar*^{au} gene flanked by *loxP* target sites (*triangles*) and the spectinomycin resistance (*aadA*) gene. Excision of the *bar*^{au} gene leaves behind a *loxP* target site in the Nt-pSS42-Δ *bar*^{au} plastid DNA (below). On top we show part of the wild-type Nt-wt ptDNA and the position of *trnI*, *rrn16*, *trnV*, and *rps12/7* plastid genes. Position of PCR primers (P) listed in Subheading 2.4, **item 13** is also shown. Primers P1, P2, and P7 are not included in vector-targeting region; P8 is within the ptDNA targeting region of the vector. (**b**) T-DNA in plasmid pKO31 encodes a plastid-targeted Cre recombinase engineered for expression in the plant nucleus. Shown are Cre fused with the Rubisco small subunit transit peptide (ssuTP) expressed in a P2′ promoter and Tnos terminator cassette, a gentamicin resistance (*aacC1*) gene expressed in the 35S promoter-terminator cassette (P35S-T35S), and the T-DNA left and right borders

injected plant are dark green, because the aurea leaf phenotype is exhibited only by rapidly growing young leaves. The Agrobacterium-injected plants directly yield plastid marker-free seed progeny (Fig. 2d). This protocol is based on the original publication by Tungsuchat-Huang and Maliga in 2012 [17].

Fig. 2 Excision of the aurea *bar*^{au} gene by Cre yields *green* sectors in leaves. (**a**) Injection of a decapitated Nt-pSS42 plant with Agrobacterium at the axillary bud site. Note that the genetically aurea leaves of this plant are *green* due to age. (**b**) *Green* spots in central aurea leaf indicate excision of the *bar*^{au} gene. Note bright *yellow* (aurea) color of adjacent, rapidly growing leaves. (**c**) Variegated shoot with *dark green* (*bar*^{au} excised) and aurea sectors (*bar*^{au} present). (**d**) Marker-free plastids are transmitted to seed progeny in Nt-pSS42 plants. Shown is a segregating population of seedlings from one capsule. *Green* and variegated seedlings grow faster than their aurea sibs

2 Materials

2.1 Plant Materials

1. Transplastomic Nt-pSS42 plants with *loxP*-flanked aurea *bar*^{au} gene [17] (*see* **Note 1**).
2. ProMix General Purpose Growing Medium (Code 0432, Premier Horticulture Inc., Grower Services, Quakertown, PA 18951) or any soil containing peat moss.
3. Plastic flats with drain holes (10 in. × 10 in.; T.O. Plastic, Inc., Clearwater, MN 5532).
4. Blu-Grow 10-10-10 General Purpose Liquid Fertilizer (Plant Food Company, Cranbury, NJ 08512).

5. Osmocote Classic 14-14-14 Controlled Release Fertilizer (Scotts-Sierra Horticultural Products Co., Marysville, OH 43401).
6. Plastic pots (7.5 in; Dillen products, Middlefield, OH 44062).

2.2 Agrobacterium Culture and Induction of T-DNA Transfer

1. *Agrobacterium tumefaciens* strain EHA 101 carrying binary plasmid pKO31 encoding a plastid-targeted Cre and a plant-selectable gentamicin resistance gene [9].
2. Agrobacterium Minimal A medium supplemented with 100 mg/l spectinomycin HCl and 50 mg/l kanamycin sulfate. Agar-solidified Agrobacterium Minimal A medium is assembled from three stock solutions. Stock A contains 10.5 g K_2HPO_4, 4.5 g KH_2PO_4, 1.0 g $(NH_4)_2SO_4$, and 0.5 g sodium citrate in 490 ml of deionized water. Stock B contains 15 g Bacto Agar in 500 ml deionized water. Stock C is obtained by mixing 1 ml of 1 M $MgSO_4{\cdot}7H_2O$ and 10 ml of 40 % sucrose. Stocks A and B are autoclaved separately, cooled to about 60 °C, mixed with filter-sterilized Stock C, and poured into Petri dishes (100 × 15 mm).
3. Agrobacterium YEB medium supplemented with 100 mg/l spectinomycin HCl and 50 mg/l kanamycin sulfate. Agrobacterium liquid YEB medium contains 5 g beef extract, 1 g yeast extract, 5 g Bacto Peptone, and 5 g sucrose per liter. The pH is adjusted to 7.4 with 1 M KOH prior to autoclaving.
4. 0.5 M 2-(N-morpholino)ethanesulfonic acid (MES) buffer. Prepare 0.5 M stock solution in water and then adjust to pH 5.6 with KOH and filter-sterilize.
5. 100 mM 3′-5′-Dimethoxy-4′-hydroxyacetophenone (acetosyringone) (Sigma, St. Louis, MO). Prepare stock solution in DMSO and filter-sterilize using DMSO-resistant membrane.
6. Agrobacterium MMA medium (MS salts, in 10 mM MES buffer) [18]. The Agrobacterium MMA medium contains 100 ml 10× macronutrients, 10 ml 100× micronutrients, 5 ml 1 % Fe-EDTA, and 20 g sucrose per liter. The pH is adjusted to 5.6 with KOH and the medium is sterilized by autoclaving. Prepare MMA medium by adding filter-sterilized acetosyringone (200 μM, freshly prepared) and 2-(N-morpholino)ethanesulfonic acid buffer (MES; 0.5 M) before use. 10× macronutrient solution of the MMA medium per liter contains 19 g KNO_3, 3.7 g $MgSO_4{\cdot}7H_2O$, 4.4 g $CaCl_2{\cdot}2H_2O$, 1.7 g KH_2PO_4, and 16.5 g $(NH_4)NO_3$) [19]. 100× Micronutrient solution of the MMA medium per liter contains 1.69 g $MnSO_4{\cdot}2H_2O$, 620 mg H_3BO_3, 860 mg $ZnSO_4{\cdot}7H_2O$, 83 mg KI, 25 ml $Na_2MoO_4{\cdot}2H_2O$ (1 mg/ml), 2.5 ml $CuSO_4{\cdot}5H_2O$ (1 mg/ml), and 2.5 ml $CoCl_2{\cdot}6H_2O$ (1 mg/ml) [19].
7. Polypropylene centrifuge bottle (250 ml) (Nalgene Labware, Rochester, NY 14625).

2.3 Bud Injection

1. 3 ml syringe with a 25G × 1½ in. needle (BD disposable syringe and PrecisionGlide Needle BD®, Becton Dickinson and Co., Franklin Lakes, NJ 07417).
2. Silwet L-77 nonionic surfactant.
3. Plastic sheet.

2.4 PCR Analyses

1. CTAB DNA extraction buffer contains 2 % C-tetradecyl-trimethyl-ammonium bromide (CTAB, Sigma T4762), 1.4 M NaCl, 200 mM EDTA pH 8, 100 mM Tris–HCl, pH 8, and 100 mM β-mercaptoethanol.
2. 1.5-ml disposable pellet pestle (Fisher Scientific Cat. No. 12-141-367, Pittsburgh, PA 15275).
3. Liquid nitrogen.
4. Stainless steel forceps (7 in.).
5. Thermo Scientific Dry Block Heater (Fisher Scientific, Pittsburgh, PA 15275).
6. Chloroform ($HCCl_3$)/isoamyl alcohol (24:1).
7. Fisher Scientific Heavy-Duty Vortex Mixer (Fisher Scientific, Pittsburgh, PA 15275).
8. Isopropanol.
9. Microcentrifuge.
10. 70 % ethanol.
11. Oven set to 65 °C.
12. Double-distilled sterile water.
13. PCR primers:
    P1: 5′-GGTAACGACTTCGGGCATGGCC-3′
    P2: 5′-TCGCCACTCCCTTTGGCAGCATCC-3′
    P3: 5′-CCTGCCGGCCCAGTATCAGCCC-3′
    P4: 5′-GGCTCCGCAGTGGATGGCGGCCTG-3′
    P5: 5′-GGGCTGATACTGGGCCGGCAGG-3′
    P6: 5′-CACCACAAACAGAGAGCCCA-3′
    P7: 5′-ACTCCGACAGCATCTAGGGT-3′
    P8: 5′-GTTCATTTGGAATCTGGGTTCTTC-3′
14. PCR tube, 0.2 ml.
15. Commercial PCR kits containing 10 mM dNTPs (2.5 mM each dNTP), 10× PCR Buffer and DNA polymerase. For fragments up to 5-kb amplicons, use Expand High Fidelity PCR System (Roche Applied Science, Indianapolis, IN). For fragments larger than 5 kb, use TaKaRa LA Taq DNA polymerase (Clontech Laboratories, Inc., Mountain View, CA).
16. PCR machine.
17. Loading dye (5×) consists of 40 mM EDTA, pH 8.0; 1 % SDS, 15 % Ficoll, bromophenol blue, and xylene cyanole FF.

18. Agarose.
19. TAE electrophoresis buffer (50× stock) contains 242 g Tris base, 57.1 ml glacial acetic acid per liter, and 100 ml 0.5 M EDTA adjusted to pH 8.
20. Gel box with power supply.
21. GelGreen Nucleic Acid Stain (Biotium, Hayward, CA 94545).
22. DNA molecular weight marker.
23. Imaging equipment set up for detection in blue light. We use an AlphaImager S-220, Alpha Ease TM (Alpha Innotech Corporation, San Leandro, CA 94577). DNA in gels is stained with GelGreen (Biotium Inc., Hayward, CA 94545), and the image is captured on a Dark Reader blue light transilluminator (DR88X, Clare Chemical Research, Inc., Dolores, CO 81323).

2.5 DNA Gel Blot Analyses

1. Total cellular DNA, used also as template for PCR amplification (*see* Subheading 3.4, **steps 1–12**).
2. Primers 1 and 8 for the amplification of DNA probe (*see* Subheading 2.4, **item 13**).
3. dCTP Ready-to-go DNA labeling Beads (GE Healthcare, Piscataway, NJ 08854).
4. ^{32}P-dCTP (50 μCi).
5. Thermo Scientific Dry Block Heater (Fisher Scientific, Pittsburgh, PA 15275).
6. Microspin G-50 column for the purification of labeled DNA fragments (GE Healthcare).
7. *Bam*HI restriction endonuclease.
8. Loading dye (*see* Subheading 2.4, **item 17**).
9. Agarose.
10. Southern hybridization 20× SSPE solution (per liter add: 175.3 g NaCl, 27.6 g $NaH_2PO_4{\cdot}H_2O$, 40 ml 0.5 M EDTA, pH to 7.4).
11. Southern hybridization depurination solution (0.25 M HCl).
12. Southern hybridization denaturation solution (0.5 M NaOH, 0.8 M NaCl).
13. Southern hybridization neutralization solution (0.5 M Tris–HCl, pH 7.0, 0.5 M NaCl).
14. Southern hybridization wash solution (0.1× SSPE, 0.2 % SDS).
15. GelGreen Nucleic Acid Stain (Biotium, Hayward, CA 94545).
16. Pyrex glass baking dish.
17. Platform shaker operating at room temperature.
18. Whatman 3MM filter paper.

19. Nitrocellulose membrane (GE Healthcare, Piscataway, NJ 08854).
20. Stratagene PosiBlot Pressure Blotter.
21. Stratagene Stratalinker UV DNA Crosslinker 1800.
22. Hybridization mesh (GE Healthcare, Piscataway, NJ 08854).
23. 50-ml Falcon tube.
24. Church hybridization buffer contains 0.5 M phosphate buffer pH 7.2, 7 % (w/v) SDS, 10 mM EDTA [20].
25. Hybridization oven (rotating) with glass tubes.
26. Shaking temperature-controlled water bath.
27. X-ray cassettes and screens to amplify radioactive signal.
28. X-ray film.
29. Developer for X-ray film.

3 Methods

3.1 Cultivation of Plants in the Greenhouse

1. Germinate seed in 10 in. × 10 in. flats filled with ProMix General Purpose Growing Medium. Mix tobacco seed (about 300, 0.1 ml) with dry sand in 1.5-ml Eppendorf tube and sprinkle seed on surface. Water seed with fine mist sprayer head. To prevent seed desiccation, cover flat with plastid dome until seedlings reach 5 mm to 10 mm in size.
2. Transfer seedlings (about 1 in.) to 7.5 in. plastic pots and grow plants until they reach 5–10 in. in size.
3. After germination fertilize seedlings weekly with liquid fertilizer and sprinkle fresh Osmocote pellets monthly on soil surface. Regular fertilizer applications are important to ensure rapid plant growth, the condition required for the manifestation of aurea phenotype.

3.2 Agrobacterium Culture and Induction of T-DNA Transfer

Acetosyringone treatment of Agrobacterium described here is modified from the transient agroinfiltration protocol [9, 18].

1. Grow *Agrobacterium tumefaciens* strain EHA 101 carrying plasmid pKO31 on Minimal A plates containing 100 mg/l spectinomycin HCl (to select for the maintenance of binary plasmid) and 50 mg/l kanamycin sulfate (to select for the maintenance of the virulence plasmid in Agrobacterium strain EHA 101) for 1–2 days at 30 °C. Agrobacterium binary plasmids can be lost or rearranged. Check plasmid in miniprep before experiment.
2. Inoculate *Agrobacterium* from plate into a tube containing 3 ml YEB medium (no MES) containing 100 mg/l spectinomycin HCl and 50 mg/l kanamycin and grow for 16 h at 30 °C.

3. Inoculate 1 ml of overnight culture into 100 ml YEB medium containing 10 mM MES, pH adjusted to 5.6, 20 μM acetosyringone, 100 mg/ml spectinomycin HCl, and 50 mg/l kanamycin, and grow culture for 16–18 h at 30 °C on a shaker.
4. Measure the optical density at 600 nm (OD_{600}). OD_{600} value should be above 1.
5. Centrifuge liquid culture at 4000 × *g* at room temperature for 15 min to collect cells.
6. Resuspend cell pellet in MMA medium to obtain a final OD_{600} = 2.4. To calculate volume (ml) of MMA medium, use formula = (OD_{600} of overnight culture/2.4) × 100.
7. Inoculate tubes at room temperature for 2–3 h by gently rocking the tubes on shaker.
8. Add 0.02 % Silwet (optional) to the culture before injecting plants.

3.3 Agrobacterium Bud Injection to Induce Marker-Free Green Sectors

1. Heavily water the transplastomic Nt-pSS42 plants before injection.
2. Decapitate shoot leaving only a short stem with three to four leaves. Remove all buds from nodes. It is important to remove all axillary buds to force regeneration of new shoot apices from the wounded surface at the axillary bud site.
3. Fill 3 ml syringe with acetosyringone-induced *Agrobacterium* culture.
4. Gently inject *Agrobacterium* suspension one to three times at axillary bud site. The injected nodes should be 1 in to 1.5 in above the soil.
5. Cover plants with a plastic sheet to prevent desiccation. Injected plants should be stored at room temperature (20–22 °C) for 48 h, and not exposed to direct sunlight. We store our plants in an air-conditioned room adjacent to the greenhouse. Note that T-DNA transfer is temperature dependent and the transfer efficiency is significantly reduced above 22 °C [21].
6. Transfer plants to the greenhouse and resume normal cultivation practices. New shoots develop in 1–2 weeks at the injection site. Excision of the aurea *bar^au* gene yields green sectors, readily detectable by the change of leaf color in 30–40 % of the injected plants. Typically, a cluster of shoots forms at the injection site, most of which do not have green sectors. The shoots should be actively managed. Shoots that do not contain green sectors should be removed to encourage the development of new shoots, some of which possibly will have sectors. The size of green sectors varies from leaf to leaf. If a sector is present in a single leaf, the marker-free plastids are absent in the shoot apex. In this case the stem should be cut off above the sectored leaf

to promote the growth of a new shoot that potentially carries marker-free plastids. Shoots developing from the axillary buds often contain marker-free plastids in the shoot apex (Fig. 2b), which are then transmitted through seed.

7. Seeds can be collected from green branches about 2 months after the injection. The aurea color disappears as growth slows down. Therefore, it is important to tag the green branches while the aurea and green sectors are still visually distinguishable from each other.
8. Plastid marker-free seedlings can be readily identified in the seed progeny by their green color. The seed progeny shown in Fig. 2d segregates for the plastid types: some of the seedlings are smaller, aurea types, and other larger and green carrying marker-free plastids, confirmed by DNA gel blot analyses [17]. Methods for identification of transplastomes are described below.

3.4 PCR Analyses to Confirm Transgene Integration

The PCR analyses described here are for rapid screening of putative transplastomic clones. The protocol comprises isolation of total cellular DNA from leaves by a modified CTAB protocol [22], amplification of ptDNA, and separation of amplicons in agarose gels.

1. Collect 20–100 mg leaf tissue, place it in a 1.5-ml microfuge tube, close tube, and freeze in liquid nitrogen. Store samples in liquid nitrogen until you start DNA extraction (*see* **Note 2**).
2. Remove tubes from liquid nitrogen with forceps, open the tubes, and grind frozen samples using a 1.5-ml pestle until the leaves are powdered. If grinding is insufficient, no DNA will be obtained, but overgrinding of DNA will cause DNA breakage.
3. Add 0.4 ml of CTAB buffer and incubate the tube in a Dry Block Heater at 60 °C for 30 min.
4. Remove tube from the Dry Block Heater, homogenize the sample with 1.5-ml disposable pellet pestle, and add another 0.4 ml CTAB buffer to the tube.
5. Add 0.7 ml chloroform ($HCCl_3$)/isoamyl alcohol (24:1), vortex vigorously for 60 s, and spin tube in a Microcentrifuge at maximum speed (14,000 × *g*) for 8 min. Centrifugation separates DNA-containing aqueous phase (upper layer) from chloroform (lower layer).
6. Transfer upper layer to a clean 1.5-ml microfuge tube by suction with pipette. Do not pipette close to interphase to avoid sucking up protein. Add 0.6 ml chloroform ($CHCl_3$)/isoamyl alcohol (24:1) and repeat DNA extraction. If white precipitate is present at the interphase, repeat chloroform/isoamyl alcohol extraction.

7. Transfer supernatant (~0.5 ml) to a clean 1.5-ml microfuge tube and precipitate DNA by adding 2/3 volume (0.4 ml) isopropanol.
8. Sediment precipitated DNA by spinning in a Microcentrifuge for 30 min at room temperature.
9. Remove the supernatant and add 1 ml 70 % ethanol to wash the DNA. Mix by inverting the tube two to three times and then sediment DNA by spinning the tube in a Microcentrifuge for 10 min at 14,000 × *g*.
10. Repeat washing the DNA with 70 % ethanol (*see* **step 9**).
11. Remove 70 % ethanol with pipette and place the open tube in 65 °C oven for 10 min to dry DNA pellet.
12. Solubilize DNA pellet in 20 μl to 50 μl sterile, double-distilled water (if used immediately) or in 10 mM Tris–Cl, pH 8.5 (if DNA is stored in freezer).
13. To screen for transplastomic clones, PCR amplify left junction fragment (P1, P3), *aadA* (primers P4, P5), and right junction fragment (primers P6, P7; *see* Subheading 2.4, **item 13** for primer sequences). The targeted region in wild-type and transgenic ptDNA can be amplified with primers P1 and P2, which are absent in the vector-targeting region (*see* Figs. 1a and 3). Assemble the reaction mix in a 0.2-ml PCR tube to contain 100 ng DNA (1 μl from **step 12** above), 4 μl 10 mM dNTPs (2.5 mM each dNTP), 2.5 μl 10× PCR Buffer, 1 μl of each primer (5 pmol/μl or 10 pmol/μl, as recommended by manufacturer), and 15.5 μl water in a total volume of 25 μl. Prepare master mix with dNTPs, buffer, and DNA and distribute the aliquots to PCR tubes; add PCR primers and start the reaction by adding 0.2–0.5 μl DNA polymerase, as recommended by manufacturer. For choice of DNA polymerase *see* Subheading 2.4, **item 15**.
14. Program for 0.6-kb *aadA* amplicon, P4–P5 primers, and Roche Expand High Fidelity PCR System (*see* lane 7, Fig. 3). Place 0.2-ml tubes in PCR machine and run program of 2 min at 94 °C, followed by 40 cycles of 30 s at 94 °C, 45 s at 56 °C, and 3 min at 72 °C (1 min per kb) and 1 cycle of 5 min at 72 °C. Store at 4 °C.

    Program for 2.5-kb right junction fragment, P6 and P7 primers, and Roche Expand High Fidelity PCR System (*see* lane 8, Fig. 3). Place 0.2-ml tubes in PCR machine and run program of 2 min at 94 °C, followed by 11 cycles of 20 s at 94 °C, 45 s at 55 °C, and 4 min at 68 °C; 40 cycles of 20 s at 94 °C, 30 s at 45 °C, and 5 min at 68 °C; and 1 cycle of 15 min at 68 °C. Store at 4 °C.

    Program for 2.2-kb left junction fragment, P1 and P3 primers, and Roche Expand High Fidelity PCR System

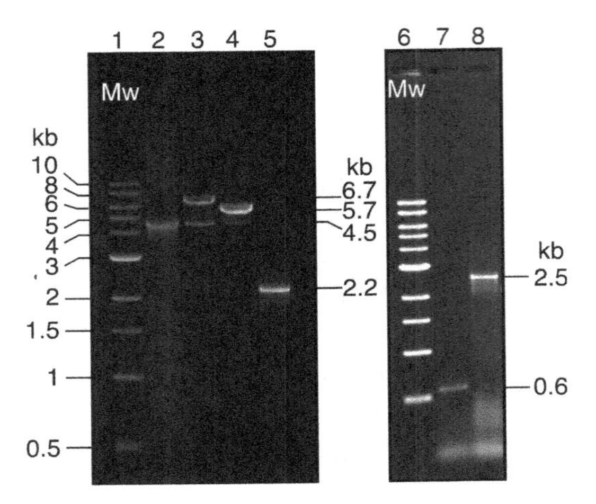

Fig. 3 PCR assay confirms transgene integration in the Nt-pSS42 ptDNA. For primer position and amplified fragments, *see* Fig. 1a. *Lanes 1* and *6*, 1-kb DNA ladder (New England Biolabs); *lane 2*, 4.5-kb fragment, primers P1 and P2, wild-type ptDNA template; *lane 3*, 6.7-kb fragment, P1 and P2, Nt-pSS42-11A DNA template; *lane 4*, 5.7-kb fragment, P1 and P2, Nt-pSS42-bar excision template; *lane 5*, 2.2-kb fragment, P1 and P3, Nt-pSS42-11A template; *lane 7*, 0.6-kb *aadA* fragment, P4 and P5, Nt-pSS42-11A template; *lane 8*, 2.5-kb right junction fragment, P6 and P7, Nt-pSS42-11A template

(*see* lane 5, Fig. 3). Place 0.2-ml tubes in PCR machine and run program of 2 min at 94 °C, followed by 11 cycles of 20 s at 94 °C, 30 s at 60 °C, and 4 min at 68 °C; 40 cycles of 20 s at 94 °C, 30 s at 55 °C, and 5 min at 68 °C; and 1 cycle of 15 min at 68 °C. Store at 4 °C.

Program for 4.5-, 5.7-, and 6.7-kb amplicons (P1–P2 primers, TaKaRa LA Taq enzyme): Place 0.2-ml tubes in PCR machine and run program of 2 min at 95 °C, followed by 24 cycles of 30 s at 95 °C, 30 s at 60 °C, and 9 min at 68 °C and 1 cycle of 10 min at 72 °C. Store at 4 °C. Products shown in lanes 2, 3, and 4, Fig. 3.

15. Test amplicon size by running 5 μl PCR reaction mixed with 1 μl 5× loading dye in 0.8 % agarose gel in TAE buffer. Stain gel with GelGreen Nucleic Acid Stain and visualize DNA under blue light (*see* Fig. 3). Calculate fragment size based on the mobility of molecular weight markers. Successful amplification of junction fragments and *aadA* indicates transplastomic clone.

3.5 DNA Gel Blot Analyses to Confirm Marker Excision

1. Prepare DNA fragment to be used as probe by PCR amplification using total cellar DNA as template and primers 1 and 8. Check amplicon size in gel. Column-purify DNA fragment for labeling. For an alternative source of probe *see* **Note 3**.
2. Denature 50 ng DNA in 20 μl volume by inserting tube in 95 °C 3 water bath for 3 min, then transfer tube to ice.

3. Suspend one dCTP Ready-to-go DNA labeling Bead in 27 μl water. Mix denatured DNA, the DNA labeling Bead, and 2.5 μl ^{32}P-dCTP (50 μCi) and incubate tube at 37 °C for 45 min. Remove unincorporated isotope by spinning in a Microspin G-50 column.
4. Denature labeled DNA probe (3 min at 95 °C) in water bath or Dry Block Heater and store DNA at −20 °C. The probe can be used up to 2 weeks.
5. To prepare the DNA blot, digest 1 microgram total cellular DNA with the *Bam*HI restriction endonuclease in a 15 μl reaction volume for at least 5 h.
6. Mix loading dye (2 μl) to digested DNA sample and load it onto 0.8 % agarose gel along with a DNA molecular weight standard. Run gel until dye runs close to the end.
7. Stain the gel with GelGreen and take a photo with a ruler next to DNA ladder.
8. Depurinate DNA by placing the gel in 250 ml of Southern blot depurination solution and incubate tray for 15 min at room temperature on a shaker.
9. Pour off the depurination solution, rinse the gel with water, and then incubate gel at room temperature in Southern blot denaturation solution on a shaker for 30 min.
10. Pour off the denaturation solution, rinse the gel with water, and then incubate gel in Southern blot neutralization solution on a shaker for 30 min. Handle with care; the gel will be slippery.
11. Transfer the DNA from the gel onto the nitrocellulose membrane using the Stratagene PosiBlot pressure blotter. First cut Whatman 3MMChr filter paper and nitrocellulose membrane to fit the size of the gel. Wet filter paper and nitrocellulose membrane briefly in 2× SSPE solution and lay sequentially on blotter: 2 layers of the filter paper, one nitrocellulose membrane, the gel, one layer of filter paper, and a sponge wetted in 800 ml 2× SSPE solution. Transfer for 1 h.
12. Rinse the nitrocellulose membrane in 2× SSPE and crosslink DNA to membrane in a Stratagene Stratalinker UV DNA Crosslinker 1800 using the 1,200 μJ setting. This is a pausing point. Membrane can be stored at 4 °C at least a month.
13. To wet membrane, submerge briefly in a tray containing 2× SSPE solution in a glass baking dish.
14. Place wet membrane between 2 sheets of hybridization mesh and insert the sandwich into a 50-ml Falcon tube. Pipette 7.5 ml Rapid Hybridization Buffer in the Falcon tube. Place Falcon tube in glass hybridization tube and rotate glass tube in a hybridization oven at 65 °C for 0.5–1 h.

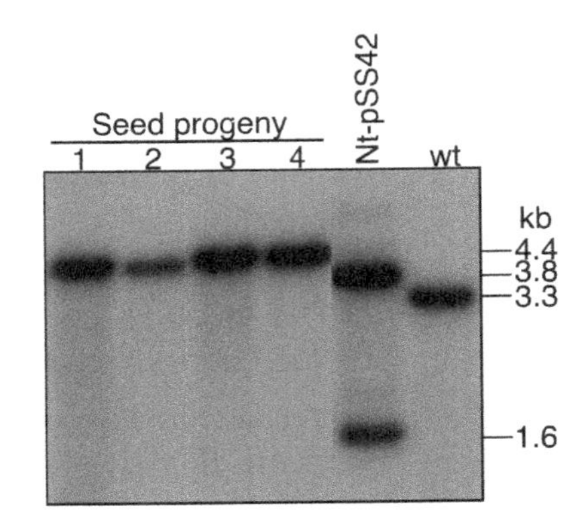

Fig. 4 DNA gel blot (Southern) analysis confirms *bar*^{au} excision in the Nt-pSS42-Δ*bar*^{au} ptDNA. Total cellular DNA was digested with the *Bam*HI restriction endonuclease and probed with the 1.9-kb *Apa*I–*Bam*HI targeting region fragment (*see* Fig. 1a). The probe detects a 3.3-kb fragment in the wild-type ptDNA (wt), 3.8- and 1.6-kb fragments in the Nt-pSS42 ptDNA, and a 4.4-kb fragment in the Nt-pSS42-Δ*bar*^{au} ptDNA after the excision of the *bar*^{au} gene. For fragment sizes, *see* Fig. 1a

15. Add 10 μl denatured, labeled probe to the tube containing the membrane. Incubate tube overnight in hybridization oven at 65 °C.
16. Carefully remove the membrane from the Falcon tube and rinse the filter briefly in 150 ml 2× SSPE in glass baking dish.
17. Wash membrane in 250 ml of Southern hybridization wash solution by incubating a shaking water bath at 55 °C for 30 min.
18. Repeat washing (**step 17**).
19. Rinse the membrane briefly by submerging in 2× SSPE at room temperature in glass baking dish.
20. Wrap moist membrane in Saran Wrap and expose to X-ray film. Exposure time at −80 °C may vary between 1 h and 3 days, dependent on signal strength (Fig. 4). For an alternative to detect radioactivity, *see* **Note 4**. Wild-type ptDNA yields 3.3-kb, the transplastome 3.8-kb and 1.6-kb, and the *bar*-excised transplastome 4.4-kb fragments. Homoplastomic plants lack the wild-type fragment.

4 Notes

1. The aurea leaf phenotype of the Nt-pSS42 plants is manifested only in rapidly growing plants. When the growth rate slows down, the leaves turn green. For injection, select 8–10 in tall, rapidly growing plants, which have not yet flowered.

2. Be careful handling liquid nitrogen. You may float the tubes in a portable liquid nitrogen container, but use forceps to handle the tubes. Protect your hand with a cloth or knitted glove under the latex glove when grinding DNA.
3. As an alternative to the PCR probe, you may use the 1.9-kb *Apa*I–*Bam*HI fragment (Fig. 1) obtained by digesting an appropriate plasmid with *Apa*I and *Bam*HI restriction endonucleases and isolating the DNA fragment from gel.
4. Phosphorimager may be used as alternative to X-ray film.

Acknowledgments

This work was supported by grants from the USDA National Institute of Food and Agriculture Biotechnology Risk Assessment Research Grant Program Award No. 2005-33120-16524, 2008-03012, and 2010-2716.

References

1. Iamtham S, Day A (2000) Removal of antibiotic resistance genes from transgenic tobacco plastids. Nat Biotechnol 18:1172–1176
2. Kode V, Mudd E, Iamtham S, Day A (2006) Isolation of precise plastid deletion mutants by homology-based excision: a resource for site-directed mutagenesis, multi-gene changes and high-throughput plastid transformation. Plant J 46:901–909
3. Klaus SMJ, Huang FC, Eibl C, Koop HU, Golds TJ (2003) Rapid and proven production of transplastomic tobacco plants by restoration of pigmentation and photosynthesis. Plant J 35: 811–821
4. Klaus SMJ, Huang FC, Golds TJ, Koop H-U (2004) Generation of marker-free plastid transformants using a transiently cointegrated selection gene. Nat Biotechnol 22:225–229
5. Ye GN, Colburn S, Xu CW, Hajdukiewicz PTJ, Staub JM (2003) Persistence of unselected transgenic DNA during a plastid transformation and segregation approach to herbicide resistance. Plant Physiol 133:402–410
6. Corneille S, Lutz K, Svab Z, Maliga P (2001) Efficient elimination of selectable marker genes from the plastid genome by the CRE-*lox* site-specific recombination system. Plant J 72: 171–178
7. Hajdukiewicz PTJ, Gilbertson L, Staub JM (2001) Multiple pathways for Cre/*lox-mediated* recombination in plastids. Plant J 27:161–170
8. Kittiwongwattana C, Lutz KA, Clark M, Maliga P (2007) Plastid marker gene excision by the phiC31 phage site-specific recombinase. Plant Mol Biol 64:137–143
9. Lutz KA, Bosacchi MH, Maliga P (2006) Plastid marker gene excision by transiently expressed CRE recombinase. Plant J 45:447–456
10. Lutz KA, Maliga P (2007) Construction of marker-free transplastomic plants. Curr Opin Biotechnol 18:107–114
11. Lutz KA, Azhagiri AK, Tungsuchat-Huang T, Maliga P (2007) A guide to choosing vectors for transformation of the plastid genome of higher plants. Plant Physiol 145:1201–1210
12. Maliga P, Bock R (2011) Plastid biotechnology: food, fuel and medicine for the 21st century. Plant Physiol 155:1501–1510
13. Maliga P (2012) Plastid transformation in flowering plants. In: Bock R, Knoop V (eds) Genomics of chloroplasts and mitochondria. Springer, Berlin, pp 393–414
14. Lutz KA, Svab Z, Maliga P (2006) Construction of marker-free transplastomic tobacco using the Cre-loxP site-specific recombination system. Nat Protocols 1:900–910
15. Lutz KA, Knapp JE, Maliga P (2001) Expression of *bar* in the plastid genome confers herbicide resistance. Plant Physiol 125:1585–1590
16. Tungsuchat-Huang T, Slivinski KM, Sinagawa-Garcia SR, Maliga P (2011) Visual spectinomycin resistance gene for facile identification of

transplastomic sectors in tobacco leaves. Plant Mol Biol 76:453–461
17. Tungsuchat-Huang T, Maliga P (2012) Visual marker and Agrobacterium-delivered recombinase enable the manipulation of the plastid genome in greenhouse-grown tobacco plants. Plant J 70:717–725
18. Kapila J, De Rycke R, Van Montagu M, Angenon G (1997) An *Agrobacterium-mediated* transient gene expression system for intact leaves. Plant Sci 122:101–108
19. Murashige T, Skoog F (1962) A revised medium for the growth and bioassay with tobacco tissue culture. Physiol Plant 15: 473–497
20. Church GM, Gilbert W (1984) Genomic sequencing. Proc Natl Acad Sci U S A 81: 1991–1995
21. Dillen W, De Clercq J, Kapila J, Zambre J, Van Montagu M, Angenon G (1997) The effect of temperature on *Agrobacterium tumefaciens-mediated* gene transfer to plants. Plant J 12: 1459–1463
22. Murray MG, Thompson WF (1980) Rapid isolation of high molecular weight plant DNA. Nucleic Acids Res 8:4321–4325

Chapter 13

Determination of the Half-Life of Chloroplast Transcripts in Tobacco Leaves

Sithichoke Tangphatsornruang and John C. Gray

Abstract

The amounts of specific transcripts that accumulate in chloroplasts are determined by the rates of synthesis and degradation of the transcripts. The 3′ untranslated region of transcripts is a major determinant of the stability of transcripts in chloroplasts. The half-lives of specific transcripts can be determined by northern blot analysis of a time course of transcripts in detached tobacco leaves incubated with actinomycin D, a potent transcription inhibitor. This analysis may be applied to transcripts of endogenous genes or of transgenes introduced into the chloroplast genome in transplastomic plants. Sequence determinants of transcript stability can be identified by analysis of transplastomic plants containing constructs of the green fluorescent protein (*gfp*) reporter gene fused to the sequences of interest.

Key words Actinomycin D, Chloroplast, Green fluorescent protein, RNA-gel blot, RNA stability, Tobacco

1 Introduction

Plastid gene expression can be controlled at several levels to ensure coordination between plastid genes and nuclear genes during different stages of development and in different tissues [1–3]. During plant development, overall plastid transcription rates and accumulation of plastid transcripts can vary enormously; however, the relative transcription rates of many plastid genes are constant [4–6]. This observation indicates that posttranscriptional mechanisms must play an important role in the differential expression of chloroplast genes. Changes in transcription activity alone cannot account for the large changes in the amounts of transcripts accumulated [6, 7]. Transcripts may undergo a large number of posttranscriptional processing steps, including 5′- and 3′-end modification [8, 9], cleavage of polycistronic transcripts [10–12], and RNA editing [13, 14] and splicing [15–17], which may also affect the stability of the transcripts [18–20].

Pal Maliga (ed.), *Chloroplast Biotechnology: Methods and Protocols*, Methods in Molecular Biology, vol. 1132,
DOI 10.1007/978-1-62703-995-6_13, © Springer Science+Business Media New York 2014

RNA-gel blot analysis (or northern blot analysis) [21] has been widely used to detect the amounts of plastid transcripts in a wide variety of plant tissues under a variety of experimental conditions. Relative rates of transcription of individual plastid genes have been determined using "run-on" transcription protocols developed with isolated plastids [4, 22]. Rates of RNA degradation have been measured as half-lives of individual transcripts by northern blot analysis of leaves treated with actinomycin D, a potent inhibitor of transcription in plastids [7, 23, 24]. Methods for the measurement of transcript half-lives are described in this chapter. They have been used to assess the role of 3′ untranslated regions on the stability of *gfp* transcripts in transplastomic tobacco plants [23].

The first line of evidence that inverted repeat sequences at the 3′-end affect plastid RNA stability came from studies with chloroplast extracts [25]. Later, a number of studies in vivo and in vitro confirmed that the 3′ UTR of plastid transcripts plays an important role in mRNA stability [23, 25–35]. Primary transcripts with stem-loops near their 3′-ends are initially processed by specific endoribonuclease cleavage downstream of the stem-loop [11]. Subsequently, a 3′–5′ exonuclease trims the processed intermediate to create the mature 3′-end of the transcript [9, 27, 36]. Large effects of different 3′ UTR sequences on the accumulation of transcripts have been demonstrated in Chlamydomonas and tobacco chloroplasts [23, 29, 32, 37]. Measurements of half-lives of chimeric transcripts, consisting of *gfp* fused to 3′ UTRs from different plastid and *E. coli* genes, in chloroplasts of transplastomic tobacco leaves have provided direct evidence for a role of 3′ UTRs in mRNA stability [23]. The half-life of the transcripts was strongly correlated with the amount of the transcript accumulated, with transcripts containing the *E. coli rrnB* 3′ region having the longest half-life (~40 h) and accumulating to the highest amount [23].

This chapter describes RNA-gel blot analysis of RNA extracted from tobacco leaves and its use in determining the half-life of chloroplast transcripts in detached leaves treated with actinomycin D to inhibit transcription.

2 Materials

2.1 Plant Material

1. Seeds of tobacco (*Nicotiana tabacum*) var. Petit Havana or seeds of transplastomic tobacco (*Nicotiana tabacum*) var. Petit Havana containing reporter genes with different 3′ UTR sequences (*see* **Note 1**).
2. Ethanol, 70 % (v/v).
3. Sterile deionized water.
4. Murashige-Skoog (MS) medium: 4.3 g MS salts (Duchefa Biochemie, Haarlem, the Netherlands), 30.0 g sucrose, and

7.0 g Phyto Agar (Duchefa Biochemie, Haarlem, the Netherlands) dispersed in ~900 mL deionized water, adjusted to pH 6.0 with 1 M NaOH, and made up to a final volume of 1 L. Autoclave at 15 psi (1.0×10^5 Pa) for 15 min.

5. Magenta GA-7 boxes, 77 mm × 77 mm × 97 mm (V8505, Sigma-Aldrich, St. Louis, MO).

2.2 Actinomycin D Treatment

1. Razor blade.
2. Actinomycin D (200 μg/mL) (A4262, Sigma-Aldrich) (*see* **Note 2**).
3. Sterile deionized water.
4. Laminar flow hood with continuous air flow at the rate of 0.5 m/s.

2.3 RNA Extraction

1. Liquid nitrogen.
2. Mortar and pestle.
3. TriPure isolation reagent (Roche Applied Science, Indianapolis, IN).
4. Chloroform (Fisher Scientific, Loughborough, UK).
5. Isopropanol (Fisher Scientific, Loughborough, UK).
6. Diethyl pyrocarbonate-treated deionized water (*see* **Note 3**).
7. Ethanol, 75 % (v/v) in DEPC-treated deionized water.

2.4 Formaldehyde-Agarose Gel Electrophoresis

1. Agarose SeaKem GTG (Cambrex Bioscience, Rockland, ME).
2. Running buffer: 40 mM MOPS-NaOH pH 7.0, 50 mM sodium acetate, and 5 mM EDTA.
3. 37 % formaldehyde (F1635, Sigma-Aldrich) (*see* **Note 4**).
4. RNA loading buffer: 48 % (v/v) deionized formamide (F9037, Sigma-Aldrich), 6 % (v/v) formaldehyde, 5 % (v/v) glycerol, 0.002 % (w/v) bromophenol blue in 40 mM MOPS-NaOH pH 7.0, 50 mM sodium acetate, 5 mM $EDTA.Na_2$, and ethidium bromide (20 ng/μL; ICN Biomedical, Aurora, OH) (*see* **Note 5**).
5. Single-stranded RNA ladder (0.5–9 kb; New England Biolabs, Ipswich, MA 01938).

2.5 RNA-Gel Blotting

1. DEPC-treated deionized water.
2. Positively charged nylon membrane, such as GeneScreen Plus membrane (NEN Research Products, Boston, MA), Hybond N+ (GE Healthcare Life Sciences, Piscataway, NJ), or similar.
3. Whatman 3MM filter paper (Whatman International, Maidstone, Kent, UK).
4. 10× SSC: 1.5 M NaCl and 0.15 M sodium citrate, adjusted to pH 7.0 with HCl.

5. Flat paper towels.
6. 2× SSC: 0.3 M NaCl and 30 mM sodium citrate, adjusted to pH 7.0 with HCl.

2.6 Radiolabeling of DNA Probes

1. Screw-capped 1.5-mL tubes.
2. DNA probes (10 ng DNA/μL) (*see* **Note 6**).
3. Sterile deionized water.
4. Random oligonucleotide 9-mer primers (27 A_{260} units/μL) (Agilent Technologies, Santa Clara, CA).
5. Oligo-labeling buffer: 1 M HEPES-KOH pH 6.6; 200 mM Tris–HCl pH 8.0; 25 mM $MgCl_2$; 100 μM each dCTP, dGTP, and dTTP; and 5 mM 2-mercaptoethanol.
6. [α^{32}P]dATP (specific activity 3,000 Ci/mmol; GE Healthcare Life Sciences, Piscataway, NJ).
7. Klenow fragment of DNA polymerase I (2 units/μL; Agilent Technologies, Santa Clara, CA).
8. MicroSpin S-200 HR column (GE Healthcare Life Sciences, Piscataway, NJ).

2.7 Probe Hybridization

1. Hybridization oven and hybridization bottles (Hybaid, Thermo Fisher Scientific, Waltham, MA).
2. 20× SSC: 3 M NaCl and 300 mM sodium citrate, adjusted to pH 7.0 with HCl.
3. 50× Denhardt's solution: 1 % (w/v) polyvinylpyrrolidone (MW 40,000) (PVP40, Sigma-Aldrich), 1 % (w/v) bovine serum albumen (B4287, Sigma-Aldrich), and 1 % (w/v) Ficoll-400 (Fisher Scientific, Loughborough, UK).
4. Pre-hybridization buffer: 10 mL 20× SSC, 5 mL deionized formamide (F9037, Sigma-Aldrich), 2 mL 50× Denhardt's solution, 1 mL 10 % (w/v) SDS, and 100 μg denatured salmon sperm DNA (10 mg/mL, ICN Biomedical, Aurora, OH).
5. Hybridization buffer: 10 mL of 20× SSC, 5 mL of deionized formamide (F9037, Sigma-Aldrich), 2 mL of 50× Denhardt's solution, 2 mL of 10 % (w/v) SDS, 2 g dextran sulfate, and 20 μL of denatured salmon sperm DNA (10 mg/mL, ICN Biomedical, Aurora, OH). Heat at 65 °C for 3 h.
6. Radioactively labeled probe.
7. 2× SSC: 0.3 M NaCl and 30 mM sodium citrate, adjusted to pH 7.0 with HCl.
8. 2× SSC, 1 % (w/v) SDS.
9. Handheld Mini Monitor Series 900 detector (Mini Instruments, Burnham-on-Crouch, Essex, UK).

10. Typhoon 8600 high-performance laser scanning system, with storage phosphor screens and ImageEraser (GE Healthcare Life Sciences, Piscataway, NJ).
11. 0.1× SSC, 1 % (w/v) SDS: 15 mM NaCl, 1.5 mM sodium citrate, 1 % (w/v) SDS, adjusted to pH 7.0 with HCl.

3 Methods

The half-lives of transcripts can be determined by RNA-gel blot analysis of the amounts of specific transcripts in detached tobacco leaves treated with actinomycin D to prevent transcription of the genes [7, 23]. Control leaves treated with water provide an indication of the balance of synthesis and degradation in the detached leaves. In all of our experiments, there is little change in the amounts of transcripts of *psbA*, *rbcL*, or *gfp* in detached control leaves over a 48 h period ([23], and *see* Fig. 2). The transcripts examined may include transcripts of endogenous chloroplast genes or transcripts from reporter gene constructs inserted into the chloroplast genome in transplastomic tobacco plants.

3.1 Preparation of Plant Materials

1. Surface sterilize tobacco seeds in a 1.5-mL microcentrifuge tube by adding 1 mL of 70 % ethanol and vortexing vigorously for 1 min. Briefly centrifuge the seeds and discard the supernatant. Wash the seeds with 1 mL of sterile deionized water and vortex for 1 min. Pellet the seeds by centrifugation and discard the supernatant. Repeat washing with sterile deionized water three times.
2. Place tobacco seeds on the surface of MS medium in a Petri dish. Seal the dish with Parafilm and place at 4 °C for 48 h.
3. Transfer tobacco seeds from the Petri dish to Magenta boxes by placing one seed on the surface of MS medium in each Magenta box. Tobacco plants are grown in a growth room for 5 weeks at 25 °C with photosynthetically active radiation of 100 μmol photons/m^2 s for a photoperiod of 16 h light: 8 h dark.

3.2 Actinomycin D Treatment

1. Cut fully expanded young leaves (5 cm long) from 5-week-old tobacco plants with a sharp blade at the base of the petiole.
2. Immediately place the leaf with the petiole submerged in 1 mL of either actinomycin D (200 μg/mL) or water in 2.0-mL microcentrifuge tubes. To increase transpiration and uptake of actinomycin D solution, the experiment is carried out in a laminar flow hood with continuous air flow at the rate of 0.5 m/s.
3. Refill the actinomycin D solution and water every 5 h.
4. Extract RNA from leaves over a period of time: 0, 10, 24, and 48 h (*see* **Notes** 7 and **8**).

3.3 RNA Extraction

1. Grind tobacco leaves to a fine powder in liquid nitrogen in a mortar with a pestle that had been precooled with liquid nitrogen.
2. Transfer the powder into a 1.5-mL tube, add 1 mL TriPure isolation reagent, and vortex to mix.
3. Add 200 μL of chloroform and mix by vortexing.
4. Incubate the solution at room temperature for 10 min, and centrifuge at 11,600 × *g* for 15 min at 4 °C.
5. Transfer the upper aqueous layer to a fresh 1.5-mL tube without disturbing the interphase layer.
6. Add 0.5 mL of isopropanol, mix by inversion, and incubate at room temperature to precipitate the RNA.
7. Pellet the RNA by centrifugation at 11,600 × *g* for 10 min at 4 °C, and wash by adding 1 mL of 75 % ethanol (75 % ethanol in DEPC-treated deionized water).
8. Centrifuge the RNA pellet and discard the supernatant.
9. Air-dry the pellet and resuspend in 50–200 μL DEPC-treated deionized water depending on the pellet size. Pipetting the RNA pellet in DEPC-treated water up and down several times can help dissolving the pellet.
10. RNA concentration and purity can be determined by spectrophotometric measurement using a Perkin Elmer Lambda 9 spectrophotometer (Perkin Elmer, Waltham, MA). The concentration can be calculated on the assumption that an absorbance of 1.0 at 260 nm corresponds to 40 μg single-stranded RNA/mL. The ratio of absorbances at 260 nm/280 nm provides an estimate of the extent of contamination with protein (preparations without any protein contamination should give a ratio of 1.8–2.0).

3.4 Formaldehyde-Agarose Gel Electrophoresis

1. To prepare a 1.2 % (w/v) agarose gel in a tray size of 150 mm × 100 mm × 7.5 mm, 1.2 g of agarose is dispersed in 100 mL of running buffer by heating in a microwave.
2. Cool down the gel mix to 55 °C, add 5.4 mL of 37 % formaldehyde in a fume hood, and mix.
3. Cast the gel in a tray with the slot comb and allow the gel to solidify for 40 min.
4. Place the gel tray into the gel tank. Fill up the gel tank with running buffer and remove the slot comb.
5. Prepare the RNA sample by mixing 5 μL of the RNA (2 μg/μL) with 15 μL of RNA loading buffer. Heat the sample at 90 °C for 5 min and incubate on ice for 5 min.
6. Flush the sample well with running buffer to remove air bubbles and load the RNA samples with a molecular weight marker next to the samples.

7. Run the gel at 70 V for 6 h or until the bromophenol blue dye reaches the bottom of the gel.
8. Inspection of the gel under UV light should identify the two major RNA bands corresponding to the cytosolic 25S and 18S rRNAs, which migrate at 3.5 kb and 1.8 kb, respectively, plus additional chloroplast rRNA bands (*see* **Note 9**).

3.5 RNA-Gel Blotting

1. Cut off the lower right gel corner for orientation. Soak the gel in 500 mL of DEPC-treated water for 15 min to remove excess formaldehyde and ethidium bromide.
2. Cut the GeneScreen Plus membrane to the exact size of the gel and soak in DEPC-treated water for 1 min. Equilibrate the membrane by submerging in 10× SSC for 2 min.
3. Set up a capillary blot with a Whatman 3MM paper wick with both ends submerged in 10× SSC solution (Fig. 1).
4. Place the gel on the wet paper wick without trapping air bubbles. To avoid short circuiting of the 10× SSC flow, strips of Parafilm are placed along the edges of the gel (Fig. 1a). Wet the gel with a few mL of 10× SSC and carefully place the nylon membrane on top of the gel without trapping air bubbles.

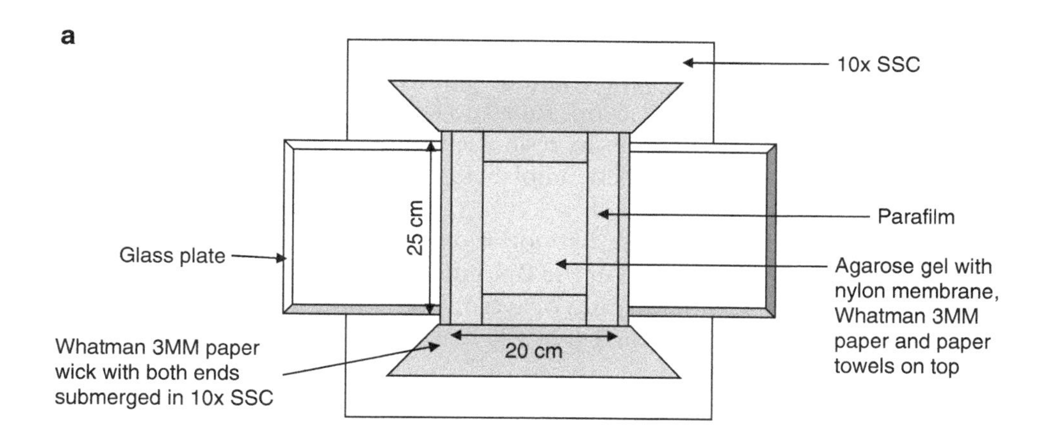

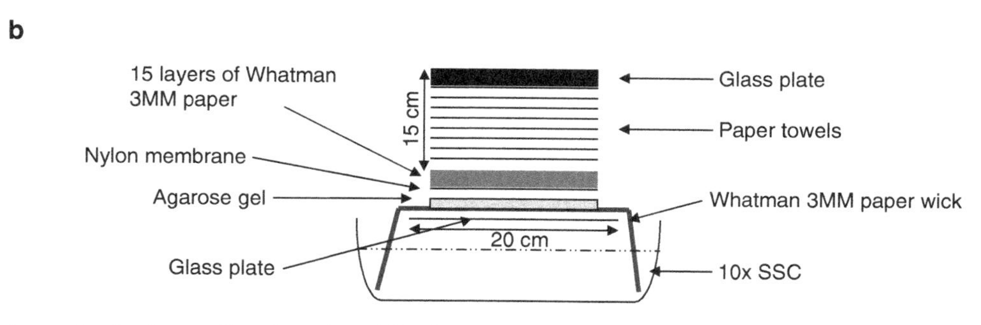

Fig. 1 Apparatus for the capillary transfer of RNA from the agarose gel to a nylon membrane. (**a**) Top view and (**b**) side view of the RNA-gel blot apparatus

5. Place 15 layers of wet Whatman 3MM paper on top of the nylon membrane (place each layer of Whatman 3MM paper at a time). Place a stack of paper towels (10–15 cm tall when compressed) and heavy glass plates on top of the Whatman 3MM paper layer. Allow the transfer of RNA overnight.
6. Wash the membrane in 2× SSC to remove any residual agarose gel. Fix the RNA to the membrane by air-drying for 2 h at room temperature.

3.6 Radiolabeling of DNA Probes

1. Mix 4 μL DNA (10 ng/μL), 19 μL deionized water, and 10 μL of random oligonucleotide primers and heat at 95 °C for 5 min in a screw-capped 1.5-mL tube.
2. Immediately place the tube on ice and centrifuge at 300 × *g* for 5 s.
3. Sequentially add 10 μL of oligo-labeling buffer, 4 μL (40 μCi) [α^{32}P]dATP (specific radioactivity 3,000 Ci/mmol), and 1.2 μL (2.4 units) of Klenow fragment of DNA polymerase I to the tube. Incubate at 37 °C for 1 h.
4. Purify radiolabeled probe from unincorporated [α^{32}P]dATP by passage through a MicroSpin S-200 HR column. First, agitate the resin in the column by vortexing for 2 min and snap the bottom closure of the column off. Place the column in a 1.5-mL screw-cap microcentrifuge tube for support and centrifuge at 735 × *g* for 1 min to remove the buffer. Place the column into a new 1.5-mL tube and add the radiolabeled probe (50 μL) to the top of the resin. Centrifuge at 735 × *g* for 2 min and collect the purified radiolabeled probe in the tube.

3.7 Probe Hybridization

1. Warm pre-hybridization buffer at 65 °C in a water bath for 10 min. Put the RNA membrane in the Hybaid bottle together with 20 mL of warm pre-hybridization buffer. Incubate at 42 °C for 2 h in a Hybaid oven.
2. Denature the ^{32}P-labeled probe by heating at 95 °C for 5 min and place the probe on ice for 10 min.
3. Discard the pre-hybridization buffer and add 20 mL of hybridization buffer to the bottle. Add the denatured ^{32}P-labeled probe directly to the hybridization buffer in the bottle. Avoid spotting the probe directly on the membrane. Incubate the bottle in the Hybaid oven at 42 °C overnight.
4. Remove the membrane from the bottle and agitate in 200 mL of 2× SSC for 30 min followed by washes with 2× SSC, 1 % (w/v) SDS at 65 °C until the background radioactivity is reduced to 50 cps. Wrap the membrane with cling film to prevent the membrane dry out. It is difficult to strip off the probe from a dry membrane.

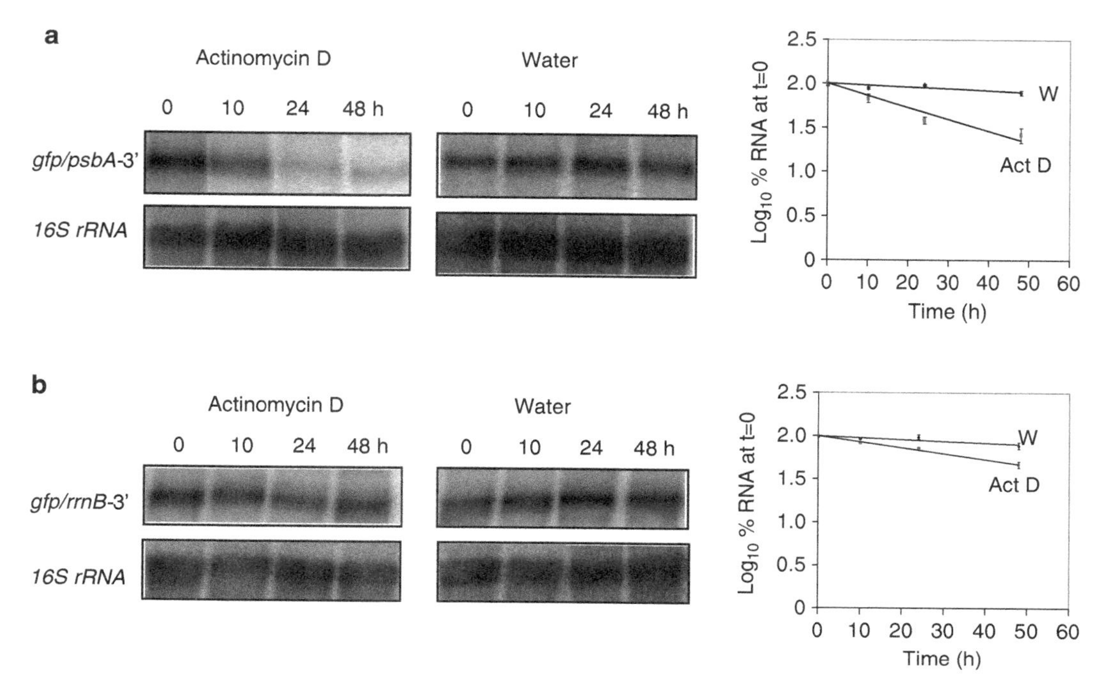

Fig. 2 RNA-gel blot analysis of *gfp* RNA stability. Transplastomic tobacco leaves, containing *gfp* constructs with different 3′ regions (from tobacco *psb*A and *E. coli rrn*B), were placed with their petioles in either water or actinomycin D. RNA was extracted at various times up to 48 h after initiation of the treatment. Equal amounts (10 μg) of total RNAs for each treatment and time point were separated by electrophoresis on formaldehyde-agarose gels, blotted on GeneScreen Plus nylon membrane, and hybridized with a ^{32}P-labeled *gfp* probe, followed by stripping the blot and hybridizing with a ^{32}P-labeled 16S rRNA probe. The radioactivity of the hybridized *gfp* and 16S rRNA probes was detected with a Typhoon 8600 phosphorimaging system. The graphs show semilogarithmic plots of $\log_{10}$ values of the amounts of the 1.0-kb *gfp* transcripts, standardized to 16S rRNA, against time. Standard error bars indicate the degree of variation between two experiments. (**a**) *gfp* transcripts containing the *psbA* 3′ UTR and (**b**) *gfp* transcripts containing the *E. coli rrnB* 3′ region. *ActD* actinomycin-treated leaves, *W* water-treated control leaves

5. Prepare a blank storage phosphor screen by exposing the screen to the ImageEraser light for 20 min.
6. Place the membrane on top of the blank storage phosphor screen in a cassette for 3 h to allow the exposure.
7. Scan the screen using a Typhoon 8600 high-performance laser scanning system (*see* **Note 10**). Typical results are shown in Fig. 2.
8. To hybridize the membrane with different probes, the radioactive probe hybridized to the membrane can be stripped off by washing with a boiling solution of 0.1× SSC, 1 % (w/v) SDS for 15 min, and rinsing twice with 2× SSC for 15 min. The membrane can be re-probed immediately or wrapped in Saran film for storage.

3.8 Analysis of Transcript Degradation and Determination of RNA Half-Lives

1. The band intensity of the images obtained by scanning with the Typhoon 8600 is determined using ImageQuant software. Use the image rectangle tool to demarcate each transcript band and read the band intensity to quantify each transcript band.
2. Standardize the amount of RNA of interest against the amount of an internal RNA control, such as chloroplast 16S rRNA, for each time point (*see* **Note 11**).
3. Relate the standardized amount of RNA of interest over a time course of treatments (such as 0, 10, 24, and 48 h after treatment with actinomycin D or water) by setting 0 h after treatment as the reference (100 %).
4. Take logarithms to base 10 of the relative amounts of the transcript to transform degradation rate of transcript into a linear relationship with time (h) (Fig. 2).
5. Calculate the best-fit equation for the relationship between $\log_{10}$ values of the amount of RNA against time (h). The half-life of a transcript, which is the time point when the transcript amount is reduced to 50 %, can be calculated by substituting the $\log_{10}$ value of the amount of RNA in the best-fit equation with $\log_{10}(50)$.

4 Notes

1. The methods described will enable the determination of half-lives of transcripts of endogenous genes in wild-type plants, or transcripts of chimeric reporter gene constructs in transplastomic plants. Protocols for the generation of transplastomic tobacco plants are given in Chapter 30.
2. Actinomycin D is a potent inhibitor of transcription in prokaryotic and eukaryotic organisms and is highly toxic, particularly if ingested. Wear gloves and protective clothing when handling actinomycin D solid or solutions. On opening the bottle, avoid breathing the dust. If there is skin contact, wash immediately with plenty of soapy water. If you feel unwell, seek medical advice immediately. Prepare a stock of 8 mg/mL actinomycin D in ethanol and dilute to a working concentration of 200 μg/mL actinomycin D with water. Dilute solutions of actinomycin D are very light sensitive and should be made up immediately before use.
3. To prepare DEPC-treated deionized water, add 1 mL of DEPC (D5758, Sigma-Aldrich) to 1,000 mL deionized water and mix using a magnetic stirrer overnight at room temperature. Autoclave to hydrolyze excess DEPC. DEPC reacts with active-site histidine residues and inhibits ribonuclease activity. It is a potential carcinogen. Always handle with caution, wear gloves, and work in a fume hood.

4. Formaldehyde is toxic by inhalation, in contact with skin, or if swallowed. Wear gloves, protective clothing and eye/face protection when handling formaldehyde solutions, and use in a well-ventilated room or fume hood. If formaldehyde is splashed in the eyes, rinse with water and seek medical advice.
5. Ethidium bromide intercalates into double-stranded regions of nucleic acids and is a potential mutagen, carcinogen, and teratogen. Extreme care should be taken to avoid contact with the skin and splashes in the eyes. Wear gloves, protective clothing, and eye/face protection when handling ethidium bromide solutions. Solutions should be made up using pre-weighed compressed tablets (806113, ICN Biomedical, Aurora, OH), rather than ethidium bromide powder, to decrease the risk of inhalation. Alternatively, ethidium bromide solutions (10 mg/mL) are available commercially (802511, ICN Biomedical, Aurora, OH).
6. The method described here will produce a double-stranded ^{32}P-labeled probe from a DNA fragment generated by PCR or by restriction digestion of a plasmid containing the gene sequence of interest. In general, probes of 200–1,500 bp are most useful. A 740-bp fragment of the *gfp* coding region, amplified by PCR, was used as a template for probe labeling to detect *gfp* transgene expression in transplastomic tobacco leaves by the methods described in this chapter [23]. The same study used a 1.4-kb fragment of the 16S rRNA gene amplified by PCR from a cloned *Bam*HI fragment of tobacco chloroplast DNA to produce a double-stranded ^{32}P-labeled probe to quantify the 16S rRNA as an internal control. For genes located in regions of the chloroplast genome showing complicated patterns of transcription, it may be necessary to produce single-stranded ^{32}P-labeled probes to identify sense transcripts, rather than antisense transcripts from the opposite strand [23].
7. Half-lives of chloroplast transcripts vary enormously depending on plant species or different transcripts. For example, the half-life of the *psbA* mRNA in young spinach leaves was estimated to be around 10 h [7], whereas the *psbA* mRNA in barley leaves had a half-life of more than 40 h [18]. A preliminary time course over 48 h with 4 different time points (0, 10, 24, and 48 h) is recommended for initial estimation of mRNA half-lives in chloroplasts. However, if the half-life of the transcript appears shorter than 10 h, a time course over the first 10 h with four different time points (0, 2.5, 7.5, and 10 h) is recommended.
8. For an experiment observing RNA degradation over 48 h, one leaf is required for RNA extraction at each time point (0, 10, 24, and 48 h) for actinomycin D and for water treatments. Select one leaf about 5-cm long from each 5-week-old tobacco plant grown in a magenta box. Normally, the most suitable-sized

leaf is the fifth leaf from the bottom. It is important to use leaves of the same developmental age from different plants, because expression of transgenes in tobacco chloroplasts is likely to decrease as the leaf ages [38]. An experiment detecting transcripts in actinomycin D- and water-treated leaves at four different time points with two replicates will therefore require 24 leaves from 24 tobacco plants.

9. The band intensity of the cytosolic 25S and 18S rRNAs in each track provides an indication of the amount of total RNA loaded and allows comparison between samples to identify unequal loading. The ratio of the band intensity of the 25S and 18S rRNAs should be ~2:1 in all samples. If the ratio is less than 2:1, it indicates that the RNA sample is partially degraded. The ratio of band intensities of the chloroplast 23S and 16S rRNAs cannot easily be used to determine the quality of the RNA preparation, because of nicking of the 2.9-kb 23S rRNA *in planta* producing fragments of 1.3, 1.1, and 0.5 kb [39–41]. However, the 1.5-kb 16S rRNA is not affected by such nicking and can be used to standardize the amount of chloroplast RNA loaded in each track. This is best accomplished by hybridization of the RNA-gel blot with a ^{32}P-labeled 16S rRNA probe (*see* **Note 6**).
10. During Typhoon 8600 scanning, check the image in the ImageQuant window for saturation. If the image appears saturated (pixels appear in red), expose the phosphor storage screen to the blot for a shorter period of time and then rescan.
11. The chloroplast 16S rRNA appears particularly stable in detached tobacco leaves over an incubation time of 48 h [23] and therefore appears to be a good internal control for standardizing amounts of transcripts of interest.

Acknowledgments

We thank Sue Aspinall for help in writing this article and for laboratory assistance during the development of the methods described. S.T. was supported by a scholarship from the Thai Government during the development of the methods described.

References

1. Barkan A, Goldschmidt-Clermont M (2000) Participation of nuclear genes in chloroplast gene expression. Biochimie 82:559–572
2. Gruissem W (1989) Chloroplast gene expression: how plants turn their plastids on. Cell 56:161–170
3. Gruissem W, Tonkyn JC (1993) Control mechanisms of plastid gene expression. Crit Rev Plant Sci 12:19–55
4. Deng XW, Gruissem W (1987) Control of plastid gene expression during development: the limited role of transcriptional regulation. Cell 49:379–387

5. Deng XW, Gruissem W (1988) Constitutive transcription and regulation of gene expression in non-photosynthetic plastids of higher plants. EMBO J 17:3301–3308
6. Rapp JC, Baumgartner BJ, Mullet J (1992) Quantitative analysis of transcription and RNA levels of 15 barley chloroplast genes. Transcription rates and mRNA levels vary over 300-fold; predicted mRNA stabilities vary 30-fold. J Biol Chem 267:21404–21411
7. Klaff P, Gruissem W (1991) Changes in chloroplast mRNA stability during leaf development. Plant Cell 3:517–529
8. Nickelsen J, Link G (1993) The 54 kDa RNA-binding protein from mustard chloroplasts mediates endonucleolytic transcript 3′ end formation *in vitro*. Plant J 3:537–544
9. Hayes R, Kudla J, Schuster G, Gabay L, Maliga P, Gruissem W (1996) Chloroplast mRNA 3′-end processing by a high molecular weight protein complex is regulated by nuclear encoded RNA binding proteins. EMBO J 15: 1132–1141
10. Westhoff P, Herrmann RG (1988) Complex RNA maturation in chloroplasts. The *psbB* operon from spinach. Eur J Biochem 171:551–564
11. Yang J, Schuster G, Stern DB (1996) CSP41, a sequence-specific chloroplast mRNA binding protein, is an endoribonuclease. Plant Cell 8: 1409–1420
12. Kuroda H, Maliga P (2002) Overexpression of the *clpP* 5′-untranslated region in a chimeric context causes a mutant phenotype, suggesting competition for a *clpP*-specific RNA maturation factor in tobacco chloroplasts. Plant Physiol 129:1600–1606
13. Bock R, Hermann M, Kossel H (1996) *In vivo* dissection of *cis*-acting determinants for plastid RNA editing. EMBO J 15:5052–5059
14. Knoop V (2010) When you can't trust the DNA: RNA editing changes transcript sequences. Cell Mol Life Sci 68:567–586. doi:10.1007/s00018-010-0538-9
15. Barkan A (1989) Tissue-dependent plastid RNA splicing in maize: transcripts from four plastid genes are predominantly unspliced in leaf meristems and roots. Plant Cell 1:437–445
16. Perron K, Goldschmidt-Clermont M, Rochaix JD (1999) A factor related to pseudouridine synthases is required for chloroplast group II intron *trans*-splicing in Chlamydomonas reinhardtii. EMBO J 18:6481–6490
17. Rivier C, Goldschmidt-Clermont M, Rochaix JD (2001) Identification of an RNA-protein complex involved in chloroplast group II intron *trans*-splicing in *Chlamydomonas reinhardtii*. EMBO J 20:1765–1773
18. Kim M, Christopher DA, Mullet JE (1993) Direct evidence for selective modulation of *psbA*, *rpoA*, *rbcL* and *16S* RNA stability during barley chloroplast development. Plant Mol Biol 22:447–463
19. Boudreau E, Nickelsen J, Lemaire SD, Ossenbuhl F, Rochaix JD (2000) The *Nac2* gene of *Chlamydomonas* encodes a chloroplast TPR-like protein involved in *psbD* mRNA stability. EMBO J 19:3366–3376
20. Vaistij FE, Goldschmidt-Clermont M, Wostrikoff K, Rochaix JD (2000) Stability determinants in the chloroplast *psbB/T/H* mRNAs of *Chlamydomonas reinhardtii*. Plant J 21:469–482
21. Thomas PS (1980) Hybridization of denatured RNA and small DNA fragments transferred to nitrocellulose. Proc Natl Acad Sci U S A 77: 5201–5205
22. Deng XW, Stern DB, Tonkyn JC, Gruissem W (1987) Plastid run-on transcription. Application to determine the transcriptional regulation of spinach plastid genes. J Biol Chem 262: 9641–9648
23. Tangphatsornruang S, Birch-Machin I, Newell CA, Gray JC (2010) The effect of different 3′ untranslated regions on the accumulation and stability of transcripts of a *gfp* transgene in chloroplasts of transplastomic tobacco. Plant Mol Biol 76:385–396. doi:10.1007/s11103-010-9689-1
24. Crouse EJ, Bohnert HJ, Schmitt JM (1984) Chloroplast RNA synthesis. In: Ellis RJ (ed) Chloroplast biogenesis. Cambridge University Press, Cambridge, pp 83–136
25. Stern DB, Gruissem W (1987) Control of plastid gene expression: 3′ inverted repeats act as mRNA processing and stabilizing elements, but do not terminate transcription. Cell 51: 1145–1157
26. Stern DB, Radwanski ER, Kindle KL (1991) A 3′ stem/loop structure of the *Chlamydomonas* chloroplast *atpB* gene regulates mRNA accumulation *in vivo*. Plant Cell 3:285–297
27. Stern DB, Kindle KL (1993) 3′ end maturation of the *Chlamydomonas reinhardtii* chloroplast *atpB* mRNA is a two-step process. Mol Cell Biol 13:2277–2285
28. Rott R, Drager RG, Stern DB, Schuster G (1996) The 3′ untranslated regions of chloroplast genes in *Chlamydomonas reinhardtii* do not serve as efficient transcriptional terminators. Mol Gen Genet 252:676–683
29. Rott R, Liveanu V, Drager RG, Stern DB, Schuster G (1998) The sequence and structure of the 3′-untranslated regions of chloroplast

transcripts are important determinants of mRNA accumulation and stability. Plant Mol Biol 36:307–314

30. Monde RA, Greene JC, Stern DB (2000) The sequence and secondary structure of the 3′-UTR affect 3′-end maturation, RNA accumulation, and translation in tobacco chloroplasts. Plant Mol Biol 44:529–542
31. Zerges W (2000) Translation in chloroplasts. Biochimie 82:583–601
32. Eibl C, Zhou Z, Beck A, Kim M, Mullet J, Koop H-U (1999) *In vivo* analysis of plastid *psbA*, *rbcL* and *rpl32* UTR elements by chloroplast transformation: tobacco plastid gene expression is controlled by modulation of transcript levels and translation efficiency. Plant J 19:333–345
33. Goldschmidt-Clermont M, Rahire M, Rochaix JD (2008) Redundant cis-acting determinants of 3′ processing and RNA stability in the chloroplast rbcL mRNA of Chlamydomonas. Plant J 53:566–577
34. Maliga P (2002) Engineering the plastid genome of higher plants. Curr Opin Plant Biol 5:164–172
35. Staub J, Maliga P (1994) Translation of *psbA* mRNA is regulated by light via the 5′-untranslated region in tobacco plastids. Plant J 6:547–553
36. Lin-Chao S, Chiou NT, Schuster G (2007) The PNPase, exosome and RNA helicases as the building components of evolutionarily-conserved RNA degradation machines. J Biomed Sci 14:523–532
37. Barnes D, Franklin S, Schultz J, Henry R, Brown E, Coragliotti A, Mayfield SP (2005) Contribution of 5′- and 3′-untranslated regions of plastid mRNAs to the expression of *Chlamydomonas reinhardtii* chloroplast genes. Mol Genet Genomics 274:625–636
38. Birch-Machin I, Newell CA, Hibberd JM, Gray JC (2004) Accumulation of rotavirus VP6 protein in chloroplasts of transplastomic tobacco is limited by protein stability. Plant Biotechnol J 2:261–270
39. Leaver CJ, Ingle J (1971) The molecular integrity of chloroplast ribosomal ribonucleic acid. Biochem J 123:235–243
40. Atchison BA, Bourque DP, Wildman SG (1973) Preservation of 23S chloroplast RNA as a single chain of nucleotides. Biochim Biophys Acta 331:382–389
41. Nishimura K, Ashida H, Ogawa T, Yokota A (2010) A DEAD box protein is required for formation of a hidden break in Arabidopsis chloroplast 23S rRNA. Plant J 63:766–777

Chapter 14

Quantification of Organellar DNA and RNA Using Real-Time PCR

Andreas Weihe

Abstract

Quantitative (real-time) polymerase chain reaction (PCR) allows the measurement of relative organellar gene copy numbers as well as transcript abundance of individual mitochondrial or plastidial genes. Requiring only minute amounts of total DNA or RNA, the described method can replace traditional analyses like Southern or Northern hybridization which require large amounts of organellar nucleic acids and usually provide only semiquantitative data. Here we describe prerequisites, reaction conditions, and data analysis principles, which should be applicable for a wide range of plant species and experimental situations where comparative and precise determination of gene copy numbers or transcript abundance is requested. Sequences of amplification primers for qPCR of organellar genes from Arabidopsis are provided.

Key words Gene copy number, Mitochondrial genes, mtDNA, Plastid genes, ptDNA, Real-time quantitative PCR, Transcript accumulation

1 Introduction

Determination of the copy number of mitochondrial and plastidial genes and their transcript accumulation is traditionally done by Southern and Northern hybridization, respectively. These methods require relatively large amounts of DNA or RNA. Relying on the quantification of hybridization signals, they often provide rather inaccurate data. Quantitative real-time PCR (qPCR) provides a simple, much more sensitive, and accurate alternative for the determination of both organellar gene copy number and transcript accumulation (*see* **Note 1**). The method does not require the isolation of organelles and rather uses total DNA or RNA (cDNA), respectively, as template.

qPCR measures amplicon concentration by monitoring fluorescence signals during the exponential phase of amplification. Probe-based assays, also known as TaqMan PCR, use, in addition to the amplification primers, a double-labeled fluorogenic probe, designed to bind inside the amplified region during primer

Pal Maliga (ed.), *Chloroplast Biotechnology: Methods and Protocols*, Methods in Molecular Biology, vol. 1132,
DOI 10.1007/978-1-62703-995-6_14,

extension. 5′ exonuclease activity of the Taq polymerase releases a quencher dye resulting in fluorescence of the reporter dye on the probe [1]. Alternatively, intercalating dyes, such as SYBR Green I, can be used instead of fluorescent probes. SYBR Green produces a fluorescent signal when bound to double-stranded DNA. Although the latter method lacks the high specificity of the TaqMan assay, it can be applied to any amplification reaction, thus providing more flexibility and avoiding the relatively high costs of TaqMan probes.

We describe a simple method for relative quantification of organellar gene copy number and transcript accumulation by qPCR which can be used for a wide range of comparative analyses. The protocol requires careful design of primers and, besides basic laboratory equipment, a thermocycler with an optical detection system, available from a variety of suppliers.

2 Materials

2.1 DNA Isolation

1. 2× CTAB: 2 % CTAB (cetyltrimethyl ammonium bromide), 100 mM Tris–HCl, pH 8.0, 20 mM EDTA, 1.4 M NaCl.
2. 5× CTAB: 5 % CTAB, 0.35 M NaCl.
3. 1× CTAB: 1 % CTAB, 50 mM Tris–HCl, pH 8.0, 10 mM EDTA.
4. HS-TE: 10 mM Tris–HCl, pH 8.0, 1 mM EDTA, 1 M NaCl.
5. 1× TE: 10 mM Tris–HCl, pH 8.0, 1 mM EDTA.
6. RNase A/T1 mix (Thermo Scientific, catalogue # EN0551).
7. 3 M sodium acetate, pH 5.2.
8. Chloroform–isoamyl alcohol (24:1, v/v).
9. Phenol/chloroform (1:1, v/v).
10. Ethanol, 96 and 70 %.
11. Mortar, pestle, liquid nitrogen.

2.2 RNA Isolation

1. TRIzol (Life Technologies, catalogue # 15596026).
2. Chloroform–isoamyl alcohol (24:1 v/v).
3. Isopropyl alcohol.
4. Ethanol, 70 %.
5. DNase I, amplification grade (Life Technologies, catalogue # 18068-015).
6. Mortar, pestle, liquid nitrogen.

2.3 DNA-qPCR

1. Power SYBR Green PCR Master Mix (Life Technologies, catalogue # 4367659) or

   TaqMan Fast Universal PCR Master Mix (for probe-based assays, Life Technologies, catalogue # 4352042).

2. Amplification primer, 20 μM.
3. LNA fluorescent probe (for probe-based assays only), from Universal Probe Library (Roche, catalogue # 04683633001 and # 04869877001), 10 μM.
4. Microtiter plates, optical grade.

2.4 Reverse Transcription and RT-qPCR

1. QuantiTect Reverse Transcription Kit (QIAGEN, catalogue # 205310) including gDNA Wipeout buffer, Quantiscript RT buffer, RT Primer Mix, and Quantiscript Reverse Transcriptase.
2. Power SYBR Green PCR Master Mix (Life Technologies, catalogue # 4367659) or
   TaqMan Fast Universal PCR Master Mix (Life Technologies, catalogue # 4352042).
3. Amplification primer, 20 μM.
4. LNA fluorescent probe (for probe-based assays only), from Universal Probe Library (Roche, catalogue # 04683633001 and # 04869877001), 10 μM.

3 Methods

3.1 DNA Isolation

DNA isolation follows the protocol of Rogers and Bendich [2].

1. 2–3 g of plant tissue is grinded in a mortar under liquid nitrogen.
2. The homogenate is transferred into a Falcon tube, resuspended in 3 vol 2× CTAB, and incubated for 15 min at 65 °C in a water bath.
3. 1 vol chloroform/isoamyl alcohol is added and the tube vigorously shaken for 1–2 min.
4. Centrifuge tube for 10 min at 6,000 × *g*.
5. Transfer the upper phase to a new tube, add 0.2 vol of 5× CTAB, and incubate the mixture for 10 min at 60 °C.
6. Repeat chloroform extraction (**step 3**).
7. Centrifuge tube for 10 min at 6,000 × *g*.
8. Mix the upper phase with 1.2 vol of 1× CTAB and incubate at least for 1 h at room temperature.
9. Pellet the CTAB–nucleic acid complexes by centrifugation at 6,000 × *g* for 30 min.
10. Dissolve the pellet in HS-TE.
11. Precipitate the nucleic acids by addition of 2.5 vol 96 % ethanol. Incubate tube for 10 min at room temperature.
12. Pellet DNA by centrifugation (20 min at 6,000 × *g*).

13. Wash DNA with 70 % ethanol, and repeat pelleting by centrifugation (10 min at 6,000 ×*g*).
14. Dissolve DNA in 1× TE.
15. The DNA still contains considerable amounts of RNA. Remove RNA by RNAse A/T1 treatment for 60 min at 37 °C.
16. Remove the enzymes by phenol/chloroform extraction.
17. Precipitate the DNA using 2.5 vol 96 % ethanol and 0.1 vol 3 M sodium acetate (10 min, room temperature), and then sediment DNA by centrifugation (6,000 ×*g* for 20 min).
18. Dissolve DNA in ddH_2O. Determine DNA concentration spectrophotometrically. DNA integrity (there should be no smear towards lower molecular weight) is checked by agarose gel electrophoresis.

3.2 RNA Isolation

TRIzol [3] is used for isolation of total RNA (*see* **Note 2**).

1. Grind plant tissue in a mortar under liquid nitrogen. Add 5 vol of TRIzol directly to the mortar and continue grinding until the homogenate is thawed.
2. Transfer homogenate to a Falcon tube. Add 0.2 vol of chloroform/isoamyl alcohol (24:1) and shake tube vigorously for 1–2 min.
3. Centrifuge tube at 6,000 ×*g* for 15 min.
4. Transfer the upper phase to a clean tube and precipitate the RNA by adding 0.7 vol (per initial volume of TRIzol) of isopropyl alcohol.
5. Pellet the RNA by centrifugation at 6,000 ×*g* for 30 min, and then wash the pellet two times with 70 % ethanol and dissolve the precipitate in ddH_2O.
6. To remove residual genomic DNA, treat the sample with RNase-free DNase for 30 min at 37 °C.
7. Repeat TRIzol extraction (see above).
8. Dissolve RNA in ddH_2O. Determine RNA concentration with spectrophotometer and RNA quality by denaturing formaldehyde-agarose gel electrophoresis (rRNA bands should be clearly visible, no smear towards lower molecular weight).

3.3 qPCR

3.3.1 Chemistries for the Detection of Amplicons and Primer-Probe Design

Both SYBR Green- and probe-based assays may be performed. If SYBR Green detection is used, specificity of amplification is solely dependent upon the amplification primers. Therefore, in this case, a dissociation curve should be generated to control whether only the specific amplicons are produced. Probe-based qPCR requires an additional, double-labeled oligonucleotide probe annealing to the target sequence between the two amplification primers, thus providing extra specificity. For probe-based assays, in the author's laboratory, the Universal Probe Library Set (Roche) is used, consisting of a set of predesigned short (8–9 nucleotides) locked nucleic acid (LNA) oligos labeled at the 5′ end with fluorescein and at the 3′ end with a dark quencher dye. The sequences of the probes in the library detect target sequences that are prevalent in the transcriptome, ensuring optimal coverage of the transcripts in a given organism (*see* **Note 3**). Specificity is maintained by the high melting temperature of the LNA oligos in combination with the amplification primers. Primer design and probe selection can be conveniently performed on the Universal Probe Library web site (http://www.roche-applied-science.com/shop/CategoryDisplay?catalogId=10001&tab=&identifier=Universal+Probe+Library#tab-0). For both SYBR Green- and probe-based qPCR, amplification primer pairs have to be designed to yield amplification products of 70–120 bp. The primers have to be carefully checked for primer dimer formation which must be avoided. For SYBR Green assays, primers can be designed using the Primer Express Software (Life Technologies) or any other primer design software, e.g., Primer3 (http://www.ncbi.nlm.nih.gov/tools/primer-blast/index.cgi?LINK_LOC=BlastHome). Table 1 lists amplification primer pairs for mitochondrial and plastidial as well as nuclear reference genes for *Arabidopsis thaliana* used in the author's laboratory (*see* **Note 4**).

3.4 DNA-qPCR

SYBR Green-based PCR reactions are carried out in a real-time PCR thermocycler (we use a 7500 Applied Biosystems Real-Time PCR System), in microtiter plates, employing the Power SYBR Green PCR Master Mix (for SYBR Green-based assays) or the TaqMan Fast Universal PCR Master Mix (for probe-based assays). Each reaction contains 0.1 ng DNA, 1 μM of each amplification primer, and 100 nM of the LNA oligo (for probe-based assays only). Biological and technical replicates are analyzed in triplicates per experiment. No template controls (replace DNA by ddH_2O) should be included for each primer pair.

Set up reactions in a total volume of 50 μl (for 3 × 15 μl replicates + 10 % for possible pipetting inaccuracy) consisting of 25 μl 2× Master Mix (Power SYBR Green Master Mix and TaqMan Fast Universal PCR Master Mix, respectively), 0.33 ng total DNA, 2.5 μl of each amplification primer, 0.5 μl LNA probe (for probe-based assay only), and ddH_2O *ad* 50 μl. Dispense 3 × 15 μl per reaction well of the microtiter plate. The cycling protocol consists of an initial denaturation step at 95 °C for 10 min (activation of the polymerase) followed by 40 cycles of 15 s at 95 °C and 1 min at 60 °C. To verify the specificity of the amplification in the SYBR Green assays, a dissociation curve is generated after the cycles are completed (denaturation over a temperature gradient from 60 to 95 °C at 0.03 °C/s). The quantitations are normalized to the DNA amount of a nuclear-encoded (usually single-copy) reference gene (*see* **Note 5**).

Table 1
Amplification primers for real-time qPCR of *Arabidopsis thaliana* genes [4, 6]

Primer	Sequence
RpoTm (nuclear, reference for DNA-qPCR)	(fw) AGCCTGTGCGTAATGCTATTCA (rv) GCCATCTTATCAGCCGGTAACT
RpoTp (nuclear, reference for DNA-qPCR)	(fw) TGGAAGCCGTCTGCTAGAACTA (rv) TGTCTGAATGCAGGTCGAAAC
UBQ (nuclear, reference for RT-qPCR)	(fw) CTTATCTTCGCCGGAAAGC (rv) GAGGGTGGATTCCTTCTGG
18S (nuclear, reference for RT-qPCR)	(fw) AAACGGCTACCACATCCAAG (rv) ACTCGAAAGAGCCCGGTATT
nad6 (mitochondrial)	(fw) AGGATGTATTCCGACGAAATGC (rv) CGTGAGTGGGTCAGTCGTCC
atp1 (mitochondrial)	(fw) CTTAGAAAGAGCGGCTAAACGA (rv) GGGAATATAGGCCGATACGTCT
rps4 (mitochondrial)	(fw) CCCATCACAGAGATGCACAGA (rv) GGAGACGAAGCGGAATAACGT
cox1 (mitochondrial)	(fw) GCCATGATCAGTATTGGTGTCTT (rv) CTACGTCTAAGCCCACAGTAAACA
clpP (plastid)	(fw) TATGCAATTTGTGCGACCC (rv) TTGGTAATTGCTCCTCCGACT
psbA (plastid)	(fw) CATCCGTTGATGAATGGCTAT (rv) AACTAAGTTCCCACTCACGACC
ndhH (plastid)	(fw) CATACCGGTGGCAGCTTCGAA (rv) TCATCTGTTATGGCTCGGCCC

3.5 RT-qPCR

3.5.1 Reverse Transcription

Reverse transcription is performed using the QuantiTect Reverse Transcription Kit (QIAGEN, catalogue # 205310). To verify removal of genomic DNA from cDNA samples, negative controls have to be included, in which reverse transcriptase is replaced by ddH_2O.

1. Set up the genomic DNA elimination reaction by adding 2 μl of 7x gDNA Wipeout buffer to up to 2 μg of total RNA in a reaction volume of 14 μl. Incubate for 2 min at 42 °C and then place on ice.
2. Add 4 μl Quantiscript RT buffer, 1 μl RT Primer Mix, and 1 μl Quantiscript Reverse Transcriptase. Incubate for 15 min at 42 °C.
3. Inactivate enzyme for 3 min at 95 °C.
4. Place on ice and proceed with real-time PCR.

3.5.2 qPCR

For RT-qPCR, an amount of cDNA produced from 50 ng RNA is used per single 15 μl reaction. Other components of the reaction,

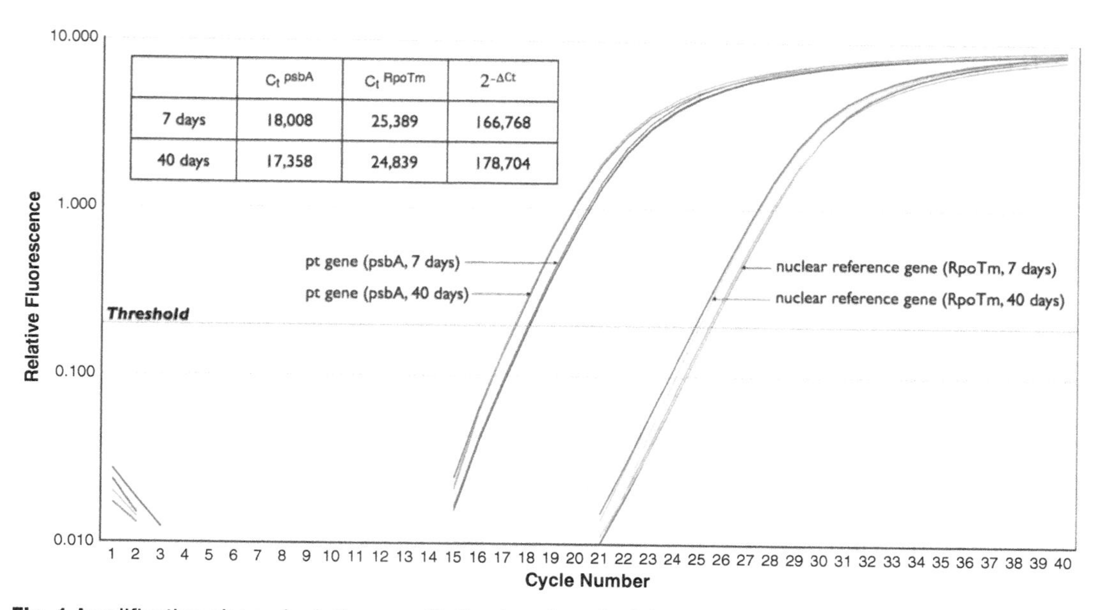

Fig. 1 Amplification plot and relative quantitation (*see insert*) of the copy numbers of a plastid gene (*psbA*) in 7- and 40-day-old Arabidopsis leaves. The nuclear single-copy gene *RpoTm* encoding the mitochondrial RNA polymerase was used as the reference gene. SYBR Green-based assay, amplification primers *see* Table 1

setup, and cycling conditions are exactly as described for the DNA-qPCR (*see* Subheading 3.4). The quantitations are normalized to the cDNA amount of a nuclear-encoded reference (housekeeping) gene (*see* **Note 6**).

3.5.3 Data Analysis

Data are analyzed by the instrument's software (Sequence Detection Software v2, Life Technologies). The quantitations are expressed as copies (transcripts) per nuclear-encoded single-copy gene (DNA-qPCR) or reference (housekeeping) gene (RT-qPCR) using the ΔC_t method.

C_t values are the cycle thresholds measured by the instrument during the exponential phase of amplification:

$$\text{Relative amount of gene copies} = 2^{-\Delta C_t}$$

where $\Delta C_t = C_t^{og} - C_t^{ref}$, og = organellar gene, ref = reference/housekeeping gene and

$$\text{Relative transcript abundance} = 2^{-\Delta C_t}$$

where $\Delta C_t = C_t^{og} - C_t^{ref}$, og = organellar gene, ref = reference gene.

Figure 1 provides an example of relative quantification of plastidial gene copy numbers in Arabidopsis leaves of different ages. Both the amplification plot and the final data derived from the assay are shown. The mitochondrial gene copy numbers are expressed as copies per nuclear reference gene; possible changes in the nuclear DNA content were not regarded.

4 Notes

1. The described method provides relative quantification of organellar genes and transcript accumulation rather than an absolute quantification. If the latter is required, individual data have to be compared to a standard curve which can be constructed by measuring defined quantities of DNA or cDNA samples, respectively. For this purpose, the vector construct described for controlling the efficiency of amplification (*see* **Note 4**) can be used. In most cases, however, relative quantification will provide reliable and exact data for comparative analyses.
2. For certain plants/tissues, an alternative RNA isolation method may be necessary. Thus, we found RNA isolated from etiolated Arabidopsis seedlings using TRIzol to be of insufficient quality. In such cases, the RNeasy Plant Mini Kit (QIAGEN, catalogue # 74904) is recommended. RNA should be isolated according to the manufacturer's protocol with buffer RLT (kit component).
3. The search algorithm for primers/probe allows the input of any target sequence (Arabidopsis is available as default), and in most cases, appropriate probe sequences will be found among the oligo set of the library.
4. New amplification primer pairs, both for SYBR Green- and probe-based assays, should be tested for efficiency of amplification which should be equal for all genes to be analyzed [4]. For this purpose, the amplicon regions should be cloned into an appropriate vector, e.g., pBluescript. The recombinant plasmids are then used as templates for real-time PCR in standard reactions as described above. The ratio between two genes is calculated as $2^{-\Delta C_t} \left(\Delta C_t = C_t^{gene1} - C_t^{gene2} \right)$
5. Organellar gene copy number data obtained by qPCR are expressed as copies per a (usually) nuclear reference gene. Copy numbers of organellar DNA *per cell* can be calculated by multiplying the relative copy number as obtained by the qPCR by the nuclear ploidy level. Experimental conditions, under which the nuclear DNA content may vary, have to account for the endoploidy level of the analyzed samples. To measure changes in nuclear DNA content due to endopolyploidization, ploidy levels have to be determined by flow cytometry [5, 6].
6. Special care should be taken in choosing the reference gene for transcript accumulation measurements. No single gene can be recommended in general. It will depend upon the plant species, particular tissue, and application. One probably needs to try several genes before choosing those whose expression remains constant over the conditions used in the experiment. For discussion of internal controls for plant gene expression studies using real-time PCR, *see* e.g. [7]. Starting with nuclear

18S rRNA or housekeeping genes like *GAPDH* or ubiquitin (*UBQ*) might be a good recommendation. When using rRNA references, the cDNA in the reference reaction should be diluted 1:10,000.

Acknowledgement

The author thanks R. Reile for providing unpublished data (Fig. 1).

References

1. Heid CA, Stevens J, Livak KJ, Williams PM (1996) Real time quantitative PCR. Genome Res 6:984–986
2. Roger SO, Bendich AJ (1985) Extraction of DNA from milligram amounts of fresh, herbarium and mummified plant tissues. Plant Mol Biol 5:702–707
3. Chomczynski P, Sacchi N (1987) Single-step method of RNA isolation by acid guanidinium thiocyanate-phenol-chloroform extraction. Anal Biochem 162:156–159
4. Preuten T, Cincu E, Fuchs J, Zoschke R, Liere K, Börner T (2010) Fewer genes than organelles: extremely low and variable gene copy numbers in mitochondria of somatic plant cells. Plant J 64:948–959
5. Barow M, Meister A (2006) Endopolyploidy in seed plants is differently correlated to systematics, organ, life strategy and genome size. Plant Cell Environ 26:571–584
6. Zoschke R, Liere K, Börner T (2007) From seedling to mature plant: Arabidopsis plastidial genome copy number, RNA accumulation and transcription are differentially regulated during leaf development. Plant J 50:710–722
7. Brunner AM, Yakovlev IA, Strauss SH (2004) Validating internal controls for quantitative plant gene expression. BMC Plant Biol 4:14

Chapter 15

Plastid Transformation for Rubisco Engineering and Protocols for Assessing Expression

Spencer M. Whitney and Robert E. Sharwood

Abstract

The assimilation of CO_2 within chloroplasts is catalyzed by the bi-functional enzyme ribulose-1, 5-bisphosphate carboxylase/oxygenase, Rubisco. Within higher plants the Rubisco large subunit gene, *rbc*L, is encoded in the plastid genome, while the Rubisco small subunit gene, *RbcS* is coded in the nucleus by a multi-gene family. Rubisco is considered a poor catalyst due to its slow turnover rate and its additional fixation of O_2 that can result in wasteful loss of carbon through the energy requiring photorespiratory cycle. Improving the carboxylation efficiency and CO_2/O_2 selectivity of Rubisco within higher plants has been a long-term goal which has been greatly advanced in recent times using plastid transformation techniques. Here we present experimental methodologies for efficiently engineering Rubisco in the plastids of a tobacco master-line and analyzing leaf Rubisco content.

Key words CABP binding, Immuno-detection, *Nicotiana tabacum*, Rubisco enzyme, Rubisco master-line, *rbc*L gene, Tobacco

1 Introduction

Rubisco catalyzes the rate-limiting step of CO_2-fixation within the photosynthetic carbon reduction cycle. This enzyme is hampered by its competition between substrates CO_2 and O_2 and its slow carboxylation rate [1]. This necessitates a significant nitrogen investment by higher plants to synthesize large amounts of Rubisco to achieve adequate photosynthetic CO_2-assimilation rates for carbohydrate synthesis that are used for plant growth and to store and transport energy [2, 3]. The poor catalytic qualities of Rubisco have been a long standing target for kinetic improvement with recent advances in structure–function detail and genetic manipulation strategies reinvigorating efforts to engineer catalytic improvements in Rubisco in higher plants [4, 5].

The disparate genetic locations of the large (*rbc*L—plastome) and small (*RbcS*—nucleus) Rubisco subunit genes in higher plants have complicated comprehensive engineering of the entire enzyme,

Pal Maliga (ed.), *Chloroplast Biotechnology: Methods and Protocols*, Methods in Molecular Biology, vol. 1132,
DOI 10.1007/978-1-62703-995-6_15, © Springer Science+Business Media New York 2014

possibly even complicating its natural evolution [4]. Manipulation of Rubisco in higher plants plastids has focused on altering its content using nucleus directed anti-*RbcS* and sense-*RbcS* applications; supplementing the plastome with additional copies of foreign Rubisco genes or tobacco *RbcS* genes; and replacing the tobacco *rbcL* with foreign or altered forms using plastome transformation [1].

The recent construction of the tobacco-*rubrum* master-line (called ^{cm}trL) for Rubisco manipulation provides unique opportunities for higher-throughput mutagenic studies on Rubisco in chloroplasts [6]. Unwanted recombination events are avoided in this tobacco line due to the sequence variability of the ^{cm}*rbcM* transgene in ^{cm}trL that enables efficient and reliable insertion of Rubisco transgenes into the native tobacco *rbcL* plastome region under the control of the tobacco *rbcL* regulatory 5′ and 3′ sequences (Fig. 1). The efficient transformation capacity of ^{cm}trL has been exploited to engineer tobacco producing large (L) subunit mutated tobacco Rubisco [7, 8], entirely foreign Rubisco [9, 10]; hybrid forms comprising foreign L and tobacco small (S) subunits [5, 8, 11], and unique complexes comprising fused L and S-subunits [12].

As Rubisco is the major protein in leaves, determining the amount present and its catalysis in wild-type and transformed plants is often undertaken. Methodologies that are used to measure Rubisco catalysis are detailed elsewhere [13]. This chapter presents protocols used to manipulate Rubisco form and function in the ^{cm}trL master-line and details how quantitative measurements of leaf Rubisco content is made using ^{14}C-labelled carboxyarabinitol 1,5-bisphosphate (CABP), a universal and tight binding inhibitor specific of Rubisco that binds stoichiometrically to each active site [14]. Also described are qualitative methodologies often used to measure the relative Rubisco content in non-purified leaf samples using densitometry measurements of Coomassie blue binding or immunoblot signals. A comparative analysis of the practical limitations and reliability of these qualitative approaches for measuring leaf Rubisco content is made. An overview of the ^{cm}trL transformation and Rubisco analysis processes are shown in Fig. 2.

2 Materials

2.1 Rubisco Gene Amplification and Plasmid Cloning

1. Plant total genomic DNA: Isolate DNA from plant material using a DNeasy Plant Mini Kit (Qiagen).
2. TA-cloning; PCR amplified *rbcL* products are cleaned up using the Wizard® SV Gel and PCR Clean-Up system (Promega) and then cloned into pGEM-T-Easy (Promega) according to supplier instructions.
3. DNA sequencing; using T7, SP6 and *rbcL* specific primers the cloned transgenes are fully sequenced using BigDye

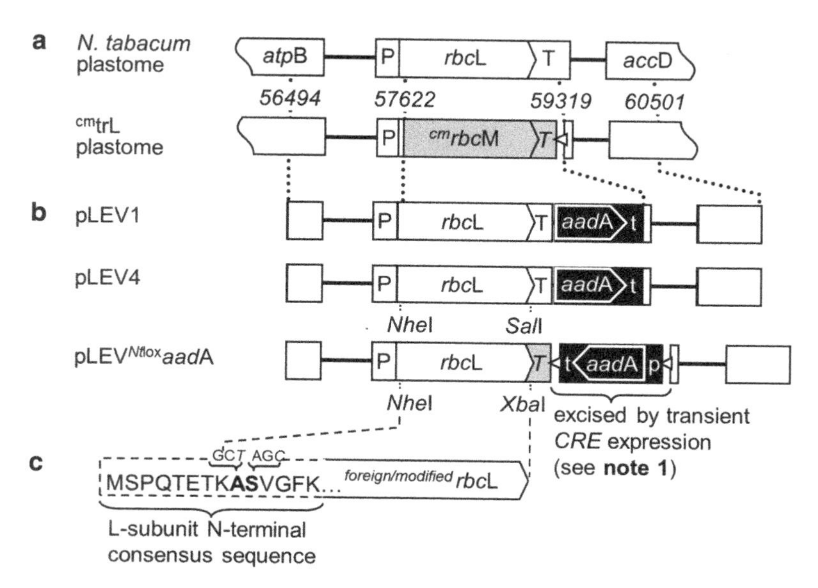

Fig. 1 Genetic tools for manipulating Rubisco in tobacco plastids. (**a**) Comparative features of the "wild-type" *Nicotiana tabacum* plastome around *rbc*L relative to that in the tobacco master-line ^{cm}trL where the *rbc*L gene (excluding the first 42 bp of coding sequence) and 285 bp of its 3′ sequence (T) has been replaced with a codon modified Rubisco gene (^{cm}*rbc*M; coding the *Rhodospirillum rubrum* bacterial L_2 Rubisco), 222 bp of the *psb*A 3′ sequence (*T*) and 34 bp of *lox*P sequence (*white triangle*) ([6], *see* **Note 1**). (**b**) Transforming plasmids derived from pLEV1 [7] that can be used to introduce (**c**) foreign *rbc*L, modified tobacco *rbc*L or entire *rbc*L-S operons into the ^{cm}trL plastome [5, 6]. The unique *Nhe*I site introduced into plasmids pLEV4 and pLEV^{Nflox}*aad*A introduces silent changes to the Ala-9 and Ser-10 codons within the N-terminal sequence that is highly conserved amongst vascular plant Rubisco L-subunits. These mutations do not affect tobacco Rubisco expression [8]. Incorporating either *Sal*I or *Xba*I sites after the *rbc*L (or *rbc*S) stop codon (TAA) allow for cloning into pLEV4 and pLEV^{Nflox}*aad*A respectively. The p-*aad*A-t gene cassette (*p* 16S rDNA *rrn* promoter and 5′untranslated region, *t rps*16 3′-untranslated region [18]) in pLEV^{Nflox}aadA transformed lines can be excised by transient CRE recombinase expression [6]. The plastome sequence incorporated into the transforming plasmids used to direct homologous recombination is indicated by the *dotted lines* and numbered according to GenBank accession no. Z00044

terminator sequencing on an ABI 3730 sequencer (Biomolecular Resource Facility, JCSMR, ANU).

4. Site directed mutagenesis; is performed using the QuickChange® Multi Site-Directed Mutagenesis Kit (Stratagene) according to supplier instructions.
5. LB-Amp; Luria–Bertani medium (10 g Tryptone, 5 g yeast, 5 g NaCl, pH 7 per liter, 1 L water, adjust to pH 7.0 with 0.1 M NaOH) containing ampicillin (0.2 mg/mL). For LB-Amp media, agar is added to 1.5 % (w/v). Autoclave to sterilize.
6. TE buffer; autoclave sterilized 5 mM Tris–Cl, pH 8.0, 0.5 mM EDTA.

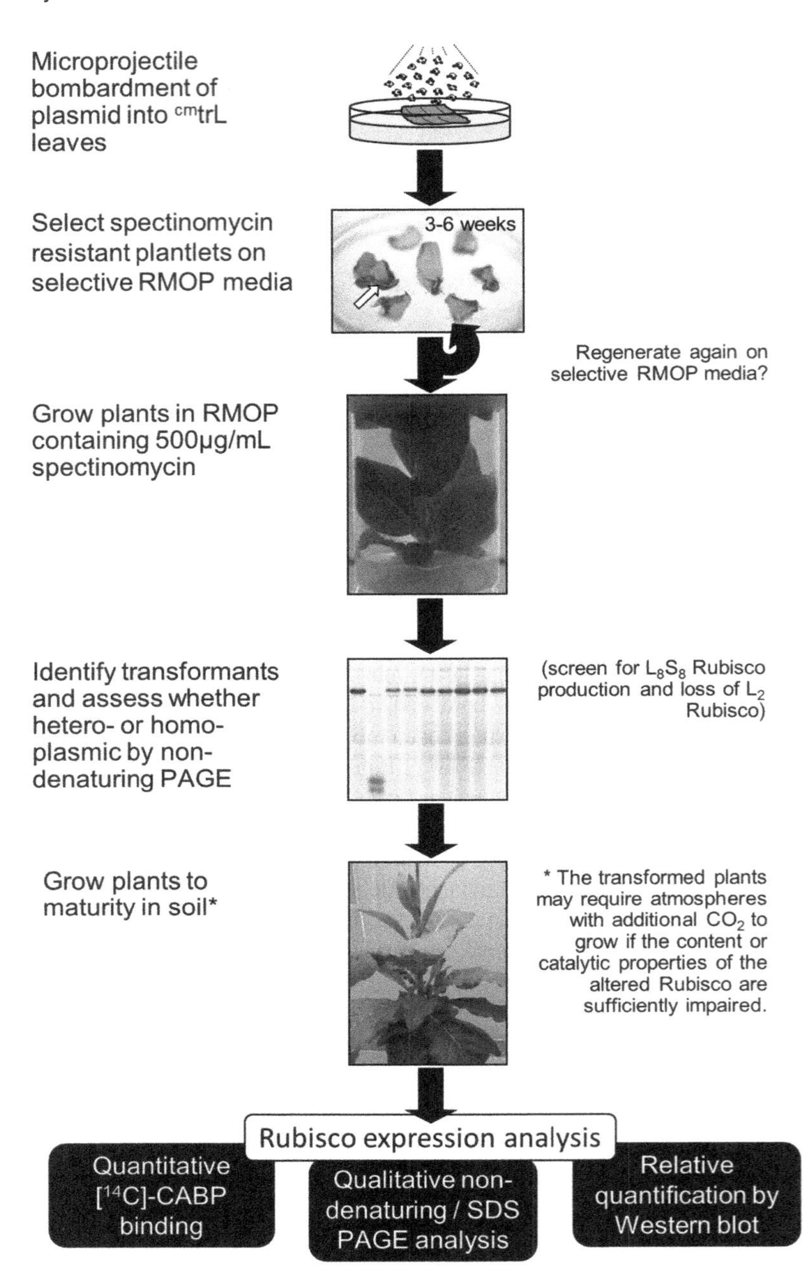

Fig. 2 Engineering Rubisco via plastomic manipulation of the tobacco master-line ^{cm}trL. Summary of the processes for biolistic transformation, selection and identification of transformants, and measuring the levels of Rubisco expression

2.2 Plant Tissue Culture

1. Leaves from ^{cm}trL plants are used for transformation (*see* **Note 1** and Fig. 2).
2. Bleach solution; in a sterile 10 mL plastic tube mix 9 mL autoclaved Milli-Q water H_2O with 1 mL sodium hypochlorite (1.2 % v/v).

3. RMOP medium, agar-solidified Murashige–Skoog salts (Sigma) containing 3 % (w/v) sucrose and 0.6 % (w/v) agar (Difco). Autoclave to sterilize and cool to 50 °C in a water bath before pouring. Filter-sterilize (0.22 μm) a 10 mL solution containing 0.1 g myo-inositol and 25 μL thiamine-HCl (40 mg/mL in water) and add to 1 L of autoclaved RMOP media cooled to 50 °C. In a laminar flow cabinet add 25 μL 6-benzylaminopurine (BAP; 40 mg/mL in DMSO) and 25 μL 1-Naphthaleneacetic acid (NAA; 4 mg/mL in DMSO), swirl to mix and dispense 50 mL of the media into twenty 90 mm × 25 mm petri dishes.
4. Germination plates, 90 mm × 8 mm plastic petri dishes containing 20 mL RM medium (same as RMOP medium, but no sucrose, BAP, NAA or thiamine).
5. Autoclaved cylindrical polypropylene growth containers (90 mm diam × 130 mm height, *see* panel 3 of Fig. 2).
6. Selective $RMOP^{spec}$ medium. Same as RMOP medium, but include 0.5 g spectinomycin in the 10 mL filter-sterilized (0.22 μm) solution containing, 0.1 g myo-inositol and 25 μL thiamine-HCl (40 mg/mL in water) and add to 1 L of autoclaved RMOP media cooled to 50 °C. In a laminar flow cabinet add 25 μL 6-benzylaminopurine (BAP; 40 mg/mL in DMSO) and 25 μL 1-Naphthaleneacetic acid (NAA; 4 mg/mL in DMSO), swirl to mix and dispense 50 mL of the media into twenty 90 mm × 25 mm petri dishes.

2.3 Leaf Protein Extraction and Polyacrylamide Gel Electrophoresis (PAGE)

1. Protein extraction buffer: 50 mM Tris-Cl, pH7.8, 10 mM $MgCl_2$, 20 mM $NaHCO_3$, 1 % (w/v) polyvinyl polypyrrolidone, 2 mM DTT. Keep on ice. Add 1 % (v/v) plant protease inhibitor cocktail (Sigma-Aldrich) to buffer just prior to protein extraction.
2. PAGE: Assumes use of 4–12 % Novex® Tris–Glycine (non-denaturing) and Bis-Tris-buffered 4–12 % NuPAGE Novex® (SDS PAGE) mini gels (Invitrogen).
3. Non-denaturing electrophoresis buffer (10×): 0.24 M Tris, 1.91 M glycine. Do not adjust pH. Dilute 100 mL with 900 mL water for use and store at 4 °C.
4. Non-denaturing sample buffer (5×): 80 % (v/v) glycerol containing 0.05 % (w/v) bromophenol blue. Store at room temperature.
5. SDS PAGE electrophoresis buffer (10×): 0.5 M MES (2-(*N*-morpholino)ethanesulfonic acid; Sigma), 0.5 M Tris, 8 mM EDTA, 1 % (w/v) sodium dodecyl sulfate (SDS). Do not adjust pH. Store at room temperature.
6. SDS PAGE Sample buffer (5×): 0.25 M Tris-Cl, pH 6.8, 8 % (w/v) SDS, 20 % (v/v) glycerol, 0.1 % (w/v) bromophenol

blue. Store at room temperature. Add 0.02 (v/v) 2-β-mercaptoethanol (i.e., ~0.3 M) just prior to use.

7. Gel fixation buffer: 45 % (v/v) Milli-Q water, 5 % (v/v) glacial acetic acid, 50 % (v/v) methanol.
8. Coomassie blue stain: GelCode Blue Stain Reagent (Pierce), 20 mL per gel.
9. Non-denaturing PAGE molecular weight markers: HMW Native Marker Kit (GE Healthcare). Dissolve the protein standards in 0.32 mL of 1× non-denaturing buffer and 0.08 mL of non-denaturing sample buffer. Load 5 μL per lane.
10. SDS PAGE molecular weight markers: Precision Plus Protein Standards (Bio-Rad). Load 5 μL per lane.

2.4 Plant Growth in Soil

1. Growth facilities: Plants are grown in temperature controlled glasshouses or growth cabinets at 25 °C.
2. Illumination and humidity in growth cabinets typically set at 200–600 μE and 60 %, respectively.
3. Growth at elevated CO_2; Atmospheric CO_2 levels in the growth cabinets are controlled by Infrared gas analyzers (IRGA's: Vaisala). ^{cm}trL plants grown in soil require an atmosphere with >0.5 % (v/v) CO_2 for adequate growth rates.
4. The soil around the plants (not the foliage) are watered daily and sufficient liquid (Hoagland nutrient mix: [15]) and slow release fertilizer (Oscmocote) are routinely applied to ensure plant growth is not nutrient limited.

2.5 [^{14}C]-CABP Synthesis

1. AG50W-X8 (H+ form, Bio-Rad) resin is prepared by adding 5 bed volumes of 0.1 N HCl to the resin for 1 min, decanting off the acid and washing the resin three times with Mill-Q water (5–10 bed volumes each) to remove excess HCl.
2. Potassium cyanide, [^{14}C] (K^{14}CN) ((*see* **Note 2**). Is supplied as a solid as 1 mCi (37 MBq) from PerkinElmer (NEC079H001MC).
3. RuBP is made from ribose 5-phosphate and purified according to [16].

2.6 Gel Filtration Chromatography

1. Separation of Rubisco bound—[^{14}C]-CABP from excess unbound [^{14}C]-CPBP is performed using Sephadex G50 (fine, GE Biosciences) chromatography.
2. Sephadex G50 resin; prepare by swelling the resin (1 g/10 mL bed volume) for at least 3 h in column buffer (20 mM EPPS-NaOH, pH 8.0, 75 mM NaCl).
3. Size exclusion columns: Degas the resin suspension under vacuum for 5 min then pour into 0.7 × 30 cm glass Econo-Columns (Bio-Rad) till the bed volume is 8.5 mL (~22 cm in height).

4. Column equilibration: Pass through >5 column volumes of column buffer.
5. Column storage buffer: Milli-Q water containing 0.05 % (w/v) chlorhexidine (Sigma).
6. Ultima Gold (PerkinElmer) is a versatile liquid scintillation cocktail that can accommodate 0.75 mL of aqueous buffer per mL of scintillant.

2.7 Coomassie and Western Blot Densitometry for Detecting Rubisco Expression

1. Densitometry: assumes use of an appropriate imaging system (e.g., VersaDoc or Pharos Plus imaging systems from Bio-Rad) to quantify banding and signal intensities using software such as Quantity One (Bio-Rad).
2. Coomassie blue stain. GelCode Blue Stain Reagent (Pierce), 20 mL per gel.
3. Protein (Western) blot apparatus: assumes use of the XCell II™ Blot Module (Life Technologies).
4. Transfer buffer (10×): 0.25 M Bicine, 0.25 M Bis–Tris, 8 mM EDTA. Do not adjust pH. Dilute 20 mL with 160 mL water and 20 mL methanol for use.
5. Tris-buffered saline (TBS, 20× stock): 2.7 M NaCl, 0.5 M Tris-Cl, pH 7.4.
6. Blocking buffer: 5 % (w/v) low fat (skim) dry milk powder in TBS (prepare fresh).
7. Primary antibody: commercially prepared against purified tobacco Rubisco in Rabbits.
8. Secondary antibody: Alkaline phosphatase conjugated anti-rabbit IgG (Bio-Rad).
9. AttoPhos® AP Fluorescent Substrate System from Promega. Dissolve each vial of AttoPhos® Substrate (36 mg) in 6 mL of AttoPhos® buffer to make a 10 mM stock (10×) and store as 0.1 mL aliquots at −80 °C. Dilute with 0.9 mL AttoPhos® buffer for use in immuno-detection.

3 Methods

3.1 Cloning Rubisco Genes into pLEV Plastome Transforming Plasmids

1. A variety of *L*arge-subunit *E*ncoding *V*ectors (pLEV) exist for cloning modified Rubisco genes (*see* **Note 3**). As shown in Fig. 1, $^{\text{foreign/modified}}$*rbc*L transgenes of interest are PCR amplified with primers (or commercially synthesized) that incorporate an in frame *Nhe*I site at codons Ala-9 and Ser-10 in the *rbc*L and either a *Sal*I (**TAA**AAACA*GTCGAC*) or *Xba*I (**TAA**AAGC*TCTAGA*) site downstream of the *rbc*L stop codon (shown in bold).

2. The *rbc*L products PCR amplified from plant gDNA are cloned into pGEM-T-Easy (Promega) by T-A cloning. The subsequent pGEM-T-Easy-*rbc*L plasmids provide useful templates for *rbc*L mutagenesis.
3. The cloned transgenes are fully sequenced to ensure correct then cloned into pLEV4 (as *Nhe*I-*Sal*I products) or pLEV^{Ntlox}aadA (as *Nhe*I-*Xba*I fragments) (*see* Fig. 1).
4. The *E. coli* stain XL1-Blue (Novagen) is used to propagate plasmids and their integrity checked by restriction digestion to ensure no transposon insertion in the *rbc*L 5′untranslated (UTR) sequence (*see* **Note 4**).
5. Plasmid DNA (pDNA) for transforming is typically prepared using commercial pDNA purification kits (e.g., Qiagen or Promega) from 100 mL cultures of pLEV-XL1-Blue cells grown at 22–37 °C (*see* **Note 4**) in LB-Amp media.
6. The pDNA is eluted in 0.2–0.5 mL TE buffer and diluted with nuclease-free water to 1 μg /μL and stored at −20 °C.

3.2 Culture of ^{cm}trL and Transformed Tobacco

1. Sterilize ~0.1 mL bed volume of ^{cm}trL seed in a sterile 1.5 mL microfuge tube by treatment with 0.8 mL of bleach solution for 10 min.
2. Remove the bleach by pipette and wash the seed four times with 0.8 mL changes of sterile Milli-Q water to remove residual bleach.
3. Using flame sterilized forceps place 10–20 seeds onto germination plates.
4. Germinate the seed at 25 °C in a growth cabinet with cool white fluorescent illumination (~10–30 μmol photon m^2/s, 16:8 h light–dark cycle) and an atmosphere containing 2.5 % (v/v) CO_2.
5. After germination (8–12 days) the plantlets (cotyledons plus an emerging leaf) are transferred using aseptic forceps to polypropylene growth containers containing 150 mL of agar solidified RMOP media.
6. When 8–12 cm in height (7–9 leaf stage, *see* panel 3 of Fig. 2 for example), the leaves are ready to transform biolistically (*see* Chapter 8 that describes an appropriate plastome transformation protocol in tobacco) with a pLEV transforming plasmid (*see* Subheading 3.1).
7. Transformed plants are selected for by repeated propagation on selective RMOP^{spec} media.

3.3 Identification of Transformed Plants by Non-denaturing PAGE

1. Leaf material (>0.2 cm^2) from spec^R tissue growing on selective media is lysed in 0.5 mL of ice-cold extraction buffer in a 2 mL glass homogenizer (Wheaton) and the lysate centrifuged (35,000 × *g*, 5 min, 4 °C).

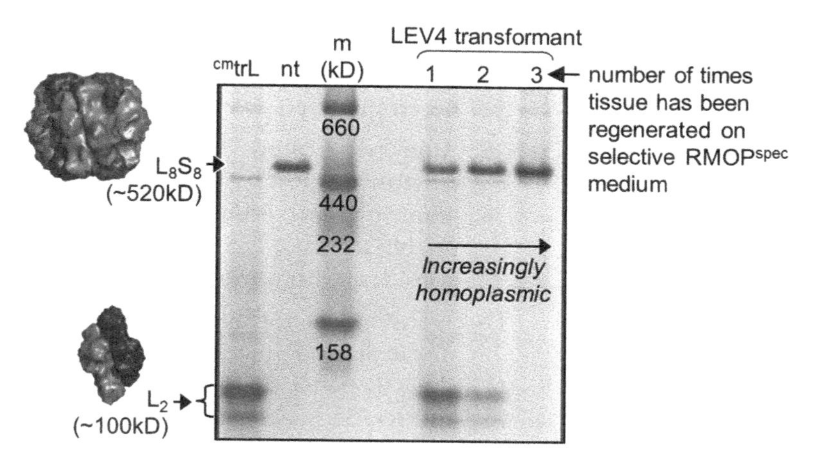

Fig. 3 Non-denaturing PAGE screening for transformants and relative homoplasmicity. Transformants are identified from the production of Rubisco hexadecamers (L_8S_8; the S constituting the endogenous tobacco small subunits synthesized in the cytosol and imported into the chloroplast) that migrates slower through the non-denaturing PAGE than the smaller *R. rubrum* L_2 Rubisco (which is the only Rubisco produced in the ^{cm}trL line). With successive rounds of regeneration on selective media (RMOP^{spec}) the transformed tissue produces less L_2 Rubisco as the population of ^{cm}*rbc*M containing plastome copies diminish. When homoplasmic, no L_2 Rubisco is present (even in Rubisco deficient transformants where the introduced *rbc*L gene does not enable L_8S_8 synthesis [6]). *nt* non-transformed *N. tabacum* leaf protein, *m* molecular mass markers with sizes shown

2. Aliquots of the soluble protein (160 μL) are diluted with 40 μL of 5× non-denaturing sample buffer.
3. Load 20 μL of the protein sample onto non-denaturing PAGE (along with non-denaturing PAGE marker proteins) and slowly separate at 60 V for 16 h (or at 100 V for 4 h) at 4 °C using non-denaturing electrophoresis buffer (Fig. 2, panel 4 and Fig. 3).
4. The gels are transferred to 10 cm^2 square culture dishes (Sarstedt), rinsed in 50 mL of Milli-Q water and then fixed in 50 mL of Gel fixation buffer for 15–30 min with slow oscillation (~60 opm).
5. The acetic acid and methanol are then removed with six successive washes (5 min each) of 100 mL Milli-Q water (at ~100 opm).
6. The protein bands are visualized by Coomassie blue staining.
7. After 1 h remove excess stain with repeated changes of Milli-Q water.
8. Transformed plants are identified from the production of L_8S_8 Rubisco (heteroplasmic tissue will still contain *R. rubrum* L_2 Rubisco; Fig. 3). Plants that produce only the L_2 enzyme are

either non-transformed or indicate assembly of the transplanted Rubisco into L_8S_8 complexes is impaired and the plant still heteroplasmic.

9. After two rounds of regeneration on selective media non-denaturing PAGE analysis typically show the transformed plants produce little, or no, L_2 Rubisco (Fig. 3) even in plants unable to make L_8S_8 complexes (*see* **Note 5**). These plants are considered homoplasmic (or close to).
10. Transfer leaf forming sections of plant material to 150 mL sterile RMOP medium in polypropylene growth containers (Fig. 2, panel 3) to promote plantlet formation (i.e., shoots and roots).
11. Once the plants have formed roots the RMOP media is gently washed away with tap water and the plants transferred to pots of soil and grown at elevated CO_2.
12. A transparent plastic container is placed over the plant for 3–5 days to maintain high humidity and enable the plant to become established in soil. Growth is dependent on sufficient amounts of suitably functional recombinant L_8S_8 Rubisco being produced to support plant growth.

3.4 Preparation of Leaf Samples for Rubisco Analysis

1. Accurate and representative measurements of leaf Rubisco content are best made using T_1 (or subsequent) progeny using leaves of comparable physiological age (*see* **Note 6**).
2. Leaf disc punches of known area (typically 0.5 cm^2) are taken using sharpened cork boring tools (Cole-Palmer) and immediately frozen in liquid N_2 and stored at −80 °C for up to 6 months without noticeable activity loss.
3. Leaf protein is extracted rapidly in 0.4–1 mL of protein extraction buffer in an ice-cold 2 mL glass homogenizer (Wheaton) and the lysate centrifuged (35,000 × *g*, 1 min, 4 °C) to remove insoluble matter.
4. The soluble protein (supernatant) is transferred to a new tube and then either reacted with ^{14}C-CPBP (Subheading 3.5) and the Rubisco content quantified by ^{14}C-CABP binding (Subheading 3.6) or its relative amount determined by Coomassie stain or immunoblot densitometry measurements of Rubisco following separation by PAGE (Subheading 3.7).

3.5 ^{14}C-CABP Synthesis

1. In a sealable 7 mL glass vial in the fume hood dissolve 18.2 μmol ^{14}C-KCN (i.e., 1 mCi with a specific activity of ~55 mCi/mmol = 122,100 dpm/nmol) in 0.2 mL of 0.1 M Tris-acetate buffer (pH 8.5).
2. Add to the vial a 20 % molar deficiency (~15 μmol) of RuBP (i.e., 750 μL of a 20 mM RuBP stock made up in 3 mM HCl buffer [16]).

3. Seal the vial and leave at room temperature in the fume hood for at least 48 h to ensure complete hydrolysis of the resulting nitriles.
4. Pass the reaction through a column containing 1 mL of AG50W-X8 (H+ form, Bio-Rad) resin (the ^{14}C-KCN binds to the resin) and collect the filtrate into a clean 20 mL glass vial.
5. Wash the resin with two 1 mL Milli-Q water washes and collect into the same vial.
6. The reaction is completely dried in a fume hood with a gentle stream of N_2 gas to remove NH_3. The dried ^{14}C-CPBP residue comprises an isomeric mix of ^{14}C-carboxyarabinitol-P_2 (^{14}C-CABP) and ^{14}C-carboxyribotol-P_2 (^{14}C-CRBP) (*see* **Note 7**).
7. Suspend the ^{14}C-CPBP mixture in ~7 mL of 50 mM Bicine-NaOH (pH 9.3) to a working concentration of ~2 mM and store at −80 °C in 0.2 mL aliquots.
8. Before use the ^{14}C-CPBP mixture is left at 4 °C overnight to ensure complete saponification of the lactone forms of ^{14}C-CABP and ^{14}C-CRBP.
9. The ^{14}C-CPBP mixture can be stably stored at 4 °C for up to 5 years.

3.6 Quantifying Leaf Rubisco Content by ^{14}C-CABP Binding

3.6.1 Forming the ^{14}C-CABP-Rubisco Complexes

1. As ^{14}C-CABP binds tightly and specifically to each active site in Rubisco [14] it provides a versatile method for quantifying the enzymes concentration in total cellular protein extract.
2. Duplicate 50 μL samples of soluble leaf protein (from **step 2** of Subheading 3.5) are incubated for 10–30 min at room temperature (22 °C) to fully activate Rubisco before incubating with either 0.5 or 1 μL (i.e., 20 or 40 μM) of [^{14}C]-CPBP (*see* **Note 7**).
3. The samples are incubated a further 15–60 min at 22 °C to ensure stoichiometric binding of the higher affinity and tighter binding [^{14}C]-CABP isomer to all eight active sites in L_8S_8 Rubisco.
4. Samples can be left on ice for up to 20 h on ice with minimal (<5 %) loss of [^{14}C]-CABP-Rubisco complexes recovered by Sephadex G50 (fine) chromatography (Subheading 3.6.2).

3.6.2 Quantify Rubisco Content by Gel Filtration Separation of [^{14}C]-CABP Bound Enzyme

1. Each protein-[^{14}C]-CPBP sample is carefully layered onto the top of a Sephadex G50 column matrix in each column (*see* Subheading 2.6 for column preparation) and the column tap opened to start sample separation by gravity flow.
2. The sampled is carefully washed into the resin with 200 μL of column buffer before applying three successive additions of column buffer (750 μL). Do not collect the eluent.

3. Collect the following six successive applications of column buffer (750 μL each) into separate 6 mL polyethylene Pony scintillation vials (PerkinElmer). The first of these samples is a blank, with >95 % of the large molecular weight Rubisco-[^{14}C]-CABP complexes eluting in the next two fractions, <5 % in the fourth, little or no [^{14}C] in the fifth and the beginning of the unbound [^{14}C]-CPBP peak eluting in the sixth fraction.
4. Add 1 mL of Ultima Gold scintillant to each sample, mixed by vortexing and quantify the amount [^{14}C] in each sample using a scintillation counter.
5. The amount of Rubisco-[^{14}C]-CABP recovered in the duplicate samples should be identical despite incubating with different amounts of [^{14}C]-CPBP (*see* **Note 7**).
6. From the specific activity of the [^{14}C]-CPBP (122,100 dpm/nmol) and adjusting for the counting efficiency of the counter (usually >80 %, ours is 87 % making the [^{14}C]-CPBP specific activity 106,200 cpm/nmol) the leaf Rubisco content (in units of μmol active sites per m^2) is quantified using the following equation;

$$\frac{\text{cpm recovered in Rubisco} -^{14}\text{C} - \text{CABP fractions}}{^{14}\text{C} - \text{CPBP specific activity (cpm / nmol)}} \times \frac{\text{ml of buffer leaf extracted in}}{\text{ml loaded onto sephadex G50}} \times \frac{10{,}000\ \text{cm}^2}{\text{cm}^2\ \text{of leaf sampled}} \div 1{,}000.$$

7. In young healthy leaves of a tobacco plant growing in a glasshouse at 25 ± 3 °C, the Rubisco content is typically between 30 and 50 μmol active sites per m^2 (Fig. 4d).

3.7 Densitometry Analysis of Rubisco Expression

3.7.1 Relative Rubisco Content from Protein Staining with Coomassie Blue

1. For non-denaturing PAGE analysis of Rubisco L_8S_8 expression, 160 μL of the protein sample from **step 4** in Subheading 3.4 is diluted with 40 μL ice-cold 5× non-denaturing PAGE Sample buffer.
2. Load 10 μL per lane (this equates to 0.4 mm^2 of leaf if 0.5 cm^2 of leaf is extracted in 1 mL of protein extraction buffer; Fig. 4a) on a non-denaturing PAGE. Separate and visualize the proteins by Coomassie staining as described in Subheading 3.3 (**steps 3–6**).
3. For SDS PAGE analysis of Rubisco L-subunit (52 kDa) and S-subunit (14.6 kDa) expression, 160 μL of the protein sample (*see* Subheading 3.5, **step 2**) is diluted with 40 μL 5× SDS PAGE Sample buffer.
4. Load 10 μL per lane (0.4 mm^2 of leaf; Fig. 4b) and separate the protein by SDS PAGE at 200 V for 40 min in SDS PAGE electrophoresis buffer.

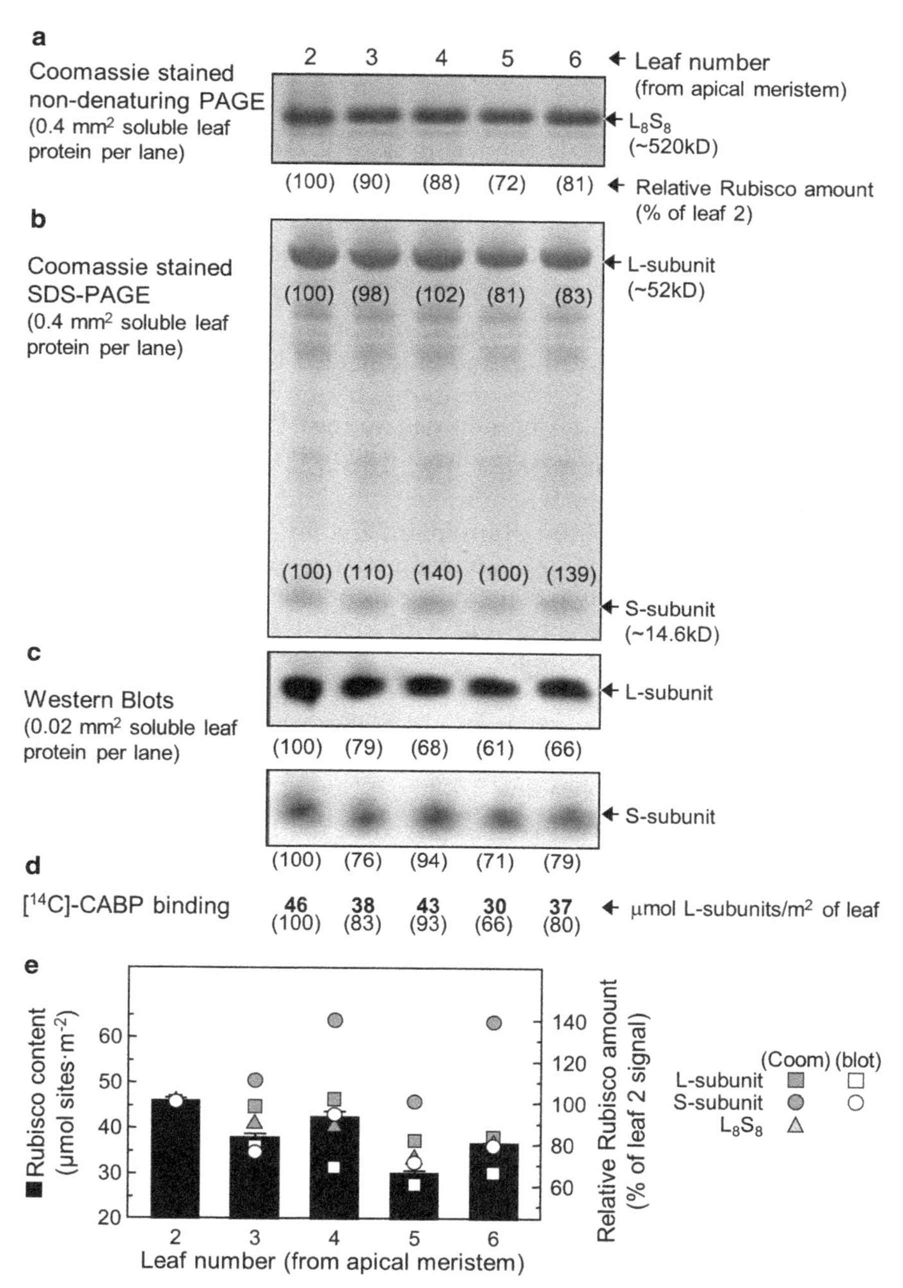

Fig. 4 Comparative methods for measuring Rubisco expression in soluble leaf protein. Rubisco content in the soluble protein from leaves in the upper canopy of a glasshouse-grown LEV4 transformed plant during exponential growth (12 leaves, 45 cm in height) were analyzed by (**a**) non-denaturing PAGE, (**b**) SDS PAGE, (**c**) Western blot analysis using an antibody raised against the tobacco Rubisco large (L) and small (S) subunits and (**d**) quantified by [^{14}C]-CABP binding. The relative levels of Coomassie and immuno-detection banding determined by densitometry relative to the signals measured for leaf 2 (set at 100 %) are shown in parenthesis. (**e**) Summary that compares Rubisco expression levels measured by the different methods

5. Visualized the proteins bands by Coomassie staining (Subheading 3.3 following **steps 4–6**).
6. Obtain a digital scanned image (at >300 dpi) of each gel using a VersaDoc imaging system (Bio-Rad) or equivalent machine.
7. Quantify the relative densitometry of the Coomassie stained L_8S_8 complex (Fig. 4a) or the L- and S-subunit bands (Fig. 4b) using Quantity One or analogous (e.g., ImageQuant), software.
8. As shown in Fig. 4e, measuring the relative Rubisco expression from Coomassie blue stained bands is most accurately determined from densitometry measurements of L_8S_8 complexes resolved by non-denaturing PAGE (shaded triangles) or the L-subunits resolved by SDS-PAGE (shaded squares).
9. Accurate densitometry measurements of relative Rubisco content need to ensure the amount of sample is within the dynamic range for quantifying relative expression (Fig. 5). Quantifying Rubisco content by densitometry is disadvantaged by the need for comparisons with known amounts of the target Rubisco (*see* **Note 8**).

3.7.2 Western Blot Analysis of Rubisco Expression

1. After separation, the SDS polyacrylamide gel is removed from the plastic casing and rinsed briefly with western transfer buffer (Subheading 2.7).
2. The gel is then placed in the transfer apparatus and proteins transferred to nitrocellulose membrane at a constant 25 V for 60 min.
3. The membrane is then incubated for 45 min in excess Blocking buffer (~50 mL) and then gently rinsed in TBS buffer.
4. The primary Rubisco specific antibody is diluted to 1:5,000 with TBS and incubated with the membrane for 1 h followed by three successive 10-min washes with TBS buffer.
5. The secondary antibody conjugated to alkaline phosphatase is diluted 1:5,000 with TBS buffer and incubated for 30 min with the membrane and washed as above.
6. The Rubisco peptides are detected by layering the membrane onto ~1 mL of 1 mM AttoPhos® AP Fluorescent Substrate. In the example blots shown in Fig. 4c, the L-subunit and S-subunit signals were obtained after incubating the respective sections of the membrane with the AttoPhos® AP substrate for 10 and 60 s respectively.
7. To terminate the reactions excess Attophos substrate is removed rapidly by blotting dry the membranes between two sheets of Whatman Paper.
8. After 1 min drying the membranes are scanned using a Pharos Plus imaging system (Bio-Rad).

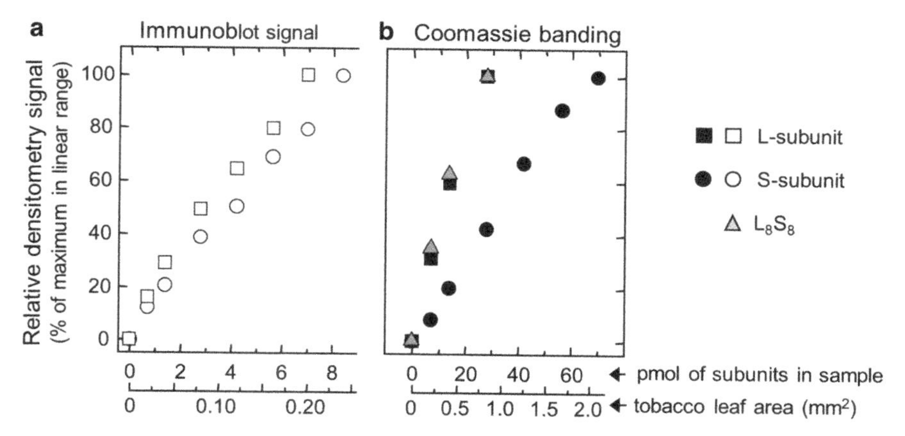

Fig. 5 Comparative efficiencies of the methods for measuring Rubisco expression. Different amounts of soluble *N. tabacum* leaf protein was separated by non-denaturing PAGE and SDS-PAGE and densitometry measurements made on their Coomassie stained L_8S_8, L-subunits and S-subunits as well as their Western Blot signals (*see* Fig. 4 for examples). The amount of Rubisco (shown as picomoles of subunits) in each sample was quantified by [^{14}C]-CABP binding. The linear range for extrapolating relative Rubisco levels varies for each technique highlighting the necessity for suitable calibration curves when using such qualitative measurements

9. The densitometry of the immune-reactive peptides are examined using Quantity-One software. As shown in Fig. 4e, measuring relative Rubisco content by immune-blotting is most accurately determined from densitometry measurements of the SDS PAGE separated S-subunits (open circles) and not the L-subunit (open squares).

10. It is important to ensure the immune-blot signals are within a linear range (i.e., not saturating). As shown in Fig. 5, the greater sensitivity of immune-blot detection reduces the effective linear range for quantifying Rubisco content by more than threefold relative to Coomassie banding densitometry measurements.

4 Notes

1. The tobacco master-line ^{cm}trL is a plastome transformed *Nicotiana tabacum* L. cv Petit Havana [*N,N*]) line where the tobacco *rbc*L gene has been replaced by a codon modified (optimized for plastid expression) version of the *rbc*M gene (^{cm}*rbc*M) from the bacterium *Rhodospirillum rubrum* and the *aad*A selectable marker gene has been excised by CRE-*lox* recombination following transient CRE expression [6] (Fig. 1). To ensure the ^{cm}trL plants are devoid of nuclear mutations, leaves from T_3 and subsequent progeny that have been successively crossed with non-transformed tobacco pollen are used for transformation. Unwanted recombination events during

transformation in ^{cm}trL do not occur due to the significant sequence divergence between ^{cm}*rbc*M and all other *rbc*L genes.

2. Potassium cyanide, [^{14}C] (K^{14}CN) is toxic. Only deal with in fume hood.
3. The introduced Rubisco genes can either be foreign or altered tobacco *rbc*L genes, or comprise a Rubisco operon coding both an *rbc*L and *rbc*S [6].
4. Transposon insertion in the tobacco *rbc*L 5′UTR sequence within the pLEV plasmids can occur as it is constitutively active in *E. coli* and accumulation of Rubisco can reduce cell viability [10]. Slowing cell growth by culturing at lower temperatures (e.g., 22 °C) can reduce transposition frequency. Restriction digestion with *Nco*I (located towards the 5′ end of the *rbc*L 5′UTR) and an enzyme that cuts within *rbc*L can be used to screen for transposon insertion in purified transforming plasmids.
5. The isolation of homoplasmic transformants can still be readily obtained using the ^{cm}trL tobacco master-line even if the introduced Rubisco is assembly deficient or catalytically impaired or non-functional [6]. As expected of Rubisco deficient plants, these lines show a bleached leaf phenotype and can only be grown in tissue culture media containing a carbon source such as sucrose.
6. Care needs to be taken to select leaves of comparable developmental stages and growth conditions when comparing Rubisco contents between different genotypes (such as non-transformed controls) as leaf Rubisco content varies significantly with leaf age and canopy position (for example, in Fig. 4 leaf 5 was more shaded than leaf 6 in its growth position within the glasshouse). Ideally samples should be from young leaves at similar locations in the canopy of healthy exponentially growing plants (20–65 cm in height when lateral branch formation is minimal). For non-transformed tobacco this phase generally constitutes a 5–7 day period during exponential growth.
7. The leaf protein samples are incubated with two different concentrations of ^{14}C-CPBP to confirm stoichiometric binding of ^{14}C-CABP to each Rubisco active site. As ^{14}C-CABP binds essentially irreversibly (and with 100-fold greater avidity than ^{14}C-CRBP) identical amounts of ^{14}C-CABP-Rubisco complexes should be recovered in both samples following gel filtration.
8. In contrast to the quantitative ^{14}C-CABP-binding method applying densitometry and immune-blot methodologies to quantify leaf Rubisco content necessitates the availability of suitable antibodies and specific Rubisco standards of known concentration. Without the availability of ^{14}C-CABP-binding quantifying the concentration of a Rubisco standard necessitates

its purification and quantification using derived extinction coefficients. If the introduced subunits are insoluble (thus negating the applicability of the ^{14}C-CABP binding technique) quantification is only feasible by immunoblot and Coomassie banding densitometry comparisons against known amounts of the target Rubisco [17].

Acknowledgments

This work was supported by the Australian Research Council grants DP0984790 and FT0991407.

References

1. Andrews TJ, Whitney SM (2003) Manipulating ribulose bisphosphate carboxylase/oxygenase in the chloroplasts of higher plants. Arch Biochem Biophys 414:159–169
2. Andrews TJ, Hudson GS, Mate CJ, von Caemmerer S, Evans JR, Arvidsson YBC (1995) Rubisco: the consequences of altering its expression and activation in transgenic plants. J Exp Bot 46:1293–1300
3. Hudson GS, Evans JR, von Caemmerer S, Arvidsson YBC, Andrews TJ (1992) Reduction of ribulose-1,5-bisphosphate carboxylase/oxygenase content by antisense RNA reduces photosynthesis in transgenic tobacco plants. Plant Physiol 98:294–302
4. Whitney SM, Houtz RL, Alonso H (2011) Advancing our understanding and capacity to engineer nature's CO_2 sequestering enzyme. Rubisco Plant Physiol 155:27–53
5. Sharwood RE, von Caemmerer S, Maliga P, Whitney SM (2008) The catalytic properties of hybrid Rubisco comprising tobacco small and sunflower large subunits mirror the kinetically equivalent source Rubiscos and can support tobacco growth. Plant Physiol 146:83–96
6. Whitney SM, Sharwood RE (2008) Construction of a tobacco master line to improve Rubisco engineering in chloroplasts. J Exp Bot 59:1909–1921
7. Whitney SM, von Caemmerer S, Hudson GS, Andrews TJ (1999) Directed mutation of the Rubisco large subunit of tobacco influences photorespiration and growth. Plant Physiol 121:579–588
8. Whitney SM, Sharwood RE, Orr D, White SJ, Alonso H, Galmés J (2011) Isoleucine 309 acts as a C_4 catalytic switch that increases ribulose-1, 5-bisphosphate carboxylase/oxygenase (rubisco) carboxylation rate in *Flaveria*. Proc Natl Acad Sci U S A 108:14688–14693
9. Alonso H, Blayney MJ, Beck JL, Whitney SM (2009) Substrate-induced assembly of Methanococcoides burtonii D-ribulose-1,5-bisphosphate carboxylase/oxygenase dimers into decamers. J Biol Chem 284:33876–33882
10. Whitney SM, Andrews TJ (2001) Plastome-encoded bacterial ribulose-1,5-bisphosphate carboxylase/oxygenase (RubisCO) supports photosynthesis and growth in tobacco. Proc Natl Acad Sci U S A 98:14738–14743
11. Zhang X, Webb J, Huang Y, Tang R, Liu A (2010) Hybrid Rubisco of tomato large subunits and tobacco small subunits is functional in tobacco plants. Plant Sci 180: 480–488
12. Whitney SM, Kane HJ, Houtz RL, Sharwood RE (2009) Rubisco oligomers composed of linked small and large subunits assemble in tobacco plastids and have higher affinities for CO2 and O-2. Plant Physiol 149:1887–1895
13. Kubien DS, Brown CM, Kane HJ (2010) Quantifying the amount and activity of Rubisco in leaves. Methods Mol Biol 684:349–362
14. Blayney M, Whitney S, Beck J (2011) NanoESI mass spectrometry of rubisco and rubisco activase structures and their interactions with nucleotides and sugar phosphates. J Am Soc Mass Spectr 22:1588–1601
15. Hoagland DR, Snyder WC (1933) Nutrition of strawberry plants under controlled conditions: (a) effects of deficiencies of boron and

certain other elements; (b) susceptibility to injury from sodium salts. P Am Soc Hortic Sci 30:288–294

16. Kane HJ, Wilkin JM, Portis AR, Andrews TJ (1998) Potent inhibition of ribulose-bisphosphate carboxylase by an oxidized impurity in ribulose-1,5-bisphosphate. Plant Physiol 117:1059–1069
17. Whitney SM, Baldet P, Hudson GS, Andrews TJ (2001) Form I Rubiscos from non-green algae are expressed abundantly but not assembled in tobacco chloroplasts. Plant J 26:535–547
18. Svab Z, Maliga P (1993) High-frequency plastid transformation in tobacco by selection for a chimeric *aadA* gene. Proc Natl Acad Sci U S A 90:913–917

Part III

Crop Specific Plastid Transformation Protocols

Chapter 16

Plastid Transformation in Tomato

Stephanie Ruf and Ralph Bock

Abstract

Tomato (*Solanum lycopersicum*) is one of the most important vegetable crops and has long been an important model species in plant biology. Plastid biology in tomato is especially interesting due to the chloroplast-to-chromoplast conversion occurring during fruit ripening. Moreover, as tomato represents a major food crop with an edible fruit that can be eaten raw, the development of a plastid transformation protocol for tomato was of particular interest to plant biotechnology. Recent methodological improvements have made tomato plastid transformation more efficient and facilitated applications in metabolic engineering and molecular farming. This article describes the basic methods involved in the generation and analysis of tomato plants with transgenic chloroplast genomes and summarizes current applications of tomato plastid transformation.

Key words Biolistic transformation, Biotechnology, Chloroplast, *Lycopersicon esculentum*, Metabolic engineering, Molecular farming, Particle gun, Plastid, Plastid transformation, *Solanum lycopersicum*, Tomato

1 Introduction

Tomato, *Solanum lycopersicum* (formerly *Lycopersicon esculentum*), is one of the world's most important vegetable crops. Moreover, tomato has been one of the classical model species of plant genetics [1, 2] and the sequencing of its nuclear genome has recently been completed (http://solgenomics.net/; [46]). The sequence of the tomato chloroplast genome (plastome) has also been determined [3]. Its size is 155,460 bp with an average AT content of 62.14 %. The tomato plastid genome contains 114 genes and conserved open reading frames (*ycf*s) and displays a high overall similarity to the plastid DNAs of other members of the nightshade family (Solanaceae), such as tobacco (*Nicotiana tabacum*) and deadly nightshade (*Atropa belladonna*). The expression of the plastid genome in different tissues and developmental stages and its regulation at the transcriptional and posttranscriptional levels have been extensively investigated [3–7]. Fruit development has received particular attention [5, 7], because the tomato fruit has long been

Pal Maliga (ed.), *Chloroplast Biotechnology: Methods and Protocols*, Methods in Molecular Biology, vol. 1132,
DOI 10.1007/978-1-62703-995-6_16, © Springer Science+Business Media New York 2014

a paradigm for chloroplast-to-chromoplast conversion [8–11]. The chromoplast of ripe red tomato fruits accumulates large amounts of storage carotenoids, especially the red pigment lycopene.

Recently, it has become possible to engineer the plastid genome of tomato plants by stable transformation [12]. Initially, the procedures were quite laborious and time consuming, but a combination of several small improvements [13, 14] have made them considerably more efficient. Nonetheless, tomato plastid transformation remains significantly more demanding than transformation in the well-established tobacco system, and it is, therefore, highly recommended to switch to tomato only after full construct optimization has been done in tobacco.

The relatively demanding procedures involved in the generation of plastid-transformed (transplastomic) tomato plants are the main reason why the technology has so far mainly been used for applications in biotechnology. For most questions in fundamental research, it is much more convenient to use tobacco plastid transformation, due to the significantly higher transformation frequency and the availability of more efficient tissue culture systems (facilitating much faster selection and plant regeneration) in tobacco.

Biotechnological applications of tomato plastid transformation that have been pursued so far include engineering of the carotenoid and tocopherol metabolic pathways and expression of antigens for subunit vaccines [13–16]. To enhance the nutritional quality of tomatoes, lycopene β-cyclase genes have been expressed from the tomato plastid genome. The best-performing construct, a lycopene β-cyclase gene from daffodil (*Narcissus pseudonarcissus*) driven by the strong ribosomal operon-derived P*rrn* promoter and the 5′-untranslated region (5′-UTR) from *gene 10* of bacteriophage T7, resulted in highly efficient conversion of lycopene into provitamin A (β-carotene). Unexpectedly, it also led to a 50 % increase in total carotenoid content of the tomatoes, thus providing an additional health benefit [15]. To explore the potential of transplastomic tomato plants as production platform for new vaccines, the HIV antigens P24 and Nef were expressed from the tomato plastid genome as a translational fusion [14]. Very high levels of foreign protein accumulation were obtained in green leaves (approximately 40 % of the total soluble protein). Expression levels in fruits were much lower, illustrating that plastid transgene expression in nongreen and nonleafy tissue does not normally reach the high protein accumulation levels needed for most applications in molecular farming [7, 17]. However, recent success with the construction of chimeric expression elements (promoter and 5′-UTR combinations) that confer high-level foreign protein accumulation in nongreen plastid types, including tomato chromoplasts [18, 19], suggests that this problem can be overcome.

Here we describe the procedures involved in the selection of transplastomic tomato plants and their characterization and provide an updated working protocol [20] for plastid transformation by particle gun-mediated transformation.

2 Materials

All chemicals used in synthetic media for tomato tissue culture must be of highest purity and, ideally, should be certified for use in plant cell and tissue culture.

2.1 Plant Material and Growth Conditions

Young leaves from tomato plants (*Solanum esculentum*) grown under aseptic conditions should be used for chloroplast transformation experiments. Sterile plants are readily obtained from surface-sterilized seeds. To this end, 800 μL ethanol, 1 drop of Tween 20, and 800 μL 6 % bleach (sodium hypochlorite solution) are added to a sample of 20 tomato seeds in a 2 mL Eppendorf tube. Following vigorous shaking for 2–4 min, the liquid is removed with a sterile pipet and the seeds are washed at least six times with 1.5 mL sterile distilled water (each time completely removing the liquid). Seeds are soaked in sterile water overnight, then transferred to sterile containers (e.g., Magenta boxes), and germinated on MS medium (see below). When the first true leaves have opened, the stem is cut directly above the cotyledons and the upper part of the plantlet is transferred to a sterile container (Magenta box) with MS medium. The plants are then grown in a chamber at 22–25 °C under a 16 h light/8 h dark cycle until they reach a height of approximately 10–15 cm. The light intensity should be between 50 and 70 $\mu E/m^2$ s. For bombardment, only the youngest 3–4 leaves are used.

Similar to nuclear transformation, the efficiency of plastid transformation in tomato is also dependent on the genotype. Two red-fruited South American varieties (bred in Brazil) have so far been successfully used for biolistic plastid transformation: Santa Clara and IPA-6. IPA-6 displays somewhat better regeneration properties and, therefore, is currently the preferred genotype in the authors' laboratory. However, recent testing of other well-regenerating tomato varieties indicates that they are transformable at similar efficiency (our unpublished data). The tissue culture and selection protocols described below are optimized for the cultivar IPA-6, but recently have been demonstrated to work equally well for two commercial green-fruited tomato varieties, Dorothy's Green and Green Pineapple [16]. If other genotypes are to be used, small modifications may be required to optimize regeneration and selection efficiency.

2.2 Chemicals and Stock Solutions for Plant Culture Media

1. 10× Macro (1 L): 19.0 g KNO_3, 16.5 g NH_4NO_3, 4.4 g $CaCl_2 \cdot 2\ H_2O$, 3.7 g $MgSO_4 \cdot 7\ H_2O$, 1.7 g KH_2PO_4.
2. 100× Micro (100 mL): 169 mg $MnSO_4 \cdot H_2O$, 86 mg $ZnSO_4 \cdot 7\ H_2O$, 62 mg H_3BO_3, 8.30 mg KI, 2.50 mg $Na_2MoO_4 \cdot 2\ H_2O$, 0.25 mg $CuSO_4 \cdot 5\ H_2O$, 0.25 mg $CoCl_2 \cdot 6\ H_2O$.
3. FeNaEDTA: 1 g/100 mL.
4. Thiamine HCl: 1 mg/mL.
5. Glycine: 1 mg/mL.
6. Nicotinic acid: 1 mg/mL.
7. Pyridoxine HCl: 1 mg/mL.
8. Indole 3-acetic acid (IAA): 1 mg/mL (in 0.1 M NaOH).
9. N6-benzylamino purine (BAP): 1 mg/mL (in 0.1 M HCl).

10× Macro and 100× Micro stock solutions are sterilized by autoclaving, and all other stock solutions by sterile filtration. As an alternative to the preparation of these nutrient and vitamin stock solutions, various salt mixtures and ready-to-use media are commercially available (e.g., from Duchefa, Haarlem, the Netherlands).

2.3 Plant Culture Media

1. MS medium [21] for growth of plants in sterile culture (1 L): 10× Macro (100 mL), 100× Micro (10 mL), FeNaEDTA (3.67 mL), 20 g sucrose, pH 5.6–5.8 (adjust with 0.2 M KOH).

   The liquid medium is supplemented with 5.6 g/L agar (certified for plant cell and tissue culture, e.g., Micro Agar, Duchefa, Haarlem, the Netherlands; *see* **Note 1**), autoclaved, and poured in Petri dishes or sterile containers for plant growth.
2. RMOL5 medium for plant regeneration from tissue explants.

   Regeneration of shoots from tissue explants is induced by addition of the plant hormones IAA (the natural auxin) and BAP (a cytokinin analog) to the basic MS medium.

   RMOL5 medium (1 L): 10× Macro (100 mL), 100× Micro (10 mL), FeNaEDTA (3.67 mL), 20 g sucrose, thiamine HCl (1 mL), glycine (2 mL), nicotinic acid (0.5 mL), pyridoxine HCl (0.5 mL), IAA (0.2 mL), BAP (3 mL), 100 mg myo-inositol, pH 5.8 (0.2 M KOH). After addition of 5.4 g/L agar (for plant cell and tissue culture; Duchefa; *see* **Note 1**), the medium is sterilized by autoclaving and chilled to approximately 60 °C. Subsequently, antibiotics are added if required and the medium is poured into Petri dishes (approximately 50 mL medium per Petri dish of 9 cm diameter and 2 cm height).

2.4 Materials for Particle Preparation

1. Gold particles (0.6 µm, Bio-Rad).
2. Ethanol 100 %.
3. Spermidine (free base).
4. $CaCl_2$.

3 Methods

Tobacco represents by far the most efficient and least time-consuming experimental system for plastid transformation. Therefore, it is highly recommended to at least initially use tobacco to set up the technology and acquire the basic expertise. Particle gun-mediated transformation, also referred to as "biolistic" transformation (ref. 22), represents the currently most effective method of delivering foreign DNA into higher plant plastids [23–25]. Polyethylene glycol (PEG) treatment of protoplasts in the presence of transforming DNA provides an alternative method [26, 27] and has also been used in tomato [28], but the cell culture and transformation procedures involved in PEG-mediated plastid transformation are considerably more demanding and time-consuming than biolistic transformation.

3.1 Vectors for Tomato Plastid Transformation

The principles of vector design for tomato plastid transformation are similar to tobacco and other higher plants [12–14]. So far, only spectinomycin selection has been successfully used to obtain transplastomic tomato plants. Spectinomycin resistance is conferred by chimeric *aadA* genes similar to the ones established previously for tobacco plastid transformation [12, 24]. Alternatively, mutant alleles of the 16S rRNA gene have been used as selectable marker [28]. However, extrapolating from tobacco plastid transformation [23, 24], they are probably considerably less efficient selectable markers than the chimeric *aadA* genes. For most applications (e.g., simple transgene expression studies), the construction of species-specific vectors for tomato plastid transformation is not necessary due to the high sequence conservation of the plastid genome among Solanaceous species. Standard plastid transformation vectors developed for tobacco have been successfully used for tomato chloroplast transformation [12, 14]. However, for reverse genetics analyses in tomato plastids and/or targeting of plastome regions of relatively low sequence conservation or regions containing non-conserved RNA editing sites [3, 6], it is advisable to use homologous flanking regions from the tomato plastid genome.

3.2 Gold Particle Preparation for Biolistic Plastid Transformation

All steps should be performed on ice or in the cold room (at 4 °C). Water and 100 % ethanol must be ice-cold.

1. Use 1.4 mg of gold particles (0.6 μm diameter) per seven shots.
2. Prepare a gold suspension: $n \times 1.4$ mg in $n \times 140$ μL of 100 % EtOH.
3. Vortex at least 1 min at maximum power.
4. Centrifuge 1 s at $11{,}000 \times g$ in a microfuge.
5. Remove the supernatant completely and redissolve the particles in 1 mL of sterile distilled water.

6. Centrifuge 1 s at 11,000×*g* in a microfuge, remove and discard the supernatant.
7. Resuspend the particles very carefully in *n*×175 μL sterile water.
8. Prepare 172 μL aliquots (per seven shots) in Eppendorf tubes.
9. Add to each aliquot in the following order: 20 μg of DNA (plastid transformation vector; concentration 1–2 μg/μL), 175 μL of 2.5 M $CaCl_2$, 35 μL of 100 mM spermidine. (Vortex briefly immediately after addition of each component.)
10. Incubate on ice for 10 min with brief vortexing every minute.
11. Centrifuge 1 s at 8,100×*g* in a microfuge and remove the supernatant completely.
12. Add 600 μL of 100 % EtOH and carefully resuspend the particles by pipetting and vigorous vortexing.
13. Centrifuge 1 s at 8,100×*g* in a microfuge and remove the supernatant completely.
14. Repeat this washing step.
15. Resuspend the particles very carefully in 50 μL 100 % EtOH by pipetting and vigorous vortexing (*see* **Note 2**).
16. Use 6.5 μL per shot and carefully resuspend particles immediately before use.

3.3 Biolistic Bombardment

Use young leaves from tomato plants (*Solanum esculentum*) grown under aseptic conditions on synthetic medium (MS medium). Place the leaves in a Petri dish onto a sterile piece of filter paper on top of a thin layer of RMOL5 medium without antibiotics. Bombard the abaxial side of the leaves (*see* **Note 3**) with the plasmid DNA-coated gold particles. For the Bio-Rad PDS-1000/He biolistic gun, suitable settings for the transformation of tomato chloroplasts (using the Hepta Adaptor setup of the Bio-Rad PDS-1000/He particle gun) are as follows:

- Helium pressure at the tank regulator: 1,300–1,400 psi.
- Rupture disks: 1,100 psi (Bio-Rad).
- Macrocarrier (flying disk) assembly: level three from the top.
- Petri dish holder: level one from the bottom.
- Vacuum (at the time of the shot): 27–28 in. Hg.

A typical chloroplast transformation experiment involves the bombardment of six Petri dishes covered with tomato leaves using the Hepta Adaptor setup of the Bio-Rad PDS-1000/He particle gun. With the Mono Adaptor setup, only a circular area of ~5 cm diameter is bombarded and the number of Petri dishes should be increased five to tenfold.

To adjust the settings of the particle gun, optimize the particle preparation procedure and control for its quality, transient nuclear

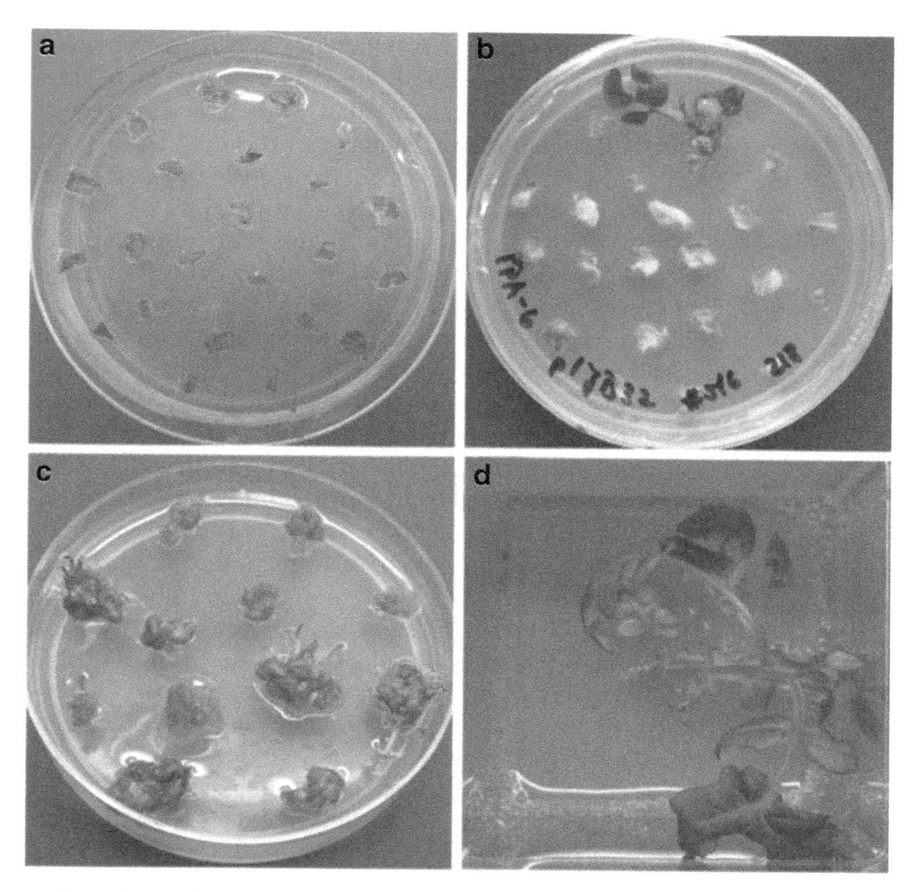

Fig. 1 Generation of transplastomic tomato (*Solanum lycopersicum*) plants. (**a**) Tomato leaf material bombarded with a helium-driven particle gun and exposed to selection on spectinomycin-containing regeneration medium. (**b**) Regeneration of a transplastomic tomato line (green callus with emerging shoot) after 3 months of selection. (**c**) Additional regeneration round under spectinomycin selection to obtain homoplasmic tissue. (**d**) Transfer of a homoplasmic transplastomic shoot for rooting on phytohormone-free medium

transformation assays using a GUS (β-glucuronidase) reporter gene construct should be used. Any plant transformation vector containing a GUS reporter gene under the control of a strong constitutive promoter can be used for these assays (*see* **Note 4**). Assessment of transient gene expression by histochemical GUS staining [20] provides a suitable proxy also for stable plastid transformation efficiency.

3.4 Selection of Transplastomic Tomato Plants

Following biolistic bombardment, the leaves are cut into small pieces (approximately 2 × 3 mm) and placed adaxial side up onto the surface of a selective regeneration medium (RMOL5 containing 500 μg/mL spectinomycin) (Fig. 1a). Resistant calli or shoots typically appear after 6–12 weeks of incubation (light intensity: approximately 7 μE/m^2 s; 16 h light/8 h dark cycle) (Fig. 1b). Note that not all of the antibiotic-resistant plant lines are true plastid transformants. Spontaneous spectinomycin resistance can be conferred by specific point mutations in the chloroplast 16S rRNA gene, which arise spontaneously at low frequency [29, 30]. True chloroplast transformants can be identified by DNA gel blot analysis or PCR assays, which test for the physical presence of the resistance gene in the plastid genome [31, 32].

Plant cells are highly polyploid with respect to their plastid genome, and a single tomato leaf cell contains thousands of identical copies of the plastid DNA distributed among at least dozens of chloroplasts. For this reason, primary plastid transformants are usually heteroplasmic, meaning that they contain a mix of wild-type and transformed genome copies. In order to obtain a genetically stable transplastomic plant, residual wild-type genome molecules must be completely eliminated. This is achieved by subjecting the primary plastid transformant to additional cycles of regeneration on selective medium [30] (Fig. 1c). Typically after one to two rounds, homoplasmic transplastomic lines are obtained, in which the wild-type genome is no longer detectable. Homoplasmy is assessed by either DNA gel blot analysis [12] or sensitive PCR assays that selectively favor amplification of residual wild-type genomes [32]. Verification of homoplasmy is particularly important when the transplastomic lines can be expected to display a mutant phenotype [31].

It is important to note that homoplasmic transplastomic lines sometimes still show faint wild-type-like hybridization signals in Southern blots (for two recent examples, *see* refs. 33, 34). These bands are not necessarily indicative of the presence of residual wild-type copies of the plastid genome. Instead, they often represent the so-called promiscuous DNA: nonfunctional copies of the plastid DNA residing in the nuclear or mitochondrial genomes [35–37]. To distinguish between residual heteroplasmy and promiscuous DNA, either Southern blots with purified plastid DNA [38, 39] or inheritance assays (seed tests; see below) should be conducted.

Finally, homoplasmic transplastomic tomato shoots regenerated on RMOL5 medium are transferred to boxes with MS medium (Fig. 1d). The phytohormone-free MS medium will induce rooting, and the plants can then be taken out of the sterile environment, transferred to soil (*see* **Note 5**), and grown to maturity in the greenhouse (*see* **Note 6**).

3.5 Analysis of Transplastomic Tomato Plants

As chloroplasts are maternally inherited in tomato (like in the vast majority of angiosperm plant species; refs. 40, 41), plastid transgene localization and homoplasmy can be easily verified genetically by conducting reciprocal crosses. To this end, wild-type and transformed plants are transferred to soil and grown to maturity in the greenhouse. Tomato fruits and seeds are collected from selfed plants and from reciprocal crosses of the transplastomic lines with the wild type. Surface-sterilized seeds are then germinated on spectinomycin-containing (100 mg/L) MS medium and analyzed for uniparental inheritance of the plastid-encoded antibiotic resistance. Selfed transformants and crosses with the plastid transformant as maternal parent give rise to uniformly green (i.e., antibiotic-resistant) progeny, whereas seeds collected from wild-type plants and crosses with the plastid transformant as paternal parent yield white (i.e., antibiotic-sensitive) seedlings that typically show strong anthocyanin accumulation in their hypocotyl [12, 30]. If white seedlings turn up in the inheritance assay with seeds from the selfed transplastomic plant, the plant was still heteroplasmic and should be purified further until homoplasmy is reached. If homoplasmy is not attainable, an essential plastid gene may have been inactivated by the transformation construct [42–44].

Standard methods are used for the molecular analysis of transplastomic tomato plants. Southern blot analyses should be performed for confirmation of plastid genome transformation and tests for homoplasmy. Northern blot experiments are suitable to assess RNA accumulation levels and transcript patterns and to analyze RNA stability. Western blots can be used to determine foreign protein accumulation or assess the consequences of mutations introduced into endogenous plastid genes. As none of these methods requires any tomato-specific modifications, they are not described here. Instead, the reader is referred to the literature and standard protocol books.

4 Notes

1. The exact amount of agar is dependent on the autoclave used (and, in particular, on its heating and cooling speeds) and, therefore, must be determined empirically. The solidified medium should be soft enough to allow pushing of leaf pieces 1–2 mm into the medium (for optimum medium contact to induce regeneration), but hard enough to stably hold an inserted stem cutting in a vertical position.
2. DNA-coated gold particles tend to aggregate. Careful resuspension of the particles is crucial to the success of the transformation experiment. Clumpy particles hit fewer target cells and will cause severe cell damage upon penetrating the leaf, thus greatly reducing both transformation efficiency and plant regeneration frequency.
3. The abaxial (lower) side of the leaf usually has a softer epidermis and cuticle than the adaxial (upper) side and thus is more efficiently penetrated by the particles.
4. Vectors for stable *Agrobacterium*-mediated plant transformation (e.g., pBI121) are often relatively large plasmids and, therefore, may give somewhat lower transient transformation

frequencies than smaller vectors for biolistic transformation or protoplast transformation. A reasonably small GUS gene-containing vector suitable for transient transformation assays is, for example, vector pFF19G [45].

5. Residual culture medium should be removed from the roots of the plantlets as completely as possible (by washing). As the medium is rich in minerals and also contains high amounts of sucrose, it otherwise will be rapidly colonized by soil microbes, including plant root-infecting pathogens.
6. Plants raised in sterile culture containers are adapted to 100 % relative humidity and, therefore, are highly sensitive to dry air. To allow for gradual adaptation to greenhouse conditions, plants should be kept under a transparent plastic cover for at least a few days, which is then stepwise lifted.

Acknowledgments

Work on plastid transformation in the authors' laboratory is supported by the Max Planck Society and by grants from the Deutsche Forschungsgemeinschaft (DFG), the Bundesministerium für Bildung und Forschung (BMBF), and the European Union (Framework Programs 6 and 7).

References

1. Alba R, Fei Z, Payton P, Liu Y, Moore SL, Debbie P, Cohn J, D'Ascenzo M, Gordon JS, Rose JKC, Martin G, Tanksley SD, Bouzayen M, Jahn MM, Giovannoni J (2004) ESTs, cDNA microarrays, and gene expression profiling: tools for dissecting plant physiology and development. Plant J 39:697–714
2. Fernandez AI, Viron N, Alhagdow M, Karimi M, Jones M, Amsellem Z, Sicard A, Czerednik A, Angenent G, Grierson D, May S, Seymour G, Eshed Y, Lemaire-Chamley M, Rothan C, Hilson P (2009) Flexible tools for gene expression and silencing in tomato. Plant Physiol 151:1729–1740
3. Kahlau S, Aspinall S, Gray JC, Bock R (2006) Sequence of the tomato chloroplast DNA and evolutionary comparison of solanaceous plastid genomes. J Mol Evol 63:194–207
4. Piechulla B (1988) Plastid and nuclear mRNA fluctuations in tomato leaves - diurnal and circadian rhythms during extended dark and light periods. Plant Mol Biol 11:1988
5. Piechulla B, Chonoles Imlay KR, Gruissem W (1985) Plastid gene expression during fruit ripening in tomato. Plant Mol Biol 5:373–384
6. Karcher D, Kahlau S, Bock R (2008) Faithful editing of a tomato-specific mRNA editing site in transgenic tobacco chloroplasts. RNA 14: 217–224
7. Kahlau S, Bock R (2008) Plastid transcriptomics and translatomics of tomato fruit development and chloroplast-to-chromoplast differentiation: chromoplast gene expression largely serves the production of a single protein. Plant Cell 20:856–874
8. Fray RG, Grierson D (1993) Molecular genetics of tomato fruit ripening. Trends Genet 9:438–443
9. Bramley PM (2002) Regulation of carotenoid formation during tomato fruit ripening and development. J Exp Bot 53:2107–2113
10. Moore S, Vrebalov J, Payton P, Giovannoni J (2002) Use of genomics tools to isolate key ripening genes and analyse fruit maturation in tomato. J Exp Bot 53:2023–2030
11. Giovannoni JJ (2007) Fruit ripening mutants yield insights into ripening control. Curr Opin Plant Biol 10:283–289
12. Ruf S, Hermann M, Berger IJ, Carrer H, Bock R (2001) Stable genetic transformation

of tomato plastids and expression of a foreign protein in fruit. Nat Biotech 19:870–875
13. Wurbs D, Ruf S, Bock R (2007) Contained metabolic engineering in tomatoes by expression of carotenoid biosynthesis genes from the plastid genome. Plant J 49:276–288
14. Zhou F, Badillo-Corona JA, Karcher D, Gonzalez-Rabade N, Piepenburg K, Borchers A-MI, Maloney AP, Kavanagh TA, Gray JC, Bock R (2008) High-level expression of HIV antigens from the tobacco and tomato plastid genomes. Plant Biotechnol J 6:897–913
15. Apel W, Bock R (2009) Enhancement of carotenoid biosynthesis in transplastomic tomatoes by induced lycopene-to-provitamin A conversion. Plant Physiol 151:59–66
16. Lu Y, Rijzaani H, Karcher D, Ruf S, Bock R (2013) Efficient metabolic pathway engineering in transgenic tobacco and tomato plastids with synthetic multigene operons. Proc Natl Acad Sci U S A 110:E623–E632
17. Ma JK-C, Barros E, Bock R, Christou P, Dale PJ, Dix PJ, Fischer R, Irwin J, Mahoney R, Pezzotti M, Schillberg S, Sparrow P, Stoger E, Twyman RM (2005) Molecular farming for new drugs and vaccines. EMBO Rep 6:593–599
18. Zhang J, Ruf S, Hasse C, Childs L, Scharff LB, Bock R (2012) Identification of cis-elements conferring high levels of gene expression in non-green plastids. Plant J 72:115–128
19. Caroca R, Howell KA, Hasse C, Ruf S, Bock R (2012) Design of chimeric expression elements that confer high-level gene activity in chromoplasts. Plant J 73:368–379
20. Bock R (2004) Studying RNA editing in transgenic chloroplasts of higher plants. Methods Mol Biol 265:345–356
21. Murashige T, Skoog F (1962) A revised medium for rapid growth and bio assays with tobacco tissue culture. Physiol Plant 15: 473–497
22. Altpeter F, Baisakh N, Beachy R, Bock R, Capell T, Christou P, Daniell H, Datta K, Datta S, Dix PJ, Fauquet C, Huang N, Kohli A, Mooibroek H, Nicholson L, Nguyen TT, Nugent G, Raemakers K, Romano A, Somers DA, Stoger E, Taylor N, Visser R (2005) Particle bombardment and the genetic enhancement of crops: myths and realities. Mol Breed 15:305–327
23. Svab Z, Hajdukiewicz P, Maliga P (1990) Stable transformation of plastids in higher plants. Proc Natl Acad Sci U S A 87: 8526–8530
24. Svab Z, Maliga P (1993) High-frequency plastid transformation in tobacco by selection for a chimeric aadA gene. Proc Natl Acad Sci U S A 90:913–917
25. Bock R (1998) Analysis of RNA editing in plastids. Methods 15:75–83
26. Golds T, Maliga P, Koop H-U (1993) Stable plastid transformation in PEG-treated protoplasts of Nicotiana tabacum. Nat Biotech 11:95–97
27. O'Neill C, Horvath GV, Horvath E, Dix PJ, Medgyesy P (1993) Chloroplast transformation in plants: polyethylene glycol (PEG) treatment of protoplasts is an alternative to biolistic delivery systems. Plant J 3:729–738
28. Nugent GD, ten Have M, van der Gulik A, Dix PJ, Uijtewaal BA, Mordhorst AP (2005) Plastid transformants of tomato selected using mutations affecting ribosome structure. Plant Cell Rep 24:341–349
29. Svab Z, Maliga P (1991) Mutation proximal to the tRNA binding region of the Nicotiana plastid 16S rRNA confers resistance to spectinomycin. Mol Gen Genet 228:316–319
30. Bock R (2001) Transgenic chloroplasts in basic research and plant biotechnology. J Mol Biol 312:425–438
31. Bock R, Kössel H, Maliga P (1994) Introduction of a heterologous editing site into the tobacco plastid genome: the lack of RNA editing leads to a mutant phenotype. EMBO J 13:4623–4628
32. Bock R, Hermann M, Kössel H (1996) In vivo dissection of cis-acting determinants for plastid RNA editing. EMBO J 15:5052–5059
33. Rogalski M, Karcher D, Bock R (2008) Superwobbling facilitates translation with reduced tRNA sets. Nat Struct Mol Biol 15: 192–198
34. Oey M, Lohse M, Scharff LB, Kreikemeyer B, Bock R (2009) Plastid production of protein antibiotics against pneumonia via a new strategy for high-level expression of antimicrobial proteins. Proc Natl Acad Sci U S A 106: 6579–6584
35. Ayliffe MA, Timmis JN (1992) Tobacco nuclear DNA contains long tracts of homology to chloroplast DNA. Theor Appl Genet 85:229–238
36. Stegemann S, Hartmann S, Ruf S, Bock R (2003) High-frequency gene transfer from the chloroplast genome to the nucleus. Proc Natl Acad Sci U S A 100:8828–8833
37. Bock R, Timmis JN (2008) Reconstructing evolution: gene transfer from plastids to the nucleus. Bioessays 30:556–566
38. Hager M, Biehler K, Illerhaus J, Ruf S, Bock R (1999) Targeted inactivation of the smallest plastid genome-encoded open reading frame reveals

a novel and essential subunit of the cytochrome b6f complex. EMBO J 18:5834–5842

39. Ruf S, Biehler K, Bock R (2000) A small chloroplast-encoded protein as a novel architectural component of the light-harvesting antenna. J Cell Biol 149:369–377
40. Ruf S, Karcher D, Bock R (2007) Determining the transgene containment level provided by chloroplast transformation. Proc Natl Acad Sci U S A 104:6998–7002
41. Bock R (2007) Structure, function, and inheritance of plastid genomes. Top Curr Genet 19:29–63
42. Drescher A, Ruf S, Calsa T Jr, Carrer H, Bock R (2000) The two largest chloroplast genome-encoded open reading frames of higher plants are essential genes. Plant J 22:97–104
43. Rogalski M, Ruf S, Bock R (2006) Tobacco plastid ribosomal protein S18 is essential for cell survival. Nucleic Acids Res 34: 4537–4545
44. Rogalski M, Schöttler MA, Thiele W, Schulze WX, Bock R (2008) Rpl33, a nonessential plastid-encoded ribosomal protein in tobacco, is required under cold stress conditions. Plant Cell 20:2221–2237
45. Timmermans MCP, Maliga P, Vieira J, Messing J (1990) The pFF plasmids: cassettes utilising CaMV sequences for expression of foreign genes in plants. J Biotech 14:333–344
46. The Tomato Genome Consortium (2012) The tomato genome sequence provides insights into fleshly fruit evolution. Nature 485:635–641

Chapter 17

Stable Plastid Transformation of Petunia

Elena Martin Avila and Anil Day

Abstract

Petunia hybrida is a commercial ornamental plant and is also an important model species for genetic analysis and transgenic research. Here we describe the steps required to isolate stable plastid transformants in *P. hybrida* using the commercial Pink Wave cultivar. Wave cultivars are popular spreading Petunias sold as ground cover and potted plants. Transgenes introduced into *P. hybrida* plastids exhibit stable expression over many generations. The development of plastid transformation *in P. hybrida* provides an enabling technology to bring the benefits of plastid engineering, including maternal inheritance and stable expression of performance-enhancing trait genes, to the important floriculture and horticulture industries.

Key words Chloroplast transformation, Floriculture, Horticulture, *Petunia hybrida*, Petunia cv. Pink Wave, Plastid transformation

1 Introduction

Petunia hybrida is an established model species used to study the biology of flowering plants by transgenic approaches. *P. hybrida* is also a popular potting and bedding plant and is one of the top five selling ornamental plants in the USA with a recorded wholesale value of $94 million in 2005 [1]. Ornamental flowering plants are required for the important floriculture and horticulture industries [2]. Petunias are annual solanaceous plants that originate in South America. *P. hybrida* is the descendant of crosses involving *Petunia axillaris* and *Petunia integrifolia* which were initiated in the nineteenth century [1]. Flower morphology and plant habit form the basis for distinguishing five groups of *P. hybrida* named: grandiflora, floribunda, milliflora, multiflora, and spreading [1]. There are well over 100 commercial cultivars of *P. hybrida* [1]. *P. hybrida* is diploid with 14 chromosomes and a nuclear DNA content (2C) of 3 pg [3, 4]. Diploidy in *P. hybrida* makes it tractable to genetic analysis of nuclear genes including those with roles in plastids. *Nicotiana tabacum* is the established model for transplastomic research but it is an amphidiploid or allotetraploid ($2n = 4x$), which

Pal Maliga (ed.), *Chloroplast Biotechnology: Methods and Protocols*, Methods in Molecular Biology, vol. 1132,
DOI 10.1007/978-1-62703-995-6_17, © Springer Science+Business Media New York 2014

Table 1
Petunia hybrida **plastid DNA sequences >300 bp in the DDBJ/EMBL/Genbank database and comparison with homologous** ***N. tabacum*** **sequences[a]**

Genes	Accession number	Size (bp)	% identity *P. hybrida*/*N. tabacum*
psbA	X04974	1,186	98.4
mat K	EF439018	935	98.2
rbcL	X04976	1,434	97.4
Intergenic *trnL–trnF*	X74578	325	95.8
arsB	X03418	1,335	95.7
trnL intron	X74572	479	95.3
trnE–trnT	AF408753	1,013	94.6
arsA	X03417	512	92.8
Intergenic *psbA–trnH*	X12856	492	84.0

[a]*arsA* and *arsB* are short sequences promoting autonomous replication of yeast plasmids and the sequence identities shown are due to other plastid genes within the compared sequences

is not suitable for genetic analysis. Plastid transformation of a genetically tractable species such as *P. hybrida* allows transplastomic tools for manipulating the plastid genome to be combined with genetic screens to identify and characterize nuclear genes with a role in plastids. For example, identification of nuclear loci involved in the expression of plastid genes or maternal inheritance of plastids [5, 6] would be facilitated by using transplastomic plants containing reporter genes in plastids. The approach has been used successfully in Chlamydomonas where genetic screens of transplastomic algae have led to the identification of nuclear genes involved in chloroplast gene expression (*see* Chapter 28).

The chloroplast genome of *P. hybrida* was one of the first chloroplast genomes to be characterized using restriction enzymes and was estimated to be about 153 kb in size [7]. Despite this early work, the *P. hybrida* chloroplast genome has not been sequenced in its entirety and is not available for designing plastid transformation vectors with homologous targeting arms. *P. hybrida* sequences deposited in the NCBI data bases are listed in Table 1. These relatively short stretches of *P. hybrida* plastid sequences are useful for phylogenetic studies [8] but are not suitable for constructing plastid targeting arms by PCR-based cloning. Figure 1 shows a high degree of conservation between *P. hybrida* and *N. tabacum* plastid DNA sequences, particularly in the regions coding for RNA and protein products. This high degree of sequence conservation has allowed *N. tabacum* plastid DNA sequences to target integration of foreign genes into the plastid genome of *P. hybrida* [11]. Heterologous *N. tabacum* sequences were also used to target

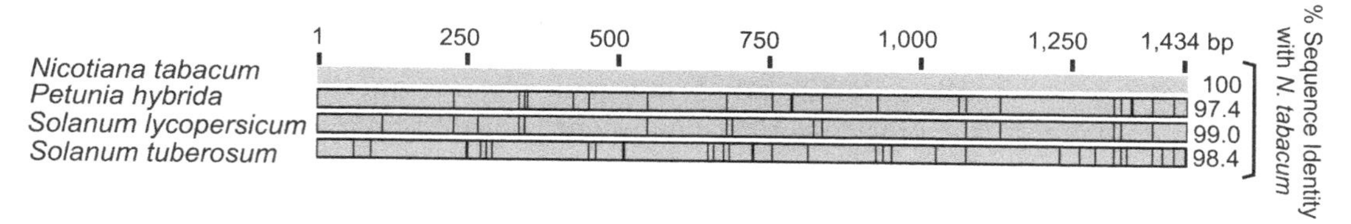

Fig. 1 Conservation of the *N. tabacum rbcL* coding region with *rbcL* genes in three Solanaceous species in which plastid transformation was established using *N. tabacum* plastid DNA targeting arms. *Vertical bars* indicate base substitutions. DDBJ/EMBL/GenBank Accession numbers: *Nicotiana tabacum* NC_001879 [14], *Petunia hybrida* X04976 [13], *Solanum lycopersicum* DQ347959 [18]; *Solanum tuberosum* DQ231562 [19]

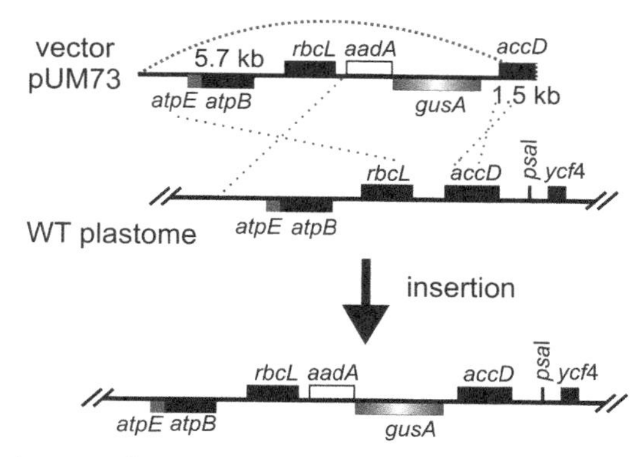

Fig. 2 Map showing integration of the *aadA* and *gusA* genes in the intergenic region between the plastid *rbcL-accD* genes. Recombination events between *N. tabacum* plastid DNA sequences in the left 5.7 kb and right 1.5 kb targeting arms and *P. hybrida* plastid DNA sequences are shown as *dotted lines*. Genes transcribed *left* to *right* (*aadA*) are placed above genes transcribed *right* to *left* (*gusA*)

transgenes into *Solanum lycopersicum* (tomato) and *Solanum tuberosum* (potato) plastids [9, 10]. Previous work had shown that sequence divergence of 2.4 % between the targeting plastid DNA sequences and resident plastid genome did not lead to an observable decrease in plastid transformation frequency [12]. Figure 1 shows conservation of the coding sequence (*rbcL* gene) for the large subunit of ribulose bisphosphate carboxylase/oxygenase. Of the three *rbcL* genes from solanaceous species, compared to the *N. tabacum rbcL* gene in Fig. 1, the *P. hybrida rbcL* gene [13] is the most distant.

The vector pUM73 used to transform *P. hybrida* plastids targets foreign genes between the *rbcL* and *accD* genes (Fig. 2). It contains 5.7 kb left and 1.5 kb right targeting arms corresponding to bases 53612–59326 and 59324–60868 of the *N. tabacum* plastid genome, respectively [14]. Recombination events in both targeting arms are required to insert the *aadA* marker gene, encoding aminoglycoside-3″-adenyltransferase, and the *gusA* reporter gene, encoding β-glucuronidase, into the *P. hybrida* plastid genome (Fig. 2).

Combined spectinomycin and streptomycin selection was used to isolate plastid transformants. The *gusA* gene is a useful reporter to follow the spread of transgenic plastids and can be used to evaluate the efficiency of containment provided by maternal inheritance of field-grown transplastomic *P. hybrida* plants [15].

The commercial Pink Wave cultivar of *P. hybrida* was used for plastid transformation experiments. The cultivar was chosen because of its relatively high regeneration and shoot formation rates from leaf explants [16]. Pink Wave can grow vertically to about 15–20 cm, is densely branched, and can spread to about 1 m. The cultivar normally starts flowering profusely 10–12 weeks after sowing. The flowers are about 7 cm in diameter. The development of plastid transformation in *P. hybrida* [11] has applications in plastid genetic engineering to improve the performance of commercial cultivars as well as providing a genetically tractable model for transplastomic research on flowering plants.

2 Materials

2.1 Plant Material

1. Seeds of *Petunia hybrida* cv. Pink Wave (*see* **Note 1**).

2.2 Culture Media Components (See Table 2)

1. Murashige and Skoog (MS) basal salt mixture [17] with vitamins and 2-(*N*-morpholino)ethanesulfonic acid (MES) buffer (Duchefa Biochemicals distributed by Melford Lab Ltd, Ipswich, UK).
2. Sucrose (Fisher Scientific, Loughborough, UK).
3. Plant growth regulators: 1 mg/mL stocks of 6-Benzylaminopurine (BA) (>99 % W/W, Fluka, Poole, UK) and indolyl-3-acetic acid (IAA) (Sigma-Aldrich, UK). Add a few drops (100 μL) of 1 M NaOH to 20 mg of BA or IAA solid powder to dissolve the powder and bring the volume up to 20 mL with Milli-Q water (*see* **Note 2**). Store the solution in disposable sterile plastic ware (*see* **Note 3**), such as 30 mL screw cap containers (Greiner Bio-One) in the dark at 4 °C. For long-term storage keep aliquots of the solutions frozen at −20 °C.
4. 1 M stock solution of KOH (Reagent Grade, Merck, Beeston, UK).
5. Gelling agents: Phytagel™ (Sigma-Aldrich, Poole, UK; *see* **Note 4**) and Agar (high strength >800 g/cm^2, Melford).
6. Antibiotics: 20 mg/mL stocks of spectinomycin dihydrochloride pentahydrate (>95 % purity W/W; Melford, Ipswich, UK) and streptomycin sulfate (100 % purity W/W; Melford, Ipswich, UK) are made up in Milli-Q water. Both aminoglycoside antibiotics are water soluble. Aqueous stocks are sterilized by passing through a 0.22 μm pore size syringe filter unit

(Millex GP, Millipore) and kept in disposable sterile plastic ware (*see* **Note 5**). Store in the dark at 4 °C and use within 10 days. For longer term storage freeze in 10 mL aliquots at 20 °C (*see* **Note 6**).

2.3 Preparation of Gold Microcarriers

1. Preparation of gold microcarriers. Weigh 60 mg of 1 μm gold particles (Bio-Rad, UK or Heraeus, Germany) into a 1.5 mL microfuge tube. Wash the gold with 1 mL of 100 % ethanol (V/V) by vortexing at maximum speed for 2 min. Then spin the particles at maximum setting (> 10,000 rpm, >8,000 × *g*) in a microfuge (Eppendorf 5415C) for 10 s to discard the supernatant. Repeat this wash step with ethanol twice. Resuspend the gold in 1 mL sterile Milli-Q water by scraping the gold pellet off the bottom of the tube carefully with a sterile disposable micropipette tip. Move the mixture up and down the micropipette tip to break up the gold clumps. Make sure that small clumps of gold do not stick to the micropipette tip. Vortex to ensure an even suspension of gold particles. An even suspension has the appearance of an opaque brown liquid without any visible grains. Store 50 μL aliquots in 1.5 mL eppendorf tubes making sure the suspension is homogeneous by vortexing the tube and withdrawing the aliquots from the center of the suspension. Store the suspension at –20 or –80 °C (*see* **Note** 7).
2. Calcium chloride. Prepare 2.5 M stock solution by dissolving $CaCl_2 \cdot 2H_2O$ (Merck) in sterile Milli-Q water (Millipore (UK) Watford). Sterilize the solution in an autoclave (*see* **Note 8**) and store 1 mL aliquots at –20 or –80 °C (*see* **Note** 7).
3. Spermidine, free base (MW 145.2, Sigma-Aldrich, *see* **Note 9**). Prepare 1 M stock solution by adding sterile water directly to the solid in the vial and making up to the correct volume in sterile disposable plastic ware. Immediately use the 1 M stock solution to make up 50 mL of the 0.1 M spermidine solution. Store 0.5 mL aliquots at –80 °C for a maximum of 3 months. Use a fresh aliquot each time.
4. Plasmid pUM73 (*see* Fig. 2). Purify the plastid transformation vector using a QIAGEN Plasmid Maxi Kit (Qiagen Ltd, Crawley, UK) and resuspend the DNA in sterile MILLI-Q water to a final concentration of 1 mg/mL. High plasmid yields from recombinant *Escherichia coli* strains are essential to ensure purity of the plastid transformation vector. Store at –20 °C.

2.4 Other Chemicals

1. 10–20 % (V/V) sodium hypochlorite (undiluted stock has >8 % active chlorine; General Purpose Grade, Fisher Scientific, Loughborough, UK).

2. 100 % (V/V) ethanol (Analytical Grade, Fisher Scientific).
3. 70 % (V/V) ethanol (96 % V/V ethanol, Analytical Grade, Fisher Scientific diluted with sterile water).

3 Methods (*See* Notes 10 and 11)

3.1 Preparation of Culture Media

1. MS media (*see* Table 2 and Subheading 2.2) is used for germination of seeds, in vitro propagation of the plants used to provide leaf explants for transformation, and for promoting the formation of roots from antibiotic resistant shoots (*see* **Note 12**). Add 4.9 g of MS salts with MES buffer and vitamins (Duchefa) together with 30 g of sucrose (3 % w/v) to 900 mL of RO water. Adjust pH with 1 M KOH to pH 6.0 and make the volume up to 1 L. Add 2.5 g of Phytagel™ (0.25 % W/V) as a gelling agent before sterilization in an autoclave. Unlike agar once solidified Phytagel™ cannot be reheated and must be poured into sterile plates and jars (*see* **Note 13**) as soon as it has cooled below 60 °C. Swirl the media to ensure even distribution of the Phytagel™ before pouring (*see* **Note 14**).
2. Solid MSB30 media [11, 16] is used for regenerating shoots from leaf explants. It is used to support leaf explants when bombarded with plasmid-coated micro-gold-projectiles and in the selection plates that also contain antibiotics. MSB30 is made by adding the plant growth regulators BA and IAA to MS media containing MES buffer, vitamins and sucrose (*see* Table 2 and Subheading 2.2). Add 1 mL of BA (1 mg/mL stock) and 0.1 mL of IAA (1 mg/mL stock) to 900 mL of MS media containing supplements. Adjust to pH 6.0 with 1 M KOH and make the volume up to 1 L with RO water. Add 6–8 g of agar (0.6–0.8 % W/V) as a gelling agent before sterilization in an autoclave. Media for microprojectile bombardment contains 0.8 % (W/V) agar (Subheading 3.4). To encourage root growth from stem sections (Subheading 3.3) we prefer 0.6–0.7 % (W/V) agar. Make sure the agar is evenly distributed by swirling the media before pouring into the sterile plates or vessels (*see* **Note 14**).
3. To prepare antibiotic-selection media, cool MSB30 media to below 60 °C with occasional swirling to avoid the agar setting at the edges of the container. Add 10 mL of 20 mg/mL filter-sterilized stocks of spectinomycin and streptomycin per liter of MSB30 medium. Mix by swirling the liquid slowly and avoid air bubbles. Pour into sterile petri dishes or Magenta™ GA 7 vessels.

3.2 Germination of Seedlings In Vitro (see Note 11)

1. Place *Petunia* seeds in 1–2 mL 100 % (V/V) ethanol for 1 min. Remove the ethanol with a sterile pipette. Dilute sodium hypochlorite (>8 % active chlorine) by three to fivefold with RO

Table 2
Composition of plant growth media[a]

No.	Chemical compounds	mg/L	Supplier
MS macronutrients			Duchefa 4.9 g/L
N1	NH_4NO_3	1,650	
N2	KNO_3	1,900	
N3	$CaCl_2{\cdot}2H_2O$	332	
N4	$MgSO_4{\cdot}7H_2O$	181	
N5	KH_2PO_4	170	
MS micronutrients			
N6	FeNaEDTA	36.7	
N7	$MnSO_4{\cdot}4H_2O$	16.9	
N8	$ZnSO_4{\cdot}7H_2O$	8.6	
N9	H_3BO_3	6.2	
N10	KI	0.83	
N11	$Na_2MoO_4{\cdot}2H_2O$	0.25	
N12	$CuSO_4{\cdot}5H_2O$	0.025	
N13	$CoCl_2{\cdot}6H_2O$	0.025	
Vitamins and amino acids			
N14	Nicotinic acid	0.5	
N15	Thiamine HCl	0.1	
N16	Pyridoxine HCl	0.5	
N17	*Myo*-inositol	100	
N18	Glycine	2	
Buffer			
N19	2-(*N*-morpholino) *ethanesulfonic acid (MES)*	500	
N20	*Sucrose*	30,000	Fisher Scientific
Growth regulators			
N21	6-benzylaminopurine (BA)	1	Fluka
N22	Indolyl-3-acetic acid (IAA)	0.1	Sigma-Aldrich
Antibiotics			
N23	Spectinomycin	200	Melford Labs
N24	Streptomycin	200	Melford Labs
Gelling agents			
N25	Agar	6,000–8,000	Melford Labs
N26	Phytagel™	2,500	Sigma-Aldrich

[a]WT plants are propagated on Murashige and Skoog (MS) salts [17] with vitamins and MES buffer, sucrose (chemicals N1–N20) solidified with Phytagel™ (N26). Transplastomic plants are grown on MS supplemented medium with antibiotic (N1–N20, N23) solidified with 0.6–0.7 % (W/V) agar (N25). MSB30 regeneration medium contains chemicals N1–N22, 0.8 % (W/V) agar (N25) and for selection two antibiotics (N23–N24) are used. 4.9 g of a mixed powder containing components N1–N19 are used per liter of medium (Duchefa, *see* Subheading 2.2). The pH is adjusted to 6.0 before sterilization

water and add 1–2 mL to the seeds (*see* **Note 15**). Place the seeds on a tube rotator (blood tube rotator, Stuart Scientific, Chelmsford, UK) for 5–10 min. Remove the sodium hypochlorite with a sterile pipette and rinse seeds with 1–2 mL of sterile RO water for 2 min (*see* **Note 16**). Remove the RO water and repeat the rinse 3–4 times with sterile RO water to remove all traces of sodium hypochlorite.

2. Mix the seeds with sterile 0.2 % (W/V) agar in RO water to facilitate their distribution on a petri dish containing solid MS Phytagel™ media containing sucrose (*see* Table 2). Seal the sides of the plates with two strips of Parafilm. Cover the plates with aluminum foil and place them at 24–25 °C (*see* **Note 17**).
3. Remove the aluminum foil as soon as germination is visible in the majority of seeds (normally 3–5 days). After 2–3 weeks transfer seedlings (Fig. 3a, *see* **Note 18**) to Magenta™ GA 7 vessels (Sigma-Aldrich, Poole, UK) containing MS media solidified with 0.25 % (W/V) Phytagel™ (*see* **Note 19**).

3.3 Propagation of Plants In Vitro

Only healthy plants should be propagated in vitro (Fig. 3b). Plants are maintained by transferring shoot tips, or stem segments containing an axillary bud, every 10 weeks to Magenta™ GA 7 vessels with fresh medium (*see* **Notes 20** and **21**). Old plants or weakly growing plants are discarded.

1. Remove the leaves and cut the stem into 2–4 cm sections making sure that each stem segment contains at least one axillary bud.
2. Place the shoot tip and stem segments vertically into MS Phytagel™ media with vitamins, MES buffer and sucrose contained in a Magenta™ GA 7 vessel.

3.4 Preparation of Leaf Explants for Particle Bombardment

1. Excise fully expanded green leaves from 9–11-week-old plants grown in vitro (*see* Fig. 3b). The new leaves at the top and old leaves at the bottom of the plant are not used. Normally 4–6 leaves of the type shown in Fig. 3c are obtained per plant. Young expanded leaves regenerate at higher efficiency (*see* Fig. 3d) than older leaves at the base of the plant (*see* Fig. 3e).
2. Remove the short petiole region and slice the leaves (*see* **Note 21**) to occupy a 4 cm diameter circle in the center of a petri dish containing MSB30 medium (*see* Fig. 3f). Place the leaf pieces abaxial side up on the MSB30 medium. Pieces from a single leaf are placed onto several plates such that each plate contains leaves from two to four different plants. Normally ten plates with leaf explants are prepared for each plastid transformation vector.

3.5 Coating Gold Microcarriers with Vector DNA

1. In a laminar flow cabinet add 5 μL of the pUM73 (*see* Fig. 2) plasmid solution (1 mg/mL) to a 1.5 mL microfuge tube containing an even suspension of the 50 μL aliquot of gold particles. The plasmid solution is slowly ejected directly into the gold suspension with the micropipette tip moving continuously below the surface of the liquid, which is at the same time agitated by finger-tapping the base of the tube. The gold particles should not settle during the procedure.

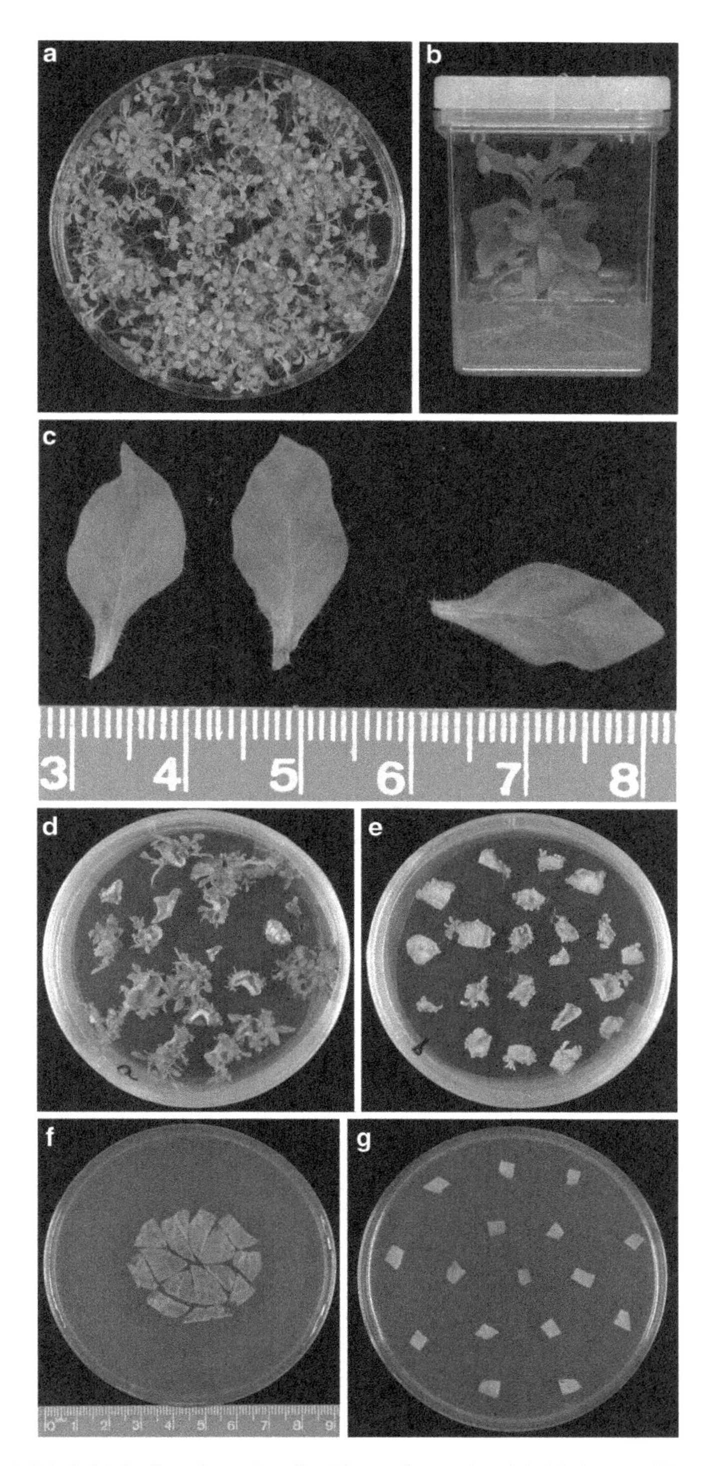

Fig. 3 Preparation of *P. hybrida* leaf explants for plastid transformation. (**a**) 14-day seedlings germinated in vitro on MS medium (*see* Table 2) with sucrose. (**b**) 10-week old plant arising from a seedling transferred to fresh MS medium with sucrose in a Magenta™ vessel. (**c**) Examples of expanded healthy leaves used for bombardment. (**d**) Regeneration from leaves from top of plant on MSB30 media. (**e**) Regeneration from older leaves near bottom of plant. (**f**). Leaf explants positioned in the middle of a 9 cm petri dish containing MSB30 media prior to particle bombardment. (**g**) Leaf explants placed on MSB30 selection plates containing antibiotics

2. Then add 50 μL of 2.5 M $CaCl_2$ (*see* **Note 22**) and mix by placing the tube on a vortex mixer for 5 s.
3. Next add 20 μL of 0.1 M spermidine solution (*see* **Note 22**), flick the tube to mix its contents and place on a vortex mixer immediately.
4. Vortex the mixture for 1 min.
5. Centrifuge the particles at 14,000 rpm, 16,000 × *g* (max setting) for 5 min in a microfuge. Discard the supernatant with a sterile micropipette tip.
6. Resuspend the pellet in 250 μL of 100 % (V/V) ethanol by breaking it up with a sterile micropipette tip and vortex briefly (<1 s) if necessary.
7. Centrifuge the suspension again for 2 min at 14,000 rpm, discard the supernatant and add 70 μL of 100 % (V/V) ethanol. Keep the tube on ice without disturbing the pellet. Just prior to use, the DNA-gold pellet is broken up with a micropipette tip. Tap the sides of the tube and move the suspension up and down a micropipette tip to obtain an even suspension of the DNA-gold particles (*see* **Note 23**). Vortex briefly (<1 s) if necessary.

3.6 Operation of Particle Delivery System

1. Particle bombardment is carried out in a laminar airflow cabinet using a Bio-Rad PDS-1000/He Particle Delivery System. The microcarrier launch assembly is placed into the top groove (shelf position 1) giving rise to a gap of 1 cm between the position of the rupture disk and macrocarrier holder. The macrocarrier holder and stopping screen are separated by two spacer rings providing a macrocarrier flight distance of 1.5 cm.
2. The PDS-1000/He Particle Delivery System is operated according to the manufacturer's instructions using 1,100 psi rupture disks and a vacuum of 28 in. Hg.
3. The microcarrier launch assembly, rupture disk holder, and macrocarrier holders are soaked in 70 % (V/V) ethanol for 10 min. before allowing them to dry in the sterile air within a laminar flow cabinet. The chamber of the PDS-1000/He Particle Delivery System is wiped with 70 % (V/V) ethanol and allowed to dry with the door open. The rupture disks (*see* **Note 24**) and macrocarriers are left in 70 % (V/V) ethanol for a minimum of 10 min. and then allowed to air-dry in a laminar flow cabinet by leaning them against the inside edges of sterile petri dishes.
4. Make sure that the DNA-gold suspension is well mixed before withdrawing 5 μL of the DNA-coated particles from the center of the suspension with a micropipette as the tube is agitated by finger taping the base. The 5 μL is immediately pipetted into

the center of a macrocarrier placed flat in the holder using the red plastic circular seating tool. DNA-gold suspension is placed on four macrocarriers at a time and once dry they are kept in a petri dish with lid closed until they are used.

5. The rupture disk holder with 1,100 psi rupture disk is placed in the PDS1000/He particle device. The metal holder containing the macrocarrier with dry DNA-gold precipitate is placed in the microcarrier launch assembly.
6. The leaf explants on media (Fig. 3f) are placed on the target shelf which is positioned in the third groove from the top (shelf position 3). This results in a 6 cm distance between the stopping screen and the target leaf explants. Remove the lid from the petri dish.
7. The vacuum is applied and the device fired according to the manufacturer's instructions. The release of vacuum is at an intermediate setting.
8. After the leaf explants have been bombarded they are placed in dim light (<30 μE/m²/s) at 25 °C for 2 days.

3.7 Antibiotic-Based Selection of Resistant Shoots

1. Forty to fifty hours after bombardment the leaf explants are cut into smaller pieces of approximately 2–5 mm wide and long, and placed abaxial side up on solid MSB30 media in a 9 cm petri dishes containing 200 mg/L spectinomycin dihydrochloride pentahydrate and 200 mg/L streptomycin sulfate (Fig. 3g, *see* **Notes 21**, **25** and **26**).
2. Incubate the plates in stacks in a plant growth cabinet (*see* **Note 17**) at 25 °C using a 12 h day (~80 μE/m²/s).
3. After 4 weeks cut the leaf explants in half and move to fresh MSB30 medium plates containing antibiotics (*see* **Note 21**).
4. Resistant green cells start appearing after about 8 weeks (*see* **Note 27**). When these are large enough (>2 mm across) transfer to fresh plates. Once shoots are formed carry out two more regeneration cycles on media with antibiotics to ensure homoplasmy of transplastomic shoots (*see* **Note 28**).
5. Transfer shoots to Magenta™ GA 7 vessels containing MS medium with 200 mg/L of spectinomycin. Once sufficient roots have formed transfer plants to soil.

3.8 Inheritance of Transgenic Plastids

1. Transplastomic plants grow well in soil (Fig. 4a) and appear indistinguishable from WT plants. They flower profusely and set seed. On ripening the seed pods open abruptly spilling seeds. To avoid loss of seeds remove the brown unopened pods from plants and place in a petri dish. Squeezing the sides of the pods releases the seeds (Fig. 4b). Plants will continue to flower profusely for the next few months allowing seeds to be collected for storage (*see* **Note 29**).

Fig. 4 Inheritance of the plastid-localized *aadA* gene in transplastomic *P. hybrida*. (**a**) Transplastomic plant flowering in soil. (**b**) Transplastomic seeds from ripe pods. Petunia seedlings from (**c**) transplastomic and (**d**), WT seeds on MS selection medium containing 200 mg/L spectinomycin. (**e**) Transplastomic and (**f**) WT Petunia seedlings growing on MS medium lacking antibiotics (control)

2. The inheritance of transgenic plastids is followed by germinating seeds on MS medium with 200 mg/L spectinomycin. Seeds collected from selfed transplastomic plants give rise to green resistant seedlings (Fig. 4c) whereas WT seedlings bleach and

are arrested at the cotyledon stage (Fig. 4d) on MS media containing 200 mg/L spectinomycin (*see* **Note 30**). Both transplastomic (Fig. 4e) and WT seedlings (Fig. 4f) grow on MS medium lacking antibiotics.

4 Notes

1. Obtained from Thompson and Morgan (Ipswich, UK). Pink wave is a popular cultivar and can be purchased from a variety of other sources.
2. We prefer Milli-Q water (Millipore UK, Watford) for making up stocks of reagents including antibiotics. RO-water (Aquaboss™, Lauer, Aquaris Ltd, Hereford, UK) is used for plant growth media.
3. Disposable plastic ware is used to reduce the possibility of contamination with trace quantities of other chemicals that might be present in shared glass and plastic containers for general use following washing with detergent solutions.
4. Phytagel™ is used to propagate plants in vitro because the soft gel matrix allows root penetration. The matrix appears to adsorb some selection agents including kanamycin, preventing its use during the selection phase of plastid transformation. Spectinomycin selection appears to be compatible with use of Phytagel™ but we prefer to use agar at 0.6–0.8 % (W/V) during the selection phase.
5. Passing 10 mL of sterile Milli-Q water through the filter unit will reduce the quantity of wetting agent on the membrane and can be done just prior to filter sterilizing the antibiotic solutions.
6. We dispose of streptomycin stock solutions after 1 month storage at −20 °C and spectinomycin stocks after 2 months storage
7. The tubes are stored frozen to minimize the risk of growth of microbial contamination and can be kept for years.
8. Autoclave settings are 121 °C, 15 psi for 20 min.
9. It is important to use the free base and not the salt. We usually order 1 g, which arrives as a sealed glass vial that can be stored at −80 °C until required.
10. Separate glassware for making media from vessels used for other laboratory methods. This minimizes the risk of contaminating chemicals such as detergents entering plant growth media. The glassware for media is washed in RO water immediately after use. Avoid washing glassware for media with detergents, which could have a negative impact on plant growth.

11. All aseptic procedures are carried out in a laminar flow cabinet preferably with a germicidal ultraviolet lamp. Sterilize the cabinet with UV light for 10 min. prior to working in the cabinet. Remove any media from the cabinet before turning on the UV lamp.
12. Media solidified with Phytagel™ is used for germination of seeds and in vitro propagation of plants without antibiotic selection. Agar is used for propagating plants on media with antibiotics and to promote root growth we reduce the agar concentration to 0.6–0.7 % (W/V) depending on the gel strength of the agar.
13. We use sterile 9 cm triple vent petri dishes (Scientific Laboratory Supplies) for shoot regeneration and germinating seedlings (Scientific Laboratory Supplies) and Magenta™ GA-7 vessels (77 mm × 77 mm × 97 mm, Sigma-Aldrich) for plant propagation in vitro.
14. If the media does not solidify this can result from the incorrect amount of agar or Phytagel™ being used or because the pH of the media was not checked and corrected to pH 6.0.
15. Higher concentrations of sodium hypochlorite solutions are needed if the seeds are contaminated with fungal spores. Fungal contamination tends to be more of a problem with seeds collected from plants grown in glass-houses compared to those grown in walk-in growth rooms. Add 0.1 % (V/V) Tween 20 (Sigma-Aldrich) as a wetting agent if fungal contamination is a problem.
16. Make sure there is a ready supply of sterile RO water by sterilizing batches of 50–500 mL in 100 mL to 1 L Duran® bottles (Schott, Stafford, UK). To avoid contamination with microbes only use once and do not reuse the bottles once opened following sterilization.
17. We use Sanyo-Gallenkamp MLR 350 illuminated growth chambers with a light setting of 3–4 and a 12 h day. Plates are evenly distributed on shelves and not placed directly adjacent to an illuminated fluorescent tube.
18. Pick only dark green and healthy seedlings. Variation in seedling appearance may reflect genetic segregation arising from heterozygous seed stocks.
19. Use Magenta™ lids with vents (10 mm vent with 0.22 μm pore size) for optimal growth and to reduce vitrification of plants. Vented lids are not required when the plants are small and occupy a limited volume of the vessel.
20. We sterilize metal scalpels with blades and forceps by placing them in 70 % (V/V) ethanol and then inserting them into a hot (250 °C) bead sterilizer (Steri 350, Sigma-Aldrich) for 30 s.

21. It is important to use a new sharp stainless steel scalpel blade. Use a rounded blade such as No. 10 (Swann-Morton, Sheffield, UK) to allow clear incisions without tearing. Process one leaf or leaf explant at a time and cut it into smaller pieces relatively quickly to ensure the leaf segment does not dry out. Ensure the source of the leaf explant does not dry out by placing the lid on Magenta™ vessels containing plants or petri dishes containing leaf pieces.
22. It is important to mix the contents of solutions previously stored frozen by inverting the tube several times once thawed.
23. The DNA-coated gold should appear as a homogeneous brown opaque solution without clumps. An even suspension of gold takes longer to settle compared to gold particles that have not been resuspended well.
24. Make sure the rupture disks are separated. Two rupture disks stacked together can sometimes be placed in the device and will not break once a pressure of 1,100 psi is achieved. A scalpel blade provides the easiest tool for separating and lifting rupture disks placed flat in a petri dish.
25. Leaf explants on one bombarded MSB 30 media plate are cut and placed onto three to four MSB30 selection plates containing antibiotics. The use of two antibiotics prevents the isolation of spontaneous resistant mutants to the single antibiotics. Spontaneous resistance mutations are most frequently located in the ribosomal RNA genes. The *aadA* gene confers resistance to both spectinomycin and streptomycin.
26. As a negative control place WT leaf explants on two to three selection plates. Bleaching of leaf pieces and eventual browning shows selection is effective. As a positive control we usually place leaf explants from transplastomic Petunia on two selection plates. Efficient regeneration of resistant shoots indicates the media composition is correct.
27. Transplastomic plants are obtained at a frequency of about one per ten plates bombarded with microprojectiles, which is about tenfold lower than the frequency obtained in *N. tabacum* [20].
28. Homoplasmy is normally verified by DNA blot analysis showing that WT bands have been replaced by transgenic plastid bands of the correct predicted size. It is important to use restriction enzymes that cleave outside the plastid DNA targeting arms present in the transformation vector.
29. Watering plants with liquid fertilizer maintains healthy growth and flowering for long periods (>4 months).
30. 100 % transmission of spectinomycin resistance from selfed transplastomic plants has been observed over five generations. All seedlings tested at each generation were GUS positive.

Acknowledgments

The authors thank Prof P. Meyer, Drs. O. Zubko (Leeds) and M. K. Zubko for their support in establishing plastid transformation in Petunia. E. M. A was the recipient of a BBSRC Ph.D. studentship and A. D. was supported by research grants (BB/E020445 and BB/I011552) from the Biotechnology and Biological Research Council (UK).

References

1. Kelly RO, Deng ZA, Harbaugh BK (2007) Evaluation of 125 petunia cultivars as bedding plants and establishment of class standards. Horttechnology 17:386–396
2. Chandler SF, Lu CY (2005) Biotechnology in ornamental horticulture. In Vitro Cell Dev Biol Plant 41:591–601
3. Conia J, Bergounioux C, Perennes C, Muller P, Brown S, Gadal P (1987) Flow cytometric analysis and sorting of plant chromosomes from *Petunia hybrida* protoplasts. Cytometry 8:500–508
4. Mishiba KI, Ando T, Mii M, Watanabe H, Kokubun H, Hashimoto G, Marchesi E (2000) Nuclear DNA content as an index character discriminating taxa in the genus Petunia sensu Jussieu (*Solanaceae*). Ann Bot 85:665–673
5. Cornu A, Dulieu H (1988) Pollen transmission of plastid DNA under genotypic control in *Petunia hybrida*. Hort J Hered 79:40–44
6. Nagata N, Saito C, Sakai A, Kuroiwa H, Kuroiwa T (1999) The selective increase or decrease of organellar DNA in generative cells just after pollen mitosis one controls cytoplasmic inheritance. Planta 209:53–65
7. Bovenberg WA, Kool AJ, Nijkamp HJJ (1981) Isolation, characterization and restriction endonuclease mapping of the *Petunia hybrida* chloroplast DNA. Nucleic Acids Res 9:503–517
8. Olmstead RG, Sweere JA (1994) Combining data in phylogenetic systematics: an empirical approach using 3 molecular data sets in the *Solanaceae*. Syst Biol 43:467–481
9. Ruf S, Hermann M, Berger IJ, Carrer H, Bock R (2001) Stable genetic transformation of tomato plastids and expression of a foreign protein in fruit. Nat Biotechnol 19:870–875
10. Sidorov VA, Kasten D, Pang SZ, Hajdukiewicz PTJ, Staub JM, Nehra NS (1999) Stable chloroplast transformation in potato: use of green fluorescent protein as a plastid marker. Plant J 19:209–216
11. Zubko MK, Zubko EI, van Zuilen K, Meyer P, Day A (2004) Stable transformation of petunia plastids. Transgenic Res 13:523–530
12. Kavanagh TA, Thanh ND, Lao NT, McGrath N, Peter SO, Horvath EM, Dix PJ, Medgyesy P (1999) Homeologous plastid DNA transformation in tobacco is mediated by multiple recombination events. Genetics 152:1111–1122
13. Aldrich J, Cherney B, Merlin E, Palmer J (1986) Sequence of the *rbcL* gene for the large subunit of ribulose bisphosphate carboxylase oxygenase from Petunia. Nucleic Acids Res 14:9534
14. Yukawa M, Tsudzuki T, Sugiura M (2005) The 2005 version of the chloroplast DNA sequence from tobacco (*Nicotiana tabacum*). Plant Mol Biol Rep 23:359–365
15. European Commission Joint Research Centre (2009) Evaluation of pollen spread of plastid located genes for the model plant *Petunia hybrida* under field condition. University of Rostock, Germany. In: Deliberate releases and placing on the EU marker of genetically modified organisms-GMO Register, Notification Number: B/DE/08/203. http://gmoinfo.jrc.ec.europa.eu/gmp_report.aspx?CurNot=B/DE/08/203. Accessed 8 Sept 2012
16. Zubko E, Adams CJ, Machaekova I, Malbeck J, Scollan C, Meyer P (2002) Activation tagging identifies a gene from *Petunia hybrida* responsible for the production of active cytokinins in plants. Plant J 29:797–808
17. Murashige T, Skoog F (1962) A revised medium for rapid growth and bio assays with tobacco tissue cultures. Physiol Plantarum 15:473–497
18. Daniell H, Lee SB, Grevich J, Saski C, Quesada-Vargas T, Guda C, Tomkins J, Jansen RK (2006) Complete chloroplast genome sequences of *Solanum bulbocastanum*, *Solanum lycopersicum*

and comparative analyses with other Solanaceae genomes. Theor Appl Genet 112:1503–1518

19. Chung HJ, Jung JD, Park HW, Kim JH, Cha HW, Min SR, Jeong WJ, Liu JR (2006) The complete chloroplast genome sequences of *Solanum tuberosum* and comparative analysis with *Solanaceae* species identified the presence of a 241-bp deletion in cultivated potato chloroplast DNA sequence. Plant Cell Rep 25:1369–1379
20. Svab Z, Maliga P (1993) High-frequency plastid transformation in tobacco by selection for a chimeric *aadA* gene. Proc Natl Acad Sci U S A 90:913–917

Chapter 18

Plastid Transformation in Potato: *Solanum tuberosum*

Vladimir T. Valkov, Daniela Gargano, Nunzia Scotti, and Teodoro Cardi

Abstract

Although plastid transformation has attractive advantages and potential applications in plant biotechnology, for long time it has been highly efficient only in tobacco. The lack of efficient selection and regeneration protocols and, for some species, the inefficient recombination using heterologous flanking regions in transformation vectors prevented the extension of the technology to major crops. However, the availability of this technology for species other than tobacco could offer new possibilities in plant breeding, such as resistance management or improvement of nutritional value, with no or limited environmental concerns. Herein we describe an efficient plastid transformation protocol for potato (*Solanum tuberosum* subsp. *tuberosum*). By optimizing the tissue culture system and using transformation vectors carrying homologous potato flanking sequences, we obtained up to one transplastomic shoot per bombardment. Such efficiency is comparable to that usually achieved in tobacco. The method described in this chapter can be used to regenerate potato transplastomic plants expressing recombinant proteins in chloroplasts as well as in amyloplasts.

Key words Amyloplast, Chloroplast, Homologous recombination, Plastid transformation, Potato, *Solanum tuberosum*

1 Introduction

Potato is by far the most important non-cereal staple food for mankind. It is the fourth crop in global production terms and a long-established part of the diet in many parts of the world [1]. Potatoes are not only a good source of carbohydrates and fibers but are also a valuable source of vitamin C, a number of minerals including iron and potassium, and other phytonutrients including carotenoids [2]. The aims of potato breeders over the last 200 years have largely been to maximize yield and to improve disease resistance. Recently, new technologies have emerged that could affect the future direction of potato breeding.

Plastid transformation offers a great potential for plant breeding as well as molecular farming, due to attractive advantages such as

Vladimir T. Valkov and Daniela Gargano contributed equally to this work.

Pal Maliga (ed.), *Chloroplast Biotechnology: Methods and Protocols*, Methods in Molecular Biology, vol. 1132,
DOI 10.1007/978-1-62703-995-6_18, © Springer Science+Business Media New York 2014

high level of transgene expression and protein accumulation, the possibility of coexpressing several transgenes in operons, the precise transgene integration by homologous recombination, and transgene containment. In the recent years, fertile transplastomic plants have been reported in several crops, such as tomato, soybean, cotton, lettuce, cabbage, and eggplant [3–8]. Interestingly, the use of species-specific flanking sequences allowed the improvement of plastid transformation efficiency (up to 1–2 transplastomic shoots per bombardment) in cotton, soybean, and lettuce. Regarding potato chloroplast transformation, early attempts performed with vectors carrying homologous tobacco flanking regions resulted in low transformation efficiency (one transplastomic shoot every 15–35 biolistic shots) [9, 10]. In 2005, the complete chloroplast genome sequence of *Solanum tuberosum* became available [11, 12], facilitating the design of species-specific vectors and the study of the effect of increasing homology on plastid transformation efficiency [13].

In this chapter, we describe an efficient and reliable protocol for potato chloroplast transformation. Optimized regeneration procedures in combination with the species-specific vectors enabled the isolation of spectinomycin-resistant green calli and, subsequently, the regeneration of transplastomic shoots with a transformation efficiency comparable to that of tobacco. The described method gives the opportunity to study transgene expression in green and nongreen plastids, broadening the available approaches for potato breeding.

2 Materials

2.1 Plant Material and Consumables

1. We used cv. Désirée potato plants (*Solanum tuberosum* L. subsp. *tuberosum*) grown in vitro.
2. Ultrapure water (*see* **Note 1**).
3. 0.1 M free base spermidine (*see* **Note 2**).
4. 2.5 M $CaCl_2$, filter sterilized (filter 0.22 μm). Stored at 4 °C for no more than 2 weeks.
5. Bombardment consumables: macrocarrier holders, rupture disks (1100 psi), stopping screens, macrocarriers (flying disks), 0.6 μm diameter gold particles (Bio-Rad, Hercules, CA, USA).
6. 1 μg/μl purified plasmid DNA.
7. 70 and 100 % ethanol.
8. Murashige and Skoog (MS) basal salt mixture; MS salts NH_4NO_3-free (Duchefa, Haarlem, the Netherlands).
9. Vitamins: Gamborg B5, Nitsch vitamin mixture (Duchefa, Haarlem, the Netherlands).
10. Plant growth regulators: 2,4-dichlorophenoxyacetic acid (2,4-D), indole acetic acid (IAA), naphthaleneacetic acid (NAA),

zeatin riboside (ZR), zeatin, gibberellic acid (GA_3), adenine hemisulfate.

11. Casein hydrolysate.
12. Sucrose, glucose.
13. Osmotic agents: sorbitol, mannitol.
14. Antibiotics: spectinomycin, cefotaxime.
15. Miscellaneous: NH_4Cl, agar.
16. Vented plastic containers and Petri dishes 100 mm×20 mm (*see* **Note 3**).
17. Forceps and scalpel.

2.2 Equipment

The biolistic transformation is carried out using a PDS-1000/He™ Helium Biolistic Particle Delivery System (Bio-Rad, Hercules, CA, USA).

2.3 Media

1. M6: MS salts with B5 vitamins, 30 g/l sucrose, 0.8 mg/l zeatin riboside (ZR), 2.0 mg/l 2,4-dichlorophenoxyacetic acid (2,4-D), 400 mg/l spectinomycin and 250 mg/l cefotaxime, solidified with 0.8 % (w/v) agar, pH 5.8 [14].
2. M6M: as above, but with 0.1 M sorbitol and 0.1 M mannitol, and without antibiotics.
3. T1: MS salts with B5 vitamins, 16 g/l glucose, 3.0 mg/l zeatin riboside (ZR), 2.0 mg/l indole acetic acid (IAA), 1.0 mg/l gibberellic acid (GA_3), 400 mg/l spectinomycin and 250 mg/l cefotaxime, 0.8 % (w/v) agar, pH 5.8 [10].
4. DH: MS salts without NH_4NO_3, 0.268 g/l NH_4Cl, Nitsch vitamin mixture, 2.5 g/l sucrose, 36.4 g/l mannitol, 0.1 g/l casein hydrolysate, 0.08 g/l adenine hemisulfate, 2.5 mg/l zeatin, 0.1 mg/l indole acetic acid (IAA), 400 mg/l spectinomycin, 0.8 % (w/v) agar, pH 5.8 [15].
5. MON: MS salts with B5 vitamins, 30 g/l sucrose, 0.1 mg/l naphthaleneacetic acid (NAA), 5 mg/l zeatin riboside (ZR), 400 mg/l spectinomycin, 0.8 % (w/v) agar, pH 5.8 [9].
6. MS rooting medium: growth regulator-free MS medium with B5 vitamins, 30 g/l sucrose, solidified with 0.8 % (w/v) agar and supplemented with 200 mg/l spectinomycin.

2.4 PCR Analysis

1. Oligonucleotides: primers can be located either outside of the homologous regions or one outside of the homologous regions and another inside of the gene of interest.
2. dNTPs.
3. Taq DNA polymerase, $MgCl_2$, and buffer.
4. DNA molecular weight marker.
5. Agarose.

3 Methods

The methods outlined below describe the treatment of potato leaves before the bombardment, the bombardment procedure, the isolation of green calli, and the regeneration of putative transplastomic shoots. Either a single- or a multistep in vitro culture procedure can be used. The latter, however, gave better results in our hands [13]. Efficient regeneration of shoots via organogenesis could be obtained by transferring green calli on shoot induction and elongation media.

3.1 Precultivation of Potato Leaves

1. Grow potato plants in vitro in vented plastic containers (*see* **Note 3**) under controlled conditions (16 h light 40 μE/m^2/s and 8 h dark at 24 °C) and on solid growth regulator-free MS media for 3–4 weeks.
2. 24 h before bombardment cut 4–5 young green leaves and place them with abaxial side on M6M medium. Incubate in the dark at 24 °C.

3.2 Microcarrier Stock Preparation

1. For 90 bombardments, mix 35 mg gold with 1.5 ml ethanol and vortex for 2–3 min.
2. Leave to sediment for 2 min.
3. Centrifuge at 4,300 × *g* for 5 s at room temperature.
4. Repeat the washing procedure three times.
5. Dissolve in 1 ml ethanol and store at −20 °C.

3.3 Sterilization and Preparation of Consumables

1. Wash stopping screens and macrocarrier holders in 70 % ethanol for 5 min and briefly flame up.
2. Wash macrocarriers in 70 % ethanol and left to dry thoroughly.
3. Wash rupture disks in 70 % ethanol briefly before the bombardment.
4. Place macrocarriers in the macrocarrier holders by using the appropriate plastic insertion tool provided with the Biolistic® PDS-1000/He Particle Delivery System.

3.4 DNA Precipitation onto the Microcarriers

1. For five bombardments, pipet 50 μl from the stock into new Eppendorf tube. Mix well by vortexing for 5 s to disrupt agglomerated particles. Leave to sediment for 2 min and centrifuge at room temperature for 2 s at 4,300 × *g*. Discard the supernatant.
2. Add 50 μl sterile cold water. Mix by vortexing for 2 min. Leave to sediment for 1 min, centrifuge for 2 s at 4,300 × *g*, and discard the supernatant. Repeat this step three times.

3. Add 50 μl of sterile cold water and mix well the gold particles by vortexing for 2 min. Add 10 μl of plasmid DNA, 50 μl cold 2.5 M $CaCl_2$, and 20 μl 0.1 M spermidine. Mix by shaking for 20 min at 4 °C.
4. Add 200 μl cold absolute ethanol and centrifuge for 2 s at 4,300 × *g*. Discard the supernatant.
5. Resuspend pellet in 200 μl cold absolute ethanol, centrifuge for 2 s at 4,300 × *g*, and discard the supernatant. Repeat **steps 4–5** four times.
6. Resuspend pellet in 30 μl cold absolute ethanol and keep on ice.

3.5 Leaf Bombardment

1. Pipet 5 μl of coated gold particles on the center of the macrocarrier.
2. Switch on the vacuum pump.
3. Open the knob of the helium gas and set up the helium pressure 200 psi higher than the desired pressure of the rupture disks (e.g., 1,300 psi with 1,100 psi rupture disks).
4. Switch on the PDS system.
5. Assemble the Biolistic gene gun following the manufacturer's instructions (Bio-Rad, Hercules, CA, USA).
6. Place the Petri dish containing the potato leaves on the shelf at a 6 cm distance from the stopping screen to the target (microcarrier flight distance).
7. Turn the vacuum button on VAC position. Allow the vacuum pressure to reach 27–28 in. of Hg, turn the vacuum button on HOLD position, and press the FIRE button until the rupture disk is bursted.
8. Release the FIRE button and turn the vacuum button to VENT position.
9. Repeat **steps 5–8** for additional samples.
10. Shut down the system following the manufacturer's instructions.
11. Incubate the bombarded leaves in the dark for 2 days at 24 °C.

3.6 Regeneration of Transplastomic Calli and Shoots

1. Cut the bombarded leaves into small pieces (3 mm × 3 mm) and transfer maximum 16 explants to M6 callus induction media supplemented with 400 mg/l spectinomycin and 250 mg/l cefotaxime. Incubate leaf pieces under dim light (about 10 μE/m^2/s) in 16 h light, 8 h dark regime for 1 month.
2. Transfer explants to T1 medium with the same antibiotics. Leaf explants should be transferred every 1–2 months to fresh selective T1 medium and incubated under the same conditions for a further 3–4 months (*see* **Note 4**).
3. Carefully excise primary spectinomycin-resistant calli from the leaf explants and transfer onto the surface of DH medium containing

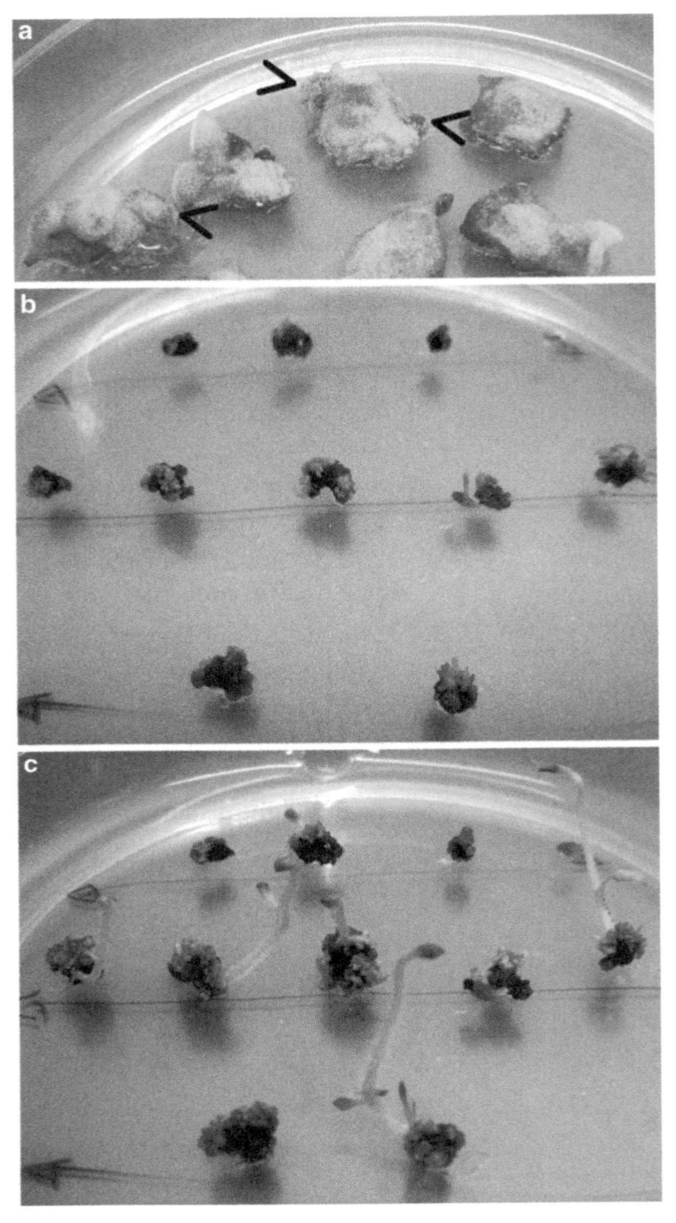

Fig. 1 Primary selection of spectinomycin-resistant potato calli and shoots. (**a**) Green calli (*arrowheads*) obtained after cultivation for 3 months on T1 medium with 400 mg/l spectinomycin. (**b**) Calli after cultivation for 3 weeks on MON medium with the same antibiotic. (**c**) The same calli with primary shoots after 6 weeks on MON medium

400 mg/l spectinomycin (*see* **Note 5** and Fig. 1a). Cultivate for 1 month for further growth of the calli.

4. Transfer all green calli to MON medium for regeneration of spectinomycin-resistant shoots (about 1–2 months) (Fig. 1b).

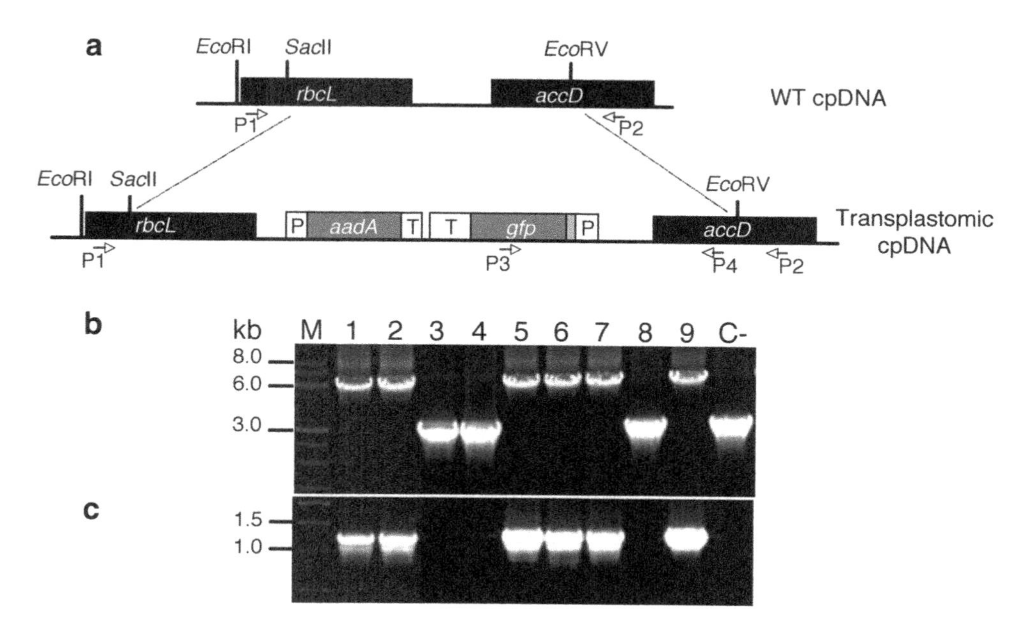

Fig. 2 PCR analyses of putative transplastomic plants. (**a**) Schematic representation of the genomic region involved in transgene integration before and after transformation [13]. (**b**) Primers located outside of the inserted sequence (P1 and P2) amplified high molecular weight products in successfully transformed plants (1, 2, 5–7, 9) and low molecular weight products in non-transformed (3, 4, 8) and wild-type control (C-) plants. The presence of a single band demonstrates that all transplastomic plants were homoplasmic. (**c**) Gene-specific primers located inside of the inserted sequence (P3 and P4) revealed a signal only in transformed plants

5. Cut the regenerated shoots and transfer to growth regulator-free MS medium with 200 mg/l spectinomycin for root formation (*see* **Note 6** and Fig. 1c).
6. Successfully rooted plants can be transferred to soil and grown under growth chamber conditions (14 h light, 200 μE/m^2/s, at 25 °C, and 10 h dark at 20 °C) for tuberization and further analyses.

3.7 PCR Analysis of Transplastomic Plants

1. Extract total leaf DNA.
2. Set PCR reaction (25 μl): 50 ng of total DNA, 2.5 μl of 10× PCR buffer, 0.75 μl of 50 mM $MgCl_2$, 0.5 μl (20 μM) of each primer, 0.5 μl of 10 mM dNTPs, 1.25 units of Taq DNA polymerase, and adjust the final volume with sterile water.
3. Set PCR cycles: initial denaturation at 94 °C for 3 min, 30 cycles of denaturation at 94 °C for 1 min, annealing at 55–60 °C for 1 min, elongation at 72 °C for 1 min/kb, final elongation at 72 °C for 5 min.
4. Run the amplification products on agarose gel. PCR products show a variable size depending on primers selected for the analysis (Fig. 2 and Table 1).

Table 1
Comparison of potato chloroplast transformation efficiency with vectors containing either tobacco- or potato-derived flanking sequences

Vectors[a]	Flanking sequence source	No. of bombarded plates	No. of spec[R] shoots	No. of shoots	
				Analyzed by PCR	PCR positive
1002	Tobacco	35	5	5	1
pVL2	Tobacco	15	3	3	0
pVL12	Tobacco	15	5	5	4
Total		65	13	13	5
pVL15	Potato	15	23	12	10
pVL14	Potato	16	20	11	11
pVL13	Potato	16	19	10	8
Total		47	62	33	29

All vectors carry the *gfp* gene under the control of different regulatory signals
[a]*See* ref. 13 for details

4 Notes

1. Purified water with 18.2 MΩ·cm resistivity was used to prepare media and all solutions.
2. Spermidine is sold in vials containing spermidine and argon. To dissolve the spermidine, heat the vial at 37 °C for 5 min. Aliquot 15.8 μl of spermidine in 1.5 ml Eppendorf tubes and store at −80 °C. Before use, thaw one tube and add 984.2 μl sterile H_2O. Mix and aliquot 23 μl of 0.1 M solution in 0.5 ml Eppendorf tubes and store at −20 °C.
3. Potato is particularly sensitive to ethylene. Hence, in order to allow better gas exchange, it is preferable to use vented containers for shoot subculture. Similarly, 20 or 25 mm high Petri dishes should be used for explant selection and regeneration.
4. We usually observe the first spectinomycin-resistant event after 2 months of incubation on T1 medium. In some cases, resistant calli are identified after 4–5 months of incubation on the same medium.
5. Big calli can be cut into 2–4 smaller pieces.
6. After rooting and some growth of regenerated shoots, we perform the molecular analyses (PCR, Southern blot) to identify the successfully transformed ones.

Acknowledgments

Work in authors' labs was supported by the European Union (FP5, project "Plastid Factory," grant no. QLK3-CT-1999-00692; FP6, project "Plastomics," grant no. LSHG-CT-2003-503238) to TC.

References

1. Solomon-Blackburn RM, Barker H (2001) Breeding virus resistant potatoes (*Solanum tuberosum*): a review of traditional and molecular approaches. Heredity 86:17–35
2. Milbourne D, Pande B, Byan GJ (2007) Potato. In: Kole C (ed) Pulses, sugar and tuber crops. Springer, Berlin, pp 205–236
3. Ruf S, Hermann M, Berger IJ, Carrer H, Bock R (2001) Stable genetic transformation of tomato plastids and expression of a foreign protein in fruit. Nat Biotechnol 19:870–875
4. Dufourmantel N, Pelissier B, Garcon F, Peltier G, Ferullo J-M, Tissot G (2004) Generation of fertile transplastomic soybean. Plant Mol Biol 55:479–489
5. Kumar S, Dhingra A, Daniell H (2004) Stable transformation of the cotton plastid genome and maternal inheritance of transgenes. Plant Mol Biol 56:203–216
6. Kanamoto H, Yamashita A, Asao H, Okumura S, Takase H, Hattori M, Yokota A, Tomizawa K-I (2006) Efficient and stable plastid transformation of *Lactuca sativa* L. cv. Cisco (lettuce). Transgenic Res 15:205–217
7. Liu CW, Lin CC, Chen JJ, Tseng MJ (2007) Stable chloroplast transformation in cabbage (*Brassica oleracea* L. var. *capitata* L.) by particle bombardment. Plant Cell Rep 26: 1733–1744
8. Singh AK, Verma SS, Bansal KC (2010) Plastid transformation in eggplant (*Solanum melongena* L.). Transgenic Res 19:113–119
9. Sidorov VA, Kasten D, Pang S-Z, Hajdukiewicz PTJ, Staub JM, Nehra NS (1999) Stable chloroplast transformation in potato: use of green fluorescent protein as a plastid marker. Plant J 19:209–216
10. Nguyen TT, Nugent G, Cardi T, Dix PJ (2005) Generation of homoplasmic plastid transformants of a commercial cultivar of potato (*Solanum tuberosum* L.). Plant Sci 168: 1495–1500
11. Gargano D, Vezzi A, Scotti N, Gray JC, Valle G, Grillo S, Cardi T (2005) The complete nucleotide sequence genome of potato (*Solanum tuberosum* cv Désirée) chloroplast DNA. In 2nd Solanaceae genome workshop 2005, Ischia, Italy, p. 107
12. Chung H-J, Jung JD, Park H-W, Kim J-H, Cha HW, Min SR, Jeong W-J, Liu JR (2006) The complete chloroplast genome sequences of *Solanum tuberosum* and comparative analysis with Solanaceae species identified the presence of a 241-bp deletion in cultivated potato chloroplast DNA sequence. Plant Cell Rep 25: 1369–1379
13. Valkov VT, Gargano D, Manna C, Formisano G, Dix PJ, Gray JC, Scotti N, Cardi T (2011) High efficiency plastid transformation in potato and regulation of transgene expression in leaves and tubers by alternative 5′ and 3′ regulatory sequences. Transgenic Res 20:137–151
14. Yadav NR, Sticklen MB (1995) Direct and efficient plant regeneration from leaf explants of *Solanum tuberosum* L. cv. Bintje. Plant Cell Rep 14:645–647
15. Haberlach GT, Cohen B, Reichert N, Baer M, Towill L, Helgeson JP (1985) Isolation, culture and regeneration of protoplasts of potato and several related species. Plant Sci 39:67–74

Chapter 19

Plastid Transformation in Eggplant

Kailash C. Bansal and Ajay K. Singh

Abstract

Eggplant (*Solanum melongena* L.) is an important vegetable crop of tropical and temperate regions of the world. Here we describe a procedure for eggplant plastid transformation, which involves preparation of explants, biolistic delivery of plastid transformation vector into green stem segments, selection procedure, and identification of the transplastomic plants. Shoot buds appear from cut ends of the stem explants following 5–6 weeks of spectinomycin selection after bombardment with the plastid transformation vector containing *aadA* gene as selectable marker. Transplastomic lines are obtained after the regenerated shoots are subjected to several rounds of spectinomycin selection over a period of 9 weeks. Homoplasmic transplastomic lines are further confirmed by spectinomycin and streptomycin double selection. The transplastomic technology development in this plant species will open up exciting possibilities for improving crop performance, metabolic engineering, and the use of plants as factories for producing biopharmaceuticals.

Key words Aminoglycoside 3″-adenylyltransferase, Eggplant, Plastid transformation, *Solanum melongena* L.

1 Introduction

Eggplant is an important solanaceous vegetable crop. It is a good source of vitamins and minerals, especially iron, making its total nutritional value comparable with that of tomato [1]. Plant genetic transformation is a core research tool in plant biotechnology and crop improvement. Nuclear transformation in eggplant has been reported by several researchers [2–8]. However, introduction of transgene(s) into the nuclear genome has led to a growing public concern of the possibility of gene escape through pollen to weedy or wild relatives of the transgenic crops or other crop plants. Plastid transformation can provide potential solution to these problems, because maternal inheritance of the chloroplast genome prevents the escape of transgene through pollen grains in most plants. Therefore, plastid transformation is an environmental friendly approach to plant genetic engineering that minimizes transgene escape to related weeds or crops through outcrossing. The very high expression level of transgene in transplastomic system is ideal for applications such as

Pal Maliga (ed.), *Chloroplast Biotechnology: Methods and Protocols*, Methods in Molecular Biology, vol. 1132,
DOI 10.1007/978-1-62703-995-6_19,

production of pharmaceuticals. Since plastids are highly polyploidy genetic system, very high expression levels of transgene can be achieved. Plastid transformation is highly a precise method of genetic engineering as the integration of transgene is mediated by homologous recombination. Hence the position effect and transgene epigenetic silencing are avoided. Improved safety coupled with high expression and the ease of selectable marker elimination in plastid transformation offers a new generation of transplastomic crop plants expressing useful agronomic traits.

The development of transplastomic technology will be useful for introducing agronomically and industrially important traits into eggplant. The production of transplastomic eggplant with maternal inheritance of the transgene could solve problems related to outcrossing between the genetically modified (GM) crop and non-transgenic crops or its wild relatives. Here we describe a protocol for generation of stable plastid transformation system for eggplant. This protocol is based on original research described from this laboratory [9].

2 Materials

2.1 Plant Materials

1. In vitro grown eggplant plants of the cultivar Pusa Uttam.
2. Green stem segments (5–7 mm in length) as explants derived from in vitro grown eggplants.

2.2 Culture Media Ingredients

1. Macro salts for MS medium (20× stock): 33 g NH_4NO_3, 38 g KNO_3, 3.4 g KH_2PO_4, 7.4 g $MgSO_4.7H_2O$, and 8.8 g $CaCl_2.2H_2O$. Make up to 1 L with distilled water. Store at 4 °C. Use 50 mL of stock solution for 1 L MS medium.
2. Micro salts for MS medium (1,000× stock): 10 g $MnSO_4$, 6.2 g H_3BO_3, 5.8 g $ZnSO_4.7H_2O$, 0.8 g KI, 0.25 g $Na_2MoO_4.2H_2O$, 0.025 g $CuSO_4.5H_2O$, and 0.025 $CoCl_2.6H_2O$. Make up to 1 L with distilled water. Store at 4 °C. Use 1 mL of stock solution for 1 L MS medium.
3. Iron solution for MS medium (500× stock): 18.65 g Na_2EDTA and 13.90 g $FeSO_4.7H_2O$. Make up to 1 L with distilled water. Stock solution should be protected from the light; store at 4 °C. Use 2 mL stock solution for 1 L MS medium.
4. Vitamins/inositol solution for MS medium (250× stock): 25 g inositol, 0.025 g thiamine HCl, 0.125 g pyridoxine HCl, 0.125 g nicotinic acid, and 0.5 g glycine. Make up to 1 L with distilled water. Store solution at 4 °C. Use 4 mL stock solution for 1 L MS medium.
5. Add 30 g sucrose to make 1 L MS medium.

6. Zeatin riboside stock solution: Add 10 mg of zeatin riboside (Sigma, St. Louis, MO, USA) to a glass container. Add 200–300 μL of 1 M NaOH to dissolve the powder. Once completely dissolved, add 10 mL of molecular biology grade water. Filter sterilize the solution with 0.22 μm filter unit and store the stock solution at −20 °C.
7. Spectinomycin stock solution: Dissolve 200 mg of spectinomycin (Sigma, St. Louis, MO, USA) in 10 mL of molecular biology grade water to obtain a final volume of 10 mL. Aliquot in 1 mL volume and store at −20 °C.
8. Streptomycin stock solution: Dissolve 1 g of streptomycin (Sigma, St. Louis, MO, USA) in 50 mL of molecular biology grade H_2O. Sterilize solution using 0.22 μm filter unit and store the stock solution at −20 °C.

2.3 Culture Media Required

1. Seed germination medium: Murashige and Skoog [10] medium solidified with 0.8 % (w/v) agar (Bacteriological grade, HiMedia, Mumbai, India).
2. Pre-culture medium: solid MS medium supplemented with 1 mg/L zeatin riboside.
3. Selection medium: solid MS medium supplemented with 300 mg/L spectinomycin and 1 mg/L zeatin riboside.
4. Double selection medium: solid MS medium supplemented with 300 mg/L each spectinomycin and streptomycin and 1 mg/L zeatin riboside.

2.4 Biolistic Delivery System

1. The PDS-1000/He Particle Delivery System (Bio-Rad).
2. Rupture disc, 1,100 psi (Bio-Rad).
3. Particle gun macro-carriers (Bio-Rad).
4. Particle gun stopping screen (Bio-Rad).
5. Particle gun macro-carrier holders.

2.5 Coating Tungsten Particles with Transforming DNA

1. Sterile tungsten particles (50 mg/mL) suspended in ethanol.
2. Plasmid DNA (pPRV111A, plastid transformation vector) 1.0 μg/μL [11] (*see* **Note 1**).
3. A 2.5 M $CaCl_2$ solution: Dissolve 3.68 g of $CaCl_2$ in 10 mL of water; filter sterilize. The solution should be stored at 4 °C for short periods and do not freeze.
4. Spermidine: Prepare a 0.1 M spermidine-free base (Sigma, St. Louis, MO, USA) solution in sterile distilled water, filter sterilize, and store at −20 °C (*see* **Note 2**).
5. Ice-cold absolute ethanol: Newly opened fresh bottle is preferred.

2.6 Total DNA Isolation from Plant Tissues

CTAB (cetyltrimethyl ammonium bromide) method.

2.7 RNA Extraction

1. RNeasy Plant Mini Kit (Qiagen, Valencia, CA, USA).
2. RNase-Free DNase set (Qiagen, Valencia, CA, USA).

2.8 Oligonucliotides

1. Primer PR1 (forward): 5′-AACTAAACACGAGGGTTGC-3′.
2. Primer PR2 (reverse): 5′-AGTATTAGTTAGTGATCCCGAC-3′.
3. Primer PR3 (forward): 5′-TTAT TTGCCGACTACCTTGGT GAT-3′.
4. Primer PR4 (reverse): 5′-ATGAGGGAAGCGGTGATCGCC-3′.

3 Methods

3.1 Plastid Transformation Vector pPRV111A

The pPRV111A plastid transformation vector is previously described in detail [10]. The plasmid pPRV111A vector contains the *aadA* gene as a selectable marker and the MCS (multiple cloning sites) flanked by *trnV* and *rps7/12* tobacco plastid DNA homologous sequences to target the insertion of the linked transgenes into the plastid genome by homologous recombination (Fig. 1). The *aadA* gene is expressed from 5′ (*PpsbA*) and 3′ (*TpsbA*) regulatory regions of the *psbA* plastid gene.

3.2 Preparation of Eggplant Stem Explants for Bombardment

1. Grow eggplant (*Solanum melongena* L., cv. Pusa Uttam) aseptically on agar solidified MS medium without any phytohormone.
2. Place green stem segments (5–7 mm in length) at the center of agar solidified MS medium supplemented with 1 mg/L zeatin riboside in a petri dish (Fig. 2a).

3.3 Preparation of Tungsten Suspension

1. Take 50.0 mg tungsten powder and add 1 mL absolute ethanol.
2. Vortex and centrifuge for 10 s and replace with 1 mL ethanol and centrifuge.
3. Remove ethanol and add 1 mL sterile distilled water.
4. Resuspend tungsten and centrifuge. Repeat three times; store at −20 °C (*see* **Note 3**).

3.4 Coating Tungsten Particles with Plasmid DNA

1. Add in order: 50 μL of 50 mg/mL tungsten particle suspension, 10 μL DNA (1.0 μg/μL), 50 μL 2.5 M $CaCl_2$, and 20 μL 0.1 M spermidine-free base (*see* **Note 4**).
2. Add 200 μl ethanol and vortex for 10 min and centrifuge for 10 s only.
3. Remove supernatant. Add 200 μL ethanol and vortex well and centrifuge for 10 s. Repeat this step three times.

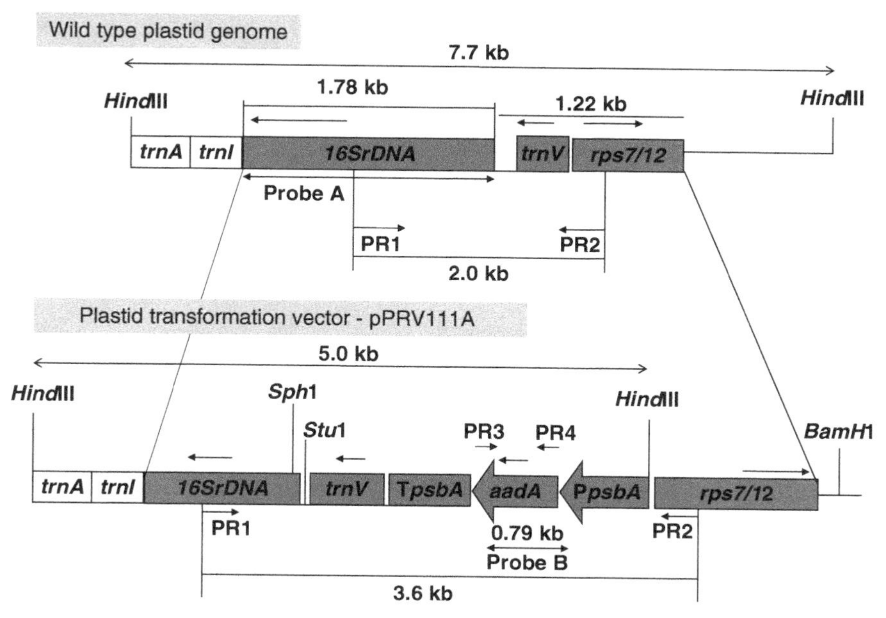

Fig. 1 Plastid transformation vector pPRV111A and the segment of the wild-type plastid genome. The *trnA* and *trnI* plastid genes indicated as two non-filled boxes in the wild-type plastid genome are not included in the pPRV111A vector. The annealing positions of primers PR1/PR2 and PR3/PR4 are shown. The PCR product of 2.0 and 3.6 kb amplified using primers PR1 and PR2, respectively, in wild-type and transplastomic plants is shown. Primer pair PR3/PR4 was used to amplify the *aadA* gene. Probe A and probe B corresponding to *16S rDNA* and *aadA* gene, respectively, used for DNA blot analysis are shown. The predicted hybridizing fragment (7.7 kb in wild-type plant and 5.0 kb in transplastomic plants) is indicated. *Arrows* above the gene names indicate the direction of transcription. The *16S rDNA*, *trnV*, and *rps7/12* are plastid genes. The P*psbA* and T*psbA* are 5′ (promoter) and 3′ (terminator) regulatory regions of the plastid *psbA* gene, respectively. Modified from Supplementary fig. S2 in ref. 9. Reproduced with permission from Springer Science + Business Media

4. Resuspend the pellet in 30 μL ethanol.
5. Place 5 μL DNA-coated tungsten particles onto each macro-carrier for bombardment.

3.5 Particle Bombardment of Stem Segments

1. Sterilize the rupture disc, macro-carriers, and stopping screen by placing these for 5 min in absolute ethanol (*see* **Note 5**).
2. Dry them by standing them up along the side of a sterile petri dish, and allow them to air-dry in a laminar flow chamber.
3. Sterilize the gene gun by wiping with 70 % ethanol (*see* **Note 6**).
4. Turn the gun on. Turn the pump on. Turn the gas cylinder on.
5. Place the macro-carrier holder in a petri plate, and into each one place a macro-carrier and press down the discs with the red plastic “caplug” so that they fit into the holders.
6. Pipette 5 μL of DNA-coated tungsten on to the surface of each disc.

Fig. 2 Generation of transplastomic plants. (**a**) Stem segments prior to bombardment. (**b**) Regeneration of spectinomycin-resistant shoot from stem explants on MS medium supplemented with 1 mg/L zeatin riboside and 300 mg/L spectinomycin after 45 days of culture. (**c**) Elongation of transplastomic shoots on MS medium containing 1 mg/L zeatin riboside, 300 mg/L spectinomycin, and 300 mg/L streptomycin. (**d**) Fertile transplastomic plant established in a pot. Modified from Supplementary fig. S1 in ref. 9. Reproduced with permission from Springer Science + Business Media

7. Place a rupture disc into the rupture disc holder using forceps. Screw tightly the holder into the vacuum chamber.
8. Place a stopping screen onto the diaphragm in the macro-carrier holder assembly. On top of this, place upside down a metal ring containing macro-carrier with DNA on it. Place the stage in the second slot from the top in the vacuum chamber (*see* **Note 7**).
9. Take a petri plate with centrally placed stem explants, and place it on the sample platform (lid off). The sample platform should be positioned in the chamber in the fourth slot from the top.
10. Press the vacuum button to begin pumping air out the shooting chamber. Allow the vacuum to reach at least 28 in. of Hg. Now press shooting button continuously.
11. After bombardment, release the shooting button and remove the vacuum by setting the vacuum button to vent.
12. Take the petri plate out of the chamber.

3.6 Selection of Spectinomycin-Resistant Lines

1. Incubate the bombarded stem segments for 7 days on MS medium containing 1 mg/L zeatin riboside, and then transfer onto selection medium (MS medium supplemented with 1 mg/L zeatin riboside and 300 mg/L spectinomycin) for induction and selection of transformed shoots. Spectinomycin-resistant shoot buds appear from cut end of explants after 6–7 weeks (Fig. 2b). Several rounds of selection are required to get spectinomycin-resistant shoots (Fig. 2c) (*see* **Note 8**).
2. Subculture the regenerated shoots at intervals of 3 weeks over a period of 9 weeks, and finally transfer to MS medium containing 0.5 mg/L zeatin riboside and 300 mg/L each of spectinomycin and streptomycin over a period of 6 weeks in order to get homoplasmy. Homoplasmic lines are obtained when spectinomycin-resistant shoots are cultured on a double selection medium for about 6 weeks (*see* **Note 9**).
3. For rooting, transfer the transformed shoots to half-strength MS medium without any phytohormones. Rooting occurs after 2–3 weeks of culture.
4. Finally, transfer the plants to soil and grow them to maturity and harvest seeds (Fig. 2d) (*see* **Note 10**).

3.7 Identification of the Transplastomic Lines

3.7.1 PCR

1. Perform polymerase chain reaction and DNA blot analysis to verify the integration of the *aadA* gene into the chloroplast genome.
2. Extract total cellular DNA from leaves of transplastomic and wild-type eggplant using CTAB (cetyltrimethyl ammonium bromide) method [12].
3. Use a pair of primers [primer PR1 (forward): 5′-AACTAAACACGAGGGTTGC-3′ and primer PR2 (reverse) 5′-AGTATTAGTTAGTGATCCCGAC-3′] to detect part of the 16S rDNA/rps7/12 sequence along with the *aadA* gene. PCR amplicon of 3.6 and 2.0 kb fragment is expected from transplastomic and wild-type plants, respectively (Fig. 3a).
4. Use one pair of gene-specific primers [primer PR3 (forward) 5′-TTATTTGCCGACTACCTTGGTGAT-3′ and primer PR4 (reverse) 5′-ATGAGGGAAGCGGTGATCGCC-3′] to verify the integration of the *aadA* gene. PCR analysis will yield a 0.79 kb fragment amplification of the *aadA* gene (Fig. 3b).
5. Use 35 PCR cycles for amplification (with denaturation at 94 °C for 1 min, an annealing at 58 °C for 1 min, and an elongation at 72 °C of 2 min and further final extension at 72 °C for 10 min), after an initial denaturation step at 94 °C for 2 min.
6. Run the PCR products on a 1 % agarose gel.

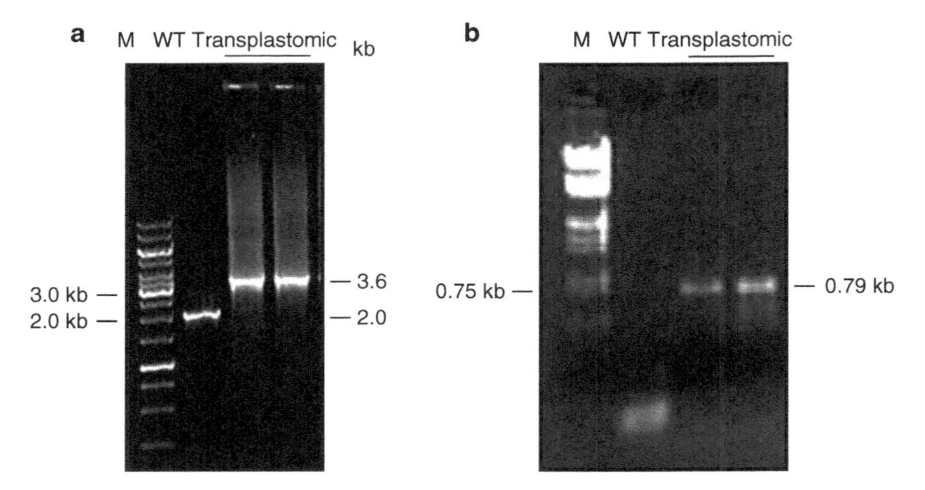

Fig. 3 PCR analysis of transplastomic plants. (**a**) PCR analysis of transplastomic plants using primer pair PR1-PR2 (as shown in Fig. 1). PCR amplification products of 2.0 and 3.6 kb were obtained, respectively, in wild-type and transplastomic plants. *Lane M*: DNA molecular weight marker (λ DNA digested with *EcoR*1 and *Hind*III). *Lane WT*: wild-type plant. *Lane transplastomic*: transplastomic lines. (**b**) PCR analysis of transplastomic plants using *aadA* gene-specific PR3 and PR4 primers (as shown in Fig. 1). A fragment of 0.79 kb corresponding to the coding region of the *aadA* gene was amplified. No PCR amplification was detected in wild-type plant. *Lane M*: DNA molecular weight marker (λ DNA digested with *EcoR*1 and *Hind*III). *Lane WT*: wild-type plant. *Lane transplastomic*: transplastomic lines. Modified from Supplementary fig. S3 in ref. 9. Reproduced with permission from Springer Science + Business Media

3.7.2 DNA Blot Analysis

1. Perform DNA blot hybridization using 10 μg of total cellular DNA digested with *Hind*III (Promega).
2. Separate the *Hind*III-digested DNA fragments by electrophoresis on a 0.8 % agarose gel.
3. Transfer the agarose gel-fractionated DNA to a nylon membrane (Hybond N+, Amersham) by capillary transfer method for 16 h.
4. Fix the DNA on the membrane by UV cross-linking.
5. Use a 1.78 kb PCR product containing 16S rDNA and 0.79 kb PCR product of the *aadA* gene as a probe for DNA blot hybridization.
6. Gel-purify the PCR product using a gel extraction kit (Qiagen, Valencia, CA, USA).
7. Label the probe with [a-^{32}P]-dCTP using the Random Primer Labeling Kit (Stratagene), and hybridize for 16 h at 65 °C with rotation in a hybridization oven.
8. Wash the membrane with 2× SSC + 0.1 % SDS, 1× SSC + 0.1 % SDS, and 0.1× SSC + 0.1 % SDS at 60 °C for 20 min.
9. Obtain an autoradiogram after overnight exposure at −80 °C. DNA blot hybridization analysis shown in Fig. 4a revealed

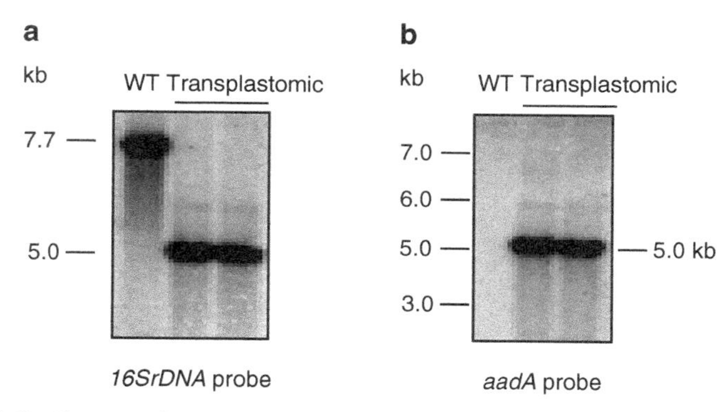

Fig. 4 Southern analysis of two transplastomic plants. Total DNA (10 μg) was digested with *Hind*III. (**a**) Homoplasmy was examined using *16S rDNA* (probe A as shown in Fig. 1) as probe. Transformed plants had one 5.0 kb fragment as expected for homoplasmic transformed plants, whereas the wild-type plant had 7.7 kb fragment. (**b**) The *aadA* gene-specific probe (probe B as shown in Fig. 1) resulted in 5.0 kb fragment in transformed plant, whereas no hybridization signal was detected in wild-type plant. *Lane WT*: wild-type plant. *Lane transplastomic*: transplastomic lines

5.0 and 7.7 kb fragments in transplastomic and wild-type plants, respectively, when a DNA blot probed with the probe A targeting sequence. Only a 5.0 kb fragment in both transplastomic lines and no hybridization signal in wild-type non-transformed plants were detected when a DNA blot is probed with probe B (Fig. 4a, b).

3.8 Analysis of Expression Level of Transgene: RNA Isolation and RT-PCR

1. Isolate total RNA from young leaves of transplastomic plants using RNeasy Plant Mini Kit (Qiagen, Valencia, CA, USA) by combining on-column DNase digestion (RNase-Free DNase set, Qiagen Valencia, CA, USA) to ensure DNA-free RNA preparations.
2. Check DNase-treated RNA samples for genomic DNA contamination by using the minus reverse transcriptase (-RT) controls in parallel to RT-PCR reactions.
3. Perform RT-PCR using a one-step RT-PCR kit (QIAGEN One-Step RT-PCR Kit). Use primers PR3 and PR4 for RT-PCR to determine the expression level of the *aadA* gene. The RT-PCR will yield an amplification of a 0.79 kb fragment in both transplastomic lines, while no amplification will be detected in the wild-type plant (Fig. 5).
4. Run the amplified PCR products on a 1 % agarose gel electrophoresis.

3.9 Genetic Stability of Plants Containing the aadA Gene

1. Germinate the T1 and F1 seed progenies of the transplastomic plants on ½ MS medium containing 300 mg/L spectinomycin in order to check the transmission of the spectinomycin

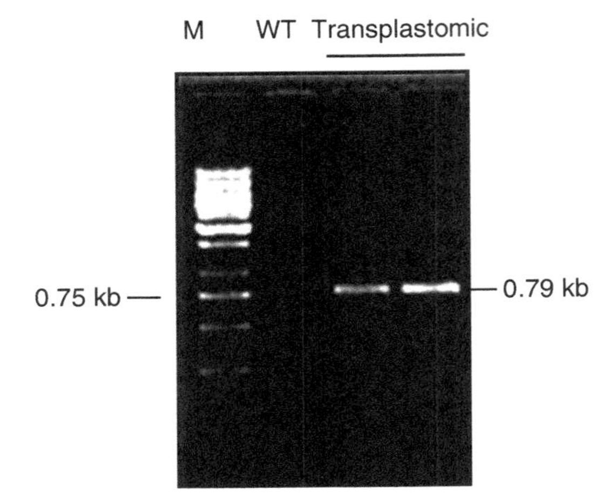

Fig. 5 RT-PCR analysis of transplastomic plants, using *aadA* gene-specific primers. A 0.79 kb fragment amplified in transformed plants, whereas no amplification was obtained in wild-type plant. *Lane M*: 1 kb DNA ladder. *Lane WT*: wild-type plant. *Lane transplastomic*: transplastomic lines

resistance trait to the seed progeny and to check the stability of the *aadA* gene in the plastid-transformed plants. The seedlings of transplastomic lines should be uniformly spectinomycin resistant (Fig. 6a).

2. Generate T1 progenies by selfing, whereas generate F1 progenies by crossing a resistant female parent with a wild-type non-transformed male parent.
3. Germinate wild-type seeds also on ½ MS medium containing 300 mg/L spectinomycin to act as controls (Fig. 6b).

4 Notes

1. Plasmid DNA should be free of proteins; otherwise it will form clumps with tungsten particles. The plasmid DNA can be purified by repeated phenol-chloroform extraction or proteinase treatment followed by ethanol precipitation.
2. Spermidine solution is very hygroscopic and oxidizable; therefore, aliquot the spermidine solution and store at −20 °C.
3. The stock of tungsten particles should be stored for short periods at −20 °C in order to avoid clumps during bombardment, which greatly reduces transformation efficiency.
4. Plasmid DNA (1 μg/μL), 50 μL of 2.5 M $CaCl_2.2H_2O$, and 20 μL of 0.1 M spermidine should be added sequentially to 50 μL of 50 mg/mL tungsten particle suspension. Adding the abovementioned components sequentially enhances efficient coating of plasmid DNA on tungsten particles.

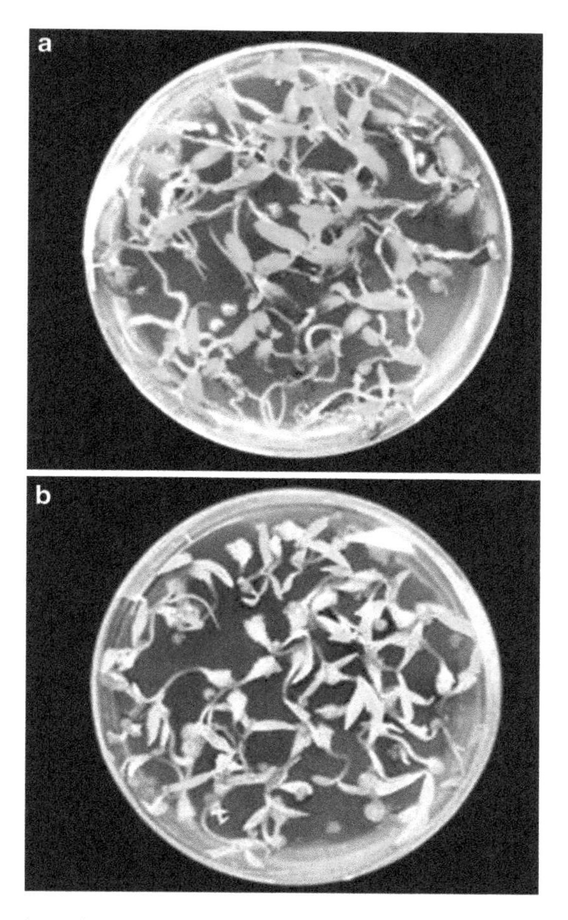

Fig. 6 Screening of seeds for resistance to spectinomycin. (**a**) Germination of transplastomic T1 seeds on ½ MS medium containing 300 mg/L spectinomycin. T1 seed progeny was raised by selfing. (**b**) Germination of wild-type seeds on ½ MS medium containing 300 mg/L spectinomycin. The bleaching of seedlings is clearly visible. Figure 2 in ref. 9. Reproduced with permission from Springer Science + Business Media

5. Macro-carrier, stopping screen, and rupture disc should be soaked in absolute ethanol for 5 min and then air dried.
6. The biolistic gene gun must be used in a laminar air flow cabinet and should be wiped with 70 % ethanol prior to use.
7. The stopping screen should be kept in place and the vacuum to be adjusted to at least 28 in. of Hg.
8. Initial selection should be performed with spectinomycin only because streptomycin selection delays shoot formation.
9. Spectinomycin-resistant shoots should be transferred on double selection medium (MS medium containing 1 mg/L zeatin and 300 mg/L each of spectinomycin and streptomycin) in order to confirm homoplasmy. Double selection also

eliminates spectinomycin-resistant shoots obtained as a result of spontaneous mutations in the 16S rDNA.

10. When plantlets are transferred from magenta boxes to pots, they should be maintained at high humidity for at least 1 week to allow hardening.

Acknowledgment

KCB is thankful to Pal Maliga and Zora Svab for the initial training on plastid transformation under the auspices of Rockefeller Career Biotechnology Fellowship.

References

1. Kalloo G (1993) Eggplant (*Solanum melongena*). In: Kalloo G (ed) Genetic improvement of vegetable crops. Pergamon, Oxford, pp 587–604
2. Guri A, Sink KC (1988) Agrobacterium transformation of eggplant. J Plant Physiol 133: 52–55
3. Filippone E, Lurquin PF (1989) Stable transformation of eggplant (*Solanum melongena* L.) by cocultivation of tissues with *Agrobacterium tumefaciens* carrying a binary plasmid vector. Plant Cell Rep 8:370–373
4. Rotino GL, Gleddie S (1990) Transformation of eggplant (*Solanum melongena* L.) using a binary *Agrobacterium tumefaciens* vector. Plant Cell Rep 9:26–29
5. Fari M, Nagy I, Csanyi M, Mityko J, Andrasfalvy A (1995) Agrobacterium mediated genetic transformation and plant regeneration via organogenesis and somatic embryogenesis from cotyledon leaves in eggplant (*Solanum melongena* L. cv. 'Kecskemeti lila'). Plant Cell Rep 15:82–86
6. Kumar PA, Mandaokar A, Sreenivasu K, Chakrabarti SK, Bisaria S, Sharma SR, Kaur S, Sharma RP (1998) Insect-resistant transgenic brinjal plants. Mol Breed 4:33–37
7. Hanyu H, Murata A, Park EY, Okabe M, Billings S, Jelenkovic G, Pedersen H, Chin CK (1999) Stability of luciferase gene expression in a long term period in transgenic eggplant *Solanum melongena*. Plant Biotechnol 16:403–407
8. Franklin G, Sita GL (2003) Agrobacterium tumefaciens-mediated transformation of eggplant (*Solanum melongena* L.) using root explants. Plant Cell Rep 21:549–554
9. Singh A, Verma SS, Bansal KC (2010) Plastid transformation in eggplant (*Solanum melongena* L.). Transgenic Res 19:113–119
10. Murashige T, Skoog F (1962) A revised medium for rapid growth and bioassay with tobacco tissue cultures. Physiol Plant 15: 473–497
11. Zoubenko OV, Allison LA, Svab Z, Maliga P (1994) Efficient targeting of foreign genes into the tobacco plastid genome. Nucleic Acids Res 22:3819–3824
12. Murray MG, Thompson WF (1980) Rapid isolation of high molecular weight plant DNA. Nucleic Acids Res 8:4321–4325

Chapter 20

Plastid Transformation in Lettuce (*Lactuca sativa* L.) by Polyethylene Glycol Treatment of Protoplasts

Cilia L.C. Lelivelt, Kees M.P. van Dun, C. Bastiaan de Snoo, Matthew S. McCabe, Bridget V. Hogg, and Jacqueline M. Nugent

Abstract

A detailed protocol for PEG-mediated plastid transformation of *Lactuca sativa* cv. Flora, using leaf protoplasts, is described. Successful plastid transformation using protoplasts requires a large number of viable cells, high plating densities, and an efficient regeneration system. Transformation was achieved using a vector that targets genes to the *trnI/trnA* intergenic region of the lettuce plastid genome. The *aadA* gene, encoding an adenylyltransferase enzyme that confers spectinomycin resistance, was used as a selectable marker. With the current method, the expected transformation frequency is 1–2 spectinomycin-resistant cell lines per 10^6 viable protoplasts. Fertile, diploid, homoplasmic, plastid-transformed lines were obtained. Transmission of the plastid-encoded transgene to the T1 generation was demonstrated.

Key words Aminoglycoside 3″-adenylyltransferase, *aadA*, *Lactuca sativa* L., Lettuce, Mesophyll protoplasts, Plastid transformation, Site specific integration, Spectinomycin resistance

1 Introduction

Plastid transformation is a technology that is being applied to an increasing number of crop species. To date, the production of fertile transplastomic plants has predominantly been carried out using biolistic DNA delivery systems. The polyethylene glycol (PEG)-mediated transformation method is considered a good alternative for DNA delivery into the plastid genome. However, it is used less frequently than biolistics mainly due to the lack of efficient single cell and protoplast regeneration protocols for many plant species. To date, the PEG system has been successfully used to produce transplastomic plants in tobacco [1–3], tomato [4], cauliflower [5], and lettuce [6].

The main steps in the procedure include enzymatic treatment of plant tissue to obtain protoplasts, collection of viable protoplasts, exposure of protoplasts to transforming DNA in the presence of PEG, stringent selection of transformed cells, and finally the

Pal Maliga (ed.), *Chloroplast Biotechnology: Methods and Protocols*, Methods in Molecular Biology, vol. 1132,
DOI 10.1007/978-1-62703-995-6_20, © Springer Science+Business Media New York 2014

regeneration of transplastomic plants. For successful PEG-mediated plastid transformation, an efficient protoplast regeneration system is a prerequisite. The main disadvantages of the system are that it requires protoplast culture experience and it takes longer to regenerate plants from single cells than from whole explants. Furthermore, a side effect of the PEG treatment is the generation of polyploid plants at a relatively high frequency. However, for those species where protoplast fusion and regeneration is feasible, the advantages of the protoplast transformation system are substantial: it is a relatively inexpensive method (no expensive equipment and materials are required) and large numbers of single cells are easily obtained and can be transformed in a single experiment.

2 Materials

2.1 Vector and Expression Cassette

Vectors used for plastid transformation typically consist of an *E. coli* plasmid backbone containing ptDNA sequences (1–2 kb) flanking cloning sites for a selectable marker gene and the gene(s) of interest. These genes are directed and integrated into the plastid genome by two homologous recombination events. The ptDNA sequences on the vector serve as targeting regions to direct this integration.

A lettuce plastid transformation vector was constructed, containing 2,253 bp of lettuce chloroplast DNA target sequence spanning the *trnI/trnA* intergenic region (GenBank AY943927) cloned into a pCR2.1 backbone (Invitrogen) [6]. The *trnI/trnA* intergenic region in the vector has unique *Pac*I and *Asc*I restriction sites for accepting transgene expression cassettes (Fig. 1).

2.2 Plants and Protoplast Isolation

1. Lettuce seeds (cultivar *Flora*).
2. MS20 medium [7] with 0.7 % microagar (Duchefa), 25 ml in 9 cm Petri dishes (Greiner).
3. MS30 medium [7] with 0.7 % microagar (Duchefa), 50 ml in 250 ml glass jars (Cataloniё B.V. Tilburg, the Netherlands).
4. PG solution. Add to mQ water 5.47 g sorbitol and 735 mg $CaCl_2.2H_2O$, adjust to a final volume of 100 ml, autoclave (*see* **Note 1**), and store at 4 °C.
5. Stock solutions required for media preparation.
   (a) 40× B5 macro. Add to mQ water 25 g KNO_3, 2.5 g $MgSO_4.7H_2O$, 1.5 g $NaH_2PO_4.H_2O$, and 1.34 g $(NH_4)_2SO_4$, adjust to a final volume of 250 ml, and freeze in 25 ml aliquots at −20 °C.
   (b) 1,000× B5 micro. Add to mQ water 25 mg $Na_2MoO_4.2H_2O$, 1 g $MnSO_4.H_2O$, 200 mg $ZnSO_4.7H_2O$, 300 mg H_3BO_3, 75 mg KI, 2.5 mg $CuSO_4.5H_2O$, and 2.5 mg $CoCl_2.6H_2O$,

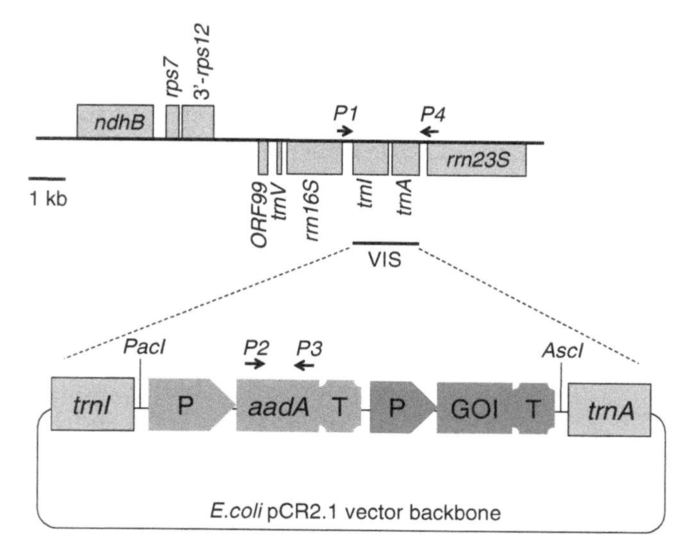

Fig. 1 Vector and expression cassettes used for lettuce plastid transformation. *Top*: gene map of the lettuce plastid genome region targeted for transformation, the *trnI*/*trnA* vector integration site (VIS) is indicated. PCR primer annealing sites flanking the VIS are indicated by P1 and P4. *Bottom*: map of the lettuce plastid transformation vector (pLCV2, 6) containing expression cassettes for *aadA* and a gene of interest (GOI). An expression cassette is a gene engineered with a promoter (P) and terminator (T) sequence suitable for plastid-based gene expression [6]. The lettuce vector has unique *Pac1* and *Asc1* restriction sites in the *trnI* and *trnA* intergenic region that are useful for cloning in transgene expression cassettes. The *aadA* gene-specific primer annealing sites are indicated by P2 and P3

adjust to a final volume of 100 ml, and freeze in 1 ml aliquots at −20 °C.

(c) 100× B5 vitamins. Add to mQ water 10 mg nicotinic acid, 10 mg pyridoxine HCL, 100 mg thiamine HCL, and 1 g meso-inositol, adjust to a final volume of 100 ml, and freeze in 10 ml aliquots at −20 °C.

(d) 1,000× NaFeEDTA. Add to mQ water 3.76 g NaFeEDTA, adjust to 100 ml, and store at 4 °C.

(e) 100× CPW salts. Add to mQ water 1.01 g KNO_3, 2.46 g $MgSO_4.7H_2O$, 272 mg KH_2PO_4, 1.6 mg KI, and 0.25 mg $CuSO_4.5H_2O$ to a final volume of 100 ml. Freeze in 25 ml aliquots at −20 °C.

6. Enzyme medium. Add to mQ water 2.5 ml 40× B5 macro, 100 μl 1,000× B5 micro, 1 ml 10× B5 vitamins, 150 mg $CaCl_2.2H_2O$, 100 μl 1,000× NaFeEDTA, 13.7 g sucrose, 1 g cellulase R10 (Brunschwig, Switzerland), and 250 mg macerozyme R10 (Brunschwig, Switzerland). Bring to a final volume of 100 ml. Adjust to pH 5.6 with 0.5 M KOH. Centrifuge the solution for 10 min at 300×*g* to remove any

particulate matter and filter sterilize (*see* **Note 2**). Freeze in 20 ml aliquots at −20 °C.

7. CPW16S solution. Add to mQ water 1 ml 100× CPW salts, 16 g sucrose, and 148 mg $CaCl_2.2H_2O$, and bring to a final volume of 100 ml. Adjust to pH 5.8 with 0.5 M KOH. Autoclave the solution and store at 4 °C.
8. W5 solution. Add to mQ water 4.5 g NaCl, 9.2 g $CaCl_2.2H_2O$, 185 mg KCl, 495 mg glucose, and 50 mg MES and bring to a final volume of 500 ml, and adjust to pH 5.8 with 0.5 M KOH. Autoclave the solution and store at 4 °C.
9. Transformation buffer. Add to mQ water 7.29 g mannitol, 304 mg $MgCl_2.6H_2O$, and 1 g MES. Bring to a final volume of 100 ml. Adjust to pH 5.8 with 10 M KOH. Autoclave the buffer and store at 4 °C.

2.3 PEG-Mediated Protoplast Transformation

1. Plasmid DNA (transformation vector) 1 μg/μl in autoclaved mQ water.
2. PEG solution. Dissolve 80 g of PEG 6000 (Sigma-Aldrich) in 100 ml of buffer (4.72 g $Ca(NO_3)_2.4H_2O$, 14.57 g mannitol in 100 ml) by gentle heating in a microwave oven, adjust the volume to 200 ml with distilled water, and divide into 2 × 100 ml aliquots. Adjust the pH of one aliquot to pH 10 with 1 M KOH and store both aliquots overnight at 4 °C. The following day, bring both aliquots to room temperature. Readjust the pH of the aliquot previously set to pH 10 to pH 8.2 using the non-adjusted aliquot, filter sterilize, and freeze in 5 ml aliquots at −20 °C.
3. Protoplast wash solution. Add to mQ water 36.45 g mannitol and 1.176 g $CaCl_2.2H_2O$ and adjust to a final volume of 500 ml. Autoclave the solution and store at 4 °C.
4. Protoplast culture medium. Add to mQ water 12.5 ml 40× B5 macro, 500 μl 1,000× B5 micro, 5 ml 100× B5 vitamins, 375 mg $CaCl_2.2H_2O$, 500 μl 1,000× NaFeEDTA, 270 mg Na succinate, 103 g sucrose, 300 μl BAP (Sigma-Aldrich, 1 mg/ml mQ water, filter-sterilized), 100 μl 2,4D (Sigma-Aldrich, 1 mg/ml mQ water, filter-sterilized), and 100 mg MES. Bring to a final volume of 1 l. Adjust the pH to 5.8 with 0.5 M KOH, filter sterilize, and store at 4 °C (do not keep longer than a week).
5. 2 % agarose/protoplast culture medium. Prepare solutions 1 and 2 separately. Solution 1: 100 ml 4 % agarose (Seaplaque, Duchefa) solution in mQ water, autoclaved. Solution 2: 100 ml 2× protoplast culture medium, filter-sterilized. Mix the two solutions together after the agarose solution has cooled to about 45 °C. Store in 10 ml aliquots at −20 °C.
6. Callus growth medium (SH2). Add to mQ water 3.184 g SH salts (Duchefa), 10 ml 100× SH vitamins (Duchefa), and 30 g

sucrose, and adjust the volume to 1 l. Adjust to pH 5.8 with 0.5 M KOH, add 5 g agarose (Sigma-Aldrich I-A, low EEO), and autoclave. Add 100 μl of filter-sterilized NAA (Sigma-Aldrich, 1 mg/ml mQ water) and 100 μl of filter-sterilized BAP (1 mg/ml mQ water).

7. Callus regeneration medium (SHREG). Add to mQ water 3.184 g SH salts, 10 ml 100× SH vitamins, 15 g sucrose, and 15 g maltose, and adjust the volume to 1 l. Adjust to pH 5.8 with 0.5 M KOH, add 5 g agarose (Sigma-Aldrich I-A, low EEO), and autoclave. Add 100 μl of filter-sterilized NAA (1 mg/ml mQ water) and 100 μl of filter-sterilized BAP (1 mg/ml mQ water).
8. Shoot outgrowth medium (SH30). Add to mQ water 3.184 g SH salts, 10 ml 100× SH vitamins, and 30 g sucrose, and bring to a final volume of 1 l. Adjust to pH 5.8 with 0.5 M KOH, add 8 g microagar (Duchefa), and autoclave.
9. Rooting medium (SH30 + IBA). SH30 medium with 1 mg/l of filter-sterilized indolebutyric acid (Sigma-Aldrich, 1 mg/ml mQ water).
10. Spectinomycin dihydrochloride. Add to mQ water 500 mg/ml spectinomycin dihydrochloride (Duchefa). Filter sterilize and freeze in single use 1 ml aliquots at −20 °C.

2.4 PCR Characterization of Transplastomic Calli and Shoots

1. AGOWA mag plant DNA Isolation Kit.
2. dNTP nucleotide mix PCR grade (Roche 25 mM each nucleotide).
3. Taq DNA polymerase (Roche 5 U/μl).
4. 10× PCR Buffer with $MgCl_2$ (Roche).
5. PCR primers specific for regions of the plastid genome flanking the vector integration site (vector dependent), for the *aadA* gene and gene(s) of interest.

2.5 Generating Homoplastomic Plant Lines

1. MS30 medium [7] with 0.7 % microagar supplemented with 500 mg/l spectinomycin (filter-sterilized), 0.1 mg/l 2,4D (1 mg/ml mQ water, filter-sterilized), and 0.3 mg/l BAP (1 mg/ml mQ water, filter-sterilized).
2. MS30 medium [7] with 0.7 % microagar supplemented with 500 mg/l spectinomycin (filter-sterilized), 0.1 mg/l NAA (1 mg/ml mQ water, filter-sterilized), 0.1 mg/l BAP (1 mg/ml mQ water, filter-sterilized), and 200 mg/l carbenicillin (Sigma-Aldrich, 200 mg/ml mQ water, filter-sterilized).
3. Shoot outgrowth medium (SH30) supplemented with 500 mg/l spectinomycin.
4. Rooting medium (SH30 + 1 mg/l IBA).

2.6 Testing Transplastomic Seedlings for Spectinomycin Resistance

1. Seeds from fertile, diploid, transplastomic lettuce plants.
2. MS20 medium [7] with 0.7 % microagar supplemented with 500 mg/l spectinomycin.

3 Methods

To isolate large numbers of viable lettuce protoplasts, it is essential to use young, fast-growing, in vitro propagated plants. The protoplasts must be handled very gently during the isolation, transformation, and plating steps as the quality of the protoplasts greatly influences the transformation rate. The PEG treatment causes disintegration of the plasma membrane, enabling uptake of the transforming DNA into the protoplasts [8]. The mechanism by which the transforming DNA is subsequently taken into the plastids is still not known [9]. If the concentration of PEG is too high, or the protoplasts are exposed to PEG for too long, the protoplasts will die and the transformation efficiency declines rapidly. It is also critical to plate the transformed protoplasts onto culture plates at an optimum density. If the plating density is too low, no protoplast division occurs, and if the density is too high, proper and stringent cell selection is hampered. One of the drawbacks of the technique is that protoplast fusion can occur during the PEG treatment, and this can give rise to cells that are tetraploid or even polyploid. After PEG treatment, diploid cells give rise to fertile plants at a frequency of about 10–20 %; however, tetraploid or polyploid cells do not result in fertile plants. Multiple plants (clones) can be obtained per cell (callus) line.

Spectinomycin-resistant cell lines, or shoots, may be true plastid transformants or spontaneous spectinomycin resistance mutants [mutations in the plastid small ribosomal RNA gene (*rrn*16) can confer spectinomycin resistance] [10, 11]. In addition, there is a small chance that spectinomycin-resistant cell lines or shoots could be the result of *aadA* integration into the nuclear genome. To confirm true plastid transformation events, three separate PCR reactions are generally used to characterize each of the primary cell lines/transformed shoots. The first reaction uses a primer pair specific for the selectable marker gene (*aadA*) or the gene of interest. The second and third reactions assess for targeted integration by using primers specific to plastid DNA flanking the integration site (left or right border primers) in combination with gene-specific primers (either *aadA* or the gene of interest). The primary regenerated lettuce shoots obtained using the PEG method may not always be homoplastomic (100 % plastid-transformed genomes), and additional rounds of regeneration under selection may be required with this method [10]. The ratio of transformed to non-transformed plastid genomes (i.e., level of heteroplasmy) may be assessed by PCR using primers that flank the vector integration site (Fig. 1) or by Southern Blot hybridization [6, 12].

3.1 Plants and Leaf Protoplast Isolation

1. Sterilize the seeds by placing them in 70 % ethanol for 30 s, followed by 0.7 % hypochlorite solution for 20 min. Wash the seeds three times in sterile deionized water, and plate on MS20 medium in 9 cm Petri dishes (Greiner). Culture the seeds at 15 °C for 2 days in the dark and then transfer to 25 °C in the light (3,000 lux, Philips Master TL-D 840, photoperiod 16 h light/8 h dark). When the first true leaves appear, transfer the shoot tips to 250 ml glass jars containing 50 ml MS30 medium and grow under the same conditions (25 °C, 16 h light/8 h dark) for a maximum of 4 weeks.
2. Remove 4–5 young, nearly expanded leaves from 4-week-old in vitro grown plants (0.5–1 g of tissue) and cut into small pieces (approximately 25–50 mm^2 in size) in 20 ml PG solution. Cut the leaves under PG solution to keep them from drying out. Seal the 9 cm plates with parafilm and leave at 4 °C, in the dark, for 2–3 h (1 h minimum) (*see* **Note 3**).
3. Remove the PG solution and discard. Add 20 ml of enzyme medium, reseal the plates, and incubate at 25 °C in the dark for 14 h (overnight).
4. The next day shake the plates (orbital shaker, 60 rpm) for 2 h at 25 °C in the dark, before generating the protoplasts.
5. Add 10 ml of CPW16S solution to the Petri dish and swirl gently to release the protoplasts. Filter the protoplast suspension through sterile nylon mesh (41 μm) and collect the flow through. Divide the flow through between three 15 ml tubes and overlay each sample with 1 ml of W5 solution (*see* **Note 4**). Centrifuge at $70 \times g$ for 8 min.
6. Collect the protoplasts from the CPW16S/W5 interface (*see* **Note 5**). Divide the protoplasts between two 15 ml tubes. Gently add 9 ml of W5 solution to each tube. Centrifuge at $70 \times g$ for 8 min.
7. Resuspend each protoplast pellet in 5 ml of W5 solution. Centrifuge at $70 \times g$ for 5 min.
8. Resuspend the pellet in transformation buffer and adjust the density to 2×10^6 protoplasts per ml (Fig. 2a). Calculate the protoplast density using a hemocytometer. The vitality of the protoplasts can be checked by FDA staining (*see* **Note 6**).
9. Divide the resuspended protoplasts into 600 μl aliquots in 15 ml tubes (Greiner). Typically 4–6 aliquots are generated per protoplast isolation.

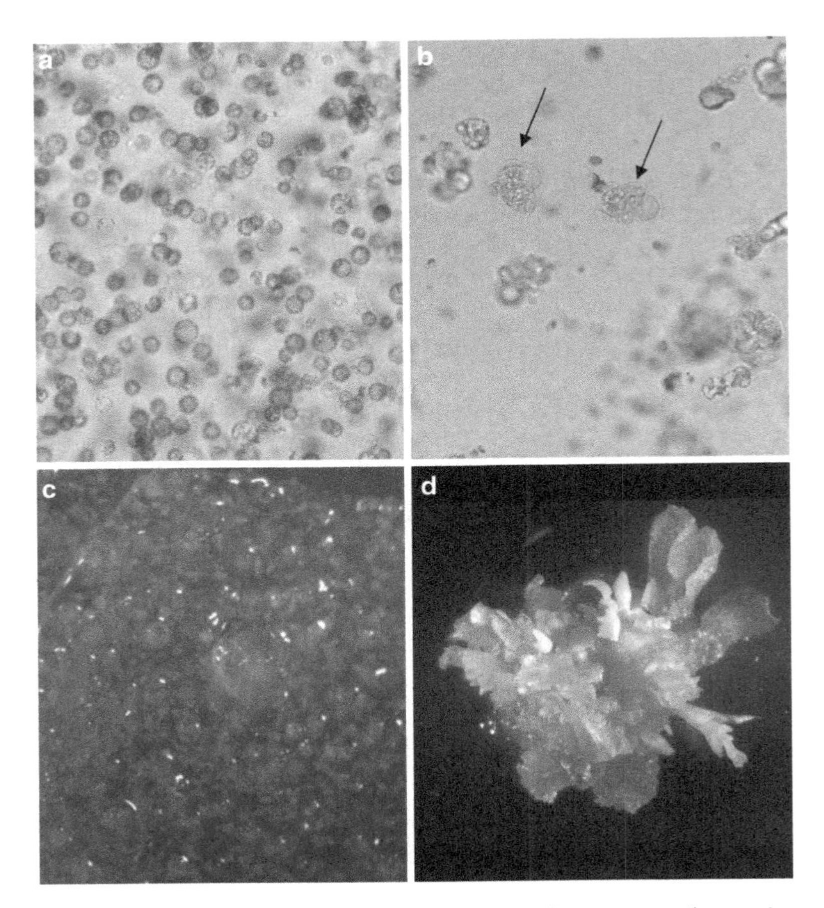

Fig. 2 Representative stages in the lettuce protoplast regeneration system. (**a**) Isolated protoplasts used for PEG-mediated transformation, viewed at 400× magnification (Olympus IX-70, inverted microscope). (**b**) Dividing protoplasts (indicated by *arrows*) viewed at 600× magnification (Olympus IX-70, inverted microscope), 6 days after transformation and plating. (**c**) A green spectinomycin-resistant micro-callus in an agarose bead seen among bleached, spectinomycin-sensitive protoplast calli. (**d**) Multiple lettuce shoots regenerating from a spectinomycin-resistant callus

3.2 PEG Leaf Protoplast Transformation

1. Add 10 μl of transforming DNA (1 μg/μl) and 400 μl of PEG solution (pH 8.2) to each 600 μl protoplast aliquot and mix gently. Incubate at room temperature for 10 min.
2. Stop the PEG treatment by adding 9 ml of wash solution and mix gently. Centrifuge at 70 × *g* for 5 min.
3. Remove the supernatant and discard. Gently resuspend the protoplast pellet to a density of 12×10^4 protoplasts/ml with protoplast culture medium (*see* **Note 7**). Mix the protoplasts with an equal volume of 2 % agarose/protoplast culture medium (*see* **Note 8**). Pipette 1.5 ml of the agarose/protoplast

mixture into 3.5 cm Petri dishes and allow the medium to solidify. Cut the solid agarose/protoplast mixture into quarters and transfer the quarters to a 6 cm Petri dish (two per plate) and overlay with 4 ml of protoplast culture medium (without spectinomycin).

4. Seal the plates with parafilm and culture at 25 °C, in the dark, for 6 days.
5. After 6 days, add spectinomycin to the solution in the plates to a final concentration of 500 mg/l and culture at 25 °C, in the dark, for one more day. Protoplast division can be observed at this time (Fig. 2b).
6. Remove 2 ml of the protoplast culture medium from the plates and replace with 2 ml of fresh solution containing 500 mg/l spectinomycin. Seal the plates and culture for 1 week at 25 °C (3,000 lux, Philips Master TL-D 840, 16 h light/8 h dark).
7. Remove the agarose quarters, and cut into small pieces of about 0.7×0.7 cm^2. Place the pieces on SH2 medium (500 mg/l spectinomycin), at a density of two quarters/plate, and culture at 25 °C (3,000 lux, Philips Master TL-D 840, 16 h light/8 h dark). Transfer the agarose pieces onto fresh SH2 medium with spectinomycin every 2 weeks until green calli appear (6–8 weeks, Fig. 2c, *see* **Note 9**).
8. When the microcalli reach about 0.5 mm in diameter, transfer them onto SHREG medium 500 mg/l spectinomycin—use deep Petri dishes for this to allow shoot regeneration to occur (Greiner) and culture at 25 °C (3,000 lux, Philips Master TL-D 840, 16 h light/8 h dark).
9. Spectinomycin-resistant shoots regenerate after approximately 6 weeks (Fig. 2d).
10. Transfer shoots onto SH30 media containing 500 mg/l spectinomycin, without hormones, in glass jars. Check the ploidy level of the plants by flow cytometry (Partec) and select the diploid plants. Tetraploid or polyploid plants will be sterile and will not set seed.
11. Once the shoots are established, transfer them to SH30 + IBA (1 mg/l) in glass jars (50 ml/jar) without spectinomycin, for rooting (*see* **Note 10**).

3.3 PCR Characterization of Putative Transplastomic Calli or Plants

1. Isolate total DNA from 50 to 100 mg of callus or leaf tissue using a GenElute plant genomic DNA mini-prep kit (Sigma-Aldrich) according to the manufacturers' instructions (*see* **Note 11**).
2. *Set* up three PCR reactions per DNA sample:
   (a) Use a primer pair specific for the *aadA* gene (or the gene of interest), e.g.:

AadA-F1 (P2, Fig. 1): 5′-TATgACgggCTgATACTgggC-3′

AadA-R1 (P3, Fig. 1): 5′-AAgTCACCATTgTTgTgCACg-3′

(b) Use a primer specific for the left border region (flanking the vector integration site) and a gene-specific primer for *aadA* (or the gene of interest), e.g.:
LCV2-FA (P1, Fig. 1): 5′-ACTggAAggTgCggCTggAT-3′
AadA-F1 (P2, Fig. 1): 5′-TATgACgggCTgATACTgggC-3′

(c) Use a primer specific for the right border region (flanking the vector integration site) and a gene-specific primer for *aadA* (or the gene of interest), e.g.:
AadA-R1 (P3, Fig. 1): 5′AAgTCACCATTgTTgTgCACg-3′
LCV2-RB (P4, Fig. 1): 5′-CTCgCCCTTAATTTTAAggC-3′

3. Combine the following reaction components in a 0.2 ml PCR tube (total volume 25 μl) for each primer set, include appropriate negative controls:

   5 μl DNA (50 ng).

   1 μl nucleotide mix (5 mM each nucleotide).

   2.5 μl 10× PCR Buffer with $MgCl_2$ (Roche).

   1 μl each of forward and reverse primer (10 pmol/μl).

   0.2 μl Taq DNA polymerase (Roche 5 U/μl).

   14.3 μl dH_2O.

   Place the tubes in a thermocycler for amplification (*see* **Note 12**). Use 35 cycles for amplification of the PCR products.

4. Run the PCR products on a 1 % agarose gel alongside an appropriate DNA ladder.
5. Stain the gel with ethidium bromide and observe and photograph the gel under UV light (Fig. 3).

3.4 Generating Homoplastomic Plant Lines

1. Use in vitro grown transplastomic lettuce plantlets.
2. Cut fully expanded leaves into pieces of about 1 cm^2 and culture, abaxial side down, on induction medium containing 2,4D, BAP, and spectinomycin (use 9 cm Petri dishes, Greiner) for 7 days at 25 °C at low light intensity (400 lux, Philips Master TL-D 840, 16 h light/8 h dark).
3. After 1 week, transfer the leaf pieces to MS30 medium containing BAP, NAA, carbenicillin, and spectinomycin in 9 cm Petri dishes. Culture the leaf pieces at 25 °C in the light (3,000 lux, Philips Master TL-D 840, 16 h light/8 h dark). Subculture every 2 weeks on the same medium until shoots appear.
4. Transfer the shoots to SH30 medium with spectinomycin.
5. Once the shoots are established, root the shoots on SH30 medium with 1 mg/l IBA, without spectinomycin.

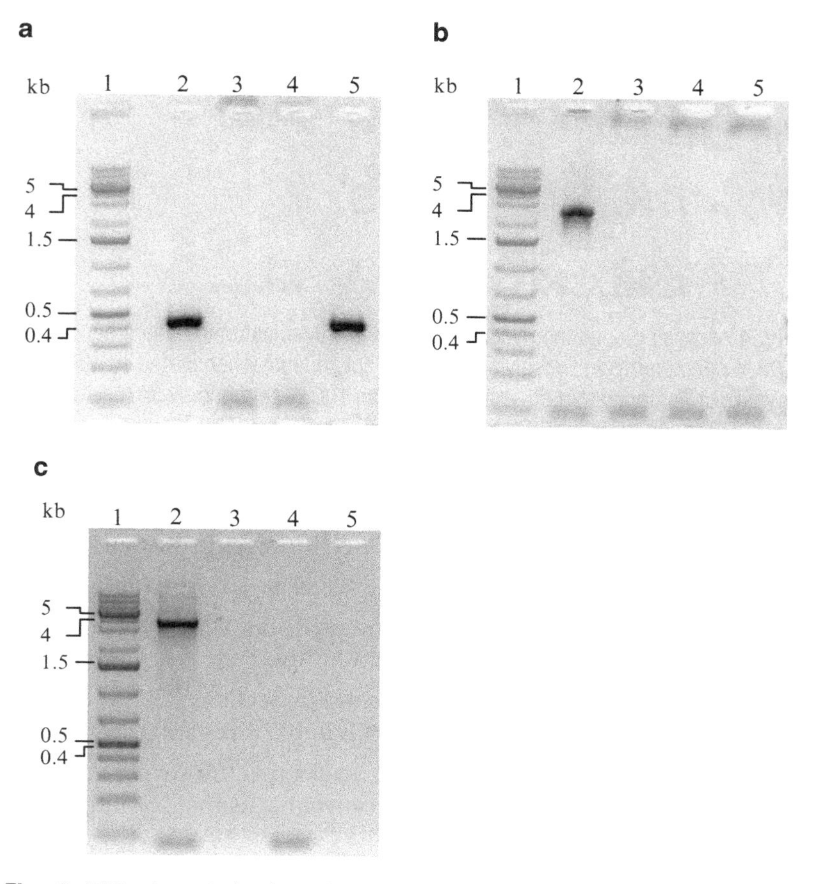

Fig. 3 PCR characterization of transplastomic regenerated lettuce shoots. (**a**) PCR amplification with *aadA* gene-specific primers. *Lane 1*, 1 kb ladder (Fermentas); *lane 2*, transplastomic lettuce line; *lane 3*, mQ water (negative control); *lane 4*, WT lettuce; *lane 5*, vector DNA (positive control). A PCR product is observed in *lane 2* indicating the presence of the *aadA* gene in the DNA of the transplastomic lettuce line. A PCR product is also generated by the positive control reaction in *lane 5*. No PCR product is generated using WT lettuce DNA as a template (*lane 4*) or in the negative control reaction (*lane 3*). (**b**) PCR amplification of the left border region following vector integration using a left border flanking primer and an *aadA* gene-specific primer (lanes indicated as above). A PCR product is observed in *lane 2* (transplastomic lettuce line), indicating targeted integration of the transgene cassette in the plastome. (**c**) PCR amplification of the right border region following vector integration using a right border flanking primer and an *aadA* gene-specific primer (lanes indicated as above). A PCR product is observed in *lane 2* (transplastomic lettuce line), indicating targeted integration of the transgene cassette in the plastome

6. Transfer rooted plants to soil (standard peat mixture, supplemented with Osmocote® Outdoor & Indoor Smart-Release® Plant Food, 16 h light/ 8 h dark, 20,000 lux greenhouse), grow to flowering, allow flowers to self-pollinate, and collect seed.

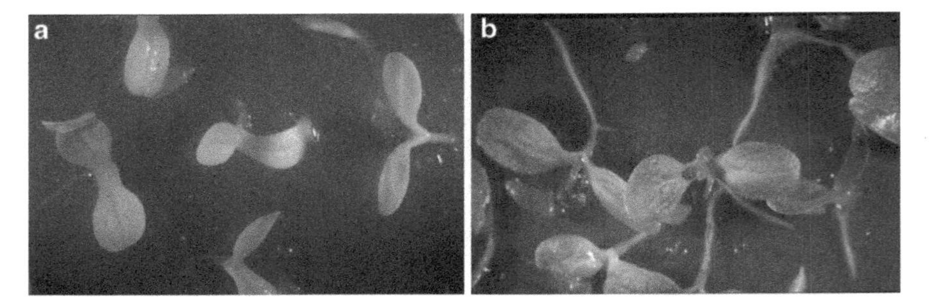

Fig. 4 Assessing spectinomycin resistance in transplastomic lettuce seedlings. (**a**) Wild-type seedlings show 100 % spectinomycin sensitivity (bleached cotyledons) when plated on MS20 medium with 500 mg/l spectinomycin. (**b**) Homoplastomic T1 seedlings show 100 % spectinomycin resistance (green cotyledons) when plated on MS20 medium with 500 mg/l spectinomycin

3.5 Testing Seedlings for Spectinomycin Resistance

1. Seed derived from homoplastomic lines should give rise to 100 % spectinomycin-resistant (green) seedlings.
2. Sterilize seeds as described in Subheading 2.2.
3. Plate the seeds on MS 20 medium with and without spectinomycin (500 mg/l).
4. Germinate the seeds at 25 °C (3,000 lux, Philips Master TL-D 840, 16 h light/8 h dark).
5. After 2 weeks spectinomycin-resistant seedlings are green and spectinomycin-sensitive seedlings are bleached (Fig. 4).

4 Notes

1. In general, all solutions and media are autoclaved at 121 °C for 20 min. The pH of solutions/media is adjusted before autoclaving. Autoclaved media with agar is stored at 15 °C and sterile liquid media is stored at 4 °C (do not keep any media longer than 4 weeks). Stocks stored at −20 °C may be kept for up to 1 year.
2. For filter sterilization use a 0.2 μm pore size filter (Nalgene).
3. Leaf age can significantly affect protoplast yield, so preferably use the youngest 2–3 leaves of plants. The protoplast yield should be about $10–15 \times 10^6$/g of fresh leaf tissue. 0.5–1 g of leaf material will give sufficient protoplasts for one transformation experiment. If the number of isolated protoplasts is low, repeat the isolation using younger in vitro propagated leaf material.
4. The enzyme treatment removes the cell walls. After enzyme treatment handle the protoplasts gently or they will burst.
5. If there is a large pellet at the bottom of the tube, rather than a thick band of protoplasts at the CPW16S/W5 interface, the protoplasts have burst and the isolation needs to be repeated.

6. Take 50 μl of protoplast suspension in wash solution and stain with a 0.002 % (w/v) FDA solution in acetone. When viewed with a microscope under UV light (excitation 450–490 using Filter set U-MNIB-2, Olympus IX-70) viable cells will fluoresce yellow/green. Normally >50 % of the protoplasts will show fluorescence with FDA.
7. Between 75 and 85 % of the protoplasts are lost, or burst, following the PEG treatment. About 8–12 Petri dishes can be plated per protoplast sample. Accurate determination of protoplast number is essential. If the protoplast plating density is too low, the spectinomycin-resistant calli will not grow.
8. Heat the 2 % agarose solution to 45–50 °C, add to the resuspended protoplasts, and swirl gently.
9. If no spectinomycin-resistant calli are obtained, the transforming DNA is not delivered to the plastids. Check the number and viability of the protoplasts, check the plasmid DNA, and check the concentration of spectinomycin used for selection.
10. Getting the regenerated shoots to root on spectinomycin is difficult, usually this step is done without selection on spectinomycin.
11. Callus material, generated using the PEG method, can be assessed for plastid transformation before shoots regenerate from the callus. Any plant DNA isolation kit/method will suffice as long as the DNA is clean enough to do PCR.
12. Annealing temperature is dependent on primer Tm. Extension time is dependent on the amplicon length (usually ≈500 bp/s). If multiple PCR products are obtained, raise the annealing temperature and/or adjust the salt concentration of the reaction. If a smear of PCR products is obtained, dilute the template DNA and repeat the PCR reaction.

Acknowledgements

This work was initially funded under the EU Fifth Framework initiative, grant number QLK-CT-1999-00692.

References

1. Golds T, Maliga P, Koop H-U (1993) Stable plastid transformation in PEG-treated protoplasts of *Nicotiana tabacum*. Bio/Technology 11:95–97
2. O'Neill C, Horváth GV, Horváth E, Dix PJ, Medgyesy P (1993) Chloroplast transformation in plants: polyethylene glycol (PEG) treatment of protoplasts is an alternative to biolistic delivery systems. Plant J 3:729–738
3. Koop H-U, Steinmuller K, Wagner H, Rossler C, Eibl C, Sacher L (1996) Integration of foreign sequences into the tobacco plastome via polyethylene glycol-mediated protoplast transformation. Planta 199:193–201

4. Nugent GD, ten Have M, van der Gulik A, Dix PJ, Uijtewaal BA, Mordhorst AP (2005) Plastid transformants of tomato selected using mutations. Plant Cell Rep 24:341–349
5. Nugent GD, Coyne S, Nguyen TT, Kavanagh TA, Dix PJ (2006) Nuclear and plastid transformation of *Brassica oleracea* var. *botrytis* (cauliflower) using PEG-mediated uptake of DNA into protoplasts. Plant Sci 170:135–142
6. Lelivelt CLC, McCabe MS, Newell CA, de Snoo CB, van Dun KMP, Birch-Machin I, Gray JC, Mills KHG, Nugent JM (2005) Stable plastid transformation in lettuce (*Lactuca sativa* L.). Plant Mol Biol 5:763–774
7. Murashige T, Skoog F (1962) A revised medium for rapid growth and bioassays with tobacco cell cultures. Physiol Plant 15:473–497
8. Maliga P (2004) Plastid transformation in higher plants. Annu Rev Plant Biol 55: 289–313
9. Kofer W, Eibl C, Steinmuller K, Koop H-U (1998) PEG-mediated plastid transformation in higher plants. In Vitro Cell Dev Biol Plant 34:303–309
10. Svab Z, Hajdukiewicz P, Maliga P (1990) Stable transformation of plastids in higher plants. Proc Natl Acad Sci U S A 87:8526–8530
11. Svab Z, Maliga P (1993) High-frequency plastid transformation in tobacco by selection for a chimeric aadA gene. Proc Natl Acad Sci U S A 90:913–917
12. Southern EM (1975) Detection of specific sequences among DNA fragments separated by gel electrophoresis. J Mol Biol 98:503–517

Chapter 21

Plastid Transformation in Lettuce (*Lactuca sativa* L.) by Biolistic DNA Delivery

Tracey A. Ruhlman

Abstract

The interest in producing pharmaceutical proteins in a nontoxic plant host has led to the development of an approach to express such proteins in transplastomic lettuce (*Lactuca sativa* L.). A number of therapeutic proteins and vaccine antigen candidates have been stably integrated into the lettuce plastid genome using biolistic DNA delivery. High levels of accumulation and retention of biological activity suggest that lettuce may provide an ideal platform for the production of biopharmaceuticals.

Key words *aadA* gene, Antibiotic selection, Biolistic transformation, Intergenic spacer, *Lactuca sativa*, Lettuce, Particle bombardment, Plastome, Plant growth regulators, Spectinomycin resistance, Tissue culture

1 Introduction

With the potential to accumulate large quantities of foreign protein within plant cells, plastid transformation technology provides a platform for recombinant expression of human therapeutic proteins and vaccine antigens [1, 2]. While the early exploration of this potential was executed in the model plant tobacco (*Nicotiana tabacum*), ultimately clinical relevance of the technology favors the expression of pharmaceutical proteins in a nontoxic system. Furthermore, if oral delivery of intact plant tissues/cell is to be implemented, an edible species that can be consumed without cooking is desired. Recent work has demonstrated the feasibility of lettuce (*Lactuca sativa*) as a suitable plant host for expression of foreign proteins with immunogenic and therapeutic properties.

Lelivelt et al. reported stable integration of foreign genes in lettuce plastids [3] some 15 years after the first land plant plastid transformation was accomplished in tobacco [4]. Targeting the intergenic spacer (IGS) between *trnI* and *trnA* of the plastid ribosomal operon, a lettuce-specific transformation vector was delivered to protoplast cultures by polyethylene glycol incubation as

Pal Maliga (ed.), *Chloroplast Biotechnology: Methods and Protocols*, Methods in Molecular Biology, vol. 1132,
DOI 10.1007/978-1-62703-995-6_21,

described in Chapter 20. Shortly thereafter, biolistic DNA delivery was reported to efficiently produce stable lettuce (cv. Cisco) transformants [5]. Regeneration of transplastomic lines from bombarded leaf pieces yielded plants that accumulated green fluorescent protein (GFP) to up to 36 % of total soluble protein. The plasmid vectors used in the experiments were designed to target the *rbcL-accD* intergenic region using either lettuce (pRL1000; *aadA* only) or tobacco (pRL1001; *aadA* and *gfp*) sequence elements to regulate expression of foreign genes lettuce. Importantly, this report describes several parameters of the transformation and regeneration protocol essential to the successful generation of transplastomic lettuce plants. These include the optimization of growth regulator concentration and cultivar selection, placement of leaf pieces for bombardment and regeneration opposite to that of tobacco (adaxial vs. abaxial exposure) and a tenfold reduction (relative to tobacco) in the concentration of the selective agent, spectinomycin, in the regeneration medium. Stable expression of human thioredoxin-1 (hTrx1) was reported recently using a vector similar to pRL1001 for targeting and transgene regulation in another lettuce cultivar, Romana [6]. Although stable expression of biologically active hTrx1 was reported, foreign protein accumulation was limited to 1 % of TSP. Apart from the gene of interest, the hTrx1 construct was essentially identical to that used in the GFP lines described above where GFP accounted for more than a third of the TSP. Accumulation of foreign protein in plastids is likely dependent on complex interactions including those influencing the stability of the mRNA and the foreign protein [7–9].

Following the work of Kanamoto et al. [5], in 2007 the generation of transplastomic lettuce (cv. Simpson Elite) lines expressing the cholera toxin B-proinsulin fusion protein (CTB–Pins) was reported [10]. By 2010, highly efficient plastid transformation of lettuce cultivar Simpson Elite had been established, and transplastomic lines were shown to accumulate CTB–Pins or anthrax protective antigen to about 25 % of the extractable protein [8]. The use of "long flanking" target sequence (pLSLF; [10]) to direct transgene integration in the *trnI–trnA* intergenic region of the plastid ribosomal operon, adoption of native regulatory elements, and cultivar-specific optimization of the regeneration medium to produce transplastomic shoots by direct organogenesis have contributed to a robust and reproducible plastid transformation protocol for lettuce. Subsequent studies using the Simpson Elite cultivar and transformation vectors based on pLSLF have produced transplastomic lettuce accumulating CTB fusions with malaria antigens AMA1 (apical membrane antigen-1) and MSP1 (merozoite surface protein-1) up to 7.3 and 6.1 %TSP, respectively [11], and more than 50 % of TSP comprised CTB–Pins engineered to contain three furin cleavage sites (CTB-PFx3; [12]).

Continued research exploring the expression of dengue-3 serotype polyprotein (prM/E; [13]) suggests virus-like particles can assemble in transplastomic lettuce broadening the range of clinical applications for this technology.

2 Materials

2.1 Construction of Lettuce-Specific Transformation Vector

1. Total cellular DNA isolated from lettuce.
2. Lettuce-specific plastid transformation/targeting vector.
3. Oligonucleotides designed to amplify sequences for desired regulatory elements from the lettuce plastome (NC_007578 and DQ383816) incorporating restriction sites to facilitate cloning steps.
4. Oligonucleotides designed to amplify gene of interest (GOI) with appropriate restriction sites for in-frame ligation to regulatory elements.
5. Enzymes: high-fidelity DNA polymerase, restriction enzymes, T4 DNA polymerase, T4 DNA ligase, alkaline phosphatase (*see* **Note 1**).
6. dNTPs; gel extraction kit, PCR purification kit; cloning vector used for assembly of expression cassette.

2.2 Growth of Wild-Type Lettuce Plants

1. Autoclavable culture vessels: Magenta® boxes or glass canning jars (wide mouth).
2. Basal medium: Murashige and Skoog salts (MS; Caisson, N. Logan, UT; cat. no. MSP001), sucrose, Phytoblend® (Caisson, cat. no. PTC 001).
3. Wild-type lettuce seeds (*see* **Note 2**).
4. Commercial chlorine bleach solution; Tween-20.
5. Sterile water (*see* **Note 3**).
6. Growth chamber: 24 ± 2 °C at 40 μE/m^2/s photon density; 16 h light/8 h dark.

2.3 Optimization of Regeneration and Selection Media

1. Basal medium.
2. Benzylaminopurine stock solution (BA; 1 mg/mL) and naphthaleneacetic acid stock solution (NAA; 1 mg/mL) (*see* **Note 4**).
3. Spectinomycin dihydrochloride, polyvinylpyrrolidone (PVP).
4. Deep culture dishes (100 × 25 mm).
5. Sterile forceps and scalpel blades.
6. Growth chamber: 24 ± 2 °C at 40 μE/m^2/s photon density; 16 h light/8 h dark.

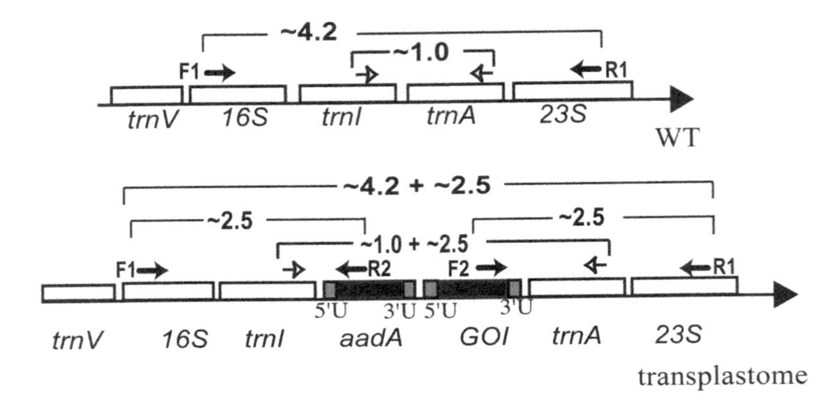

Fig. 1 Schematic representation of a lettuce plastome integration site. The transformation vector pLSLF targets transgene insertion between the *trnI* and *trnA* genes of the ribosomal operon of the lettuce plastome. *Black arrows* represent annealing sites useful to investigate plastomic integration of the transgene cassette. *Open arrows* represent alternative annealing sites that could be useful to gauge homoplasmy in confirmed transformants. *Lines with numbers* above indicate expected amplification products in kilobases (approximate). *Empty bars*, plastome flanking sequence; *black bars*, transgenes; *gray boxes*, engineered 5′ and 3′ regulatory elements (U). Plastome genes (not to scale) are presented in the order they are transcribed

2.4 Biolistic DNA Delivery

1. PDS-1000/He Biolistic® device.
2. 0.6 μM gold particles, 900 psi rupture disks, macrocarrier disks, macrocarrier disk holders, stopping screens (BioRad, Hercules CA, USA).
3. Glycerol 50 % (autoclaved in 1 mL aliquots), absolute ethanol, 70 % ethanol, 2.5 M calcium chloride solution (filter sterilized), 0.1 M spermidine free base (filter sterilize, store frozen in 20 μL aliquots).
4. Lettuce plastid transformation vector (~1 μg/μL).

2.5 Regeneration and Confirmation of Transplastomic Lines

1. Optimized culture media with selective agent in deep culture plates.
2. Sterile forceps and blades.
3. Oligonucleotides designed to amplify across integration endpoints; dependent on integration site and transgene (GOI) cassette sequence. *See* Fig. 1 for example.

   F1: CAGCAGCCGCGGTAATACAGAGGA.
   R2: CCGCGTTGTTTCATCAAGCCTTACG.
   R1: GCTGTTTCCCTCTCGACGATGAAG.
   F2: specific to gene of interest.
4. Qiagen DNeasy Plant Mini Kit (cat. no.69104).
5. PCR reagents including Taq polymerase, polymerase buffer and dNTPs.

6. Autoclavable culture vessels: Magenta® boxes or glass canning jars (wide mouth).
7. Half-strength basal medium.
8. Growth chamber: 24±2 °C at 40 μE m^2/s photon density; 16 h light/8 h dark.

2.6 Transfer to Soil, Cultivation, and Seed Production

1. Substrate for hardening rooted plants, such as Jiffy-7® peat pellets (Jiffy Products, Shippagan, N.B., CAN).
2. Plastic film or clear plastic bags.
3. Potting soil, pots, staking material (poles, ties).
4. Climate-controlled greenhouse facility.
5. Balanced plant fertilizer product such as Miracle-Gro® (The Scotts Co. Marysville OH, USA).
6. Paper bags.

3 Protocols

Aseptic technique should be employed at all times when handling plant culture media, additives, utensils and vessels, plant tissues, etc., to avoid microbial contamination. Limit access, if possible, to areas where sterile plant media is prepared and tissues are cultured.

3.1 Construction of Lettuce-Specific Transformation Vector

As mentioned in the introduction, two plasmids have been successfully employed in lettuce plastid transformation experiments using the biolistic method. These plasmids are designed to introduce foreign genes into different intergenic regions: *rbcL-accD* (pRL200; [5]) and *trnI–trnA* (pLSLF and pLsDV; [8, 10]). Investigators involved in basic research may acquire these plasmids through material transfer agreement executed between institutions. The general methods for amplification and cloning of regulatory elements and foreign genes are well established (*see* ref. 14). Investigators choosing to use lettuce native expression elements should design oligonucleotides according to the lettuce plastome sequence (NC_007578 and DQ383816). It is strongly advised to employ a high-fidelity DNA polymerase in all amplification steps. The final transformation plasmid/transgene cassette should be sequenced to confirm accuracy of amplifications and subsequent cloning reactions.

3.2 Growth of Wild-Type Lettuce Plants

1. Preparation of 1 L basal medium: In a large (≥1 L) beaker, combine one 4.33 g packet of MS salts and 30 g of sucrose in approximately 700 mL water. Stir until completely dissolved; adjust pH to 5.8 and volume to 1 L. Place 5.8 g Phytoblend® into a 2 L flask and add liquid media; cover the flask with foil.

Thoroughly clean and rinse culture vessels, place covers on magenta boxes or use foil to cover jars. Autoclave the vessels along with the media solution on liquid cycle 30 min at 121 °C. Remove from autoclave and place all items inside the clean bench or laminar flow hood immediately. If more than 1 L of medium is needed, it may be prepared in bulk but divided into several 2 L flasks prior to autoclaving such that none exceeds 1 L of medium.

2. Dispense approximately 100 mL molten medium into vessels. At this time a portion of the basal medium may be dispensed into deep culture plates. Allow media to solidify in the hood before capping and sealing. Wide mouth canning jars can be capped with the top or, preferably, the bottom of a shallow culture plate and sealed with parafilm.
3. Sterilize lettuce seeds: Prepare a 10 mL solution of 30 % commercial bleach (commercial bleach contains about 5 % sodium hypochlorite) with a few drops of Tween-20 in sterile water. Place seeds (~25) in a microcentrifuge tube. Add ~1 mL 70 % ethanol and incubate with vortexing for 1 min. Aspirate ethanol and rinse seeds twice with sterile water. Add ~1 mL bleach solution to seeds and incubate on a rocker for ≤5 min. Aspirate solution and rinse seeds four times with sterile water. Seeds may be distributed directly onto medium in culture vessels. Alternatively, seeds may be placed onto medium in culture plates then transferred to vessels upon germination.
4. Culture under long-day conditions: 24 ± 2 °C, 16 h light and 8 h dark. About three weeks following germination, the apical, fully expanded leaves may be used for optimization of regeneration or bombardment experiments.

3.3 Optimization of Culture Conditions for Regeneration from Leaf Explants

Studies have shown that response to plant growth regulators (PGRs) is highly genotype dependent in lettuce [15, 16]. Furthermore, different groups have observed variable results in shoot regeneration under replicated conditions [15, 16]. This suggests that subtle differences may affect tissue culture response [17] and highlights the need to optimize for regeneration in independent labs. Described here is a factorial experiment to evaluate regeneration efficiency of lettuce cultivars in response to the cytokinin BA and auxin NAA. Other factors such as light intensity, tissue age, explant size and water quality may further influence experimental outcome and may have to be addressed empirically. Note that much of the published literature on lettuce regeneration has utilized cotyledon explants, which may be of limited utility for the generation of transplastomic lines as primary regenerants typically require repeated regeneration to achieve homoplasmy, i.e., there will be no cotyledon available for subsequent regenerations.

Table 1
A simple factorial experiment to evaluate growth regulator response in a single cultivar[a]

PGR, μg/mL	BA 0.1	BA 0.2	BA 0.4	BA 0.8	BA 1.0	BA 1.5
NAA 0.05						
NAA 0.1						
NAA 0.25						
NAA 0.5						
NAA 0.75						
NAA 1.0						

[a]In each cell record, the number of shoots regenerated divided by the number of explants cultured under each condition as an indicator of regeneration efficiency

1. Preparation of media for regeneration experiment: The basal medium is modified by addition of 500 mg/L of PVP and PGRs as specified by the experiment (Table 1 is given as an example; BA and NAA added from 1 mg/mL stock solution). The pH is adjusted to 5.8, Phytoblend is added, and media is autoclaved for 30 min. Molten media (~25 mL) is dispensed into deep culture plates and allowed to solidify. Carefully label plates according to PGR condition.
2. Using a sharp, sterile blade, excise young leaves (~4 cm^2) from the wild-type lettuce plants. Working on a sterile surface, such as a Petri dish, prepare explants of ~0.25 cm^2 and distribute, adaxial side down, onto the regeneration medium. Continue to prepare explants placing 6–8 explants per plate allowing five plates for each cultivar and each condition. Label carefully and seal plates with parafilm. Incubate cultures at 24 ± 2 °C, 16 h light and 8 h dark, 40 μE/m^2/s. Shoots should appear in 21–28 days.
3. Following selection of cultivar and optimal regeneration medium, the appropriate level of spectinomycin is determined. The spectinomycin is dissolved in water to obtain a 100 mg/mL solution. The solution is filter sterilized and stored in single-use aliquots at −20 °C. Prepare the optimized regeneration medium, divide into flasks, and autoclave. When medium is cooled to ~50 °C, add spectinomycin stock to each flask to yield final concentrations ranging from, for example, 10 to 200 μg/mL. Swirl flasks, dispense media into plates, cool to solidify, and label accordingly (*see* **Note 4**).
4. Prepare explants as described and distribute to various media. The ideal concentration of antibiotic is the least amount that completely inhibits regeneration of plantlets.

3.4 Preparation of Consumables for Biolistic DNA Delivery

For the bombardment of lettuce leaves, 0.6 μm gold particles are used. The stock suspension of gold particles is prepared in advance and stored at −20 °C. Stopping screens and macrocarrier holders may be autoclaved. Macrocarrier disks and rupture disks should be covered in absolute ethanol for 5 min. Disks are allowed to dry in the hood on autoclaved filter paper or paper towels.

3.4.1 Preparation of Gold Suspension

1. Take 50 mg gold particles in a siliconized tube (1.5 mL).
2. Add 1 mL absolute ethanol.
3. Vortex for 2 min.
4. Centrifuge at 10,000 × *g* for 3 min; aspirate supernatant.
5. Add 1 mL 70 % EtOH; vortex for 1 min.
6. Incubate at RT for 15 min with intermittent mixing by gentle shaking.
7. Centrifuge at 5,000 × *g* for 2 min; aspirate.
8. Wash with 1 mL sterile H_2O; incubate for 1 min at room temperature.
9. Centrifuge at 5,000 × *g* for 2 min.
10. Repeat (**steps 8** and **9**) a total of four times.
11. Resuspend particles in 1 mL sterile 50 % glycerol.

3.4.2 Preparation of Lettuce Leaves for Bombardment

1. Carefully excise the youngest fully expanded leaves from wild-type lettuce plants. The leaves should be no larger than ~4 cm^2.
2. Place the leaves abaxial side down onto antibiotic-free regeneration medium. If smaller leaves are used, several (~6) may be placed on a plate in a single, centered layer. The DNA precipitation procedure (below) allows gold suspension for five bombardments; therefore, five plates of leaves should be prepared.
3. Seal the plates with parafilm and keep in dark until bombardment.

3.4.3 Precipitate Plasmid DNA onto Gold Particles

1. Vortex gold stock suspension thoroughly. Transfer 50 μL of gold suspension to a siliconized tube. Perform subsequent additions quickly in the exact order given and pulse vortex with each addition.
2. Add 5 μg plasmid stock; vortex.
3. Add 50 μL of 2.5 M $CaCl_2$, vortex.
4. Add 20 μL 0.1 M spermidine stock, vortex.
5. Vortex/mix for 20 min at 4 °C.
6. Centrifuge at 10,000 × *g* for 1 min; aspirate.
7. Wash with 200 μL 70 % ethanol.
8. Wash with absolute ethanol.
9. Resuspend with 50 μL absolute ethanol. Place on ice.

3.4.4 Loading Gold Suspension onto Macrocarrier Disk

1. Seat the macrocarrier disk into the steel macrocarrier holder using forceps and the red "thimble" tool. Ascertain that the edges of the disk are inserted under the inside lip of the holder.
2. Vortexing thoroughly each time, dispense 10 μL of the DNA-gold suspension onto the center of each disk. Do not apply suspension to disk beyond the outline of the opening in the holder.
3. Allow the ethanol to evaporate completely in the hood. You should see a uniform, opaque gold coating when drying is complete.

3.5 Bombardment Procedure

1. Remove all parts from the inside of the PDS-1000/HE device and place on autoclaved paper towels inside the hood. Clean all pieces and inside of device chamber with 70 % ethanol.
2. Attach vacuum tubing and turn on the PDS-1000/HE device and the vacuum pump.
3. Open the main valve (like a faucet handle) on the helium tank. Note that the pressure gauge responds accordingly indicating sufficient helium gas to execute the bombardment.
4. Turn the second (T-bar) valve clockwise until the second pressure gauge indicates a psi ~200 above the rupture pressure of the selected disk (i.e., 1,100 psi).
5. Using forceps, place a rupture disk in the disk retaining cap; assure the disk is seated properly in the bottom of the housing.
6. Screw the cap onto the gas acceleration tube and tighten using the torque tool provided with the device.
7. Place a stopping screen into the stopping screen support ring of the microcarrier launch assembly.
8. Place the loaded macrocarrier disk in its holder into the launch assembly with the DNA-gold facing down toward the stopping screen. Secure the macrocarrier cover lid.
9. Insert the assembly into the device chamber immediately below the disk retaining cap.
10. Place an uncovered plate with leaf sample on the target shelf and insert into the device chamber at a desired level (6 or 9 cm below the launch assembly); close and secure the chamber door.
11. Move the vacuum switch to the VAC position. When the vacuum gauge indicates 28 in. Hg has been reached, move the switch to the HOLD position.
12. Depress the FIRE switch and monitor the helium pressure gauge atop the device. The rupture disk should give way when the appropriate psi is reached.

Fig. 2 Production of transplastomic lettuce. The time line from primary transplastomic regenerant to harvest from T1 progeny is approximately 20–24 weeks. Under optimal conditions, cultivar Simpson Elite yielded ~75 g fresh weight per plant, per harvest; loose-leaf cultivars can be harvested at least twice. (**a**) Green spectinomycin resistant soot (primary regenerant). (**b**) Second selective regeneration. (**c**) Bolting. (**d**) Achene with pappus. (**e**) Confirming spectinomycin resistance by germinating T1 seed on spectinomycin medium (*right*). Sensitive (bleached) wild-type seedlings (*left*). (**f**) Mature T1 transplastomic plant

13. Immediately release the FIRE switch and move the vacuum switch to the VENT position.
14. When the vacuum is completely released, open the chamber door and remove and cover sample.
15. Seal plate with bombarded sample and hold in the dark at 24 ± 2 °C for 48 h.

3.6 Regeneration and Screening for Transplastomic Lines

Regeneration of transplastomic shoots from bombardment of young leaves of the Simpson Elite cultivar was achieved using the modified basal medium described above (Subheading 3.2) supplemented with 0.2 μg/mL BA and 0.1 μg/mL NAA (Fig. 2). A spectinomycin concentration of 50 μg/mL was sufficient to exclude non-transformed regenerants. The ideal concentration of PGRs and selective antibiotic/agent should be empirically determined.

1. Prepare fresh regeneration media with spectinomycin. Approximately 1 L of media (~40 plates) will be needed per five bombardments.
2. Working on a sterile surface, such as a Petri dish, prepare explants of ~0.25 cm^2 and distribute, adaxial side down, onto the regeneration medium. Continue to prepare explants placing

6–8 explants per plate. Incubate cultures at 24 ± 2 °C, 16 h light and 8 h dark, 40 μE/m²/s. Resistant transplastomic shoots should appear in 21–28 days.

3. When the shoot is large enough to be removed from the explant, an additional regeneration cycle under selection is initiated. Leaf tissue from the regenerated plantlet is cut into explants and placed on fresh regeneration media with antibiotic.
4. At this time, providing that sufficient tissue can be taken from the plantlet, a genomic DNA extraction using a kit such as the Qiagen DNeasy Plant Mini Kit can be performed, and DNA can be used to implement PCR screening for the integration event (*see* **Note 5**). Oligonucleotide primers are designed based on the published plastome sequence for lettuce and depend on the site of integration and the transgenes inserted. An amplification strategy that relies on primer hybridization within the foreign, inserted sequence and flanking native plastome sequence, at both sides of the insert, is recommended. Amplification using primers F1 and R2 (Fig. 1) may be executed using the following conditions: 94 °C for 5 min, 30 cycles (94 °C for 1 min, 60 °C for 30 s, 72 °C for 2 min), 72 °C for 10 min. The flanking sequence primers should be designed to anneal beyond the extent of native sequence used in the plastid transformation vector for targeting integration.
5. Numerous shoots should arise from each explant in the second regeneration cycle (Fig. 2). These plantlets (subclones) are transferred to half-strength basal media (no PGRs) with up to 100 μg/mL spectinomycin for root establishment.

3.7 Transfer to Soil and Seed Production

Once roots have developed, regenerants are transferred to soil and moved to the greenhouse for seed production (Fig. 2). At this time additional tissue samples may be taken for DNA isolation and PCR and/or Southern blotting for confirmation of transplastomic status (*see* **Note 5**).

1. Thoroughly hydrate potting substrate; several jiffy pellets can be put together in a single pan with sides high enough to accommodate plants with some headspace when the pan is covered with plastic film. If individual pots are used, these may be tented with clear plastic bags.
2. Remove media with embedded plant from the culture vessel. If necessary, the vessel can be placed in a warm bath to soften media facilitating removal without damaging the plant.
3. Carefully extract the plant with intact roots from the medium. Submerge the media block in warm water and gently massage/loosen to free the roots.
4. Transfer the plant to soil substrate covering the roots; water the plant in.

5. Cover the newly transferred plants and allow 4–6 days for hardening. After 2–3 days, small openings may be cut in the coverings (film or bags) and plants are monitored for loss of turgidity.
6. Move established plants to the greenhouse. Allow soil to approach dryness before watering. Use balanced nutrient solution according to manufacturer's direction; never apply nutrient solution to dry soil. When plants begin to bolt (internode elongation and appearance of inflorescence), they should be staked to support the stalk.
7. Harvest achenes when pappus is present. The entire inflorescence may be cut and stored in a paper bag until completely dry.
8. Sterilize and germinate seeds on spectinomycin-containing basal medium to confirm transmission of spectinomycin resistance (*aadA* gene) to progeny.

4 Notes

1. Unless otherwise indicated, chemical reagents are purchased from Sigma (St. Louis MO, USA). Molecular biology enzymes may be purchased from the investigators preferred vendor. Invitrogen (Carlsbad, CA), Fermentas (Glen Burnie, Maryland), and New England Biolabs (Ipswich, MA) are suitable examples.
2. Seeds of the Simpson Elite lettuce cultivar were purchased from New England Seed Company, Hartford, CT.
3. Herein, "water" refers to water that has been purified and deionized to a high degree (typically 18.2 MΩ cm) such as that dispensed from a Millipore water purification system.
4. The immediate use of plant media containing PGRs or selective agents is recommended for best performance. If media must be stored, seal securely, wrap with plastic film, and store at 4 °C protected from the light. Media containing spectinomycin should not be used after two weeks of storage.
5. Analysis of PCR results may be used to confirm the presence transplastomes in lettuce regenerants. It may also be possible to gauge homoplasmy using a PCR strategy. For transformants generated using the pLSLF-based vectors, which incorporate ~4.2 kb of flanking sequence, primer annealing outside the targeting sequence would generate long products in wild-type plants/plastomes. Adding the size of the GOI, regulatory elements, and the *aadA* sequence in transformants suggests the need to use a specialized long-range enzyme. Alternatively, once integration is confirmed a primer pair annealing proximal to the insertion site could be used (open arrows in Fig. 1). Regardless, the smaller product is likely to be preferentially amplified in the PCR reaction, and the possibility of amplifying

plastid sequences from other cell compartments can confound analysis of results. Recently, Ruiz et al. [18] have used a quantitative PCR approach to gauge homoplasmy in plastid transformants. Southern blot is typically used to assess homoplasmy in transplastomic lines.

Acknowledgments

Support was provided by the National Science Foundation (IOS-1027259) and the Fred G. Gloeckner Foundation.

References

1. Cardi T, Lenzi P, Maliga P (2010) Chloroplasts as expression platforms for plant-produced vaccines. Expert Rev Vaccines 9:893–911
2. Daniell H, Singh ND, Mason H, Streatfield SJ (2009) Plant-made vaccine antigens and biopharmaceuticals. Trends Plant Sci 14:669–679
3. Lelivelt CLC, McCabe MS, Newell CA, deSnoo CB, Dun KMP, Birch-Machin I, Gray JC, Mills KHG, Nugent JM (2005) Stable plastid transformation in lettuce (*Lactuca sativa* L.). Plant Mol Biol 58:763–774
4. Svab Z, Hajdukiewicz P, Maliga P (1990) Stable transformation of plastids in higher plants. Proc Natl Acad Sci U S A 87:8526–8530
5. Kanamoto H, Yamashita A, Asao H, Okumura S, Takase H, Hattori M, Yokota A, Tomizawa K-I (2006) Efficient and stable transformation of *Lactuca sativa* L. cv. Cisco (lettuce) plastids. Transgenic Res 15:205–217
6. Lim S, Ashida H, Watanabe R, Inai K, Kim Y-S, Mukougawa K, Fukuda H, Tomizawa K, Ushiyama K, Asao H, Tamoi M, Masutani H, Shigeoka S, Yodoi J, Yokota A (2011) Production of biologically active human thioredoxin 1 protein in lettuce chloroplasts. Plant Mol Biol 76:335–344
7. Zou Z, Eibl C, Koop H-U (2003) The stem-loop region of the tobacco psbA 5′UTR is an important determinant of mRNA stability and translation efficiency. Mol Genet Genomics 269:340–349
8. Ruhlman T, Verma D, Samson N, Daniell H (2010) The role of heterologous chloroplast sequence elements in transgene integration and expression. Plant Physiol 152:2088–2104
9. Waheed MT, Thönes N, Müller M, Hassan SW, Gottschamel J, Lössl E, Kaul H, Lössl AG (2011) Plastid expression of a double-pentameric vaccine candidate containing human papillomavirus-16L1 antigen fused with LTB as adjuvant: transplastomic plants show pleiotropic phenotypes. Plant Biotechnol J 9:651–660
10. Ruhlman T, Ahangari R, Devine A, Samsam M, Daniell H (2007) Expression of cholera toxin B–proinsulin fusion protein in lettuce and tobacco chloroplasts – oral administration protects against development of insulitis in non-obese diabetic mice. Plant Biotechnol J 5:495–510
11. Davoodi-Semiromi A, Schreiber M, Nallapali S, Verma D, Singh ND, Banks RK, Chakrabarti D, Daniell H (2010) Chloroplast-derived vaccine antigens confer dual immunity against cholera and malaria by oral or injectable delivery. Plant Biotechnol J 8:223–242
12. Boyhan D, Daniell H (2011) Low-cost production of proinsulin in tobacco and lettuce chloroplasts for injectable or oral delivery of functional insulin and C-peptide. Plant Biotechnol J 9:585–598
13. Kanagaraj AP, Verma D, Daniell H (2011) Expression of dengue-3 premembrane and envelope polyprotein in lettuce chloroplasts. Plant Mol Biol 76:323–333
14. Sambrook J, Russell DW (2001) Molecular cloning: a laboratory manual. CSHL Press, Woodbury, NY
15. Xinrun Z, Conner AJ (1992) Genotypic effects on tissue culture response of lettuce cotyledons. J Genet Breed 46:287–290
16. Ampomah-Dwamena C, Conner AJ, Fautrier AG (1997) Genotypic response of lettuce cotyledons to regeneration in vitro. Sci Hort 71:137–145
17. George EF (1993) Plant propagation by tissue culture: Part 1. The technology, 2nd edn. Exegetics, Edington
18. Ruiz ON, Alvarez D, Torres C, Roman L, Daniell H (2011) Metallothionein expression in chloroplasts enhances mercury accumulation and phytoremediation capability. Plant Biotechnol J 9:609–617

Chapter 22

Plastid Transformation in Soybean

Manuel Dubald, Ghislaine Tissot, and Bernard Pelissier

Abstract

The biotechnological potential of plastid genetic engineering has been illustrated in a limited number of higher plant species. We have developed a reproducible method to generate plastid transformants in soybean (*Glycine max*), a crop of major agronomic importance. The transformation vectors are delivered to embryogenic cultures by the particle gun method and selection performed using the *aadA* antibiotic resistance gene. Homoplasmy is established rapidly in the selected events without the need for further selection or regeneration cycles, and genes of interest can be expressed at a high level in green tissues. This is a significant step toward the commercial application of this technology.

Key words Chloroplast transformation, Embryogenic callus, *Glycine max*, Homoplasmy, Particle inflow gun, Soybean

1 Introduction

Soybean (*Glycine max* L. Merr.) has recently become a dominant source of protein and oil for both human consumption and animal feed. Much effort has been devoted to the genetic engineering of insect resistance or herbicide tolerance. There is a growing need to optimize productivity and further enhance the nutritional value of this crop. The nuclear and plastid genetic modification of soybean will certainly contribute to meet these objectives.

As in all other species, chloroplast engineering is significantly more difficult to achieve than nuclear transformation. One important step is tissue culture. Owing to their ability to regenerate whole plants, embryogenic cultures are, in soybean, the most common starting material for transformation by particle bombardment [1]. As a drawback, these undifferentiated cells contain fewer and much smaller plastids than for instance chloroplasts in leaf cells. Therefore, transformation of soybean plastids is a rare event and presents a real technical challenge. Nevertheless, the protocol that we describe here reliably allows the isolation of homoplasmic and fertile soybean plastid transformants [2–4].

Pal Maliga (ed.), *Chloroplast Biotechnology: Methods and Protocols*, Methods in Molecular Biology, vol. 1132,
DOI 10.1007/978-1-62703-995-6_22, © Springer Science+Business Media New York 2014

2 Materials

2.1 Plant Material

1. Immature seeds for embryo culture are obtained from greenhouse-grown plants. Seeds of explant donor plants, soybean cv. Jack (Ohio State University), are planted in bedding plant substrate TS3 (Klasmann-Deilmann GmbH, Geeste, Germany) and cultivated in the greenhouse with the following conditions: 18 h day photoperiod at 170 $\mu E/m^2/s$, at 24 °C (22 °C at night), and hygrometry at 60 %. Plants are watered with tap water and supplemented with fertilizer Triabon HPK (16/8/12; BASF), added to the pots when plants start flowering. Pods are harvested from healthy plants when immature seeds are about 3–4 mm in length.
2. Embryo induction medium I containing MS basal salt medium [5], supplemented with B5 vitamins [6], 6 % sucrose, 40 mg/L of 2,4D, 3 g/L of Gelrite, at pH 7 before autoclaving. Tissue culture media were purchased from Sigma. Gelrite was purchased from Duchefa.
3. Embryo induction medium II containing MS basal salt medium, with B5 vitamins, 3 % sucrose, 20 mg/L of 2,4D, 3 g/L of Gelrite, pH 5.7 [7].
4. FNL medium [8], consisting of MS basal salt medium, B5 vitamins, 1 % sucrose, 0.67 g/L asparagine, 10 mg/L of 2,4D, 3 g/L of Gelrite, pH 5.7.

2.2 Biolistic Transformation

1. Plasmid DNA is isolated from recombinant bacteria using Qiagen miniprep extraction kits. The backbone of soybean transformation high-copy plasmid vectors, such as pCLT312 (GenBank AY575999) or pCLT323 (GenBank DQ459070), targets the transgenes in the inverted repeat of the plastid genome, upstream of the *trnV* and *16S rRNA* genes (Fig. 1a). The nucleotide length of the flanking plastid sequences (left/1,370 and right/1,762) is important to allow recombination and integration. Those sequences were cloned from the cultivar to be transformed (cv. Jack), in order to maximize the efficiency of recombination with the target genome as described in [2]. Other integration sites have not yet been explored. The *aadA* selective marker gene, encoding resistance to spectinomycin, is under the control of tobacco plastid regulatory elements as described by [9] (*see* **Note 1**). Cassettes encoding genes of interest, such as herbicide tolerance in pCLT323, are cloned in tandem in the same orientation as the selection gene, between the plastid flanking sequences (Fig. 1a).
2. The particle influx gun described in this protocol (Fig. 2a) is identical to the one published by Finer et al. [10]. The DNA-coated particles are accelerated at a high speed through the

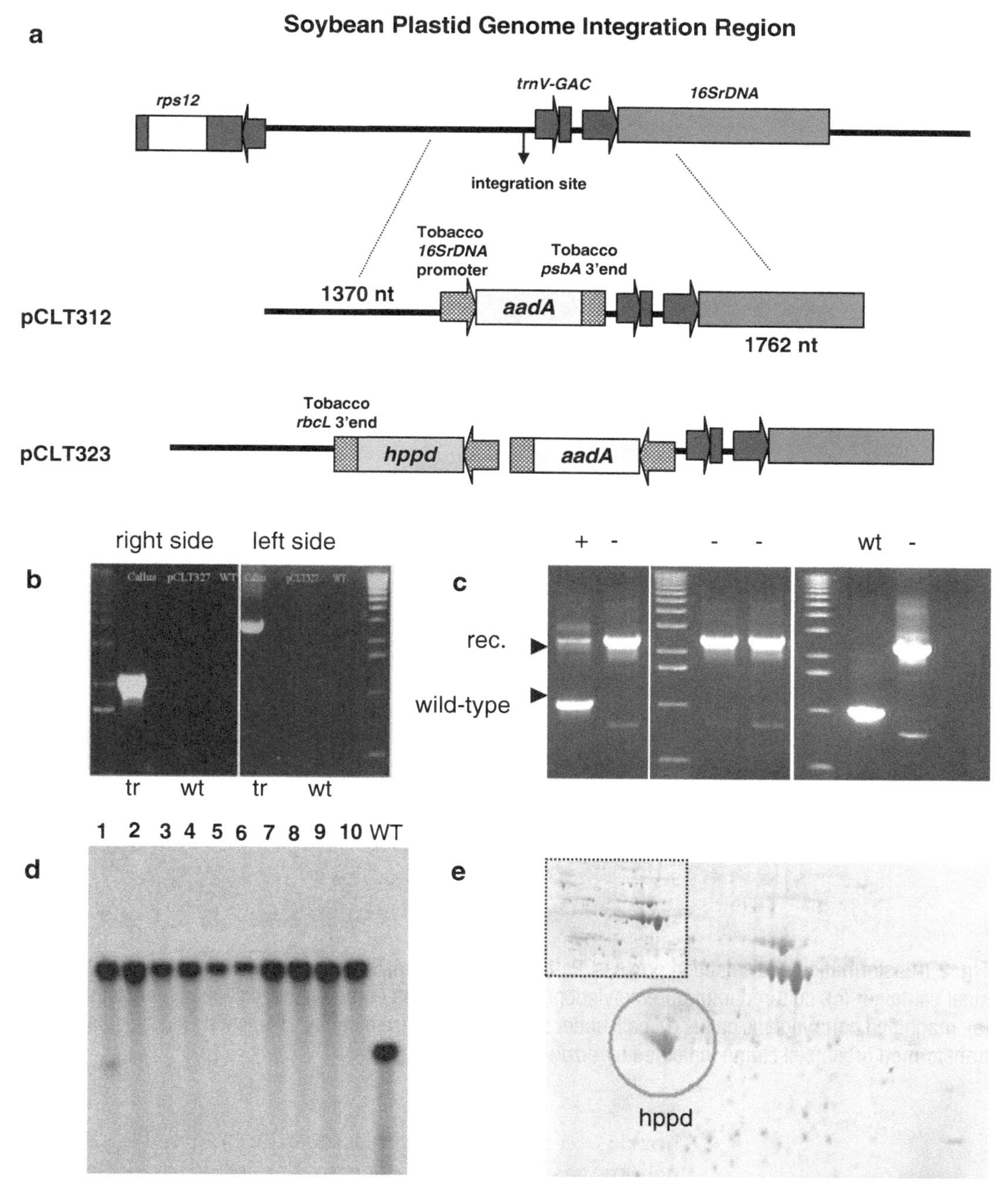

Fig. 1 Molecular analysis of transplastomic plants. Structure of integration site and two transforming vectors (**a**); example of PCR analysis on *left* and *right* sides showing integration for the transgenic event (tr) at the expected site (**b**); example of PCR evaluation of homoplasmy, revealing homoplasmic (–) and heteroplasmic (+) material, using primers encompassing the integration site (**c**); Southern blot analysis of 10 independent events (**d**); and 2D separation of wild-type (box in *upper left* corner) and transgenic leaf proteins from an event expressing hydroxyphenylpyruvate dioxygenase (hppd) *circled* (**e**)

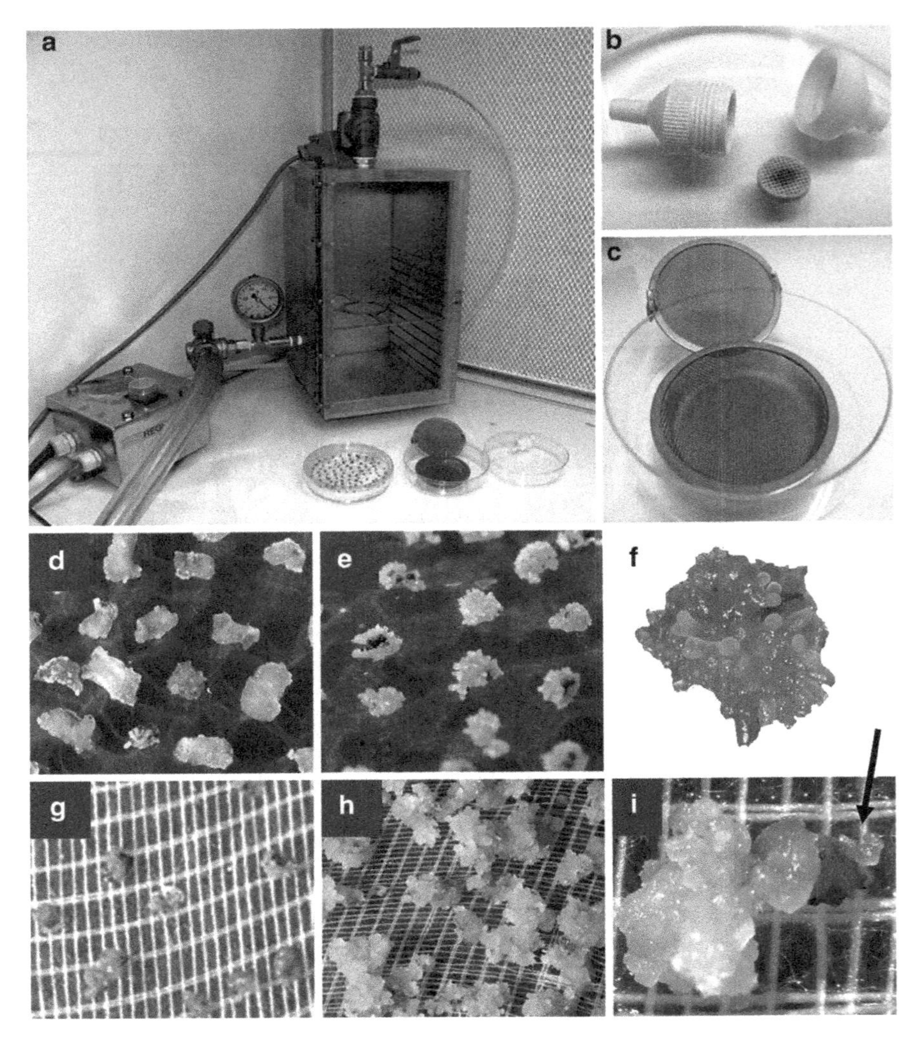

Fig. 2 Transformation and selection process. Particle influx gun (**a**), Swinnex filter holder (**b**), meshed stainless steel container (**c**), cultured immature cotyledons (**d**), proliferating embryogenic calli ready for transformation (**e**), magnified embryogenic callus (**f**), calli under selection on gauze mesh (**g**), bleached wild-type calli (**h**), and transformed green cell clump indicated by *arrow* (**i**)

nozzle of a Swinnex filter/macrocarrier holder (Fig. 2b) using helium gas pressure as energy. The pressure of helium is adjusted to 6 bars. The distance between the Swinnex and the tissues is about 17 cm (*see* **Note 2**), and the shot is performed when a partial vacuum of approximately 0.95 bar (29 in. Hg) is reached in the chamber, connected to a vacuum pump.

3. A special meshed container made of stainless steel to accommodate embryogenic calli at the time of bombardment, in order to diminish tissue dispersion due to the deflagration (Fig. 2c).
4. Gold particles (0.6 μm; Biorad).

5. $CaCl_2$ (2.5 M).
6. Spermidine (0.1 M).

2.3 Selection of Plastid Transformants

1. FNL medium containing 200 mg/L of spectinomycin.
2. SBP6 liquid medium [11] with 200 mg/L of spectinomycin.
3. Embryo maturation medium consisting of MS basal salt medium, B5 vitamins, 6 % maltose, 2 g/L of Gelrite, pH 5.7 for 2 months, and 200 mg/L of spectinomycin.
4. Embryo germination medium containing half-strength basal MS salts with B5 vitamins, 1.5 % sucrose, 2 g/L of Gelrite, pH 5.7.

2.4 Analyses of Plastid Transformants

1. PCR primers for identification of transplastomic clones transformed with vector pCLT323 or pCLT327 [3] on the right flanking side, as in Fig. 1b, are the following:

   5′-GTTAAGGTAACGACTTCGGCATGG-3′ and 5′-CTCAG TACTCGAGTTATTTGCCGACTACCTTGGTGATCT CGCC-3′.
2. PCR primers for evaluation of the homoplasmy level are located in the soybean plastid genome outside the transformation vector sequences [2] (Fig. 1c) as follows:

   5′-CATGGGTTCTGGCAATGCAATGTG-3′ and 5′-CAGGA TCGAACTCTCCATGAGATTCC-3′.
3. Southern blot analysis (Fig. 1c) is performed with 1 μg of total DNA digested by *EcoRI*. The P^{32} labeled probe covering the left flanking region was amplified by PCR with the following primers [2]:

   5′-CTAGTGGTACCGATCCAATCACGATCTTCTAATAA GAAC-3′,

   5′-GAACCTCCTTGCTTCTCTCATGTTACAATCCTCTT GCCGC-3′.

3 Methods

3.1 Preparation of Plants Material

1. Immature seed preparation. Sterilize pods with a 20 % (v:v) bleach containing solution of Domestos (Unilever) for 20 min, then wash 3 times with sterile distilled water. All subsequent manipulations should be carried out under a sterile laminar hood. Immature seeds should be excised (are extracted) from the pods using suitable (adapted set of) tweezers and scalpel. The seeds are cut transversally on the radicle side to allow the separation of the two independent cotyledons by practicing a light pressure on the seed. No additional wounding is performed.

2. Induction of somatic embryogenesis. The cotyledons are transferred to Petri dishes, abaxial side down, on embryo induction medium I. After 3 weeks, the proliferating tissues are subcultured once to fresh medium (Fig. 2d). Explants are incubated at 24 °C in a Sanyo-type growth chamber, with a photoperiod of 16 h, under a light intensity of 27 $\mu E/m^2/s$. One month after, primary somatic embryos are excised from the cotyledons and transferred to embryo induction medium II and subcultured every 2 weeks. This protocol is modified from the publication of Santarem and Finer [7].
3. Somatic embryo propagation. Somatic clumps are then transferred to the FNL medium [8] and subcultured every 2 weeks for 2–4 months prior to transformation.
4. Explant preparation for transformation. Ten to 12 days before transformation, embryogenic calli (Fig. 2e, f) are transferred to fresh FNL medium. The night before bombardment, Petri dishes containing the tissues to bombard are placed in a fridge at 4–5 °C. 30 to 40 green calli are placed in the meshed metallic container (Fig. 2c) and partially dehydrated for 5–10 min in the laminar hood.

3.2 Biolistic Transformation

1. Preparation of DNA-coated gold particles. A stock of gold particles is first prepared by adding 5 mL of ethanol to 250 mg of particles. This stock is distributed in 1 mL and can be kept at room temperature for months before use [1]. All following steps are performed with sterile solutions and material in a lamina flow hood.
2. For two shots, 50 μL of particles is taken from the stock and washed with 500 μL sterile water and centrifuged $2 \times g$ in an Eppendorf tube. The final gold pellet is resuspended in 50 μL of water, and 10 μg of DNA of the transforming vector is added, from a stock solution in water at a concentration of at least 1 μg/μL. Then, 50 μL of $CaCl_2$ (2.5 M) is added and mixed, followed rapidly by 20 μL of spermidine (0.1 M).
3. The mixture is energetically mixed during 10 s using a micropipette tip and left at least 10 min to allow sedimentation of the particles.
4. The majority of the supernatant is discarded and the particles resuspended using a micropipette tip in remaining 5 μL of the solution.
5. For each shot, approximately 2.5 μL of this particle suspension is then deposited from above in the middle of the macrocarrier grid in the Swinnex filter holder and bombarded within a few minutes.
6. The partially dehydrated embryogenic calli (*see* Subheading 3.1, **step 4** above) are bombarded twice with the transformation

plasmid. Each shot is performed as described in references [1] and [10] at a 6 bar helium pressure with as much as 5 μg of adsorbed DNA.

3.3 Selection of Plastid Transformants

1. After bombardment, embryogenic calli are transferred to fresh FNL medium and subcultured for 3 days, then cut in small pieces, and transferred on a fresh FNL medium containing 200 mg/L of spectinomycin (*see* **Note 3**). For an easier process, calli are transferred on a mesh made with gauze that allows the easy and fast transfer of the totality of the tissues every 2 weeks on fresh media (Fig. 2g). Spectinomycin induces a progressive bleaching of the calli and blocks proliferation (Fig. 2h). Putative transformants are detected as yellow-green clumps after 2–3 months on selection media, on a background of bleached calli (Fig. 2i).
2. Putative transgenic events are then amplified in 30 mL of SBP6 liquid medium [11] with 200 mg/L of spectinomycin for 1 month in the same conditions as previously described.
3. Green embryogenic calli are matured by transfer to MS basal salt medium, B5 vitamins, 6 % maltose, 2 g/L of Gelrite, pH 5.7 for 2 months with 200 mg/L of spectinomycin in the same previously described conditions. The somatic embryos are then placed in empty Petri dishes with lid on and gently desiccated for 2–3 days in the laminar flow hood.
4. Somatic embryos are germinated on a half-strength basal salt MS medium with B5 vitamins, 1.5 % sucrose, 2 g/L of Gelrite, pH 5.7 under the same previously described conditions. After root and shoot elongation, the developed plantlets are transferred to an acclimatization room, before being definitively transferred in the greenhouse for plant development and seed production.

3.4 Analysis of Plastid Transformants

3.4.1 Checking Integration in the Plastid Genome by PCR

Integration of the transgenes occurs in principle precisely by homologous recombination at the target locus. In theory, resistance to spectinomycin can also result from the integration of the *aadA* gene elsewhere in the plastid or nuclear genomes or result from mutations in the target genes, such as plastid *16SrDNA* (*see* **Note 4**). Typically, integration in the plastid genome can be checked at any stage using couples of converging primers, the first oligonucleotide landing in the transgenes (*aadA* or gene of interest) and the other landing in the plastid genome, just outside of the sequences present in the transformation vector. A unique band of expected size is detected only for true plastid transformants (Fig. 1b), and this can be checked on both sides of the integrated transgenes as in [3].

3.4.2 Evaluation of Homoplasmy by PCR

In soybean, homoplasmy is generally rapidly reached during or after regeneration of the events through somatic embryogenesis (*see* **Note 5**). The less ambiguous method to check the homoplasmy level is by PCR, using convergent primers landing on both sides of the plastid genome insertion site, as shown in [2]. A small, easily amplifiable PCR product is expected for the wild-type genome. If transgenes are integrated at the targeted locus, a significantly longer PCR product is amplified (Fig. 1c). The absence of the smaller wild-type fragment in those same lanes very strongly suggests that these events are homoplasmic.

3.4.3 Southern Blot Analysis

Once the transgenic events have been transferred to the greenhouse, enough leaf material can be harvested to proceed to a Southern blot analysis, using standard protocols as described in [2]. Southern analysis has the advantage (over PCR methods) that it provides access to a more global evaluation of the structure of the recombinant plastid genome (*see* **Note 6**). Because plastid genomes represent a significant proportion of the total DNA content of leaves, low amounts of total DNA, typically 1 μg, are needed per lane (*see* **Note 7**). The wild-type and recombinant DNA are fragmented with the appropriate restriction enzymes, separated on agarose gels, blotted on nylon membrane, and analyzed separately with probes covering the transgenes and the plastid genome in the vicinity of the insertion site, as in Fig. 1d. The latter experiment can confirm or substitute the evaluation of the homoplasmy level by PCR, as described above, although the detection limit is significantly higher.

3.4.4 Recombinant Protein Expression

Expression levels achieved in soybean plastid transformants can be spectacular in green tissues such as leaves (Fig. 1e), but are significantly lower in most other tissues (*see* **Note 8**). Of course, this only occurs if appropriate expression of regulatory elements is used and if the protein has a long half-life. The *16SrDNA* and *psbA* plastid promoters are among the most active promoters in green tissues and are widely used to drive expression of the transgene. Nevertheless, this does not always result in maximal expression because the limiting factor is often the efficacy of translation initiation, which depends on the ribosome binding site.

Proteins can be readily extracted from recombinant chloroplasts by any standard procedure and the recombinant products detected and characterized by the appropriate method. Very often, the expression level of the recombinant protein is high enough (>1 % of total soluble proteins) that it can be detected simply on 1 or 2 dimensional gels by Coomassie blue staining, as described in [4] and [12].

Transmission of Trait to the Progeny. In contrast to nuclear events, genes inserted in the plastid genome should not segregate in the selfed progeny of the homoplasmic soybean transformants (*see* **Note 9**). Seeds harvested from transgenic events can be sterilized and sown in vitro on MS medium containing a high concentration of spectinomycin (500 mg/L). All seedlings should uniformly display green cotyledons and first leaves, whereas the development of wild-type seedlings stops rapidly after germination, and those plantlets get bleached.

4 Notes

1. The *aadA* marker gene can also be linked in a dicistronic unit to a green fluorescent protein as in pCLT554 (GenBank EU70886) in order to monitor the in vitro selection process, as well as later the elimination of the antibiotic resistance gene (as described in patent application WO2010079117).
2. Parameters for biolistic delivery used to generate soybean plastid transformants correspond to conditions that are also optimal for nuclear transformation.
3. Streptomycin cannot be used for selection since soybean is naturally resistant to very high doses of this antibiotic (up to at least 800 mg/L).
4. In soybean, the selection of random *aadA* integration events (nuclear or plastidial) or of spectinomycin-resistant mutants is very rare. In our hands, this happened only once, out of more than 25 independent events obtained with various constructs.
5. Before embryogenesis, at the callus stage, selected events can still be heteroplasmic [2]. But after regeneration through somatic embryogenesis, following our selection protocol, all plants that we have analyzed so far proved to be homoplasmic for the recombinant genome. This is not the case in most other species for which the technology of plastid transformation exists, including tobacco, and for which the production of homoplasmic plants can be more difficult.
6. The use of soybean plastid regulatory elements (promoters, terminators) to drive the expression of the transgenes is not recommended since this can lead to further rearrangements of the plastid genome by recombination with resident genes, which can be detected on DNA blots, as in tobacco [13].
7. Autoradiogram exposure time is also typically short, a few hours at most, compared to times required to detect single-copy genes in complex nuclear genomes. We use P^{32} radioactive labeled probes, but without any doubt, nonradioactive detection methods can certainly be used with similar success.
8. Expression levels in leaves are not always extreme. The level of expression depends on the efficient transcription and translation of the transgene but also on the stability of the recombinant protein.

We achieved very high expression levels for HPPD [4] or GFP (WO2010079117), but low levels are not exceptions, such as the product of the *aadA* selection gene.

9. We assume that the inheritance of the plastid genome is predominantly maternal in soybean, but we are not aware of precise data backing up this assumption. Concerning fertility of the plastid transformants, we did not experience any particular problem, unlike for some other species (e.g., *Arabidopsis*).

Acknowledgements

We thank Dr. Nathalie Dufourmantel who significantly contributed to developing plastid transformation in soybean.

References

1. Finer JJ, McMullen MD (1991) Transformation of soybean via particle bombardment of embryogenic suspension culture tissue. In Vitro Cell Dev Biol Plant 27:175–182
2. Dufourmantel N, Pelissier B, Garcon F et al (2004) Generation of fertile transplastomic soybean. Plant Mol Biol 55:479–489
3. Dufourmantel N, Tissot G, Goutorbe F et al (2005) Generation and analysis of soybean plastid transformants expressing *Bacillus thuringiensis* Cry1Ab protoxin. Plant Mol Biol 58:659–668
4. Dufourmantel N, Dubald M, Matringe M et al (2007) Generation and characterization of soybean and marker-free tobacco plastid transformants overexpressing a bacterial 4-hydroxyphenylpyruvate dioxygenase which provides strong herbicide tolerance. Plant Biotechnol J 5:118–133
5. Murashige T, Skoog F (1962) A revised medium for rapid growth and bioassays with tobacco tissue culture. Plant Physiol 15:473–497
6. Gamborg O, Miller R, Ojima K (1968) Nutrient requirement suspensions cultures of soybean root cells. Exp Cell Res 50:151–158
7. Santarem ER, Finer JJ (1999) Transformation of soybean (*Glycine max* L. Merrill) using proliferative embryogenic tissue maintained on semi-solid medium. In Vitro Cell Dev Biol Plant 35:451–455
8. Samoylov VM, Tucker DM, Thibaud-Nissen F et al (1998) A liquid medium based protocol for rapid regeneration from embryogenic soybean cultures. Plant Cell Rep 18:49–54
9. Svab Z, Maliga P (1993) High-frequency plastid transformation in tobacco by selection for a chimeric *aadA* gene. Proc Natl Acad Sci U S A 90:913–917
10. Finer JJ, Vain P, Jones MW et al (1992) Development of the particle inflow gun for the DNA delivery to plant cells. Plant Cell Rep 11:95–97
11. Finer JJ, Nagasawa A (1988) Development of an embryogenic suspension culture of soybean (*Glycine max* L. Merrill). Plant Cell Tissue Organ Cult 15:125–136
12. Bally J, Nadai M, Vitel M et al (2009) Plant physiological adaptations to the massive foreign protein synthesis occurring in recombinant chloroplasts. Plant Physiol 150:1474–1481
13. Nadai M, Bally J, Vitel M et al (2009) High-level expression of active human alpha1-antitrypsin in transgenic tobacco chloroplasts. Transgenic Res 18:173–183

Chapter 23

Plastid Transformation in Cabbage (*Brassica oleracea* L. var. *capitata* L.) by the Biolistic Process

Menq-Jiau Tseng, Ming-Te Yang, Wan-Ru Chu, and Cheng-Wei Liu

Abstract

Cabbage (*Brassica oleracea* L. var. *capitata* L.) is one of the most important vegetable crops grown worldwide. Scientists are using biotechnology in addition to traditional breeding methods to develop new cabbage varieties with desirable traits. Recent biotechnological advances in chloroplast transformation technology have opened new avenues for crop improvement. In 2007, we developed a stable plastid transformation system for cabbage and reported the successful transformation of the *cry1Ab* gene into the cabbage chloroplast genome. This chapter describes the methods for cabbage transformation using biolistic procedures. The following sections are included in this protocol: preparation of donor materials, coating gold particles with DNA, biolistic bombardment, as well as the regeneration and selection of transplastomic cabbage plants. The establishment of a plastid transformation system for cabbage offers new possibilities for introducing new agronomic and horticultural traits into *Brassica* crops.

Key words *aadA* gene, Biolistic DNA delivery, *Brassica oleracea* L. var. *capitata* L., Cabbage, Chloroplast transformation, *uidA* gene

1 Introduction

Cabbage (*Brassica oleracea* L. var. *capitata* L.), a member of the mustard or cruciferous family (*Brassicaceae*), is derived from the wild mustard plant. It is native from the Mediterranean coast to the North Sea [1] and is one of the important economical vegetables worldwide. Cabbage has been the most widely cultivated leafy vegetables among the *Brassica* vegetables. In 2008, the world cultivation area of cabbage and other brassicas was estimated to be 3.1 million hectares amounting to 69.7 million tonnes with the leading countries being China, India, Russia, and the Republic of Korea [2]. Cabbage is an excellent source of many important nutrients, particularly vitamin A, vitamin C, vitamin B6, vitamin K, folic acid, iron, and calcium. Moreover, cabbage contains a

Pal Maliga (ed.), *Chloroplast Biotechnology: Methods and Protocols*, Methods in Molecular Biology, vol. 1132,
DOI 10.1007/978-1-62703-995-6_23, © Springer Science+Business Media New York 2014

powerful anticancer compound known as glucosinolate [3]. Traditional breeding methods have been very successful in making major improvements in the yield and quality of cabbage over the past 50 years [4]; however, the traditional breeding methods have been limited by genetic resources.

Currently, gene transformation offers new possibilities for obtaining high-yielding, heat-tolerant cabbage cultivars and also those resistant to major diseases. Thus far, the most widely used method for cabbage genetic transformation is the *Agrobacterium-mediated* transformation system [5]. *Agrobacterium*-mediated transformation in cabbage has been reported with various genes, such as *uidA* (*gus*, beta-glucuronidase) [6–9], *gfp* (green fluorescent protein) [10], *bt* (*Bacillus thuringiensis* delta-endotoxin) [11–14], *cpt1* (cowpea trypsin inhibitor) [15], *oc-1* (cysteine proteinase inhibitor, oryzacystatin-1) [16], *betA* (choline dehydrogenase) [17], *vhb* (*Vitreoscilla* hemoglobin) [18], and *hsp101* (heat shock protein) [19].

Since the first stable chloroplast genetic engineering in tobacco was reported two decades ago [20], chloroplast transformation has become a powerful tool used for gene transformation. The most widely used selectable marker for chloroplast transformation is a gene encoding aminoglycoside 3′-adenylyltransferase (*aadA*) [21]. Transplastomic plantlets were then selected by spectinomycin and/or streptomycin resistance. Expression of foreign genes via chloroplast genomes offers several unique advantages, including high protein levels, expressing multiple genes in operons, transgene containment by maternal inheritance, as well as lack of gene silencing, position, and pleiotropic effects [21, 22]. Our laboratory has successfully employed biolistic bombardment to obtain a stable and highly efficient transformation platform for transplastomic cabbage [23]. Moreover, we have successfully transferred the *cry1Ab* gene into the "K-Y Cross" cabbage chloroplast genome [24]. The transformed lines exhibited significant resistance to the larvae of diamondback moth (*Plutella xylostella*), a ubiquitous pest of *Brassica* crops.

In this chapter we describe a method for the biolistic bombardment of young leaves of cabbage cultivar "K-Y Cross" and the recovery of transgenic plants using the *aadA* (aminoglycoside 3′-adenylyltransferase) gene as a selectable marker gene, conferring resistance to the spectinomycin and streptomycin. The plasmid pMT91GDA contains the chloroplast expression cassette and harbors a chimeric *aadA*, *daao* (D-amino acid oxidase), and *uidA* genes flanked by the cabbage 16S and 23S rRNA sequences (Fig. 1). These three genes are driven by the chloroplast promoter *Prrn*, and transcription is terminated by the *psbA* 3′ untranslated region. The protocol described here is based on the research of Liu et al. [23, 24] with slight modifications to improve several important factors for cabbage plastid transformation including preculture of explants, culture of explants, and selective agents.

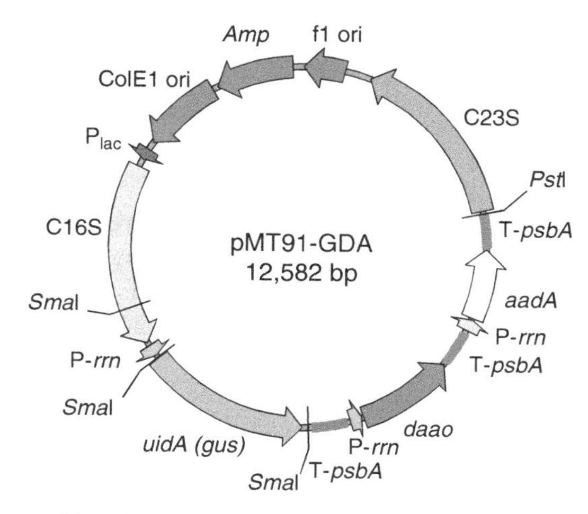

Fig. 1 Cabbage chloroplast transformation vector pMT91-GDA. The plasmid harbors chimeric *uidA*, *daao*, and *aadA* genes flanked by the cabbage *trnV–rrn16S* and *trnI–trnA–rrn23S* sequences, which serve as regions of homologous recombination after biolistic DNA delivery. The *aadA* selectable marker gene confers resistance to spectinomycin and streptomycin. The *daao* gene encodes D-amino acid oxidase. The *uidA* gene is the scorable reporter gene. These three genes are driven by the chloroplast promoter *Prrn*, and terminated by the *psbA* 3′ untranslated region

2 Materials

2.1 Chloroplast Transformation Vector

Figure 1 shows the plasmid pMT91GDA containing the *aadA*, *daao*, and *uidA* genes under the control of the tobacco plastid 16S rRNA promoter that has been successfully used for biolistic bombardment-mediated cabbage transformation. A plasmid vector that contains *aadA*, *daao*, and *uidA* genes is constructed. All of the constructed genes are driven by the *Prrn* promoter and terminated by *TpsbA* terminator. The DNA fragments of *aadA*, *daao*, and *uidA* genes are inserted into the species-specific expression cassette pASCC201 [23, 24] to produce the chloroplast transformation vector pMT91GDA (Fig. 1). The plastid transformation vector contains the *uidA*, *daao*, and *aadA* genes between the *trnV–rrn16S* and *trnI–trnA–rrn23S* regions, which is designed to be inserted into the chloroplast genome by homologous recombination after biolistic bombardment.

2.2 Plasmid DNA Preparation

1. The High-Speed Plasmid Mini Kit (Geneaid) is used to isolate the plasmid DNA according to the manufacturer's instructions. DNA is resuspended in sterile distilled water and stored at −20 °C (*see* **Note 1**).

2.3 Plant Material

1. The commercial cabbage (*Brassica oleracea* L. var. *capitata* L.) cultivar "K-Y Cross" is used in plastid transformation.
2. Uniform and plump seeds are selected.

3. Seeds are surface-sterilized using 70 % ethanol for 1.5 min and washed with autoclaved distilled water once.
4. Seeds are soaked in 2.5 % Clorox bleach with several drops of Tween-20 for 15 min with vigorous shaking at about 250 rpm and washed three times with autoclaved distilled water.
5. Seeds are germinated on 1/2 MS medium [25] supplemented with 3 % sucrose and 0.8 % agar (USB) and grown at 24 ± 2 °C in a 16-h/8-h day/night regimen at 150 μE/m^2/s light intensity for 14–21 days.

2.4 Culture Media

1. 1/2 MS medium: 1/2 MS salts plus vitamins (2.2 g/l, Duchefa Biochemie), 30 g/l sucrose (Merck), 0.8 % agar (USB), pH 5.7.
2. G1 medium (preculture medium): MS salts plus vitamins (4.4 g/l, Duchefa Biochemie), 0.05 ppm picloram (Sigma), 0.5 mg/l BA (Sigma), 0.5 mg/l $AgNO_3$ (Sigma), 3 % sucrose (Merck), 0.8 % agar (USB), pH 5.7 (*see* **Notes 2**, **3**).
3. S1 medium (shoot selection medium 1): MS salts plus vitamins (4.4 g/l, Duchefa Biochemie), 0.05 ppm picloram (Sigma), 0.5 mg/l BA (Sigma), 0.5 mg/l $AgNO_3$ (Sigma), 20 mg/l spectinomycin (Sigma), 3 % sucrose (Merck), 0.8 % agar (USB), pH 5.7.
4. S2 medium (shoot selection medium 2): MS salts plus vitamins (4.4 g/l, Duchefa Biochemie), 0.05 mg/l picloram (Sigma), 0.5 mg/l BA (Sigma), 0.5 mg/l $AgNO_3$ (Sigma), 30 mg/l spectinomycin (Sigma), 3 % sucrose (Merck), 0.8 % agar (USB), pH 5.7.
5. S3 medium (shoot selection medium 3): MS salts plus vitamins (4.4 g/l, Duchefa Biochemie), 0.05 mg/l picloram (Sigma), 0.5 mg/l BA (Sigma), 0.5 mg/l $AgNO_3$ (Sigma), 100 mg/l spectinomycin (Sigma), 3 % sucrose (Merck), 0.8 % agar (USB), pH 5.7.
6. R1 medium (rooting medium 1): MS salts plus vitamins (4.4 g/l, Duchefa Biochemie), 0.2 mg/l NAA (Sigma), 0.5 mg/l $AgNO_3$ (Sigma), 3 % sucrose (Merck), 0.8 % agar (USB), pH 5.7.
7. H1 medium (hardiness medium 1): 1/2 MS salts plus vitamins (2.2 g/l, Duchefa Biochemie), 1.5 % sucrose (Merck), 0.8 % agar (USB), pH 5.7.

2.5 Biolistic Bombardment Apparatus

1. The device used for plastid transformation is the Biolistic® PDS-1000/He Particle Delivery System supplied by Bio-Rad Laboratories, and aseptic operations are performed under a laminar flow hood.
2. The apparatus is sterilized using 70 % ethanol and allowed to air-dry.

3. The rupture disks (1,100 psi) (Bio-Rad) and macrocarrier disks (Bio-Rad) are sterilized by soaking in 100 % ethanol for 5 min and dried in a laminar flow hood.
4. The stopping screens (Bio-Rad) are sterilized by autoclaving.

2.6 Preparation of Gold Particles

1. Gold particles (50 mg) (0.6 μm, Bio-Rad) are weighed into a sterile 1.5 ml Eppendorf tube, and 1 ml of 100 % ethanol is added. The gold particles are vortex-mixed for 2 min.
2. Gold particles are pelleted by centrifuging the tube for 1 min at maximum speed in a benchtop microcentrifuge and then the supernatants are discarded.
3. The procedure of ethanol wash is repeated three times.
4. 1 ml of autoclaved distilled water is added and vortex-mixed for 1 min.
5. Gold particles are pelleted by centrifuging the tube for 1 min at maximum speed in a benchtop microcentrifuge and then the supernatants are discarded.
6. The procedure of autoclaved distilled water wash is repeated twice.
7. Gold particles are resuspended in 1 ml of autoclaved distilled water.
8. Finally, the gold particles are then divided into 50 μl aliquots in sterile 1.5 ml Eppendorf tubes. While dispensing the gold particles, care must be taken to ensure that the gold particles are kept in suspension by frequent vortex-mixing.
9. The gold particles are stored at −20 °C. The concentration of gold particles was 50 μg/μl.

2.7 Coating Gold Particles with DNA

1. One Eppendorf tube containing 50 μl of gold particle solution is removed from the freezer and allowed to thaw.
2. 10 μl of the target plasmid DNA (1 μg/μl) is added to this Eppendorf tube by pipetting the DNA onto the side of the Eppendorf tube, and then the gold and DNA are mixed immediately by vortexing for 1 min.
3. Then, 50 μl of 2.5 M $CaCl_2$ is added and vortex-mixed for 1 min (*see* **Note 4**).
4. Next, 20 μl of 0.1 M spermidine is added, and then all components are mixed immediately by vortexing for 1 min, followed by incubation at room temperature for 10 min (*see* **Notes 5, 6**).
5. DNA-coated gold particles are pelleted by centrifuging the Eppendorf tube for 30 s at maximum speed in a benchtop microcentrifuge and then the supernatants are discarded.
6. Then, 250 μl of 100 % ethanol is added and mixed by finger-tapping the bottom of the tube to ensure good mixing, and the ethanol is removed by centrifuging the Eppendorf tube for 30 s at maximum speed.

7. The procedure of ethanol wash is repeated three times.
8. Finally, 50 μl of 100 % ethanol is added to the DNA-coated gold particles and then vortex-mixed.
9. The DNA-coated gold particles are now ready for use.

2.8 Histochemical GUS Staining Solution

1. Following components are combined to prepare 10 ml of GUS histochemical staining solution [26] (*see* **Note 7**):

   0.64 ml 1 M Na_2HPO_4 (100 mM phosphate buffer).

   0.36 ml 1 M KH_2PO_4 (100 mM phosphate buffer).

   0.2 ml 0.5 M Na_2EDTA (pH 8.0) (10 mM).

   0.1 ml 50 mM $K_3Fe(CN)_6$ (0.5 mM).

   0.1 ml 50 mM $K_4Fe(CN)_6$ (0.5 mM).

   0.1 ml 10 % Triton X-100 (0.1 %).

   1 ml methanol (10 %).

   1 ml 3 % X-Gluc (MDBio) (dissolve 30 mg X-Gluc in 250 μl DMSO and then add distilled water to 1 ml) (0.3 %).

   Make up to 10 ml with distilled water, final pH 7.0.
2. The solution is filter-sterilized and stored in small aliquots at −20 °C.

3 Methods

3.1 Preparation for Bombardment

1. Set the PDS-1000/He Particle Delivery System to the following parameters:

   Rupture disk retaining cap to microcarrier launch assembly: 1/4″.

   Stopping screen to sample: 6 cm (position 3 from the base of the chamber).

   Vacuum: 28 in. Hg.
2. An alcohol-sterilized rupture disk (1,100 psi) is inserted into the rupture disk retaining cap by using a tweezer. The rupture disk has to be set flat into the rupture disk retaining cap (*see* **Note 8**).
3. Then, the rupture disk holder is tightly screwed into the gas acceleration tube of the bombardment chamber by using a torque wrench.
4. A sterile macrocarrier is placed into the macrocarrier holder by using a tweezer with its concave side facing upward. Macrocarrier insertion tool is used to push the macrocarrier in place inside the holder, if necessary.
5. 10 μl of the prepared DNA-coated gold particles is pipetted, and carefully spread it out on the center of the macrocarrier (*see* **Note 9**). Care must be taken to mix the DNA-coated gold particles by finger-tapping the bottom of the Eppendorf tube frequently to ensure that they are kept in suspension.

6. A sterile stopping screen is inserted into the microcarrier launch assembly.
7. The macrocarrier holder containing a macrocarrier prepared with dried DNA-coated gold particles is inverted and placed over the stopping screen.
8. The lid of the microcarrier launch assembly is tightened, and the whole unit is inserted into the bombardment chamber below the rupture disk holder.

3.2 Preparation of Plant Tissue for Bombardment

1. Young leaves are excised from 14- to 21-day-old aseptically propagated cabbage seedlings and precultivated in G1 medium under low-light conditions (5 μE/m²/s) for 1 week (Fig. 2a, b) (*see* **Notes 10–12**).
2. Six to eight young-green leaves are placed adaxial side up in a circle with a 2 cm radius on a G1 medium (9 cm plastic Petri dish) (Fig. 2c) (*see* **Note 13**). The leaves are now ready for biolistic bombardment.

3.3 Performing Bombardment

1. The Petri dish containing the leave samples to be bombarded is put on the target shelf that has been inserted into position 3 from the base of the chamber.
2. The lid of the Petri dish is removed and the door of the bombardment chamber is closed.
3. The "Vac" switch is flipped on to build up the vacuum until it reaches its maximum (up to 28 in. Hg), then the "Hold" switch is flipped swiftly to maintain the vacuum.
4. Now the "Fire" switch is pressed until the rupture disk is burst. Each shot carries 500 μg gold particles and 2 μg plasmid DNA (*see* **Note 14**).
5. After bombardment, the switch is toggled to "Vent" to release the vacuum, the door of the bombardment chamber is open, and the bombarded sample is removed from the chamber.

3.4 Cabbage Tissue Regeneration and Selection

1. The bombarded leaves are transferred to G1 medium and cultured in dark conditions.
2. 1 week later, bombarded leaves are cut into 9 mm² pieces (Fig. 2d) and are transplanted to S1 medium (containing 20 mg/l spectinomycin) for callus induction and primary antibiotic selection (Fig. 2e) (*see* **Note 15**).
3. Bombarded explants are subcultured onto S1 medium every 2 weeks until new shoots appear (Fig. 2f).
4. Over 4–6 weeks, the tissues are then transferred onto S2 medium (containing 50 mg/l spectinomycin) for shoot development and secondary antibiotic selection.

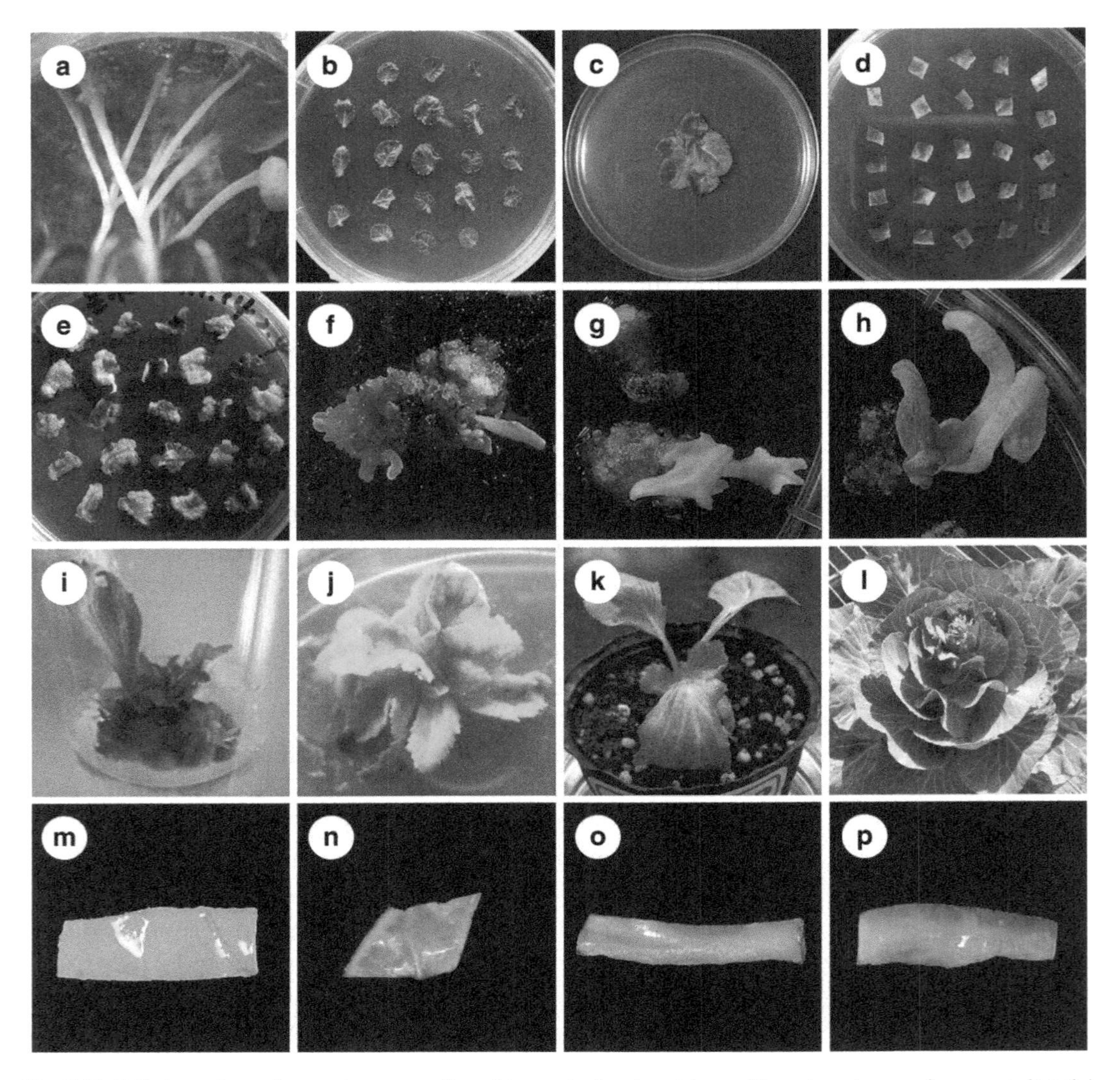

Fig. 2 Biolistic transformation and regeneration of transplastomic cabbage (*Brassica oleracea* L. var. *capitata* L.) cultivar "K-Y Cross." (**a**) Young leaves are excised from 14- to 21-day-old aseptically propagated cabbage seedlings. (**b**) Leaves are precultivated in G1 medium under low-light conditions. (**c**) Leaves are placed adaxial side up in a circle with a 2 cm radius on a GI medium for biolistic bombardment. (**d**) After cultivation in the dark for 1 week, bombarded leaves are cut into 9 mm^2 pieces and are transplanted to S1 medium. (**e–h**) Sequential selection of chloroplast transformed plantlets on the regeneration medium supplied with 20–100 mg/l spectinomycin. (i) Surviving shoots are transferred to R1 medium for induction of rooting. (**j**) Transformed plants are hardened off in H1 medium. (**k**) Transformed plants are transferred to potted soil. (**l**) Chloroplast transformed cabbage plants are grown in growth chambers. (**m**) GUS histochemical staining in mature leave of nontransformed cabbage. (**n–p**) GUS histochemical staining in mature leaves of transplastomic cabbage

5. The tissues are subcultured onto S2 medium every 2 weeks for a total of three times (Fig. 2g).
6. Then, the plantlets are transferred onto S3 medium (containing 100 mg/l spectinomycin) for shoots' growth and tertiary antibiotic selection.

7. The plantlets are subcultured onto S3 medium every 2 weeks at least twice (Fig. 2h).
8. Surviving shoots are transferred to R1 medium for induction of rooting (Fig. 2i).
9. Finally, the surviving transformants are transplanted and hardened off in H1 medium (Fig. 2j) (*see* **Note 16**).
10. Hardened plants with sufficient shoots, leaves, and roots are finally transferred to potted soil (Fig. 2k) and grown in growth chambers (Fig. 2l) (*see* **Note 17**).

3.5 Histochemical GUS Assay

1. The putative transgenic plants (spectinomycin resistant) are examined by PCR and RT-PCR, for the presence and expression of the *uidA* and *aadA* transgenes.
2. Transplastomic leaves are cut into 3 × 10 mm pieces and placed inside the Eppendorf tubes.
3. GUS staining solution is added to the Eppendorf tube until the tissues are completely covered (*see* **Note 18**).
4. Samples are vacuum infiltrated (28 in. Hg) for 5 min to enhance penetration of GUS staining solution and then incubated at 37 °C overnight in the dark.
5. The GUS staining solution is removed and replaced with 70 % ethanol several times until all the chlorophyll has been removed from the leaves (*see* **Note 19**).
6. Transformed leaves will turn deep blue, while that of the control leaves remain white.
7. Figure 2m–p shows the aspect of different transplastomic cabbage leaves after GUS staining assays.
8. GUS histochemical staining analyses show the blue-colored reactions are present in the *gus* gene of transplastomic cabbage leaves (Fig. 2n–p) but not in the nontransformed cabbage (Fig. 2m) (*see* **Note 20**).

4 Notes

1. DNA should be very pure, free of RNA or protein; otherwise, it forms clumps with gold particles. Ideally, the concentration of DNA should be at least 1 mg/ml. If the DNA concentration is too low, it is necessary to increase DNA concentration by ethanol precipitation.
2. Care must be taken not to autoclave tissue culture medium for more than 30 min. Selective agents are added prior to pouring autoclaved culture medium into Petri dishes. To avoid nonuniform distribution of selective agents, the autoclaved culture medium should be shaken well before pouring.

3. The addition of $AgNO_3$ (Sigma) in the induction medium is shown to reduce the browning of cabbage explants and improve regeneration. Picloram (Sigma) at 0.05 mg/l has been used as an alternative auxin.
4. $CaCl_2$ should be prepared fresh and the DNA-coated gold particles must be used within 3 h.
5. Spermidine deaminates with time and solutions are hygroscopic and oxidizable; therefore, they should be maintained at −80 °C and fresh stocks should be made monthly.
6. $CaCl_2$ and spermidine act to bind, stabilize, and precipitate the DNA. Strictly follow the order of adding gold, DNA, $CaCl_2$, and spermidine. This is critical for the proper coating of the gold particles.
7. The problem of endogenous plant GUS activities can be avoided by adding methanol, or increasing the GUS staining solution to pH 7.0–9.0 [26].
8. A rupture disk of 1,100 psi has been found to be optimal for the cabbage leaves reported here; a rupture disk 900 psi also results in successful transformation but with lower efficiency.
9. It is important to avoid clumps of gold particle when loading the macrocarriers because this will damage the leaf tissue during bombardment and decrease transformation efficiency.
10. Hypocotyls may also be used but they tend to produce more nontransformed shoots.
11. Fully expanded mature leaves may be used for cabbage plastid transformation. However, the concentrations of selective agents for screening the mature leaves are higher than those of young leaves.
12. The preculture period conditions the cabbage leaves for bombardment, mainly because dehydrated cells are better able to withstand the bombardment.
13. There is little or no waxy cuticle in aseptically propagated leaves. The side of the leaf to be bombarded has to be the adaxial side. This allows for better penetration of the gold particles and increases transformation efficiency.
14. The concentration of gold particles and DNA in gold/DNA mixture can vary. We recommend 250–500 μg gold particles and 1–2 μg plasmid DNA per shot to obtain a consistent efficiency of transformation.
15. When placing the bombarded leaf pieces on S1 medium, enough space is kept in between the pieces to allow the full expansion of the leaf segments.
16. The time when selection is imposed and the concentration of the selective agent are the critical factors upon which the quality and physiological status of the explants cultured depend.

One of the major problems faced in biolistic bombardment-mediated transformation of cabbage is the poor regeneration of explants after addition of spectinomycin or streptomycin. It is advisable to choose a lower concentration of selective agent at the stages of callus initiation and shoot induction, followed by a higher concentration at the stages of shoot elongation and enlargement.

17. Cover each transplastomic plant with either a plastic container or a bag immediately after transfer to soil as plants wilt quickly after removal from the culture containers. The plants are cultivated at a low light intensity.
18. Since cabbage has thick leaves, penetration of GUS staining solution may be a problem. It can be improved by slicing the leaf, infiltrating by vacuum, and adding Triton X-100.
19. The chlorophyll should be removed after GUS staining by incubating the tissues in 70 % ethanol. It is necessary to replace the ethanol several times and to incubate for overnight.
20. The procedures and growth media described here have been successfully used in our laboratory for the cabbage cultivars "K-Y Cross" and "Summer Summit." For other genotype, the growth medium and concentration of selective agent need to be optimized.

Acknowledgments

The authors sincerely wish to thank PeyJen Lee for her help with the preparation of the manuscript. This research was supported by funds from the Council of Agriculture, Taiwan, ROC (101AS-9.1.1-FD-Z1 to Menq-Jiau Tseng and Ming-Te Yang).

References

1. Dickson MH, Wallace DH (1986) Cabbage breeding. In: Bassett MJ (ed) Breeding vegetable crops. AVI Pub Co, Westport, CT
2. Food and Agricultural Organization of the United Nations (FAO) (2010) FAOSTAT. http://faostat.fao.org/site/613/default.aspx#ancor. Accessed 1 Nov 2010.
3. van Poppel G, Verhoeven DT, Verhagen H, Goldbohm RA (1999) Brassica vegetables and cancer prevention. Epidemiology and mechanisms. Adv Exp Med Biol 472:159–168
4. Fang Z, Liu Y, Lou P, Liu G (2005) Current trends in cabbage breeding. J New Seeds 6:75–107
5. Cardoza V, Stewart CNJ (2004) Brassica biotechnology: progress in cellular and molecular biology. In Vitro Cell Dev Biol Plant 40: 542–551
6. Christey MC, Sinclair BK (1992) Regeneration of transgenic kale (*Brassica oleracea* var. *acephala*), rape (*B. napus*) and turnip (*B. campestris* var. *rapifera*) plants *via Agrobacterium rhizogenes* mediated transformation. Plant Sci 87:161–169
7. Pius PK, Achar PN (2000) *Agrobacterium-mediated* transformation and plant regeneration of *Brassica oleracea* var. *capitata*. Plant Cell Rep 19:888–892
8. Tsukazaki H, Kuginuki Y, Aida R, Suzuki T (2002) *Agrobacterium*-mediated transformation of a doubled haploid line of cabbage. Plant Cell Rep 21:257–262
9. Sretenovic-Rajicic T, Ninkovic S, Miljus-Dukic J, Vinterhalter B, Vinterhalter D (2006) *Agrobacterium rhizogenes*-mediated transformation of *Brassica oleracea* var. *sabauda* and *B. oleracea* var. *capitata*. Biol Plant 50:525–530

10. Cogan N, Harvey E, Robinson H, Lynn J, Pink D, Newbury HJ, Puddephat I (2001) The effects of anther culture and plant genetic background on *Agrobacterium rhizogenes-mediated* transformation of commercial cultivars and derived doubled-haploid *Brassica oleracea*. Plant Cell Rep 20:755–762
11. Metz TD, Dixit R, Earle ED (1995) *Agrobacterium tumefaciens*-mediated transformation of broccoli (*Brassica oleracea* var. *italica*) and cabbage (*Brassica oleracea* var. *capitata*). Plant Cell Rep 15:287–292
12. Jin RG, Liu YB, Tabashnik BE, Borthakur D (2000) Development of transgenic cabbage (*Brassica oleracea* var. *capitata*) for insect resistance by *Agrobacterium tumefaciens*-mediated transformation. In Vitro Cell Dev Biol Plant 36:231–237
13. Bhattacharya RC, Viswakarma N, Bhat SR, Kirti PB, Chopra VL (2002) Development of insect-resistant transgenic cabbage plants expressing a synthetic *cryIA(b)* gene from *Bacillus thuringiensis*. Curr Sci 83:146–150
14. Paul A, Sharma SR, Sresty TVS, Devi S, Bala S, Kumar PS, Saradhi PP, Frutos R, Altosaar I (2005) Transgenic cabbage resistant to diamondback moth. Indian J Biotechnol 4:72–77
15. Fang HJ, Li DL, Wang GL, Li YH, Zhu Z, Li XH (1997) An insect resistance transgenic cabbage plant with cowpea trypsin inhibitor (CpTI) gene. Acta Bot Sinica 39:940–945
16. Lei JJ, Yang WJ, Yuan SH, Ying FY, Qiong LC (2006) Study on transformation of cysteine proteinase inhibitor gene into cabbage (*Brassica oleracea* var. *capitata* L.). Acta Hortic 706:231–238
17. Bhattacharya RC, Maheswari M, Dineshkumar V, Kirti PB, Bhat SR (2004) Transformation of *Brassica oleracea* var. *capitata* with bacterial *betA* gene enhances tolerance to salt stress. Sci Horti 100:215–227
18. Li X, Peng RH, Fan HQ, Xiong AS, Yao QH, Cheng ZM, Li Y (2005) *Vitreoscilla* hemoglobin overexpression increases submergence tolerance in cabbage. Plant Cell Rep 23:710–715
19. Rafat A, Aziz MA, Rashid AA, Abdullah SNA, Kamaladini H, Sirchi MHT, Javadi MB (2009) Optimization of *Agrobacterium tumefaciens-mediated* transformation and shoot regeneration after co-cultivation of cabbage (*Brassica oleracea* subsp. *capitata*) cv. KY Cross with *At*HSP101 gene. Sci Horti 121:447–450
20. Svab Z, Hajdukiewicz P, Maliga P (1990) Stable transformation of plastids in higher plants. Proc Natl Acad Sci U S A 87:8526–8530
21. Verma D, Daniell H (2007) Chloroplast vector systems for biotechnology applications. Plant Physiol 145:1129–1143
22. Sinagawa-García SR, Tungsuchat-Huang T, Paredes-López O, Maliga P (2009) Next generation synthetic vectors for transformation of the plastid genome of higher plants. Plant Mol Biol 70:487–498
23. Liu CW, Lin CC, Chen JJ, Tseng MJ (2007) Stable chloroplast transformation in cabbage (*Brassica oleracea* L. var. *capitata* L.) by particle bombardment. Plant Cell Rep 26:1733–1744
24. Liu CW, Lin CC, Yiu JC, Chen JJ, Tseng MJ (2008) Expression of a *Bacillus thuringiensis* toxin (*cry1Ab*) gene in cabbage (*Brassica oleracea* L. var. *capitata* L.) chloroplasts confers high insecticidal efficacy against Plutella xylostella. Theor Appl Genet 117:75–88
25. Murashige T, Skoog F (1962) A revised medium for rapid growth and bioassays with tobacco tissue culture. Physiol Plant 15:473–497
26. Martin T, Wöhner RV, Hummel S, Willmitzer L, Frommer WB, Gallagher SR (1992) The GUS reporter system as a tool to study plant gene expression. In: Gallagher SR (ed) GUS protocols: using the GUS gene as a reporter of gene expression. Academic, San Diego, pp 23–43

Chapter 24

Plastid Transformation in Sugar Beet: *Beta vulgaris*

Francesca De Marchis and Michele Bellucci

Abstract

Chloroplast biotechnology has assumed great importance in the past 20 years and, thanks to the numerous advantages as compared to conventional transgenic technologies, has been applied in an increasing number of plant species but still very much limited. Hence, it is of utmost importance to extend the range of species in which plastid transformation can be applied. Sugar beet (*Beta vulgaris* L.) is an important industrial crop of the temperate zone in which chloroplast DNA is not transmitted trough pollen. Transformation of the sugar beet genome is performed in several research laboratories; conversely sugar beet plastome genetic transformation is far away from being considered a routine technique. We describe here a method to obtain transplastomic sugar beet plants trough biolistic transformation. The availability of sugar beet transplastomic plants should avoid the risk of gene flow between these cultivated genetic modified sugar beet plants and the wild-type plants or relative wild species.

Key words Sugar beet, *Beta vulgaris* L, Biolistic transformation, Chloroplast transformation, Plant regeneration, Petioles

1 Introduction

Although improvements in various sugar beet traits, such as sugar yield [1] or disease resistance [2], have been achieved through conventional breeding, other traits such as herbicide resistance could be introduced in this crop by genetic transformation. Unfortunately, this species is still considered recalcitrant to genetic transformation even if transformation of the sugar beet genome is performed in several research laboratories [3]. Moreover, a lot of concerns could arise from the introduction of transgenic sugar beet in the field due to its well-documented cross-compatibility with the wild-relative sea beet (*B. vulgaris* ssp. *maritima*) [4, 5]. Transplastomic sugar beet lines with a maternal inheritance of the transgene could solve problems of gene flow between genetic modified varieties and wild relatives or conventional cultivars.

Pal Maliga (ed.), *Chloroplast Biotechnology: Methods and Protocols*, Methods in Molecular Biology, vol. 1132,
DOI 10.1007/978-1-62703-995-6_24, © Springer Science+Business Media New York 2014

Sugar beet chloroplast transformation, resulting in homoplasmic transgenic plants, is achieved via biolistic bombardment of leaf petioles. We select from bombarded petioles heteroplasmic sugar beet calli on spectinomycin-containing medium. Sugar beet regeneration from leaf rosette-isolated petioles is based on the method described by Mishutkina and Gaponenko [6] with few modifications concerning the selection of calli. These calli are able to regenerate by somatic embryogenesis into heteroplasmic shoots only after antibiotic removal from the culture medium. Stable homoplasmic plants can be then obtained via organogenesis after subsequent rounds of regeneration in a low spectinomycin-containing medium. These primary transformant plants (T0) can show sometimes some morphological alterations and be male sterile. To overcome this problem, the T0 plants are crossed with wild-type plants and seeds are collected from the transplastomic plants and germinated. Plants (T1) obtained from these seeds grow in a manner similar to wild-type plants and have a normal flowering phenotype.

2 Materials

2.1 Seed Sterilization and Germination

1. Seeds from sugar beet Z025 multigerm breeding line. These seeds are provided by Dr. Walther Faedi, Centro di Ricerca per le Colture Industriali, CRA-CIN, via di Corticella, 133, 40128, Bologna, Italy.
2. Ethanol 70 %.
3. Formaldehyde 35 % (v/v). Very toxic. Manipulate it under a fume hood.
4. $HgCl_2$ 0.05 % (w/v).
5. Hypochlorite 5 % (w/v).
6. Sterilized distilled water.
7. Medium 1 (1 L): 100 mL MS, Murashige and Skoog medium [7], stock solution 10×; benzylaminopurine (BA) 1 mg/L; 2,3,5-triiodobenzoic acid (TIBA) 0.5 mg/L; sucrose 30 g; pH to 5.8. The liquid medium is supplemented with 4 g of Phytagel and autoclaved.
8. MS stock solution 10×: dissolve the medium powder (Duchefa Biochemie, Haarlem, the Netherlands) in 1 L of distilled water and adjust pH between 2.0 and 3.0. Store at 4 °C.

2.2 Sample Preparation

1. Medium 2 (1 L): 100 mL MS, Murashige and Skoog medium [7], stock solution 10×; benzylaminopurine (BA) 0.25 mg/L; sucrose 30 g; pH to 5.8. The liquid medium is supplemented with 4 g of Phytagel and autoclaved.
2. Medium 3 (1 L): 100 mL MS, Murashige and Skoog medium [7], stock solution 10×; benzylaminopurine (BA) 0.5 mg/L;

α-naphthaleneacetic acid (NAA) 0.1 mg/L; sucrose 30 g; pH to 5.8. The liquid medium is supplemented with 4 g of Phytagel and autoclaved.

2.3 Gold Particle Coating

1. 0.6 μm gold particle (Bio-Rad).
2. Ice-cold HPLC absolute ethanol. The ethanol purity should be as high as possible. Do not leave open the bottle for a long time because the ethanol purity can decrease due to aqueous vapor.
3. Ice-cold sterile water.
4. 2.5 M $CaCl_2$. Freshly prepared and filter sterilized.
5. 0.1 M spermidine (cod. S4139, Sigma-Aldrich, St. Louis, MO, USA). Prepare dissolving spermidine in distilled water. Filtered sterilized small aliquots can be stored at −20 °C. Use each sample once.
6. DNA for gold coating at a concentration of 1–2 μg/μL in TE 10:1 buffer (Tris–HCl 10 mM, pH 8, EDTA 1 mM).

2.4 Biolistic Transformation

1. DuPont PDS1000He biolistic gun (Bio-Rad Laboratories, Hercules, CA, USA).
2. Medium 3 with 50 mg/mL spectinomycin.

2.5 Sugar Beet Regeneration

1. Medium 4 (1 L): 100 mL MS, Murashige and Skoog medium [7], stock solution 10×; benzylaminopurine (BA) 1 mg/L; abscisic acid (ABA) 0.5 mg/L; sucrose 30 g; pH to 5.8. The liquid medium is supplemented with 4 g of Phytagel and autoclaved.
2. MSIBA3 (1 L): 100 mL MS, Murashige and Skoog medium [7], stock solution 10×; indole-3-butyric acid (IBA) 3 mg/L; sucrose 30 g; pH to 5.8. The liquid medium is supplemented with 4 g of Phytagel and autoclaved.

3 Methods

To develop a protocol for sugar beet chloroplast transformation, as for all the plant species, is of extreme importance to have an efficient and reproducible plant regeneration methodology from tissue culture explants. Since sugar beet regeneration is strictly dependent on plant genotype and a widely applicable protocol has not yet been reported, it is a good rule to test the regeneration efficiency of the sugar beet varieties before starting biolistic bombardment experiments.

3.1 Sowing Seeds

1. Seeds of the multigerm breeding line Z025 (*see* **Note 1**) were kept at 4 °C for 1 week. This step is important to synchronize seed germination.
2. Soak the seeds overnight in tap water at 4 °C.

3. Remove water and transfer the seeds in 70 % ethanol (v/v) for 1 min.
4. Remove ethanol from the seeds and add formaldehyde 35 % (v/v) for 1 min.
5. Discard formaldehyde (into a hazardous waste container) and treat the seeds for 5 min with a solution of $HgCl_2$ 0.05 % (w/v). Shake the sample to promote sterilization of all seeds.
6. Replace the $HgCl_2$ solution with hypochlorite 5 % (w/v) for 10 min. Shake the sample to promote sterilization of all seeds.
7. Eliminate the hypochlorite solution and rinse very carefully the sterilized seeds by 10 min washing with autoclaved distilled water. Shake to facilitate the complete removal of hypochlorite. Repeat this step three times.
8. Place seeds on Petri dishes supplemented with Medium 1 and allow them to germinate in the dark at 25 °C for 1 month.

3.2 Petioles from Leaves

1. Cut with a scalpel from the germinating seedlings sugar beet apexes with the two cotyledons and a hypocotyl segment of 4–5 mm.
2. Place the apexes in glass pots filled with Medium 2, and place them in the light (16 h photoperiod under fluorescent light, 27 μE/s/m^2) at 25 °C. Sugar beet leaf rosette formation will take about 6–9 weeks. Transfer the leaf rosettes to fresh Medium 2 every 3 weeks.
3. After 6–9 weeks cut the rosette external leaves and use them to pick up petioles. Continue to propagate the internal leaves on Medium 2 as further petioles source.
4. Cut petioles into 0.5 cm long pieces and place 30–50 pieces on a Petri dish filled with Medium 3.

3.3 Gold Preparation and Coating

1. Weight 2–2.2 mg of gold particles. This gold preparation will be sufficient for 10 shots.
2. All the following steps are carried out on ice.
3. Resuspend very carefully the gold particles in 100 μL of absolute ethanol vortexing for at least 1 min.
4. Centrifuge for 1 s at 1,500×*g* in a microfuge. Simply let the centrifuge speed reach 1,500×*g* and then press stop.
5. Remove ethanol completely and resuspend gold particles in 1 mL sterile water by vigorous vortexing and pipetting.
6. Centrifuge for 1 s at 1,500×*g* and discard the supernatant.
7. Suspend gold very carefully in 250 μL of sterile water.
8. Add to the gold preparation in the following order: 25 μg DNA (*see* **Note 2**), 250 μL 2.5 M $CaCl_2$, 50 μL 0.1 M spermidine, with brief vortexing after addition of each compound.

9. Incubate on ice for 10 min. Vortex for few seconds every 1 min.
10. Add 200 μL of 100 % ethanol and mix by vortexing.
11. Centrifuge for 1 s at 1,500×*g* and remove the supernatant completely.
12. Add 600 μL of 100 % ethanol and mix by vigorous vortexing and pipetting.
13. Centrifuge for 1 s at 1,500×*g* and remove the supernatant completely.
14. Repeat this step.
15. Resuspend the coated gold in 65 μL of absolute ethanol by vigorous vortexing and pipetting.

3.4 Petiole Bombardment

1. These instructions assume the use of a DuPont PDS1000He biolistic gun. It is critical to work in a laminar flow hood to prevent contamination. Vortex for at least 3 min the gold preparation and use 6 μL for each shot.
2. Place the Petri dish containing petioles sample at a distance of 6 cm from the micro-projectile stopping screen.
3. Perform bombardment at a pressure of 1,350 psi and at a vacuum of 28″ mercury.
4. After shooting incubate the explants in the dark for 48 h, then transfer petioles to Petri dishes filled with Medium 3 containing 50 mg/mL spectinomycin.
5. Incubate in the light for 4 months changing the medium every 30 days.

3.5 Transplastomic Sugar Beet Plants

1. Transfer small friable calli developing from brown petioles on Medium 4 containing 50 mg/L spectinomycin (*see* **Note 3**).
2. Pale-green spectinomycin-resistant calli will appear approximately after 1 month of selection on Medium 4 (*see* **Note 4**).
3. To obtain regeneration from these calli, after 2–3 months the spectinomycin concentration in Medium 4 should be gradually reduced and then completely eliminated (*see* **Note 5**).
4. Leave the regenerated embryos on Medium 4 without antibiotic selection until leaflets form.
5. Transfer shoots into glass pots filled with MSIBA3 medium supplemented with 12.5 mg/L of spectinomycin to promote rooting.
6. To obtain homoplasmic sugar beet plants, make two additional rounds of regeneration from petioles on Medium 3 supplemented with spectinomycin 12.5 mg/mL (*see* **Note 6**).
7. Transfer homoplasmic sugar beet plants to soil in pots for further development in the greenhouse (*see* **Note 7**).

4 Notes

1. We used this sugar beet line because of its good regeneration frequency from petioles. The frequency of shoot formation was high both for direct regeneration (shoots arising directly from petioles) and for indirect regeneration (shoots arising from petiole-derived calli).
2. Gold particles have been coated with pSB1 or pSB2 plasmids in a recent report [8]. These vectors carry sugar beet flanking regions to assure the best homologous recombination frequency between the transgene constructs and the sugar beet plastome. Vector pSB1 targets the transgene expression cassette in the inverted repeat region of the plastid genome between the *rrn16* and *rps12* genes, while pSB2 targets the homologous recombination in the large single copy chloroplast region (between the *accD* and *rbcL* genes). Both constructs carry the *aadA* and *gfp* genes in the expression cassette, respectively, for spectinomycin resistance and for plastid transformant visual screening. No transformants were obtained with the pSB2 plasmid.
3. It doesn't matter if petioles, after long culturing on regeneration medium, turn black. It is from these petioles that friable regenerative calli arise (at least for the Z025 breeding line). Petioles in which direct shoot regeneration occur stay green and thick, whereas they become soft and deep brown before callus differentiation.
4. The presence of a *gfp* expression cassette in the pSB1 vector was very useful at this stage [8], because it allowed visual identification and isolation of the transgenic calli early on. To further confirm the site of the insertion and the presence of the entire transgenic cassette on sugar beet plastome, a PCR analysis with specific primers was performed.
5. In the sugar beet line Z025, the spectinomycin selection inhibits shoot regeneration from callus.
6. The additional rounds of regeneration were carried out via direct shoot organogenesis from petioles.
7. Sugar beet primary transformants (T0) can show sometimes some morphological alterations (Fig. 1a) and can be also male sterile. In this case, the T0 plants should be crossed with wild-type plants, and seeds should be collected from the transplastomic plants which are the female parent. Plants (T1) germinated from these seeds grow in a manner similar to wild-type plants (Fig. 1b) and have a normal flowering phenotype.

Fig. 1 Transplastomic sugar beet plants. (**a**) A sugar beet primary transformant (T0) shows a dwarf phenotype in comparison to the wild-type plant (WT). (**b**) A progeny plant (T1) obtained crossing the two plants in (**a**) and collecting the seeds from the T0 transformant

References

1. Biancardi E, Lewellen RT, De Biaggi M, Erichsen AW, Stevanato P (2002) The origin of rizomania resistance in sugar beet. Euphytica 127:383–397
2. McGrath JM, Saccomani M, Stevanato P, Biancardi E (2007) Beet. In: Kole C (ed) Genome mapping and molecular breeding in plants, vol 5, Vegetables. Springer, Berlin, pp 135–151
3. Ivic-Haymes SD, Smigocki AC (2005) Biolistic transformation of highly regenerative sugar beet (*Beta vulgaris* L.) leaves. Plant Cell Rep 23:699–704
4. Bartsch D, Cuguen J, Biancardi E, Sweet J (2003) Environmental implications of gene flow from sugar beet to wild beet – current status and future research needs. Environ Biosafety Res 2:1–11
5. Saeglitz C, Pohl M, Bartsch D (2000) Monitoring gene flow from transgenic sugar beet using cytoplasmic male-sterile bait plants. Mol Ecol 9:2035–2040
6. Mishutkina IV, Gaponenko AK (2006) Sugar beet (*Beta vulgaris* L.) morphogenesis in vitro: effects of phytohormone type and concentration in the culture medium, type of explants, and plant genotype on shoot regeneration frequency. Genetika 42:210–218
7. Murashige T, Skoog F (1962) A revised medium for rapid growth and bioassays with tobacco tissue cultures. Physiol Plant 15: 473–497
8. De Marchis F, Yongxin W, Stevanato P, Arcioni S, Bellucci M (2009) Genetic transformation of the sugar beet plastome. Transgenic Res 18: 17–30

Chapter 25

Integration and Expression of *gfp* in the Plastid of *Medicago sativa* L.

Shaochen Xing, Zhengyi Wei, Yunpeng Wang, Yanzhi Liu, and Chunjing Lin

Abstract

Here we describe a protocol of alfalfa (*Medicago sativa* L.) plastid transformation by which *gfp*, a gene encoding the green fluorescent protein (GFP), is inserted into plastid genome via particle bombardment and homoplastomic plant is obtained. Plastid engineering is likely to make a significant contribution to the genetic improvement of this crop and the production of vaccines and therapeutic proteins.

Key words Alfalfa, Green fluorescent protein, GFP, *Medicago sativa* L., Plastid transformation, Spectinomycin selection

1 Introduction

Alfalfa (*Medicago sativa* L.) is one of the most important perennial leguminous forage crops [1]. Since ancient times it has been widely cultivated in the temperate zones of the world to produce high-quality feed for livestock. It is also an excellent source of proteins, vitamins, and minerals and a source of remedy for humans. Importantly, alfalfa yields of tissue culture-derived embryos at a high efficiency, making it an ideal model legume for genetic engineering manipulation and enabling development of new value-added products of commercial interest. *Medicago sativa* belongs to the minority of plant species, in which plastids are preferentially transmitted by pollen [1, 2].

Transformation of the alfalfa nucleus by *Agrobacterium tumefaciens* has been achieved more than two decades ago [3–8]. A major limitation of conventional nuclear transformation, like gene silencing and low-level expression of foreign gene, led scientist to find new approaches to obtain transgenic plants. Plastid genetic transformation, as an attractive alternative, offers several unique advantages and has been used for improving agronomic

Pal Maliga (ed.), *Chloroplast Biotechnology: Methods and Protocols*, Methods in Molecular Biology, vol. 1132,
DOI 10.1007/978-1-62703-995-6_25, © Springer Science+Business Media New York 2014

traits of crops [9], as well as producing value-added vaccine antigens and biopharmaceuticals in flowering plants and alga [10–12]. To date, there are 14 species of flowering plants with which fertile homoplastic plants have been obtained by means of plastid genetic modification. One exception is the cereal crop rice in which only heteroplastomic plants were obtained [13].

We report the development of a successful plastid transformation system in *Medicago sativa* L. Molecular evidence and confocal fluorescent detection confirm the existence and stable expression of foreign *gfp* gene in the alfalfa plastid genome. This protocol is based on ref. [14] from this laboratory.

2 Materials

2.1 Plastid Transformation Vector Construction

1. Plastid DNA isolated from young leaves of alfalfa.
2. Primers designed to amplify flanking sequences from plastid genome and primers annealed to 5′ and 3′ ends of expression cassette.
3. *Pfu*-based DNA polymerase and deoxynucleotide triphosphates (dNTPs) for polymerase chain reaction (PCR).
4. T4 polynucleotide kinase (TaKaRa Biotechnology Co., Ltd).
5. T4 DNA ligase (TaKaRa).
6. Calf intestinal alkaline phosphatase (CIAP) (TaKaRa).
7. Plasmid pUC19 (accession number: L09137).
8. DNA restriction enzyme *Pvu* II (Promega Co.).
9. Isopropyl-1-thio-β-D-galactoside (IPTG) is dissolved at 1 M in water (*see* **Note 1**), sterilized by 0.22 μm filter, stored in aliquots at −20 °C, and then added to media as required.
10. 5-Bromo-4-chloro-3-indolyl β-D-galactopyranoside (X-gal) is dissolved at 20 mg/mL in N, *N*-dimethylformamide (DMF), stored in aliquots at −20 °C, and then added to media as required.
11. Antibiotics: ampicillin is dissolved at 100 mg/mL, sterilized by 0.22 μm filter, and stored at −20 °C in aliquots.
12. Luria Broth (LB) medium: weight 10 g of peptone, 5 g of yeast extract, and 10 g of NaCl, dissolve in 1 L of water, autoclave, and store at 4 °C. For solid medium, 15 g/L of agar powder is added before autoclave.
13. LB/Amp plate: ampicillin is added to solid LB medium at a ratio of 1 μL of antibiotics to 1 mL LB medium after autoclaved when temperature dropped to approximately 50 °C and poured into 9-cm Petri dishes.

2.2 Tissue Culture, Transformation, and Transplastomic Plant Generation

1. Phytohormones stock solutions: 2, 4-dichlorophenoxyacetic acid (2, 4-D) was dissolved in drops of 1 M KOH, and the concentration was adjusted to 4 mg/mL using water. Kinetin (KT) was dissolved in water at 0.5 mg/mL. Benzylaminopurine (BAP) was dissolved in drops of 1 M HCl and adjusted into a concentration of 2.5 mg/mL with water. Naphthalene acetic acid (NAA) was dissolved in drops of 1 M KOH and adjusted to a concentration of 0.5 mg/mL. The stocks were stored at 4 °C.
2. Antibiotic stock: spectinomycin (Ameresco, Inc.) was dissolved in water at 125 mg/mL and stored in aliquots at –20 °C.
3. Basic MS medium: 4.74 g Murashige and Skoog medium (MS) powder mixture (Hopebio Inc. Qingdao, China) and 30 g of sucrose are dissolved in 1 L double distilled water; adjust the pH to 5.8 using 1 M KOH. The medium is solidified by supplementing 8 g/L of agar.
4. Explant preparation medium (EPM): 2 g of casein hydrolysate (Ameresco Inc.), 0.5 mL of 2, 4-D stock, and 0.5 mL of KT stock are added to basic medium (*see* **Note 2**).
5. Regeneration and selection medium (RSM): 500 mg/L of spectinomycin is added to EPM. Medium is autoclaved and antibiotic is added after the medium is cooled to about 60 °C.
6. Embryo multiplication and selection medium (EMSM): 0.1 mL of BAP stock, 0.1 mL of NAA stock, and 500 mg/L of spectinomycin are supplemented to basic medium. Medium is autoclaved and antibiotic is added after the medium is cooled to about 60 °C.
7. Embryo germination and rooting medium (EGRM): 500 mg/L of spectinomycin is added to basic medium. Medium is autoclaved and antibiotic is added after the medium is cooled to about 60 °C.
8. 0.1 % $HgCl_2$ (w/v): weight 0.1 g of $HgCl_2$, dissolve in 100 mL water, autoclave, and store at room temperature.
9. 100 % ethanol.
10. 70 % ethanol: 70 mL of 100 % ethanol thoroughly mix with 30 mL of sterilized water, and store in sterilized bottle at room temperature.
11. 50 % glycerol (v/v): 50 mL of glycerol thoroughly mix with 50 mL of sterilized water. After autoclaving, store at room temperature.
12. 2.5 M $CaCl_2{\cdot}H_2O$: 36.75 g of $CaCl_2{\cdot}H_2O$, dissolve in water and make a final volume of 100 mL. After autoclaving, store at room temperature. Always prepare freshly.
13. 0.1 M spermidine-free base: 1.45 g of spermidine-free base, dissolve in water and make a final volume of 10 mL, sterilize by 0.22 μm filter, and store in aliquots at –20 °C.

14. PDS 1000/He biolistic particle delivery system (gene gun): the gene gun is sterilized by UV light in the ultraclean cabinet for 30 min after wiping with 70 % ethanol. The UV light in ultraclean cabinet is turned on for 20 min before gene gun is wiped.
15. Consumables for bombardment (Bio-Rad): autoclave the macrocarrier holders, stopping screens, filter paper, and Kimwipes. Macrocarriers, rupture disks, and macrocarrier insertion tool are sterilized by submerging them in 100 % ethanol for 10 min. Place the sterilized macrocarriers and rupture disks (1,100 psi) over autoclaved Kimwipes and dry them under the laminar flow hood. Allow the macrocarrier insertion tool to dry in laminar flow cabinet.

2.3 Molecular Analysis of Transplastomic Plants

1. Plastid DNA extraction: plant tissue plastid DNA extraction kit (Genmed Scientifics Inc.).
2. Dounce tissue grinders.
3. PCR: DNA *Taq* polymerase, dNTPs, and primers annealing to the plastid genome sequences, which are flanking the insert expression cassette.
4. Southern blot analysis: hybridization probe flanking the insert expression cassette and DIG High Prime DNA Labeling and Detection Starter Kit I (Roche, Co.).
5. Tris-Borate-EDTA (TBE) buffer stock (5×): 54 g of Tris base, 27.5 g of boric acid, and 20 mL of 0.5 M EDTA (pH 8.0) was dissolved in 800 mL water and made up to a final volume of 1 L and stored at room temperature.
6. 6× gel loading buffer: 0.25 % bromophenol blue, 0.25 % xylene cyanol FF, 15 % ficoll type 4000, and 120 mM EDTA.
7. Nucleic acid dye stock: 10,000× of GeneFinder™ dye (Bio-V Inc. Xiamen, China).

2.4 Laser Scanning Confocal Fluorescent Microscopic Analysis of Transplastomic Plants

1. CPW buffer: 27.2 mg/L KH_2PO_4, 101.0 mg/L KNO_3, 1,480.0 mg/L $CaCl_2{\cdot}2H_2O$, 246.0 mg/L $MgSO_4{\cdot}7H_2O$, 0.16 mg /L KI, 0. 025 mg/L $CuSO_4{\cdot}5H_2O$, and pH 5.8.
2. 13 % mannitol solution: CPW buffer containing 13 % mannitol.
3. Enzyme buffer: 10.8 g of glucose, 2 g of cellulase Onozuka R-10 (Yakult, Co.), and 0.5 g of pectinase PectalyaseY-23 (Yakult, Co.) dissolved in 100 mL of CPW and freshly prepared for each use.
4. Washing buffer: CPW buffer containing 10.8 % glucose.

3 Methods

3.1 Plastid Transformation Vector Construction

1. Amplification and cloning of flanking sequences. Because the complete plastid genome sequence of alfalfa is not available, we designed two primers (annealed to 16*S* rDNA and 23*S* rDNA, respectively) according to the complete plastid genome sequence of *Medicago truncatula* (accession number: AC093544.8), a close relative of alfalfa, for amplification of flanking sequences for plastid transformation vector construction. The pair of primers was MsflF, 5′-ACTCTGCTGGGCCGACACTGACACTG-3′ and MsflR, 5′-CCTGGCTGTC TCTGCACTCCTACC-3′. The flanking sequences should be high homologous (commonly >99 %) to the plastid genome so that a homologous recombination can occur, and thus we use an enzyme with high fidelity, *TransStart FastPfu* DNA polymerase (TransGen Biotech Ltd. Beijing, China), to set up a PCR reaction. The components of reaction are as follows: template DNA, 20–50 ng; buffer, 1×; dNTPs, 0.25 mM; primer MsflF, 0.2–0.4 μM; primer MsflR, 0.2–0.4 μM; and enzyme, 2.5 U. Make up the total volume to 50 μL with sterilized water. The PCR program is as follows: pre-denature at 95 °C for 5 min; then perform 30 circles of denaturation at 95 °C for 45 s, annealing at 50 °C for 40 s, elongation at 72 °C for 2 min, and finally an extension for 5 min at 72 °C. The product is subsequently treated with T4 polynucleotide kinase (*see* **Note 3**) and ligated with the help of T4 DNA ligase to *Pvu*II-digested pUC19 dephosphorylated by calf intestinal alkaline phosphatase (CIAP) according to the manufacturer's instructions. The recombinant is subsequently sequenced and named pUCMs.
2. Construction of alfalfa plastid transformation vector. We designed and synthesized the expression cassette Pprrn-T7g10-smGFP-*aadA*-Trps16-MCS-TpsbA in plasmid pUC57. In this cassette, a soluble modified green fluorescent protein gene (*smGFP*, accession number: U70495), aminoglycoside-3′-adenylyltransferase gene (*aadA*), and T7g10 represents the leader sequence of gene 10 of T7 phage, and 5′ and 3′ regulatory sequences derived from tobacco plastid genome (accession number: NC_001879) including the promoter of the rrn operon (Pprrn), the *rps16* gene 3′ region (Trps16), and the 3′-untranslated region of *psbA* gene (TpsbA) as well as a multiple cloning site (MCS) are included. Primers annealed to the 5′- and 3′-ends of the cassette. The PCR reaction is the same as the reaction for amplification of the flanking sequences except changing the amount of DNA (cassette involved in pUC57) template to 5–10 ng. The PCR parameters are as follows: at 95 °C for 5 min, 30 cycles at 95 °C for 45 s, at 55 °C for 40 s, and elongation at 72 °C for 90 s and finally an extension for 3 min at 72 °C. The product is treated with T4

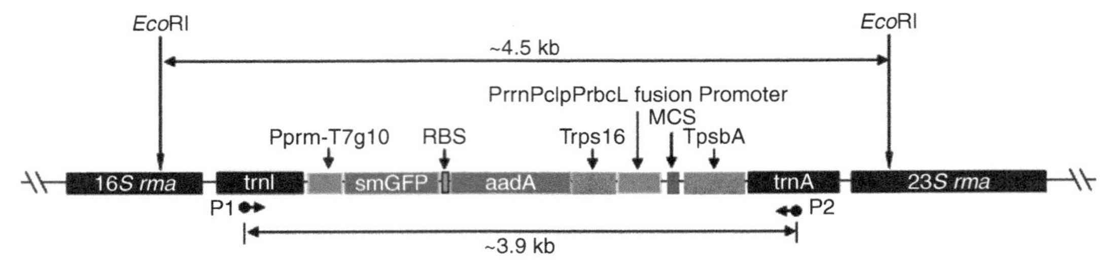

Fig. 1 The 3.9-kb pLXW-Ms01 alfalfa chloroplast vector after integration in the plastid genome. Plasmid pLXW-Ms01 carried the Pprrn-T7g10-*smGFP-aadA*-Trps16 and PrrnPclpPrbcL-MCS-TpsbA cassette. *smGFP* is soluble modified fluorescent protein; *aadA* encodes the aminoglycoside-3″-adenylyltransferase enzyme; MCS is a multiple cloning site including *XbaI* and *SalI* sites; Prrn, Trps16, and TpsbA are regulatory elements from tobacco chloroplast genome (GenBank accession number NC_001879). PrrnPclpPrbcL is a promoter. T7g10 is the T7 phage gene 10 leader sequence. RBS is the tobacco *rbcL* gene ribosome-binding site (GTAGGGAGGGATTT). The cassette was inserted into a *Pvu* II site between the *trnI* and *trnA* genes. P1 and P2 were primers used in PCR analysis. *EcoR* I was used to digest the chloroplast DNA for Southern analysis and to obtain the *trnI-trnA* fragment to be used as probe

polynucleotide kinase and inserted into dephosphorylated *Pvu*II-digested pUCMs with the help of T4 DNA ligase. Thus we obtained the plastid transformation vector pLXW-Ms01 (Fig. 1).

3.2 Preparation of Explants for Bombardment

1. *Medicago sativa* (cv. Longmu 803) seeds are surface sterilized as follows (*see* **Note 4**): seeds are first dipped in 70 % ethanol for 30 s and then dipped in 0.1 % mercuric chloride for 8 min. After that wash the seeds in sterile distilled water three times for 2 min each and allow seeds imbibe 6 h in sterile distilled water on a shaker. Dip the imbibed seeds in 0.1 % mercuric chloride again for 8 min, and wash by sterile distilled water five times for 2 min each. The sterile seeds were placed onto agar-solid 1/2 MS medium in transparent plastic bottles and cultured in a period of 16 h light and 8 h dark (*see* **Note 5**).
2. After 3-week culture, leaves of sterile plants were harvested as explants for bombardment. The leaflets of trifoliate compound leaves were placed on EPM medium with axial side down on the medium and bombarded as soon as possible. For bombardments with callus as explants, the leaflets are placed on EPM medium with the abaxial side down and cultured for 7–10 days (16 h light and 8 h dark) (*see* **Note 6**). For bombardment the explants (leaves or calli) are placed closely in a 2-cm radius (9-cm plastic Petri dish).

3.3 Sterilization of Gold Particles for Bombardment (*See* Note 7)

1. Weigh 60 mg of gold particles (0.6 μm) in a sterile 1.5-mL centrifuge tube.
2. Add 1 mL of 100 % ethanol and vortex-mix for 3 min.
3. Centrifuge the tube at 15,700×*g* for 5 min at room temperature and discard the supernatant.

4. Add 1 mL of 70 % ethanol and vortex-mix for 2 min.
5. Incubate the tube for 15 min at room temperature. Vortex-mix the contents of the tube thoroughly about three times during the incubation.
6. Centrifuge the tube at 15,700×*g* for 5 min, and then discard the supernatant.
7. Add 1 mL of distilled water and vortex until the particles are completely suspended.
8. Allow the particles to settle down for 1 min at room temperature and then centrifuge the tube at maximum speed for 2 min; discard the supernatant (*see* **Note 8**).
9. Repeat **steps** 7 and **8** two additional times.
10. Add 1 mL of 50 % glycerol to the gold particles and store −20 °C for use.

3.4 Coating of DNA on Gold Particles for Bombardment (*See* Note 7)

1. Vortex-mix the previously prepared gold particles stored at −20 °C until they are completely resuspended.
2. Pipet out 50 μL of the gold particle suspension into a sterilized 1.5-mL centrifuge tube.
3. Add 10 μL of plasmid DNA (1 μg/μL), and vortex-mix for 5 s.
4. Add 50 μL of 2.5 M $CaCl_2$ and vortex-mix for 5 s.
5. Add 20 μL of 0.1 M spermidine-free base and vortex-mix for 5 s (*see* **Note 9**).
6. Vortex the mixture for 20 min at 4 °C.
7. Add 200 μL of room temperature 100 % ethanol to the mixture, vortex for 5 s, and then centrifuge the mixture for 30 s at 800×*g*. Remove the supernatant and repeat **step** 7 four times (*see* **Note 10**).
8. After the final step, add 50 μL of 100 % ethanol and vortex to suspend the particles; keep them on ice for use.
9. The macrocarrier is placed inside the macrocarrier holder with the help of macrocarrier insertion tool. Repeat this step to make sure the number of macrocarrier-macrocarrier holder sets matches the number of bombardments you plan to carry out.
10. Resuspend the gold particles coated with plasmid DNA completely by vortex-mixing and pipetting to eliminate any clumps. Pipet 10 μL of the suspension out and spread them in the center of the macrocarrier. The DNA-coated gold particles (microcarriers) should be used within 1 h; we do not commend to use them for a second application, so just coat enough for your experiment.

3.5 Setup of Gene Gun and Particle Bombardment (See Note 7)

1. Rotate the adjustment handle of the helium regulator clockwise until the pressure is set at 1,350 psi.
2. Turn on power supply of the gene gun by turning up the left button of the gene gun.
3. Place a rupture disk (1,100 psi) into the rupture disk retaining cap, and rotate it into position in the gene gun tightly.
4. Place a stopping screen into the holder groove on the microcarrier launch assembly, and place a set of macrocarrier-macrocarrier holder into the groove with the microcarrier side downwards.
5. Place the microcarrier launch assembly in level one (three centimeter distance from rupture disk.) in the "chamber" of the gene gun.
6. Place the medium disk with explants in level 3 in the gene gun and close the door and fix it with the bolt beside.
7. Turn on the vacuum pump, press the *Vac* switch, and wait until the index of the vacuum gauge points at 27–28; hold it on by pressing the switch down quickly to "hold" position. You can either keep the vacuum pump on or turn it off during hold status.
8. Turn the right button (*Fire* button) up and hold it until the rupture disk is ruptured (you will hear a burst sound).
9. Press the *Vent* switch to release the vacuum.
10. When the vacuum gauge shows zero, open the door and take out the explants.
11. Discard the ruptured rupture disk. Take out macrocarrier-macrocarrier holder set and discard the macrocarrier.
12. Repeat **steps 3–11** until all the explants are bombarded.
13. Rotate tightly the adjustment handle of the helium regulator anticlockwise. Create some vacuum in the gene gun chamber and keep pressing the Fire button on and off until both pressure gauges show zero reading. Release the vacuum pressure and turn off the gene gun and vacuum pump.
14. The bombarded explants are recovered in dark at 26 °C for 3 days.

3.6 Selection, Regeneration, and Germination of Somatic Embryos

1. After recovery, the leaf explants (leaflets) are cut into two parts transversely and placed on RSM medium with the abaxial side closed to medium for selection. The calli explants are removed to RSM medium as well for selection. Selection medium is renewed every approximately 25 days during the selection period.
2. After approximately 2 months, resistant calli (somatic embryos) appear (Fig. 2a). Two weeks later, the somatic embryos should be transferred to EMSM medium for proliferation (Fig. 2b).

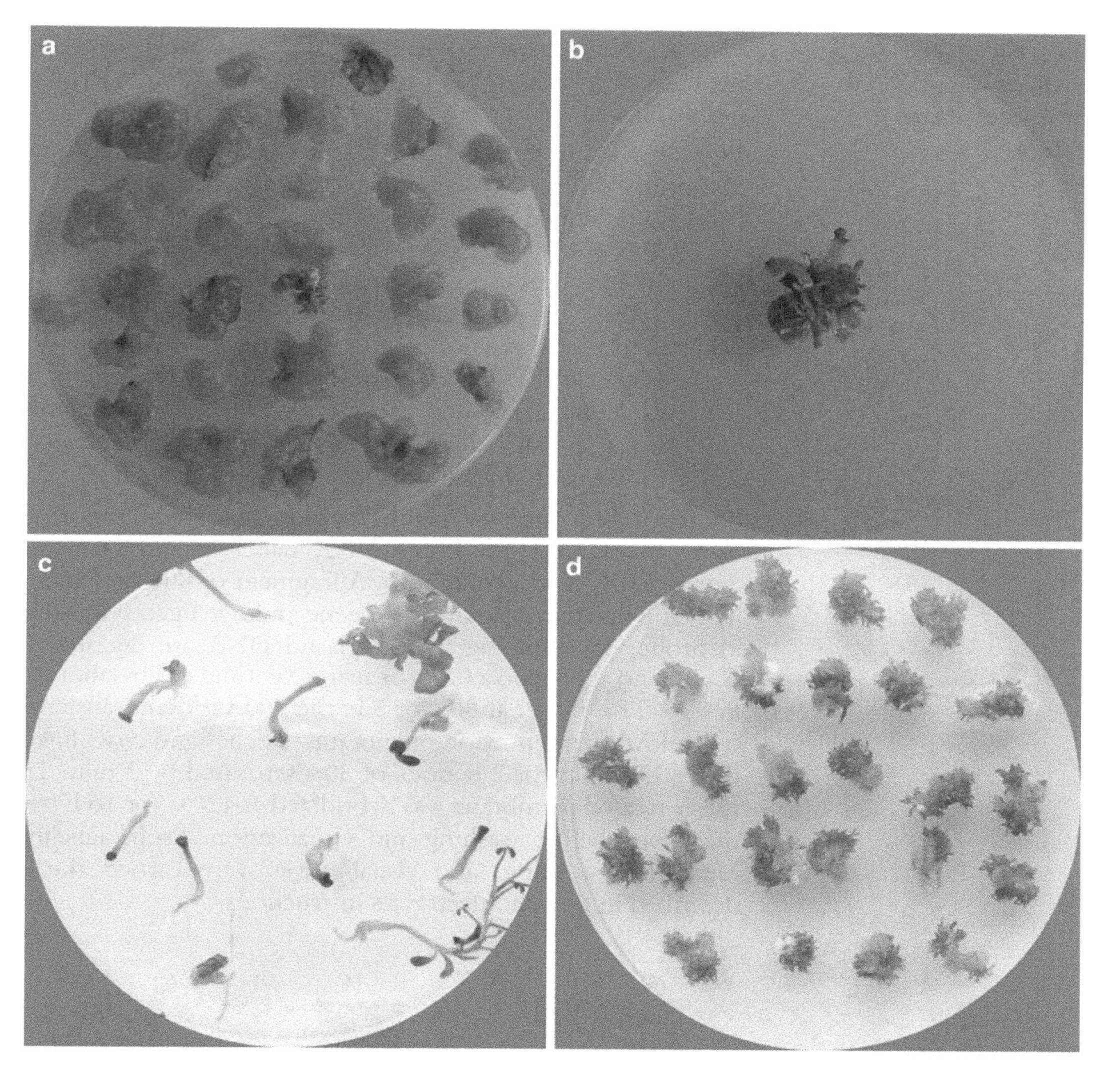

Fig. 2 Regeneration of transplastomic alfalfa. (**a**) Spectinomycin-resistant calli grew from bombarded callus explants on RSM regeneration and selection medium. (**b**) Resistant embryos proliferate on embryo multiplication and selection medium (EMSM) containing 500 mg/L spectinomycin. (**c**) Germination of resistant embryos and rooting of regenerated plantlet on embryo germination and rooting medium (EGRM) with 500 mg/L spectinomycin. (**d**) Resistant embryos during the second round of regeneration

3. When the embryos grow to a length about 1 cm, they are moved to EGRM medium for germination (*see* **Note 11**).
4. After approximately 20 days, the shoots grown from somatic embryos are excised and moved to EGRM medium in transparent plastic bottles for rooting (Fig. 2c).
5. Above is the description of one round of selection cycle. Three or more additional rounds of regeneration-selection cycles are needed starting with the leaves of regenerated plants to obtain genetically stable transplastomic plants (Fig. 2).

3.7 Molecular Analysis of Transplastomic Plants

1. Plastid DNA extraction: we extract alfalfa plastid DNA using a plant tissue plastid DNA extraction kit, following the product's user manual.
2. PCR analysis: to detect if the transplastomic plants are homoplastomic or not, PCR reactions are performed following the protocol described in 3.1, except that the DNA polymerase is *TansStart*™ *Taq* DNA polymerase (TransGen Biotech Ltd.) and PCR analysis is carried out using primers P1 and P2 shown in Fig. 1. Take the plasmid pLXW-Ms01 and plastid DNA from wild-type plants as positive control and negative control, respectively (Fig. 3a). The PCR parameters are as follows: predenature at 95 °C for 5 min; then perform 30 circles of denature at 95 °C for 45 s, annealing at 55 °C for 40 s, and elongation at 72 °C for 6 min and finally an extension for 10 min at 72 °C.
3. Southern blot analysis: southern blot analysis is performed using the DIG High Prime DNA Labeling and Detection Starter Kit I (Roche) (Fig. 3b). A fragment of 806 bp flanking insertion position (flanking *trnI* gene and *trnA* gene) was used as probe. Two micrograms of plastid DNA are digested by *Eco*R I and applied to a 0.8 % of agarose gel and electrophoresed in 0.5× TBE buffer at 80v for 3 h; the DNA is then transferred to positive charged nylon membrane (Roche) and cross-linked under UV light for 2 periods of 30 s separated by 2 min. The DNA-treated membrane was hybridized at 42 °C for 16 h with the probe before washing and visualization. Probe labeling, membrane washing, and visualization were carried out as described in the manufacturer's instructions.

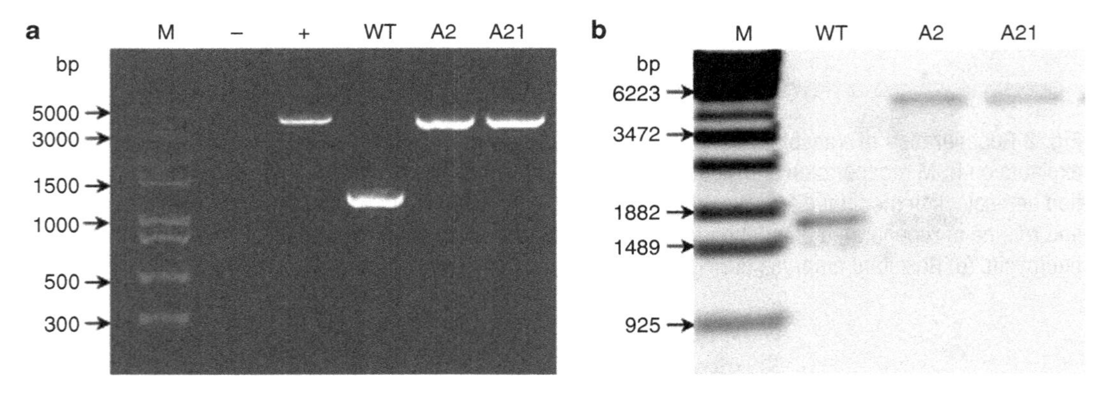

Fig. 3 PCR and Southern blot analysis of transplastomic plants. (**a**) PCR analysis using primers P1 and P2 marked in Fig. 1. A fragment of approximate 3.9 kb was amplified from resistant embryos (A2 and A21) indicating the integration of the expression cassette. Lane designation: –, double distilled water as negative control; +, pLXW Ms01 as positive control; WT, wild-type negative control; and M, DNA ladder (catalog No. BM421. TransGen Ltd., China). (**b**) Southern blot using chloroplast DNA digested by *Eco*R I and hybridized with the *trnI* and *trnA* probe flanking the insertion site (*see* Fig. 1). A 4.5-kb transplastomic fragment is present in the A2 and A21 lines and a 1.7-kb fragment in the wild type. This indicated the A2 and A21 lines were homoplastomic. WT, wild-type control; M, DNA standard (λ DNA digested by *Sty* I)

3.8 Laser Scanning Confocal Fluorescent Microscopic Analysis of Transplastomic Plants

1. In order to confirm the expression of GFP in plastid, a confocal fluorescent microscopic analysis is carried out.
2. The sterile leaves from transplastomic plants (always 0.1 g) are cut into chips and first osmotically treated with 13 % of mannitol for 1 h and subsequently treated with enzyme buffer for 8 h to release protoplasts in plastic dishes in the dark in a shaking incubator at 100 rpm.
3. The protoplasts are transferred to sterilized 1.5-mL centrifuge tube and collected by centrifuge at 100×*g* at room temperature for 6 min (*see* **Note 12**).
4. The protoplasts are washed by resuspending in washing buffer and collected as described above in **step 3**.
5. After washing, the protoplasts are resuspended in 200 μL washing buffer and then dropped to a cover glass slice (thickness ≤0.17 mm) and applied to laser scanning confocal microscopy (FV1000 IX, Olympus Co. Japan).
6. Images are scanned every 0.38 μm and captured at 1,000× magnifications using TRITC channel and FITC channel for red fluoresce and green fluoresce imaging. Serials of images from one sample are subsequently merged into one picture (Fig. 4).

4 Notes

1. "Water" in this text refers to water that has a resistivity of 18.2 MΩ cm and total organic content of less than five parts per billion [e.g., Milli Q (Millipore, Bedford, MA) grade water]. All the solutions are prepared using this standard water, unless stated otherwise. All chemicals used in this research are analytical grade.
2. Casein hydrolysate has buffering effect at about pH 5.8, so you can directly add agar powder and autoclave EPM medium without adjusting the pH value.

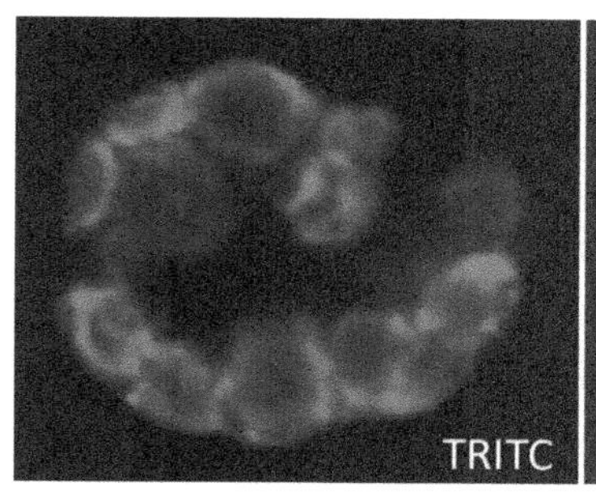

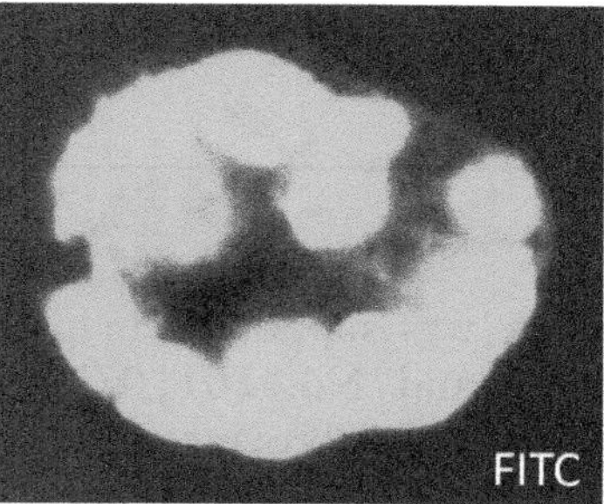

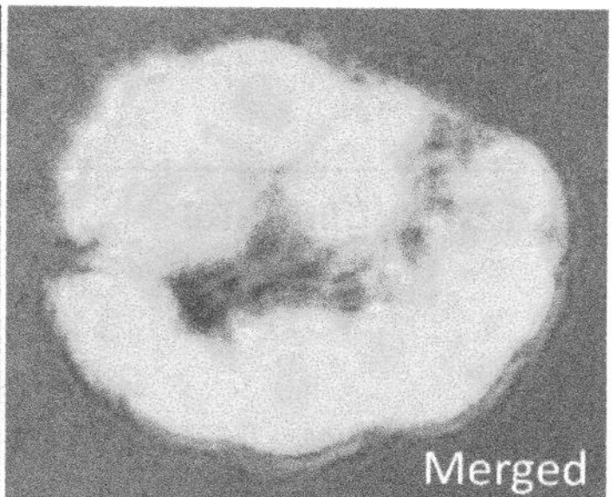

Fig. 4 Strong emission of green fluorescence from a transplastomic alfalfa protoplast observed by laser scanning confocal microscopy. This images was compounded from 45 slices of images scanned every 0.38 μm at 1,000× magnifications excited by 488 nm laser

3. This DNA polymerase produces a product without 5′ phosphate. It is best to treat the product with kinase before ligation, but you can also directly perform a ligation reaction without this treatment.
4. Alfalfa is an allotetraploid crop and self-sterile, so even though from the same cultivar, the genetic backgrounds of individuals are different from each other. A particular medium might not be suitable for all individuals, and explants from different individuals will have different regeneration frequency. In order to get uniform growth vigor, a preliminary experiment should be carried out to identify a particularly responsive individual plant and propagate this plant to meet the need in your research.
5. The protocol described here was carried out with a particular alfalfa plant selected from the cultivar Longmu 803.
6. It is better to cut the leaflets into two or more parts (depending on the size of the leaflet) transversely to make more wounds for callus induction.
7. Parts of these manipulations are adapted from ref. [15] with modifications.
8. The gold particles might float on the surface of the water because of surface tension. In that case, water can be pipetted out carefully with a 200-μL pipettor.
9. Make sure the gold particles do not form clumps. If the clumps do appear, pipet to eliminate them and consider using less spermidine next time.
10. If it is inconvenient to carry out **step 6** at 4 °C, manipulate at room temperature and use ice-cold ethanol in this step.
11. The embryos generally separate automatically when ready to germinate. In this case, just transfer the embryos to EGRM medium for germination. In some cases, the developing embryos do not separate (the length is about 1 cm). At this stage they can be separated from each other manually.
12. One can directly load the protoplasts onto cover glass slice without centrifuging and washing; in this case, the background will be much stronger.

Acknowledgements

This work was supported in part by founding of projects No. 20076021 and No. 20110753 from Science and Technology Department, Jilin Province, China.

References

1. Schumann CM, Hancock JF (1989) Paternal inheritance of plastids in *Medicago sativa*. Theor Appl Genet 78:863–866
2. Smith SE (1989) Influence of parental genotype on plastid inheritance in *Medicago sativa*. J Hered 80:214–217
3. Shahin EA, Spielmann A, Sukhapinda K, Simpson RB, Yashar M (1986) Transformation of cultivated alfalfa using disarmed *Agrobacterium tumefaciens*. Crop Sci 26:1235–1239
4. Deak M, Kiss GB, Koncz C, Dudits D (1986) Transformation of *Medicago* by *Agrobacterium* mediated gene transfer. Plant Cell Rep 5: 97–100
5. Chabaud M, Passiatore JE, Cannon F, Buchanan-Wollaston V (1988) Parameters affecting the frequency of kanamycin resistant alfalfa obtained by *Agrobacterium tumefaciens* mediated transformation. Plant Cell Rep 7: 512–516
6. D'Halluin K, Botterman J, Greef WD (1990) Engineering of herbicide-resistant alfalfa and evaluation under field conditions. Crop Sci 30:866–871
7. Kuchuk N, Komaritski I, Shakhovsky A, Gleba Y (1990) Genetic transformation of *Medicago* species by *Agrobacterium tumefaciens* and electroporation of protoplasts. Plant Cell Rep 8:660–663
8. Du S, Erickson L, Bowley S (1994) Effect of plant genotype on the transformation of cultivated alfalfa (*Medicago sativa*) by *Agrobacterium tumefaciens*. Plant Cell Rep 13:330–334
9. Daniell H, Kumar S, Dufourmantel N (2005) Breakthrough in chloroplast genetic engineering of agronomically important crops. Trends Biotechnol 23:238–245
10. Daniell H, Singh ND, Mason H, Streatfield SJ (2009) Plant-made vaccine antigens and biopharmaceuticals. Trends Plant Sci 14:669–679
11. Specht E, Miyake-Stoner S, Mayfield S (2010) Micro-algae come of age as a platform for recombinant protein production. Biotechnol Lett 32:1373–1383
12. Mata TM, Martins AA, Caetano NS (2010) Microalgae for biodiesel production and other applications: a review. Renew Sust Energ Rev 14:217–232
13. Lee SM, Kang K, Chung H, Yoo SH, Xu XM, Lee SB, Cheong JJ, Daniell H, Kim M (2006) Plastid transformation in the monocotyledonous cereal crop, rice (*Oryza sativa*) and transmission of transgenes to their progeny. Mol Cells 21:401–410
14. Wei Z, Liu Y, Lin C, Wang Y, Cai Q, Dong Y, Xing S (2011) Transformation of alfalfa chloroplasts and expression of green fluorescent protein in a forage crop. Biotechnol Lett 33: 2487–2494
15. Daniell H, Ruiz ON, Dhingra A (2005) Chloroplast genetic engineering to improve agronomic traits. Methods Mol Biol 286: 111–137

Part IV

Plastid Transformation in Algae

Chapter 26

Rapid Screening for the Robust Expression of Recombinant Proteins in Algal Plastids

Daniel Barrera, Javier Gimpel, and Stephen Mayfield

Abstract

Chlamydomonas reinhardtii has many advantages as a photosynthetic model organism. One of these is facile, targeted chloroplast transformation by particle bombardment. Functional recombinant proteins can be expressed to significant levels in this system, potentially outperforming higher plants in speed of scaling, cost, and space requirements. Several strategies and regulatory regions can be used for achieving transgene expression. Here we present two of those strategies: one makes use of the *psbD* promoter for expressing moderate levels of the recombinant protein in a photosynthetic background. The other strategy is based on the strong *psbA* promoter for obtaining high yields of the recombinant product in a non-photosynthetic strain. We herein describe the vectors, transformation procedures, and screening methods associated with these two strategies.

Key words *Chlamydomonas reinhardtii*, Chloroplast expression, Expression screening, Plastid transformation, *psbA* promoter, *psbD* promoter, Recombinant protein

1 Introduction

Chlamydomonas reinhardtii (*Chlorophyceae*) has long served as a model organism for photosynthesis, chloroplast biogenesis, and flagellar research [1]. Its chloroplast genome was first transformed over 20 years ago [2], and since then plastid transformation in algae has become a standard technique in many laboratories. The efficiency of homologous recombination in the chloroplast and the ability of cells to grow heterotrophically on acetate have allowed researchers to produce many targeted photosynthetic mutants [3–6]. *C. reinhardtii* chloroplasts are also an excellent platform for the expression of valuable recombinant proteins [7]. Algal plastid expression has many of the benefits of higher plant chloroplast expression, but is inherently faster and potentially less expensive and requires far less space when large numbers of transformants are required. Another advantage is that algae grow rapidly and easily in containment, thus making the scaling-up process faster and

Pal Maliga (ed.), *Chloroplast Biotechnology: Methods and Protocols*, Methods in Molecular Biology, vol. 1132,
DOI 10.1007/978-1-62703-995-6_26, © Springer Science+Business Media New York 2014

environmentally friendly. The time frame between transformation and scale-up can be as short as 2 months [8], considerably shorter than with higher plants because large amounts of seeds may require up to 5 months to produce in tobacco and 3 years in corn [9]. Additionally, algae can achieve up to 30–50 % protein by dry weight [10], and since they are unicellular, they don't have to maintain protein-poor differentiated structures like plants do.

Currently, particle bombardment is the best method for *C. reinhardtii* plastid transformation. Plastid expression vectors follow the same overall scheme as in plants, consisting of two flanking regions for homologous recombination, a promoter for mRNA transcription, and 5′ and 3′ untranslated regions used for translational regulation and mRNA stability. All of these control elements are required for driving the expression of selection markers as well as any recombinant genes of interest. The sites for transgene insertion differ among laboratories, but the *psbA* and *psbH* loci are commonly used. The *aadA* and *aphaVI* genes are two of the preferred selection markers conferring resistance to spectinomycin and kanamycin, respectively [11, 12]. There are also reporter genes available such as β-glucuronidase, GFP, and several luciferases (*see* ref. 8 for a review). The most widely used promoters/UTRs are from *psbA*, *psbD*, *psaA*, *atpA*, and *rbcL* [13–15]. In our laboratory we have demonstrated robust expression with the *psbA* promoter/UTR in a *psbA* knockout background [16]. It is also worth noting that *Chlamydomonas* plastid genome has a very AT-rich codon bias; thus, most of the genes have to be codon optimized and synthesized for optimal results. For example, GFP expression has been shown to increase 80-fold after codon optimization [17].

In this protocol we will show how to transform, screen, and monitor expression of transgenes by using the *psbA* promoter/UTR in a *psbA* knockout background (non-photosynthetic, highest expression) and a *psbD* promoter/UTR in a photosynthetic background (moderate expression in photosynthetic background).

2 Materials

2.1 Cell Growth and Preparation

1. *C. reinhardtii* strain W1.1 or *psbH*(−) (*see* **Note 1**).
2. Tris–acetate–phosphate (TAP) media.

2.2 DNA and Gold Particle Preparation

1. Seashell Technology S04e gold formulation kit.
2. *psbA* or *psbH* transformation vector (Fig. 1).
3. Ice-cold 100 % ethanol.
4. Macrocarrier holders (Bio-Rad).
5. Macrocarriers (Bio-Rad).

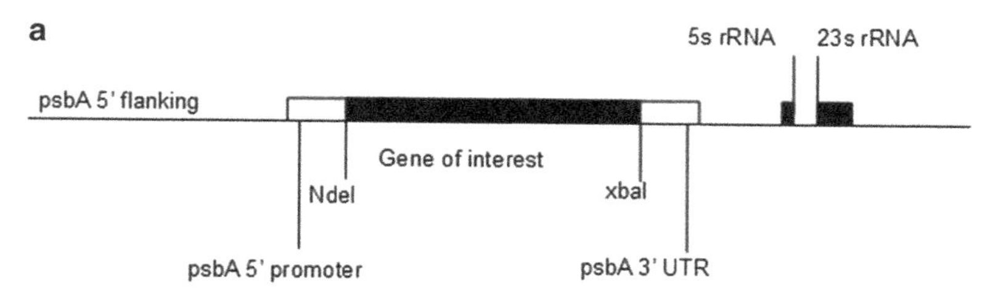

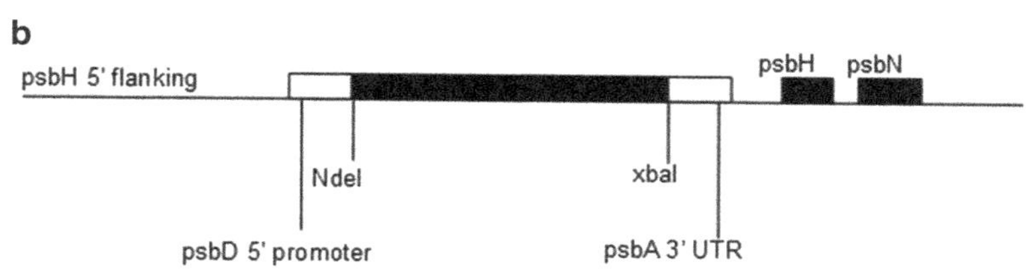

Fig. 1 *psbA* and *psbH* transformation vector schematics. (**a**) The *psbA* vector contains the gene of interest (GOI) inserted directionally with *Nde*I to *Xba*I restriction sites. The GOI has a *psbA* promoter and 5′-UTR and a *psbA* 3′-UTR. This cassette is flanked by regions of homology to the outside of the *psbA* locus. (**b**) The *psbH* vector also has *Nde*I and *Xba*I directional restriction sites. The GOI has a *psbD* promoter and 5′-UTR and a *psbA* 3′-UTR. The regions of homology flanking the cassette correspond to the *psbH* locus

2.3 Particle Bombardment

1. PDS-1000He particle delivery system (Bio-Rad).
2. Helium tank.
3. Vacuum pump.
4. Stopping screens (Bio-Rad).
5. 1,350 psi rupture discs (Bio-Rad).
6. Prepared macrocarrier assembly.
7. TAP + 100 μg/ml kanamycin 1.5 % w/v agar plates.
8. High-salt media (HSM) 1.5 % w/v agar plates.

2.4 PCR Screening

1. 10× Tris–EDTA (TE) pH 7.4 buffer.
2. Toothpicks.
3. PCR mix (*see* Tables 1 and 2).
4. Agarose gel electrophoresis and visualization equipment.

2.5 Expression Test

1. Lysis buffer: 750 mM Tris–HCl, pH8.0, 15 % sucrose (w/v), Roche Complete protease inhibitor cocktail. Store in aliquots at −20 ° Celsius after adding protease inhibitor.
2. Tip sonicator (Sonic Dismembrator 500, Fisher Scientific).
3. 4× Laemmli protein sample buffer: 8 % SDS, 40 % glycerol, 20 % 2-mercaptoethanol, 0.02 % bromophenol blue, 0.25 M Tris–HCl, pH 6.8.
4. Standard SDS-PAGE electrophoresis equipment and buffers.
5. Standard Western blot equipment and buffers.

Table 1
PCR reaction mixes for gene-positive and homoplasmic screens

	Gene positive (μl)	Homoplasmic (μl)
10x Taq ThermoPol buffer NEB	2.5	2.5
$MgSO_4$ 25 mM	1	1
dNTPs 25 μM	0.5	0.5
Primer 87 for	1.25	1.25
Primer *psbA*, *psbH*, or GOI rev	1.25	1.25
Primer 79 for	0	1.25
Primer 80 rev	0	1.25
Taq polymerase (e.g., NEB)	0.2	0.4
Water	16.3	13.6
Algae lysate	2	2
Total	25	25

Table 2
Primers for gene-positive and homoplasmic PCR reaction mixes

Primer	Sequence (5′–3′)	Working stock (μM)
79 for	CCGAACTGAGGTTGGGTTTA	3
80 rev	GGGGGAGCGAATAGGATTAG	3
87 for	GTGCTAGGTAACTAACGTTTGATTTTT	10
psbA rev	CTTGAATAGTTTCTCTAGCGTTAC	10
psbH rev	CTCCGATGCATCCTCTCTTAGG	10
GOI rev	Gene of interest specific	10

3 Methods

3.1 Cell Growth and Preparation

1. Inoculate a 50 ml starter culture of W1.1 or psbH(−) cells into liquid TAP media from an agar plate and grow under low light to a density of 1.0×10^6 to 3.0×10^6 cells/ml.
2. Transfer an aliquot of starter culture into a 250 ml flask at 2.0×10^4 cells/ml and grow in TAP to a concentration of between 8.0×10^5 and 2.0×10^6 cells/ml (*see* **Note 2**).
3. Harvest cells in 50 ml conical tubes and centrifuge at 3,000 rpm for 5 min at room temperature to pellet.
4. Decant supernatant and resuspend cells to 3.0×10^7 cells/ml.

5. Plate 1.5×10^7 cells on selective agar medium (*see* **Note 3**).
6. Allow excess liquid to absorb into the plate and perform DNA and gold preparation.

3.2 DNA and Gold Preparation

1. Place ethanol sterilized macrocarrier membranes in ethanol sterilized membrane holders.
2. Add 50 μl binding buffer to a sterile Eppendorf tube.
3. Add 10–14 μg of concentrated (midiprep) transformation plasmid (*see* **Note 4**).
4. Add 60 μl (3 mg) Seashell S550d carrier gold to the mixture and incubate on ice for 1 min.
5. Add 100 μl precipitation buffer and incubate on ice for 1 min.
6. Mix briefly by mild vortex and centrifuge for 10 s at 10,000 rpm to collect gold.
7. Remove the supernatant by pipette and wash with 500 μl ice-cold ethanol.
8. Perform 10 s spin again, remove supernatant, and add 50 μl ice-cold ethanol.
9. Resuspend pellet with a 1–2 s burst from a sonicator probe tip (minimum amplitude).
10. Immediately aliquot 10 μl portions of DNA/gold mixture onto individual macrocarrier membranes.

3.3 Particle Bombardment

1. Turn on PDS-1000He particle delivery system and open helium tank completely. The helium pressure should be at least 1,750 psi.
2. Turn on vacuum pump connected to PDS-1000He particle delivery system.
3. Sterilize all biolistics components including bombardment chamber with 100 % ethanol and allow to dry.
4. Place a 1,350 psi rupture disc into rupture disc holder and screw tightly into PDS-1000He delivery system.
5. Place a stopping screen into the center of the macrocarrier launch assembly unit.
6. Follow with prepared gold-coated macrocarrier membrane and holder placed gold side down.
7. Place assembly unit in the highest chamber slot nearest to the rupture disc.
8. Place plate holder and plate in the second slot from the bottom. Uncover plate.
9. Close chamber door and turn on the vacuum switch until pressure reaches 28 in. Hg.
10. Flip switch from vacuum to hold.

11. Press and hold fire button until rupture disc breaks.
12. Flip switch from vacuum to vent. When pressure reaches 0 in. Hg, open the door and immediately cover the plate.
13. Recover transformations in low light (<1,500 lux) until background dies off and transformed colonies appear.

3.4 PCR Screening of Transformed Colonies for Transgene Integration

1. Patch 24 colonies from the transformation plates onto selection plates by touching a colony with a toothpick and streaking it into a small patch.
2. Grow patches for 1 week.
3. Prepare cell lysates for PCR by touching a patch with a pipette tip and immersing it into 25 μl of 10× TE buffer (*see* **Note 5**). Make sure to include untransformed controls.
4. Use a thermocycler to heat cell suspensions to 95 ° C for 10 min. Cool to 4 °C.
5. Prepare the "gene-positive" PCR master mix (Table 1).
6. Distribute 23 μl of PCR master mix and use 2 μl of prepared lysates as templates. Run PCR.
7. Run an agarose gel to check for gene-positive strains.

3.5 PCR Screening for Homoplasmic Lines

1. Streak at least eight different gene-positive strains retrieved from the initial patches onto selective plates.
2. Wait for single colonies to come up (~1 week). Patch at least three colonies from each parental strain.
3. Wait for patches to grow. Repeat **steps 1–3** from at least three colonies. Always add at least two non-transformed parental strains as controls.
4. Set up a "homoplasmic PCR mix" (Table 1). Recommended extension time of 30 s and annealing temperature of 54 °C.
5. Run a 1.5 % agarose gel. A ~500 bp control band is given by primers 79 and 80. A ~300 bp (87 and *psbA* rev) band should disappear in *psbA* transgene homoplasmic strains. A ~800 bp band should disappear in *psbH* transgene homoplasmic strains (Fig. 2). Be sure that all non-transformed parental strains have both bands.
6. If no strains are homoplasmic, repeat **steps 1–5** using algae from **step 3**.
7. Pick at least four homoplasmic patches and streak them for protein extraction in selective plates (¼ of a plate each) (*see* **Note 6**).

3.6 Protein Extraction

1. Collect ¼ of a plate of algae with an inoculation loop.
2. Resuspend in 200 μl of lysis buffer.
3. Sonicate on ice for 20 s at 10 % power using the probe tip sonicator. Check in a microscope for evidence of lysis if necessary.
4. Optional: separate membrane and soluble protein by spinning at 14,000 rpm in a tabletop microcentrifuge for 10 min at

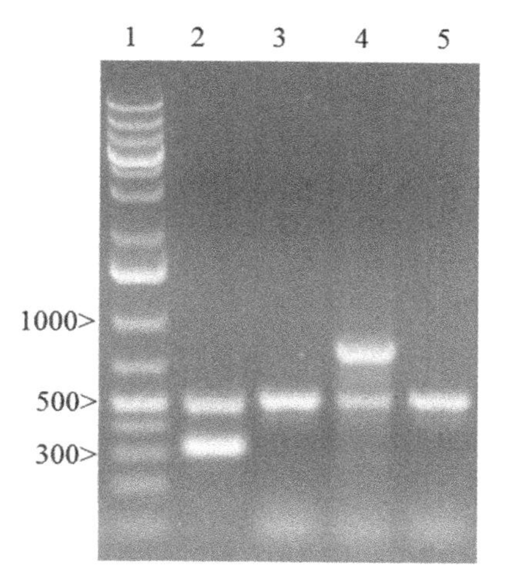

Fig. 2 Homoplasmic screen. *Lane 1*: 1 Kb Plus Ladder (Fermentas). *Lane 2*: non-homoplasmic strain transformed with the *psbA* vector. *Lane 3*: homoplasmic strain transformed with the *psbA* vector. *Lane 4*: non-homoplasmic strain transformed with the *psbH* vector. *Lane 5*: homoplasmic strain transformed with the *psbH* vector

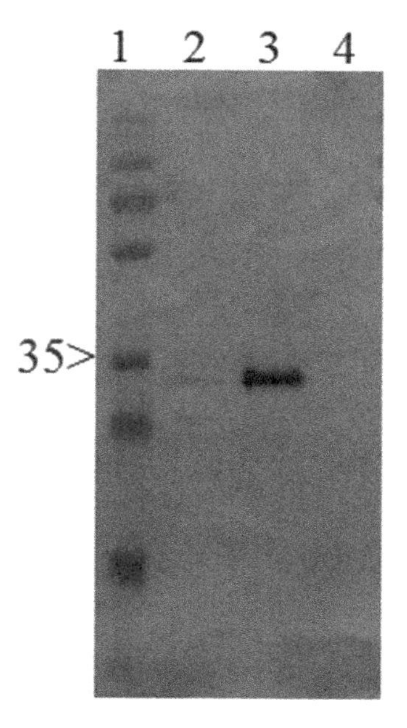

Fig. 3 Western blot of GFP expressing *C. reinhardtii* using *psbH* and *psbA* transformation vectors. *Lane 1* shows Fermentas PageRuler Plus Protein Ladder. *Lane 2* shows GFP expression under the *psbD* promoter transformed into the *psbH* locus. *Lane 3* shows GFP expression under the *psbA* promoter transformed into the *psbA* locus. *Lane 4* shows wild type with no GFP expression

4 °C (partial separation) or with an ultracentrifuge (complete separation).

5. Quantitate total, soluble, or resuspended membrane protein using the Bio-Rad DC protein assay (*see* **Note 7**).
6. Run 30 μg of protein per well on a SDS-PAGE gel (*see* **Note 8**).
7. Proceed with a standard Western blot protocol (Fig. 3).

4 Notes

1. W1.1 is a *psbA* knockout strain resistant to spectinomycin (16s rRNA mutation) (*see* ref. 16 for further details). *psbH*(−) is a non-photosynthetic strain with an early stop codon in the *psbH* gene. It has a kanamycin-resistance cassette driven by the *psbA* promoter.
2. Cells must be at mid-log phase in order to achieve high transformation efficiency. Never use cultures above 2.0×10^6 cells/ml.
3. *psbA* transformants are resistant to kanamycin thanks to the *apha-VI* cassette. *psbH* transformants recover photosynthesis by restoration of the wt *psbH* gene and thus can grow in media without acetate (HSM).
4. Plasmid concentration should be at least 1 μg/μl.
5. Use 10× TE not 1× TE.
6. Some proteins driven by the *psbA* promoter only express in liquid media upon light induction (*see* ref. 16 for further details). Do not use antibiotics in liquid media.
7. Lowry assays with absorbance at 750 nm are less prone to be affected by chlorophyll absorption.
8. 30 μg is a good initial loading amount for moderately expressed proteins. If no expression is observed proceed to load as much as possible. Heterologous proteins usually can't be observed in stained gels.

References

1. Harris EH (2001) Chlamydomonas as a model organism. Annu Rev Plant Physiol Plant Mol Biol 52:363–406
2. Boynton JE, Gillham NW, Harris EH, Hosler JP, Johnson AM, Jones AR, Randolph-Anderson BL, Robertson D, Klein TM, Shark KB (1988) Chloroplast transformation in Chlamydomonas with high velocity microprojectiles. Science 240:1534–1538
3. Fischer N, Stampacchia O, Redding K, Rochaix JD (1996) Selectable marker recycling in the chloroplast. Mol Gen Genet 251: 373–380
4. Johanningmeier U, Bertalan I, Hilbig L, Schulze J, Wilski S, Zeidler E, Oettmeier W (2006) Engineering the D1 subunit of photosystem II. Biotechnological applications of photosynthetic proteins: biochips, biosensors and biodevices, pp 46–56. ISBN 978-0-387-36672-2
5. Rochaix JD, Fischer N, Hippler M (2000) Chloroplast site-directed mutagenesis of photosystem I in Chlamydomonas: electron transfer reactions and light sensitivity. Biochimie 82: 635–645
6. Xiong L, Sayre RT (2004) Engineering the chloroplast encoded proteins of Chlamydomonas. Photosynth Res 80:411–419
7. Specht E, Miyake-Stoner S, Mayfield S (2010) Micro-algae come of age as a platform for recombinant protein production. Biotech Lett 32:1373–1383
8. Mayfield SP, Manuell AL, Chen S, Wu J, Tran M, Siefker D, Muto M, Marin-Navarro J (2007) *Chlamydomonas reinhardtii* chloroplasts as protein factories. Curr Opin Biotechnol 18: 126–133
9. Mayfield SP, Franklin SE (2005) Expression of human antibodies in eukaryotic micro-algae. Vaccine 23:1828–1832
10. Becker EW (2007) Micro-algae as a source of protein. Biotech Adv 25:207–210
11. Bateman JM, Purton S (2000) Tools for chloroplast transformation in Chlamydomonas: expression vectors and a new dominant selectable marker. Mol Gen Genet 263:404–410

12. Goldschmidt-Clermont M (1991) Transgenic expression of aminoglycoside adenine transferase in the chloroplast: a selectable marker of site-directed transformation of Chlamydomonas. Nucleic Acids Res 19: 4083–4089
13. Barnes D, Franklin S, Schultz J, Henry R, Brown E, Coragliotti A, Mayfield SP (2005) Contribution of 5′-and 3′-untranslated regions of plastid mRNAs to the expression of *Chlamydomonas reinhardtii* chloroplast genes. Mol Genet Genomics 274:625–636
14. Fletcher SP, Muto M, Mayfield SP (2007) Optimization of recombinant protein expression in the chloroplasts of green algae. Adv Exp Med Biol 616:90–98
15. Michelet L, Lefebvre-Legendre L, Burr SE, Rochaix JD, Goldschmidt-Clermont M (2010) Enhanced chloroplast transgene expression in a nuclear mutant of Chlamydomonas. Plant Biotechnol J 9: 565–574
16. Manuell AL, Beligni MV, Elder JH, Siefker DT, Tran M, Weber A, McDonald TL, Mayfield SP (2007) Robust expression of a bioactive mammalian protein in Chlamydomonas chloroplast. Plant Biotechnol J 5:402–412
17. Franklin S, Ngo B, Efuet E, Mayfield SP (2002) Development of a GFP reporter gene for Chlamydomonas reinhardtii chloroplast. Plant J 30:733–744

Chapter 27

A Simple, Low-Cost Method for Chloroplast Transformation of the Green Alga *Chlamydomonas reinhardtii*

Chloe Economou, Thanyanan Wannathong, Joanna Szaub, and Saul Purton

Abstract

The availability of routine techniques for the genetic manipulation of the chloroplast genome of *Chlamydomonas reinhardtii* has allowed a plethora of reverse-genetic studies of chloroplast biology using this alga as a model organism. These studies range from fundamental investigations of chloroplast gene function and regulation to sophisticated metabolic engineering programs and to the development of the algal chloroplast as a platform for producing high-value recombinant proteins. The established method for delivering transforming DNA into the *Chlamydomonas* chloroplast involves microparticle bombardment, with the selection of transformant lines most commonly involving the use of antibiotic resistance markers. In this chapter we describe a simpler and cheaper delivery method in which cell/DNA suspensions are agitated with glass beads: a method that is more commonly used for nuclear transformation of *Chlamydomonas*. Furthermore, we highlight the use of an expression vector (pASapI) that employs an endogenous gene as a selectable marker, thereby avoiding the contentious issue of antibiotic resistance determinants in transgenic lines.

Key words Algae, *Chlamydomonas reinhardtii*, Chloroplast, Glass beads, Homologous recombination, Transformation

1 Introduction

The green freshwater alga *Chlamydomonas reinhardtii* occupies a special niche in the history of chloroplast biotechnology, with early work using this organism confirming that chloroplasts contain their own genetic system, and screens for photosynthetic mutants resulting in the isolation of numerous strains carrying deletions in the chloroplast genome [1]. It was one such mutant that was used for the first ever demonstration of stable chloroplast transformation—a

Pal Maliga (ed.), *Chloroplast Biotechnology: Methods and Protocols*, Methods in Molecular Biology, vol. 1132, DOI 10.1007/978-1-62703-995-6_27, © Springer Science+Business Media New York 2014

strain with a deletion affecting the ATP synthase gene, *atpB* being rescued to phototrophic growth by microparticle bombardment using a plasmid carrying the wild-type *atpB* [2]. Since that time, ever more advanced tools for genetic engineering of the *Chlamydomonas* chloroplast have been developed including dominant selectable markers such as the *aadA* cassette that confers spectinomycin resistance; expression vectors and new recipient strains, and techniques for controllable gene expression [3–6].

The low cost of cultivation and fast growth rate of *Chlamydomonas*, combined with the ease of chloroplast engineering, are fuelling current interest in the use of this alga as a platform for the production of high-value recombinant products such as therapeutic proteins [7] and also as a test-bed for chloroplast metabolic engineering studies aimed at producing designer metabolites such as novel fuel molecules [8]. As part of our own research efforts in these areas, we have sought to simplify and accelerate the process of generating transplastomic lines. As a consequence, we have exploited a DNA delivery method that is considerably cheaper and easier than microparticle bombardment. The agitation of a cell suspension in the presence of glass beads and naked DNA was first developed as a nuclear transformation method for yeast, and subsequently for *Chlamydomonas* [9]. Kindle et al. then showed that the method could also be applied to chloroplast transformation of *Chlamydomonas* [10], but the low transformation rates achieved when compared with those obtained by microparticle bombardment appear to have dissuaded researchers from adopting this method. However, the recovery of only a handful of colonies is not really an issue when generating chloroplast transformant lines since all should be genetically identical. The integration of foreign DNA into the chloroplast genome occurs exclusively via homologous recombination mediated by left and right flanking sequences [2], and as a consequence, the outcome of a transformation experiment is the precise targeting of the DNA to a predetermined locus in each transformant [4].

The success of the glass bead transformation method relies on the prior removal (or significant weakening) of the proteinaceous cell wall of *Chlamydomonas*. This can be achieved either by using a cell wall-deficient mutant such as *cw15* (as illustrated in Fig. 1) or by pretreatment of walled cells with the lytic enzyme, gametolysin—an extracellular metalloprotease produced during the mating of *Chlamydomonas* gametes that mediates cell wall removal [1]. In this chapter, we cover the methodology of transformation using either walled strains pretreated with gametolysin or the *cw15* strain. In addition, we describe the use of our expression vector pASapI (unpublished; GenBank accession number KF534756) that exploits the rescue of prototropy in a non-photosynthetic strain

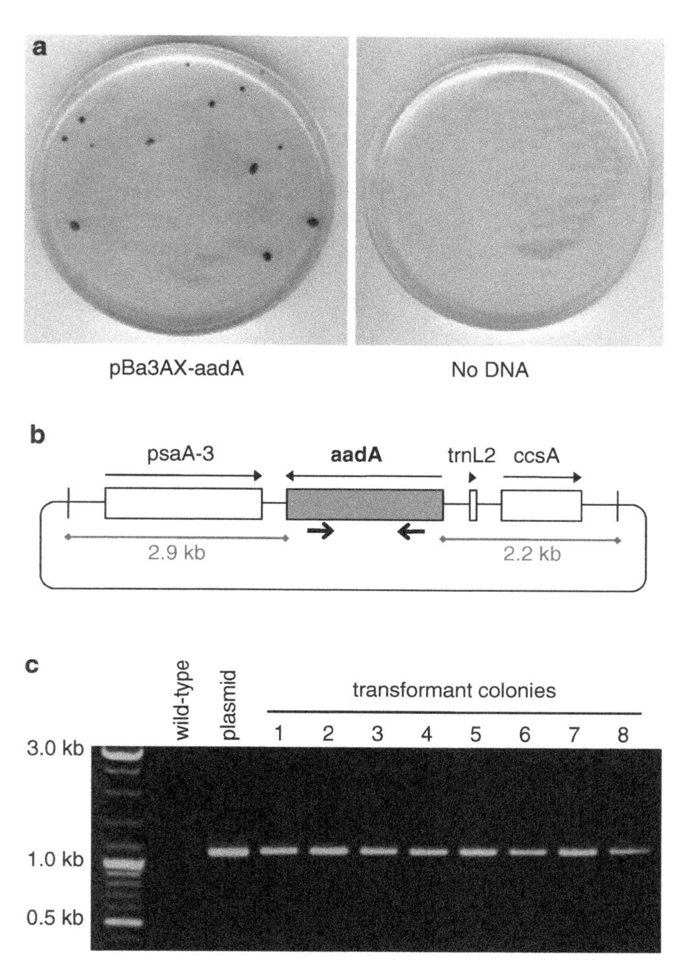

Fig. 1 Glass bead-mediated chloroplast transformation of the cell wall-deficient strain *cw15* using a plasmid carrying the *aadA* cassette that confers spectinomycin resistance. (**a**) A representative plate showing colonies growing on TAP medium containing 100 μg/ml spectinomycin 3 weeks after vortexing with plasmid pBa3AX-aadA. A control experiment without plasmid DNA yields no colonies. (**b**) Plasmid pBa3AX-aadA containing a 5.1 kb piece of the *Chlamydomonas* chloroplast genome into which the *aadA* cassette has been inserted within the *psaA-3–trnL2* intergenic region. (**c**) PCR confirmation of the presence of the cassette in each transformant line using primers designed to the *aadA* cassette, as indicated in (**b**)

carrying a mutation in the chloroplast gene, *psbH* (Fig. 2). The use of endogenous genes such as *psbH*, rather than bacterial antibiotic resistance genes, as the basis of selection is particularly attractive from the point of view of biosafety since the only foreign DNA in the resulting transgenic lines is the gene of interest.

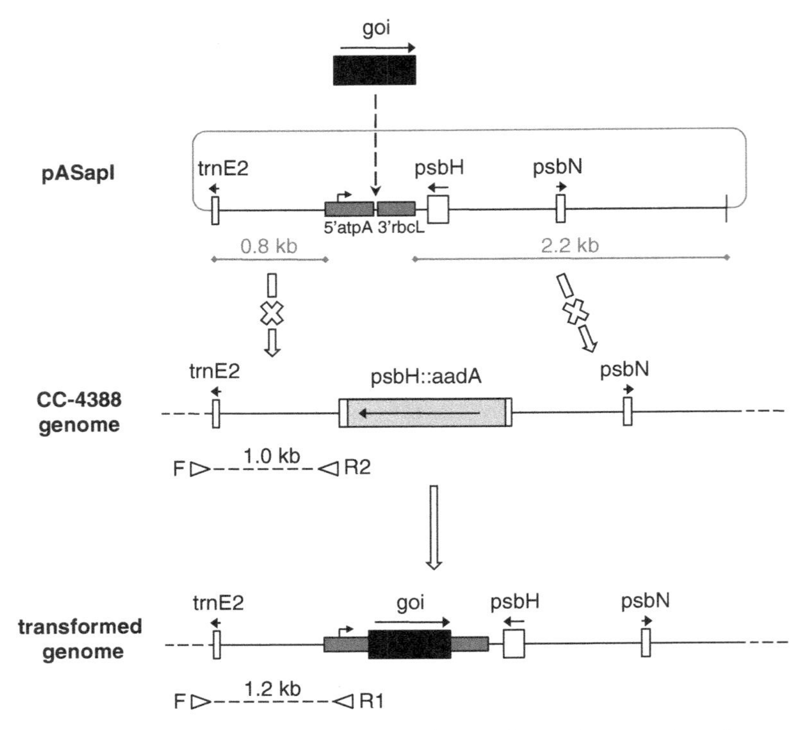

Fig. 2 The pASapI vector allows targeted integration and expression of transgenes using the essential photosynthesis gene *psbH* as a selectable marker. The vector contains an "expression cassette" within the *psbH–trnE2* intergenic region that comprises the following: the promoter, 5′ untranslated region, and start codon of *atpA*; a multiple cloning site for insertion of the coding sequence of the gene of interest (goi); and the stop codon and 3′ untranslated region of *rbcL*. Following glass bead transformation of the non-photosynthetic recipient strain (CC-4388) in which *psbH* has been disrupted with the *aadA* cassette, homologous recombination events replace the disrupted *psbH* with both a functional *psbH* and the expression cassette harboring the goi. Phototrophic transformants are checked by PCR using a mix of three primers (F, R1, and R2: *open triangles*) where homoplasmic transformants produce a 1.2 kb product only, but heteroplasmic transformants produce both the 1.2 kb product and a 1.0 kb product derived from the CC-4388 genome. Homoplasmy can also be confirmed by scoring transformants for loss of spectinomycin resistance

2 Materials

2.1 Chlamydomonas Strains

1. Wild-type and mutant strains are obtained from the Chlamydomonas Resource Center (www.chlamy.org) based at the University of Minnesota. Strains required are CC-620 (wild type, mating type +), CC-621 (wild type, mt−), CC-400 (*cw15*, mt+), and CC-4388 (*psbH*::*aadA*, mt+) (*see* **Note 1**).

2.2 Growth of Chlamydomonas

1. Use AnalaR grade chemicals (e.g., from Sigma or BDH) and purified water (distilled or reverse-osmosis quality). Two media are most commonly used for culturing of *Chlamydomonas*: TAP or "Tris–acetate–phosphate" medium and HSM or "high salt minimum" (*see* **Note 2**). Recipes for both media are at www.chlamy.org/media.html or in ref. 1. "HSM-N" is a medium lacking a source of nitrogen and is made by simply omitting the NH_4Cl from the Beijerinck salts during preparation of HSM. Liquid media are autoclaved in bottles, or in glass Erlenmeyer flasks (with the medium filling the flask to no more than 50 % volume) firmly stoppered with a cotton wool bung that is covered with foil. Sterile media can be stored at room temperature.
2. For solid media, a high-grade agar such as Bacto™ Agar (Becton Dickinson) is required. This is added to liquid media at 2 % (w/v) prior to autoclaving. Following autoclaving, the medium is poured into 90 mm Petri dishes (Sterilin) once the temperature is below ~50 °C to a depth of ~5 mm, and allowed to set. TAP and HSM plates are stored at room temperature and inspected for any bacterial or fungal contamination prior to use.
3. Hemocytometer, depth 0.1 mm, 1/400 mm^2 (Hawksley), and tincture of iodine solution, 20 mM iodine in 95 % (v/v) ethanol. Store at room temperature.

2.3 Glass Bead Transformation

1. Acid-washed glass beads, 425–600 μm (Sigma). Aliquot 0.3 g of glass beads into 5 ml test tubes, cap, and autoclave (*see* **Note 3**).
2. Gametolysin stock as prepared in Subheading 3.1. Stored at −80 °C.
3. HSM soft agar: 0.5 % (w/v), store at room temperature. Melt agar in microwave and equilibrate in a water bath for 1 h at 42 °C before use.
4. HSM plates (for selection using *psbH*) or TAP + Spc^{100} plates (for selection using *aadA*).

2.4 Genomic Extraction and PCR Analysis

1. Chelex 100 resin (Bio-Rad). 5 % (w/v) in sterile water. Store at room temperature.
2. Phusion DNA polymerase (New England Biolabs), with supplied 5× High Fidelity buffer.
3. Mix of four dNTPs at 25 mM (Sigma).
4. Oligonucleotide primers (*see* **Note 4**).
5. 1 % (w/v) agarose gel prepared in Tris–acetate–EDTA buffer and DNA size markers (New England Biolabs).

3 Methods

The introduction of exogenous DNA into the single chloroplast of the *Chlamydomonas* cell relies on the brief agitation of a suspension of cells and DNA with glass beads. It is believed that the abrasive action of the beads creates transient holes within the cell membrane allowing entry of DNA into the cell. Where the two membranes of the chloroplast are appressed against the cell membrane, the opportunity exists for entry of the DNA directly into the organelle [10]. Since the cell wall represents an additional barrier to DNA delivery, then transformation is best achieved if the cell wall is first removed using gametolysin. Alternatively, a cell wall-deficient strain can be used.

3.1 Preparation of Chlamydomonas Gametolysin

1. Re-streak the two wild-type strains (mt+ and mt−) on separate TAP plates several times over a period of about 1 week to ensure that the algae are healthy and actively growing. Incubate the plates at 25 °C in the light (~20–50 μE/m^2/s PAR).
2. Set up liquid cultures of both strains by inoculating 25 ml of HSM medium in 50 ml flasks with a large loopful of cells, and grow with shaking (~100 rpm) for 2 days at 25 °C in the light (~50 μE/m^2/s).
3. Transfer each culture to 300 ml HSM in 1,000 ml flasks and grow until the cultures reach early stationary phase (~5×10^6 cells/ml). Cell concentration is determined by adding 10 μl of tincture of iodine to a 1 ml aliquot of the culture to kill the motile cells, placing an aliquot onto the counting grid of a standard hemocytometer, overlaying the cover slip, and counting total cells within the 5 × 5 gridded area using a light microscope with ×40 objective lens (total magnification ×400). Repeat the count for the second gridded area and take the average. Cell number per ml is calculated by multiplying the count by 10^4.
4. Pellet the cells using 250 ml centrifuge bottles spun at 4,000 × *g* for 5 min. Pour off supernatant and resuspend each pellet in 5 ml of HSM-N (*see* **Note 5**).
5. Transfer each strain to 1.2 l of HSM-N in 3 l flasks.
6. Grow overnight in the light at 25 °C with shaking, as before. Pellet the cells as above and concentrate tenfold by resuspending each pellet in 120 ml of HSM-N.
7. Mix the two cell suspensions in an empty sterile 1,000 ml flask to create a mating mix.
8. Allow the cells to mate for 2 h by leaving the flask undisturbed in low light (~10 μE/m^2/s) (*see* **Note 6**).
9. Remove cells and debris from the mating mix by centrifugation using 50 ml tubes spun at 40,000 × *g* for 10 min.

10. Filter the supernatant containing the gametolysin through a 0.22 μm filter attached to a 50 ml syringe into sterile plastic tubes as 10 ml aliquots.
11. Store the tubes at −80 °C.

3.2 Preparing Cells for Transformation

1. Re-streak the recipient strain (CC-400 or CC-4388) on TAP plates several times over a period of about 1 week to ensure that the alga is healthy and actively growing. Incubate the plates at 25 °C in the light (*see* **Note 7**).
2. Use a large loopful of cells to inoculate 25 ml of TAP medium in a 50 ml flask, and grow the culture for 2–3 days under continuous light (*see* **Note 7**) with shaking (100 rpm).
3. Use 4 ml of this starter culture to inoculate 400 ml of TAP in a 1,000 ml flask and grow under continuous light to mid-log phase ($1–2 \times 10^6$ cells/ml), determining cell concentration as above (*see* **Note 8**).
4. Pellet the cells by centrifuging at $4{,}000 \times g$ at room temperature for 5 min.
5. Discard the supernatant and resuspend the cells in either fresh TAP medium (for cell wall-deficient strains) or gametolysin solution (walled strains) to a final concentration of $\sim 2 \times 10^8$ cells/ml (i.e., add ~3 ml of medium). Resuspend pellet by gently drawing up and down using a sterile pipette.
6. For cells in gametolysin solution, incubate at room temperature for 1 h prior to transformation to allow cell wall digestion.

3.3 Glass Bead Transformation

1. Transfer 300 μl of the cell suspension to ten test tubes, each containing 0.3 g of sterile glass beads (*see* **Note 3**).
2. Add 5 μg of plasmid DNA to each of eight tubes, leaving the remaining two as “no DNA” controls.
3. Agitate the cell/glass bead/DNA suspension for 15 s at top speed using a Vortex Genie-2 (Scientific Industries).
4. After vortexing all the tubes, add 3.5 ml of 0.5 % (w/v) agar that has been kept molten by incubation at 42 °C and quickly pour the mix onto the surface of selection medium plates (*see* **Note 9**). Gently tip the plates to allow the molten agar to spread across the whole plate.
5. Allow the agar to set for approximately 20 min (cover in foil or black cloth to prevent phototactic migration of cells). Invert the plates and seal with Parafilm.
6. Incubate the plates at 25 °C in very dim light (1–5 $\mu E/m^2/s$) overnight then move to higher light (40–50 $\mu E/m^2/s$) and leave for approximately 2–3 weeks until green colonies appear.

3.4 Isolating Homoplasmic Transformants (See Note 10)

1. Pick individual colonies using an inoculating loop or sterile toothpicks and re-streak to single colonies on fresh plates containing selective medium. Incubate plates as before until single colonies are visible.
2. Repeat **step 1** once more and then check for homoplasmy by extracting genomic DNA for PCR analysis using the following method [11].
3. Pick a single colony and resuspend in 20 μl sterile water.
4. Add 20 μl of absolute ethanol and incubate for 1 min at room temp.
5. Add 200 μl of a 5 % (w/v) suspension of Chelex 100 resin (Bio-Rad).
6. Vortex briefly then place in a boiling water bath for 5 min. Cool briefly on ice.
7. Pellet cell debris and resin with a 2 min spin in a microfuge and transfer the supernatant containing genomic DNA to a fresh tube.
8. Set up a standard PCR reaction using 1 μl of the genomic DNA (*see* **Note 4**).
9. Once homoplasmic lines have been obtained, then further studies into the expression of the transgene can be carried out as illustrated in Fig. 3.

4 Notes

1. The choice of mating type (mt) as recipient strain is important. During the *Chlamydomonas* sexual cycle, the chloroplast DNA is inherited uniparentally from the mt+ parent [1]. Consequently, if you plan to subsequently transfer the genetically engineered chloroplast DNA into different nuclear backgrounds, you should use an mt+ recipient. Alternatively, use of an mt− recipient increases biological containment of any foreign DNA since escaped transgenic lines will not transmit the DNA to interfertile species within the environment.
2. TAP medium contains acetate as an exogenous source of fixed carbon and is used for heterotrophic growth in the dark or mixotrophic growth in the light. HSM medium contains no fixed carbon and is used for phototrophic growth.
3. We use small (5 ml) glass test tubes capped with a loose aluminum cap. Weighing and dispensing 0.3 g of glass beads into these tubes can be tedious and frustrating because the small beads readily pick up electrostatic charge. We have found that a simple measuring scoop can be made by cutting off the bottom ~8 mm of a standard microfuge tube with a razor blade

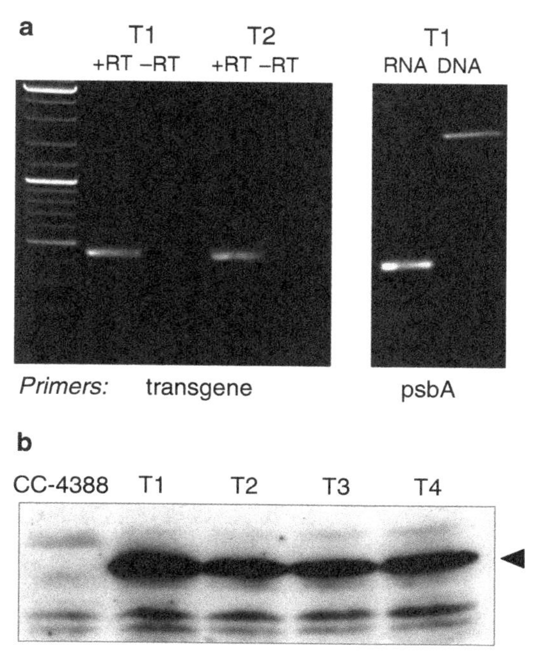

Fig. 3 Example of successful introduction and expression of a transgene achieved using the pASapI vector and the glass bead transformation method. Four transformant colonies carrying a transgene from *E. coli* were examined for the presence of mRNA from the transgene and the gene product. (**a**) Reverse transcription PCR analysis of total RNA isolated from transformant lines T1 and T2. The *left panel* shows that a transcript for the transgene can be readily detected, with the "no reverse transcriptase enzyme" (−RT) controls confirming that this band has arisen from the RNA rather than contaminating DNA. The *right panel* shows amplified transcript from an endogenous gene (*psbA*) for comparison. This gene is chosen since it contains introns, allowing a clear size distinction between RT-PCR of the spliced transcript and PCR of the gene itself from contaminating DNA. (**b**) Western blot analysis of total soluble protein from the recipient strain, psbH::aadA, and four transformant lines using antibodies to the foreign protein. A prominent band of the expected size (39 kDa) is seen in the transformants (unpublished data, T. Wannathong and S. Purton)

and heat fusing it to the end of a 1 ml pipette tip. A standard amount of beads can then be rapidly dispensed into each tube. The rack of tubes is then wrapped in foil and autoclaved.

4. When using the pASapI vector for transformation, we have devised a three-primer method that confirms both integration of the foreign gene into the chloroplast genome and homoplasmy. Primer F (5′-GTCATTGCGAAAATACTGGTGC-3′) is designed to a region immediately downstream of *trnE2*—i.e., outside the chloroplast DNA region cloned in pASapI (*see* Fig. 2). When used in conjunction with primer R1 (5′-ACGTCCACAGGCGTCGTAAGC-3′), which is designed

to the opposite strand of the *atpA* element, then a 1.2 kb PCR product is generated only if the expression cassette has integrated correctly into the genome. A third primer, R2 (5′-GATGACGTTTCTATGAGTTGGG-3′), designed to the genomic region immediately downstream of *psbH* yields a 1.0 kb band in conjunction with primer F1 if the cassette has failed to integrate. Consequently, a PCR analysis of putative transformants using the three primers together yields three possible outcomes—1.2 kb band only (cassette integrated and genome homoplasmic), 1.2 and 1.0 kb bands (cassette integrated, but genome still heteroplasmic), and 1.0 kb band only (untransformed or cassette has failed to integrate at the correct locus). For PCR analysis of transformants carrying the *aadA* cassette [3], the following primer pair can be used to generate a product of 1.1 kb as illustrated in Fig. 1: atpA.F (5′-CAAGTGATCTTACCACTCAC-3′) and rbcL.R (5′-CAAACTTCACATGCAGCAGC-3′).

5. The conversion of vegetative cells to sexually competent gametes is achieved by removal of fixed nitrogen from the medium. Depletion of N for 24 h in the light is sufficient to bring about this physiological switch, allowing mating and gamete fusion once opposite mating types are mixed.
6. Healthy gametes are motile (with two anterior flagella) and appear highly active when viewed under the light microscope. When opposite mating types are mixed, the cells recognize a mating partner through flagellar "tipping" and rapidly form "wrestling" pairs that fuse to form a quadriflagellate diploid cell.
7. While wild-type and cell wall-deficient strains are tolerant to light levels as high as 1,000 μE/m^2/s PAR and are typically grown in our lab at non-saturating levels of ~50 μE/m^2/s, photosystem II mutants such as CC-4388 are sensitive to light stress and should be grown below 10 μE/m^2/s.
8. Transformation efficiency declines if cells are used that have progressed into the later stages of exponential growth or are at stationary phase.
9. For selection of phototrophic transformants of CC-4388, both the soft agar and the plates are prepared with HSM medium. For selection of antibiotic-resistant transformants of CC-400, the soft agar is prepared using TAP medium and the plates with TAP medium containing spectinomycin at 100 μg/ml [3].
10. The *Chlamydomonas* chloroplast contains approximately 80 copies of its genome. During the initial stage of transformation, only one or a few copies of the genome are modified. It is therefore important to maintain the selection for the modification for sufficient generations such that all genome copies eventually possess the modification and a stable homoplasmic

state is reached [4]. In practice, this is best achieved by repeated re-streaking on selective medium such that single cells give rise to discrete colonies. For selection based on the rescue of prototrophy, we find that a single re-streaking is sufficient to achieve homoplasmy. For antibiotic resistance conferred by the *aadA* marker, three re-streakings on spectinomycin medium are typically required.

Acknowledgements

Research in the Purton lab into the genetic engineering of the *Chlamydomonas* chloroplast is funded by the UK's Biotechnology and Biological Sciences Research Council and the "GIAVAP" and "SUNBIOPATH" FP7 projects of the European Union. SP acknowledges the equal contribution that CE and TW have made to this chapter.

References

1. Harris EH (2008) The Chlamydomonas sourcebook, volume 1: introduction to Chlamydomonas and its laboratory use. Academic, San Diego, CA
2. Boynton JE, Gillham NW, Harris EH et al (1988) Chloroplast transformation in Chlamydomonas with high velocity microprojectiles. Science 240:1534–1538
3. Goldschmidt-Clermont M (1991) Transgenic expression of aminoglycoside adenine transferase in the chloroplast: a selectable marker for site-directed transformation of Chlamydomonas. Nucl Acids Res 19:4083–4089
4. Purton S (2007) Tools and techniques for chloroplast transformation of Chlamydomonas. Adv Exp Med Biol 616:34–45
5. Michelet L, Lefebvre-Legendre L, Burr SE et al (2011) Enhanced chloroplast transgene expression in a nuclear mutant of Chlamydomonas. Plant Biotechnol J 9:565–574
6. Surzycki R, Cournac L, Peltier G et al (2007) Potential for hydrogen production with inducible chloroplast gene expression in Chlamydomonas. Proc Natl Acad Sci U S A 104:17548–17553
7. Specht E, Miyake-Stoner S, Mayfield S (2010) Micro-algae come of age as a platform for recombinant protein production. Biotechnol Lett 32:1373–1383
8. Radakovits R, Jinkerson RE, Darzins A et al (2010) Genetic engineering of algae for enhanced biofuel production. Eukaryot Cell 9:486–501
9. Kindle KL (1990) High-frequency nuclear transformation of *Chlamydomonas reinhardtii*. Proc Natl Acad Sci U S A 87:1228–1232
10. Kindle KL, Richards KL, Stern DB (1991) Engineering the chloroplast genome: techniques and capabilities for chloroplast transformation in *Chlamydomonas reinhardtii*. Proc Natl Acad Sci U S A 88:1721–1725
11. Werner R, Mergenhagen D (1998) Mating type determination of *Chlamydomonas reinhardtii* by PCR. Plant Mol Biol Rep 16:295–299

Chapter 28

Tools for Regulated Gene Expression in the Chloroplast of Chlamydomonas

Jean-David Rochaix, Raymond Surzycki, and Silvia Ramundo

Abstract

The green unicellular alga *Chlamydomonas reinhardtii* has emerged as a very attractive model system for chloroplast genetic engineering. Algae can be transformed readily at the chloroplast level through bombardment of cells with a gene gun, and transformants can be selected using antibiotic resistance or phototrophic growth. An inducible chloroplast gene expression system could be very useful for several reasons. First, it could be used to elucidate the function of essential chloroplast genes required for cell growth and survival. Second, it could be very helpful for expressing proteins which are toxic to the algal cells. Third, it would allow for the reversible depletion of photosynthetic complexes thus making it possible to study their biogenesis in a controlled fashion. Fourth, it opens promising possibilities for hydrogen production in Chlamydomonas. Here we describe an inducible/repressible chloroplast gene expression system in Chlamydomonas in which the copper-regulated Cyc_6 promoter drives the expression of the nuclear *Nac2* gene encoding a protein which is targeted to the chloroplast where it acts specifically on the chloroplast *psbD* 5′-untranslated region and is required for the stable accumulation of the *psbD* mRNA and photosystem II. The system can be used for any chloroplast gene or transgene by placing it under the control of the *psbD* 5′-untranslated region.

Key words Chlamydomonas, Chloroplast gene, Chloroplast genetics, Chloroplast transformation, Inducible chloroplast gene expression, Inducible promoter, Trans-acting factor

1 Introduction

The chloroplast genome of Chlamydomonas exists in 80 copies, each of which contains ca. 100 genes which can be grouped in three classes [1]. The first includes genes involved in photosynthesis. These genes code for many of the subunits of the photosynthetic complexes photosystem I (PSI) and photosystem II (PSII), the cytochrome b_6f (Cyt b_6f) complex, ATP synthase, and ribulose bisphosphate carboxylase. This group also includes genes coding for two assembly factors of PSI, three proteins involved in light-independent chlorophyll synthesis, and one protein required for *c*-type cytochrome maturation. The second group comprises genes required for chloroplast gene expression such as subunits of the

Pal Maliga (ed.), *Chloroplast Biotechnology: Methods and Protocols*, Methods in Molecular Biology, vol. 1132,
DOI 10.1007/978-1-62703-995-6_28,

plastid RNA polymerase, ribosomal RNA and ribosomal proteins, tRNAs, and translation factors. The third group contains proteins involved in different processes, a subunit of an ATP-dependent protease, and several putative proteins of unknown function.

An important tool for investigating the role of these genes has been chloroplast transformation, which can be readily achieved in Chlamydomonas [2]. This is possible because, first, foreign DNA can be introduced into the chloroplast compartment by bombardment with a gene gun. Second, selectable markers for transformation are available which confer antibiotic resistance. Alternatively, chloroplast mutants can be rescued with the corresponding wild-type gene by selecting for phototrophic growth. Among all markers tested, the *aadA* gene conferring resistance to spectinomycin and streptomycin has emerged as the most efficient one, mainly because Chlamydomonas cells are highly sensitive to these antibiotics [3]. Third, an efficient plastid homologous recombination system allows one to insert the foreign DNA at any desired site in the chloroplast genome. In particular, it is possible to perform specific gene disruptions and site-directed mutagenesis of any plastid gene of interest [4]. However, it is important to note that chloroplast genomes are present in multiple copies and that usually multiple selection rounds are required for inactivating all gene copies. In this respect, attempts to inactivate a chloroplast gene can have two different outcomes. In the first case, all gene copies can be disrupted with an antibiotic resistance cassette resulting in a homoplasmic state with no remaining wild-type gene copy. The fact that all copies of a gene can be inactivated indicates that the function of this gene is no longer required under the growth conditions used. In Chlamydomonas this occurs for genes involved in photosynthesis when the cells are grown in the presence of a carbon source such as acetate. Under these conditions photosynthesis is no longer required. In the second case, it is never possible to inactivate the entire set of the gene copies because the particular gene chosen is essential for growth under all conditions. Hence, a heteroplasmic state is reached with a mixture of mutant copies which confer antibiotic resistance and wild-type copies which allow for growth and/or survival of the cells. Examples of this type include genes involved in chloroplast gene expression and the *ClpP* gene, which encodes an essential catalytic subunit of the ClpP protease complex [5]. Moreover there is a small set of chloroplast genes of unknown function which are essential based on the fact that attempts to disrupt them result in a persistent heteroplasmic state.

New approaches have been developed for studying the function of these genes. One strategy relies on attenuation. By changing the AUG start codon of a mRNA to AUU, it is possible to decrease the expression of the protein to 10–50 % of wild-type levels [6]. In some cases this is sufficient to elicit a phenotype. As an example in the case of the *ClpP* gene, it could be shown through this approach that the ClpP protease plays a role in the degradation of the Cytb_6f complex under nitrogen starvation [7]. Another strategy has been to establish an inducible/repressible chloroplast gene expression system which allows one to switch chloroplast genes on and off depending on the absence or presence of a micronutrient in the medium [8]. To understand how this system works, it is first necessary to provide some background on chloroplast gene expression.

Analysis of numerous photosynthetic mutants of Chlamydomonas has revealed a large number of nucleus-encoded factors which are involved in specific chloroplast posttranscriptional steps such as RNA processing, RNA stability, splicing, and translation [9]. Remarkably, most of these factors each act on a specific mRNA, and in many cases the target site is localized within the 5′-untranslated region (5′-UTR) of the mRNA. As an example, the nucleus-encoded Nac2 protein is specifically required for stabilizing the chloroplast *psbD* mRNA which encodes the D2 reaction center polypeptide of photosystem II [10]. The target site of Nac2 is located within the *psbD* 5′-UTR as shown by the fact that if this region is fused to a reporter gene, expression of the reporter becomes Nac2 dependent [11]. Reciprocally, when the *psbD* 5′-UTR is replaced by another chloroplast 5′-UTR, *psbD* expression is no longer dependent on Nac2. Because no inducible chloroplast promoter is known in Chlamydomonas, we have taken advantage of the Nac2 system to establish a chloroplast inducible system [8]. This was achieved by fusing the Nac2 gene to an inducible nuclear promoter. We chose the promoter of the cytochrome c_6 (*Cyc*$_6$) gene because this gene is tightly repressed when copper is present in the growth medium and strongly expressed under copper deprivation or in the presence of nickel [12, 13]. By introducing the chimeric *Cyc6-Nac2* gene into a *nac2* mutant deficient in Nac2, chloroplast *psbD* expression is repressed in the presence of copper because Nac2 is no longer expressed and *psbD* RNA is destabilized. In contrast, in the absence of copper or in the presence of nickel ions, Nac2 is expressed and *psbD* RNA is stabilized and translated. It is possible to extend this inducible expression system to any plastid gene of interest by replacing its 5′-UTR with that of *psbD* (Fig. 1a). In some cases, it may be desirable to have photosystem II present under both repressive and inducible conditions. This can easily be achieved by replacing the *psbD* 5′-UTR of *psbD* by another chloroplast 5′-UTR which does not respond to Nac2 [8].

Another important application of this inducible chloroplast gene expression system is in biotechnology [14–16]. The Chlamydomonas chloroplast gene expression system has already been used successfully for expressing proteins of commercial interest. For each foreign gene to be expressed in the chloroplast of *C. reinhardtii*, it is recommended to test different plastid 5′-UTRs for optimal expression and to use a reconstructed gene with codon optimization for the chloroplast. However, in some cases the toxicity of the expressed foreign protein can severely affect cell growth.

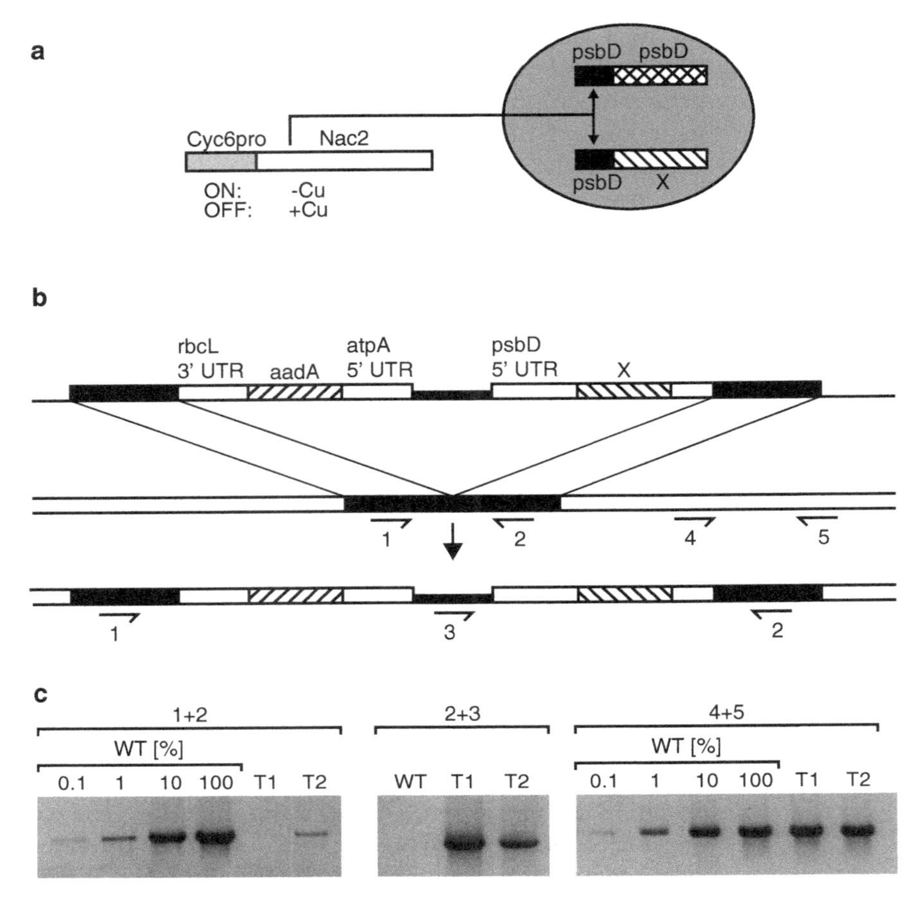

Fig. 1 A Nac2-mediated inducible/repressible chloroplast gene expression. (**a**) The nuclear Nac2 gene is driven by the Cyc6 promoter which is activated in the absence of copper and repressed in the presence of copper. The Nac2 protein is targeted to the chloroplast where it is required for *psbD* RNA stability. Because the target site of Nac2 is within the *psbD* 5′-UTR, the Nac2-dependent RNA stability can be conveyed to any gene (gene X) if its coding sequence is fused to the *psbD* 5′-UTR. (**b**) Copper-regulated expression of chloroplast gene X where X may represent any plastid gene. The *Ind41-18* strain is bombarded with a gene gun with a plasmid containing the *aadA* expression cassette and gene X driven by the *psbD* promoter and 5′-UTR (*upper line*). The plasmid contains in addition flanking sequences of gene X to promote homologous recombination in the chloroplast genome. The *small arrows* indicate primers used for assessing the homoplasmicity of the transformed strain by PCR. (**c**) Transformant T1 is homoplasmic as no wild-type copy is detectable with primers 1 and 2, whereas transformant T2 is heteroplasmic as wild-type DNA is still detectable. Control PCRs with serial dilutions of DNA show that the method is sufficiently sensitive to detect less than one wild-type copy as there are 80 copies of chloroplast genome per cell. A new PCR product arising from primers 2 and 3 is detected in T1 and T2 but not in the wild type. The PCR product obtained with primers 4 and 5, which are on the same side of the insertion site, is present as expected both in the transformants and the wild type

In this respect, the inducible chloroplast gene expression system can circumvent this problem because it is possible to induce the expression of the protein of interest once the cell culture has reached the appropriate cell density and toxicity effects are minimized [16]. This chapter provides a detailed description of the *Nac2-psbD* inducible/repressible gene expression system.

2 Materials

2.1 Growth Media

Preparation of the stock solutions is described below. The growth media are described in Table 1.

1. 4× Beijerinck's salts: 16 g NH_4Cl_2, 2 g $CaCl_2$ $2H_2O$, and 4 g $MgSO_4{\cdot}7H_2O$. Adjust to 1 l with H_2O.
2. 2× PO_4 for HSM: 14.38 g K_2HPO_4 (0.08 M) and 7.26 g KH_2PO_4 (0.05 M). Adjust to pH 6.9 with KOH to 1 l with H_2O.
3. Trace elements [21].
   (a) Dissolve in 550 ml H_2O in the order indicated and then heat to 100 °C: 11.4 g H_3BO_3, 22.0 g $ZnSO_4$ $7H_2O$, 5.06 g $MnCl_2$ $4H_2O$, 4.99 g $FeSO_4$ $7H_2O$, 1.61 g $CoCl_2$ $6H_2O$, 1.57 g $CuSO_4$ $5H_2O$, and 1.1 g (NH_4) $6Mo_7O_{24}$ $4H_2O$.
   (b) Dissolve 50 g Na_2EDTA in 250 ml H_2O by heating and add to the first solution at 100 °C.
   (c) Heat the combined solutions to 100 °C, cool to 80–90 °C, and adjust to pH 6.5–6.8 with 20 % KOH (requires <100 ml). The pH meter should first be calibrated at 75 °C. The temperature should remain above 75 °C.
   (d) Adjust to 1 l and allow for a rust-colored precipitate to form during 2 weeks at room temperature in a 2 l

Table 1
Growth media

For 1 l[a]	Tris–acetate phosphate[b] (TAP)	High salt minimal[c] (HSM)
H_2O	975 ml	925 ml
Tris	2.42 g	–
4× Beijerinck's salts	25 ml	25 ml
1 M (K)PO_4, pH 7.0	1 ml	–
2× PO_4 for HSM	–	50 ml
Trace	1 ml	1 ml
Glacial acetic acid	1 ml	–
Concentrated HCl	–	–

Copper-free medium is prepared by omitting the addition of cupper salt to the trace solution. All glassware should be carefully acid washed to remove trace amounts of copper which can inhibit the expression of cytochrome c_6 and of the *Cyc6-Nac2* gene (*see* also **Note 1**)

[a]Supplement solid media with 20 g of agar (Difco) per liter

[b]Reference [22]

[c]Reference [23]

Erlenmeyer flask loosely stoppered with cotton. The solution will change from green to purple.

(e) Filter several times through three layers of Whatman No. 1 under suction (Büchner funnel). Store the clear purple solution at −20 °C.

2.2 Materials

1. Gene gun. A skilled workshop will be able construct such a gun based on the literature [17]. Alternatively the PDS1000/He™ gene gun can be purchased from DuPont.

   Gold carrier particles can be obtained from Seashell Technology LLC (San Diego, CA) and Tungsten particles from American Elements (Los Angeles, CA).
2. Spectinomycin (Sigma-Aldrich) is dissolved in water at 10 mg/ml, stored frozen, and used at a final concentration of 100 μg/ml.
3. Paromomycin (Sigma-Aldrich) is dissolved in water at 10 mg/ml, stored frozen, and used at a final concentration of 10 μg/ml.
4. "Copper-free" TAP plates. 20 g of Difco agar is washed four times by gentle stirring at 4 °C in 2 l of 50 mM EDTA per wash, followed by four washes with 2 l Milli-Q purified water (Millipore Corp., Bedford, MA) per wash. Each wash cycle is −12 h long. The agar is allowed to settle between exchanges of wash solutions [24]. Chemicals used for the preparation of the growth media should be reagent grade.

3 Methods

3.1 Strains and Growth Conditions

1. *Chlamydomonas reinhardtii* strains are maintained on TAP (Tris–acetate–phosphate) medium supplemented with 1.5 % Bacto-agar at 25 °C either under constant light (10–60 μE/m^2/s). The strains can be grown on either TAP or HSM (minimum medium lacking acetate) on solid medium or in liquid culture. For experiments in which copper-supplemented or copper-deficient medium is used, it is critical to acid-wash the glassware and to use ultra pure chemicals (>99.99 % purity) for preparing the medium, especially for the trace element solution [18].
2. The *nac2-26* mutant strain contains a single base change within the Nac2 gene and is deficient in PSII [19]. The *cyc-6Nac2-49* strain contains a transgene consisting of the Nac2 gene in the nuclear genome of the *nac2-26* mutant (Table 2). The *Ind41* strain is a derivative of *cyc6Nac2-49* in which the *psbD* promoter and 5′-UTR are replaced by a 675 bp fragment containing the *petA* promoter and 5′-UTR. This strain grows photoautotrophically both in the presence and absence of copper in the medium. The *Ind41-18* strain derives from *Ind41* from which the *aadA* expression cassette has been excised and is therefore sensitive to spectinomycin.

Table 2
Strains used for copper-regulated chloroplast gene expression in Chlamydomonas

Strain	Genotype	Reference
nac2-26	*nac2-26*	[10]
Cyc6Nac2.49	*nac2-26::cy6proNac2*	[8]
Ind41	*nac2-26::cy6proNac2::petA-psbD, aadA*	[8]
Ind41-18	*nac2-26::cy6proNac2::petA-psbD*	[8]

3.2 Construction of Strains with Copper-Mediated Repression of Any Chloroplast Gene

The *Ind41-18* strain can be used as host for inducible expression of any chloroplast gene through the Nac2 system. It is first necessary to construct a plasmid in which the promoter and 5′-UTR of the gene of interest is replaced with a 210 bp fragment containing the *psbD* promoter and 5′-UTR (Fig. 1b). The construct needs to be surrounded on both sides by ca. 1–2 kb of flanking chloroplast DNA to ensure homologous recombination of the construct upon chloroplast transformation by bombardment. In order to avoid off-site homologous recombination events, it is best to insert the *aadA* expression cassette adjacent to the *psbD* promoter (Fig. 1b). This cassette usually consists of *aadA* flanked upstream by the *atpA* promoter and 5′-UTR and downstream by the *rbcL* 3′-UTR (*see* **Note 2**). If necessary other 5′ and 3′ UTRs can be chosen. The plasmid can then be used to transform the *Ind41-18* strain with the gene gun.

1. Inoculate *Ind41-18* from a solid agar plate into 10 ml TAP liquid medium in a 75 ml Erlenmeyer flask and put it on a shaker (120 rpm) under 60 $\mu E/m^2/s$ in the morning (day 1) (*see* **Note 3**).
2. Dilute the cells the next afternoon to 10^6 cells/ml (day 2).
3. Collect the cells the next morning (day 3). Determine cell concentration with a hemocytometer, and concentrate cells to 10^8 cells/ml.
4. Plate 0.3 ml on a TAP plate containing 100 μg/ml spectinomycin. Let the plate dry for a few hrs in a laminar flow hood (sterile conditions).
5. Coat gold or tungsten particles with plasmid engineered for chloroplast homologous recombination. The plasmid used is depicted in Fig. 1b in which X represents the chosen gene. The gold/tungsten particle-DNA solution is prepared as described by the manufacturer (http://www.seashelltech.com/chlamydomonas.shtml) and then layered on a plastic filter holder which is introduced into the gun together with the plate containing the algae. Air is removed from the gun chamber

until the pressure reaches 0.8 bar. Then a solenoid-activated valve is opened to release the pressure (8 bars) from a helium tank into the chamber so that the gold particle-DNA solution is accelerated by the helium pulse on the algae spread on the plate.

6. The plate is then removed from the chamber and placed in a growth room under 60 μE/m^2/s at 25 °C. Colonies will appear after several days (4–5 days in the case of selection for spectinomycin resistance, 7–10 days in the case of selection for photoautotrophic growth).
7. Once the colonies are large enough to be picked with a toothpick, they are streaked on fresh TAP-spectinomycin plates. This operation is repeated 3–4 times.
8. At this stage the transformed colonies should be homoplasmic which can be checked by PCR. Usually two PCRs are performed, one with primers on both sides of the chloroplast DNA region which has been replaced and one with a primer within the *psbD* 5′-UTR and one of the other two primers (Fig. 1b, c). Additionally a control PCR can be performed with two primers on the same side of the insertion (Fig. 1b, c). Because the chloroplast DNA of *C. reinhardtii* is present in 80 copies per chloroplast, it is important to check that one chloroplast gene copy can be detected under the conditions used. Thus 1/80 of the amount of wild-type DNA used for the PCR should yield a detectable signal (Fig. 1c).
9. Once the homoplasmicity of the transformed strain has been established, it can be used for induction/repression experiments (*see* **Note 4**).

3.3 Inducible/Repressible Expression of Chloroplast Genes

In principle two types of experiments are possible. One can start under growth conditions in which the chloroplast gene of interest is expressed and then repress its expression by adding copper to the medium. Alternatively one can start under conditions in which the gene of interest is repressed in the presence of copper in the growth medium followed by a transfer to copper-free medium after washing the cells with copper-free medium. Under these conditions expression of the gene is induced.

1. The strain chosen is grown in TAP medium lacking copper.
2. Once the cell concentration has reached 2×10^6 cells/ml, copper is added and aliquots of cells are collected at different time points and examined by immunoblotting or by activity measurements. An example with *psbD* 5′-UTR-*psbD* is shown in Fig. 2. It takes ca. 30 h for PSII to reach its lowest activity measured by FV/FM where FV is the variable fluorescence and FM is the maximum fluorescence.

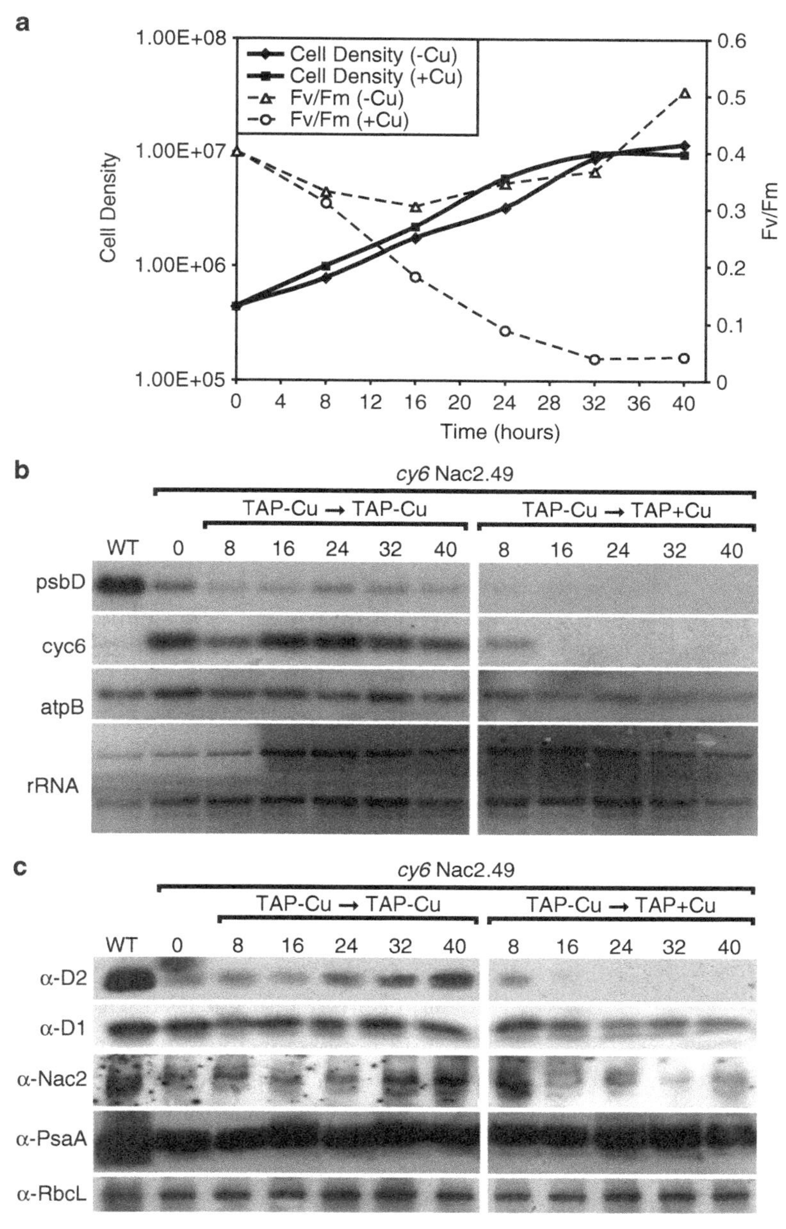

Fig. 2 Copper-mediated repression of PSII synthesis in *cy6Nac2.49*. (**a**) Cells were grown in TAP medium depleted of copper to a cell concentration of 2×10^6 cells/ml. At time 0 copper was added and F_V/F_M and cell concentration were determined at different time points. (**b**) RNA was isolated from cells at different times, fractionated by agarose gel electrophoresis, and blotted and hybridized with the gene probes indicated. *Left panel*: cells grown in TAP lacking copper. *Right panel*: cells were grown in TAP lacking copper; at time 0 copper was added to the culture. (**c**) Immunoblot analysis of proteins from *cyNac2.49* was performed as described in (**c**). The antibodies used are indicated on the *left*. (Reproduced by permission from ref. 8; Copyright 2007 National Academy of Sciences, USA)

3. In the reverse experiment, cells are grown in normal TAP medium.
4. Once the cell concentration has reached 2×10^6 cells/ml, the cells are washed and resuspended in TAP medium depleted of copper, and aliquots of cells are collected at different time points and examined by immunoblotting or by activity measurements. Figure 3 shows an example with *psbD* 5′UTR-*psbD*. PSII activity (FV/FM) started to rise after 25 h. This lag most likely reflects the time needed for the cells to deplete their internal copper reserve.

3.4 Inducible Expression of Foreign Genes in the Chloroplast

Foreign genes have been expressed in the chloroplast of Chlamydomonas with variable success ranging between a fraction of a percent and 26 % of total cell protein [14–16]. It is not clear what the limiting steps are, but clearly codon optimization, proteolytic activities, and genetic changes accompanying chloroplast transformation can greatly influence the expression levels. Several sites of the chloroplast genome of Chlamydomonas have been used for expressing foreign genes. The best results have been obtained with the insertion of the transgene within the inverted repeat downstream of the rRNA operon. In this way two copies of the transgene are inserted per chloroplast genome. The inducible expression of foreign transgenes can be achieved using the *Ind41-18* strain as described in Subheading 3.3.

4 Notes

1. A modified TAP medium with lower concentrations of trace elements has been described in which induction of Cyc6 expression is more efficient than with normal TAP medium [20].
2. It is important to avoid using a promoter and 5′-UTR or 3′-UTR flanking the *aadA* gene which is near the insertion site. Otherwise recombination may occur between the repeated elements and lead to deletions, inversions, or genomic rearrangements.
3. Because *C. reinhardtii* is a haploid organism, new mutations can easily arise. It is therefore important to check the growth properties of the strains regularly. A safe method is to freeze the strains in liquid nitrogen especially if they are not used for several months (http://www.chlamy.org/methods/freezing.html).
4. If difficulties are encountered for obtaining a homoplasmic transformed strain, it is advisable to increase the concentration of spectinomycin during the first rounds of selection. Spectinomycin concentration can be increased up to 1,000 mg/l without any important side effect.

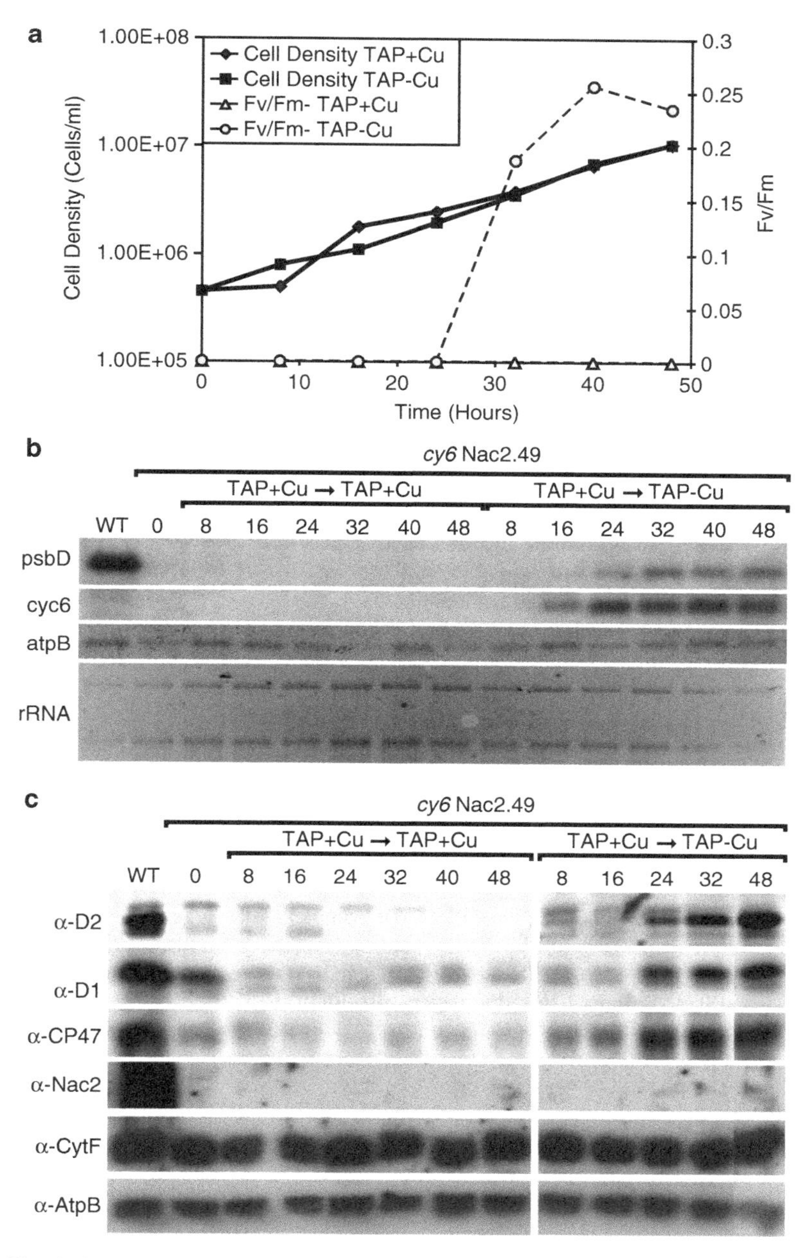

Fig. 3 Copper depletion induces PSII synthesis in *cy6Nac2.49*. (**a**) Cells were grown in TAP medium to a concentration of 2×10^6 cells/ml and centrifuged and resuspended in TAP medium depleted of copper. F_V/F_M and cell concentration were determined at different time points. (**b**) RNA blot analysis. RNA was extracted from cells at different times under either copper-replete (*left*) or copper-deficient (*right*) conditions and subjected to RNA blot analysis. The probes used are indicated on the *left*. (**c**) Immunoblot analysis of proteins from the samples described in (**b**). The antibodies used are indicated on the *left*. (Reproduced by permission from ref. 8; Copyright 2007 National Academy of Sciences, USA)

References

1. Maul JE, Lilly JW, Cui L, dePamphilis CW, Miller W, Harris EH, Stern DB (2002) The *Chlamydomonas reinhardtii* plastid chromosome: islands of genes in a sea of repeats. Plant Cell 14:2659–2679
2. Boynton JE, Gillham NW, Harris EH, Hosler JP, Johnson AM, Jones AR, Randolph-Anderson BL, Robertson D, Klein TM, Shark KB et al (1988) Chloroplast transformation in Chlamydomonas with high velocity microprojectiles. Science 240:1534–1538
3. Goldschmidt-Clermont M (1991) Transgenic expression of aminoglycoside adenine transferase in the chloroplast: a selectable marker of site-directed transformation of Chlamydomonas. Nucleic Acids Res 19:4083–4089
4. Rochaix JD (1997) Chloroplast reverse genetics: new insights into the function of plastid genes. Trands Plant Sci 2:419–525
5. Huang C, Wang S, Chen L, Lemieux C, Otis C, Turmel M, Liu XQ (1994) The Chlamydomonas chloroplast clpP gene contains translated large insertion sequences and is essential for cell growth. Mol Gen Genet 244:151–159
6. Chen X, Kindle K, Stern D (1993) Initiation codon mutations in the Chlamydomonas chloroplast petD gene result in temperature-sensitive photosynthetic growth. EMBO J 12:3627–3635
7. Majeran W, Wollman FA, Vallon O (2000) Evidence for a role of ClpP in the degradation of the chloroplast cytochrome b(6)f complex. Plant Cell 12:137–150
8. Surzycki R, Cournac L, Peltier G, Rochaix JD (2007) Potential for hydrogen production with inducible chloroplast gene expression in Chlamydomonas. Proc Natl Acad Sci U S A 104:17548–17553
9. Barkan A, Goldschmidt-Clermont M (2000) Participation of nuclear genes in chloroplast gene expression. Biochimie 82:559–572
10. Kuchka MR, Goldschmidt-Clermont M, van Dillewijn J, Rochaix JD (1989) Mutation at the Chlamydomonas nuclear NAC2 locus specifically affects stability of the chloroplast psbD transcript encoding polypeptide D2 of PS II. Cell 58:869–876
11. Nickelsen J, van Dillewijn J, Rahire M, Rochaix J-D (1994) Determinants for stability of the chloroplast psbD RNA are located within its short leader region in *Chlamydomonas reinhardtii*. EMBO J 13:3182–3191
12. Merchant S, Bogorad L (1987) The Cu(II)-repressible plastidic cytochrome c. Cloning and sequence of a complementary DNA for the pre-apoprotein. J Biol Chem 262:9062–9067
13. Quinn JM, Eriksson M, Moseley JL, Merchant S (2002) Oxygen deficiency responsive gene expression in *Chlamydomonas reinhardtii* through a copper-sensing signal transduction pathway. Plant Physiol 128:463–471
14. Michelet L, Lefebvre-Legendre L, Burr SE, Rochaix JD, Goldschmidt-Clermont M (2011) Enhanced chloroplast transgene expression in a nuclear mutant of Chlamydomonas. Plant Biotechnol J 9:565–574
15. Rasala BA, Muto M, Lee PA, Jager M, Cardoso RM, Behnke CA, Kirk P, Hokanson CA, Crea R, Mendez M, Mayfield SP (2010) Production of therapeutic proteins in algae, analysis of expression of seven human proteins in the chloroplast of *Chlamydomonas reinhardtii*. Plant Biotechnol J 8:719–733
16. Surzycki R, Greenham K, Kitayama K, Dibal F, Wagner R, Rochaix JD, Ajam T, Surzycki S (2009) Factors effecting expression of vaccines in microalgae. Biologicals 37:133–138
17. Finer JJ, Vain P, Jones MW, McMullen MD (1992) Development of the particle inflow gun for DNA delivery to plant cells. Plant Cell Rep 11:323–328
18. Quinn JM, Merchant S (1998) Copper-responsive gene expression during adaptation to copper starvation. Methods Enzymol 297:263–279
19. Boudreau E, Nickelsen J, Lemaire SL, Ossenbühl F, Rochaix J-D (2000) The Nac2 gene of Chlamydomonas encodes a chloroplast TPR-like protein involved in psbD mRNA stability. EMBO J 19:3366–3376
20. Ferrante P, Catalanotti C, Bonente G, Giuliano G (2008) An optimized, chemically regulated gene expression system for Chlamydomonas. PLoS ONE 3:e3200
21. Surzycki SJ (1971) Synchronously grown cultures of *Chlamydomonas reinhardii*. Methods Enzymol 23:67–73
22. Gorman DS, Levine RP (1965) Cytochrome f and plastocyanin: their sequence in the photosynthetic electron transport chain of *Chlamydomonas reinhardi*. Proc Natl Acad Sci U S A 54:1665–1669
23. Sueoka N, Chiang KS, Kates JR (1967) Deoxyribonucleic acid replication in meiosis of *Chlamydomonas reinhardi*. I. Isotopic transfer experiments with a strain producing eight zoospores. J Mol Biol 25:47–66
24. Quinn JM, Merchant S (1995) Two copper-responsive elements associated with the Chlamydomonas Cyc6 gene function as targets for transcriptional activators. Plant Cell 7:623–628

Part V

Plastid Transformation in Bryophyte

Chapter 29

Plastid Transformation in *Physcomitrella patens*

Mamoru Sugita

Abstract

The moss *Physcomitrella patens* performs efficient homologous recombination in both the nucleus and plastid enabling the study of individual gene function by generating precise inactivation or modification of genes. Polyethylene glycol (PEG)-mediated transformation of protoplasts is routinely used to study the nuclear gene function of *P. patens*. PEG-mediated protoplast transformation is also applied for plastid transformation of this moss. The efficiency of plastid transformation is quite reliable, and one or two homoplasmic transplastomic lines are obtained in a plastid transformation experiment (5×10^5 protoplasts) by selection for spectinomycin resistance.

Key words *aadA*, Bryophytes, Gene targeting, Homologous recombination, Moss, PEG, *Physcomitrella patens*, Polyethylene glycol (PEG)-mediated transformation, Protoplast

1 Introduction

Physcomitrella patens is a moss (Bryophyta) and its life cycle is characterized by an alteration of two generations: a haploid gametophyte that produces gametes and a diploid sporophyte where haploid spores are produced. A spore germinates and develops into a filamentous tissue termed the protonema, composed of two types of cells, chloronema with large and numerous chloroplasts and caulonema with small chloroplasts. Protonema filaments grow exclusively by tip growth of their apical cells and side branches can originate from subapical cells. Some side branch initial cells can differentiate into buds which give rise to leafy gametophores where sporophytes are produced.

The dominant haploid makes mutant isolation and genetic analysis easier than in other plant species with dominant diploidy, like Arabidopsis and rice. The most unique advantage of *P. patens* is its high-frequency homologous recombination, which enables gene targeting for studying the functions of nuclear genes [1, 2]. Therefore, *P. patens* is widely used as an excellent experimental model of choice not only for basic molecular, cytological, and

Pal Maliga (ed.), *Chloroplast Biotechnology: Methods and Protocols*, Methods in Molecular Biology, vol. 1132,
DOI 10.1007/978-1-62703-995-6_29, © Springer Science+Business Media New York 2014

developmental questions in plant biology [3, 4] but also as a bioreactor for the production of human protein [5].

The first trial for plastid transformation of *P. patens* was performed using plasmid pSBL-CtV3 and a biolistic delivery system [6]. Since the entire *P. patens* plastid genome sequence was completed [7], we then adapted the transformation procedure for nuclear genes to develop moss plastid transformation [8]. An outline of the procedures is as follows. Protoplasts to be transformed are prepared from vegetatively propagated protonemata. The linearized DNA (*see* **Note 1**) is incubated with protonemata protoplasts in the presence of polyethylene glycol (PEG) (*see* **Note 2**). The regenerated protoplasts are transferred to plates containing selection medium with spectinomycin (*see* **Note 3**) and cultured for 3 weeks ("first selection"). The moss colonies that survive are transferred to the medium without spectinomycin and incubated for one week ("release") and then transferred back to selection medium ("second selection"). The colonies grown on the selection medium are candidate plastid transformants and are used for genome analysis. Thus, the moss transplastomic lines whose plastid *trnR-CCG* gene locus was disrupted by integration of the *aadA* gene cassette were obtained [8] (*see* **Note 4**). In the future, this locus can be used as a target for overproduction of foreign protein for molecular farming.

2 Materials

2.1 Plant Material and Plasmids

1. *Physcomitrella patens* is grown in 9-cm Petri dishes containing BCDAT or BCDATG medium [9] at 25 ± 1 °C under continuous illumination (30 μmol photons/m^2/s, approximately 2,000 lux) or under a 16-h photoperiod condition (*see* **Note 5**). Protoplasts to be transformed are prepared from vegetatively propagated protonemata cultured for 4 days on BCDATG medium (*see* **Note 6**). For vegetative propagation, the protonemata are collected every 4–7 days and ground with a generator shaft NS-4 attached to a Handy Micro-homogenizer Physcotron NS-310E (Microtec Co., Ltd, Funabashi City, Chiba, Japan) at a speed setting of 10 (max). *P. patens* possesses the ability to regenerate protonema tissues from any tissues after blending the tissues with a homogenizer. The ground protonemata are soaked onto the BCDAT medium. The medium plate should be covered with a sterile cellophane sheet (PT #300, Futamura Chemical Industries Co. Ltd., Nagoya, Japan) to facilitate the transfer of the regenerating plants at subsequent stages.
2. Transforming plasmid pCSΔtrnR [8].

2.2 Stock Solutions for Culture Media

Store all stock solutions at 4 °C.

1. Stock A, 1 l (×100): 0.5 M $Ca(NO_3)_2$, 4.5 mM $FeSO_4{\cdot}7H_2O$. Cover the medium bottle with aluminum foil.
2. Stock B, 1 l (×100): 2.5 % (0.1 mM) $MgSO_4{\cdot}7H_2O$.
3. Stock C, 1 l (×100): 2.5 % (1.84 mM) KH_2PO_4. Adjust to pH 6.5 with 4 M KOH.
4. Stock D, 1 l (×100): 10.1 % (1 M) KNO_3, 0.12 % (4.5 mM) $FeSO_4{\cdot}7H_2O$. Cover the medium bottle with aluminum foil. When the stock solution is oxidized (brownish colored), then new stock should be freshly prepared.
5. Alternative TES (×1,000): 0.0055 %(0.22 mM) $CuSO_4{\cdot}5H_2O$, 0.0614 %(10 mM) H_3BO_3, 0.0055 %(0.23 mM) $CoCl_2{\cdot}6H_2O$, 0.0025 %(0.1 mM) $Na_2MoO_4{\cdot}2H_2O$, 0.0055 %(0.19 mM) $ZnSO_4{\cdot}7H_2O$, 0.0389 %(2 mM) $MnCl_2{\cdot}4H_2O$, 0.0028 % (0.17 mM)KI.
6. Ammonium tartrate stock: 500 mM diammonium (+)-tartrate.
7. $CaCl_2$ stock: 50 mM $CaCl_2{\cdot}2H_2O$. Sterilize the stock by autoclaving.

2.3 Culture Media

BCD medium [9] and Knop medium [10] are the most commonly used media for standard growth of *Physcomitrella* plants. This protocol uses BCD medium below. Prepare culture media by dilution of stock solutions. For the preparation of solid medium, add 0.8 % (w/v) agar (Sigma A9799 Plant Cell Culture Tested). Autoclave and pour the medium into Petri dishes. The plates can be stored in a sealed bag up to one month at room temperature.

1. BCD medium: 1× stock B, 1× stock C, 1× stock D, 1× Alternative TES, 1 mM $CaCl_2{\cdot}2H_2O$.
2. BCDAT medium: BCD medium supplemented with 5 mM ammonium tartrate.
3. BCDATG medium: BCDAT medium supplemented with 0.5 % glucose.

2.4 Solutions and Media for Transformation

1. 8 % mannitol: Dissolve 8 g D-mannitol in 100 ml H_2O and sterilize by autoclaving. Store at room temperature.
2. Driserase solution (2 % (w/v), make fresh as required): Dissolve 0.5 g driserase-20 (Kyowa Hakko Kirin Co., Ltd., Japan) in 25 ml of 8 % D-mannitol in a 50-ml Falcon tube. Centrifuge at 2,700×*g* for 5 min and filter sterilize the supernatant using a 0.45-μm filter.
3. MMM medium (10 ml): Dissolve 0.91 g D-mannitol in 8.85 ml H_2O and add 0.15 ml of 1 M $MgCl_2$ and 1 ml of 1 % (2-(*N*-morpholino)-ethanesulfonic acid (MES). Final concentrations are 0.5 M D-mannitol, 15 mM $MgCl_2$, 0.1 % MES. Filter sterilize using a 0.22-μm filter. Make fresh as required.

4. PEG/T solution: 40 % polyethylene glycol 6000 (PEG 6000), 0.1 M $Ca(NO_3)_2$, 0.01 M Tris–HCl (pH 8.0), 7.2 % D-mannitol. Weigh out 2 g PEG 6000 in a 20-ml glass vial with a screw cap and sterilize by autoclaving. Add 5 ml of the filter-sterilized solution (0.1 M $Ca(NO_3)_2$, 0.01 M Tris–HCl (pH 8.0), 7.2 % D-mannitol) and dissolve it with a stir bar. Make fresh as required.
5. DNA solution: Digest the transforming plasmid (pCSΔtrnR) with appropriate restriction enzymes and purify the linearized DNA by standard phenol extraction/ethanol precipitation, and dissolve at a final concentration of 1 μg/μl TE solution. The linearized DNA fragment contains the target gene of interest and the chimeric *aadA* gene cassette from pZS197 [11], which consists of the tobacco 16*S* rDNA promoter, the *aadA* coding region, and the 3′-region of tobacco *psbA*.
6. Protoplast liquid solution (100 ml): 1× stock A, 1× stock B, 0.1× stock C, 0.005 % diammonium (+)-tartrate, 6.6 % (0.36 M) mannitol, 0.5 % glucose. Sterilize by autoclaving. Make fresh as required.
7. PRM/T solution: 1× stock B, 1× stock C, 1× stock D, 1× Alternative TES, 5 mM diammonium (+)-tartrate, 10 mM $CaCl_2 \cdot 2H_2O$, 8 % mannitol, and 0.8 % agar (Sigma A9799, Plant Cell Culture Tested). Sterilize by autoclaving and incubate at 45 °C as required.
8. PRM/B medium (for regeneration of protoplasts): 1× stock B, 1× stock C, 1× stock D, 1× Alternative TES, 5 mM diammonium (+)-tartrate, 10 mM $CaCl_2 \cdot 2H_2O$, 6 % D-mannitol, 0.8 % agar (Sigma A9799). Autoclave and pour into 9-cm Petri dishes.
9. Selection medium: Prepare standard solid BCDAT medium and autoclave. Let cool down to 60 °C and add spectinomycin stock solution (500 mg/ml) to reach a final concentration of 500 μg/ml. Pour the medium into Petri dishes. The plates can be stored up to 4 weeks at 4 °C.
10. DNA extraction buffer (×10): 0.75 M Tris–HCl (pH 8.8), 0.2 M $(NH_4)_2SO_4$, 0.1 % Tween-20. Autoclave and store at room temperature.
11. PCR primers to confirm integration of transforming DNA.

3 Methods

3.1 Preparation of Moss Protoplasts

Perform on a clean bench as best possible.

1. The moss protonemata cultured for 3 days are harvested from 10 to 12 Petri dishes using sterilized forceps and transferred to a 50-ml glass centrifuge tube (IWAKI TE-32 PYREX) with 25 ml of 2 % (w/v) driserase solution (Subheading 2.4, **item 2**).

2. Cover the glass centrifuge tube with aluminum foil and incubate for 30 min at room temperature (25 °C) while gently rolling the tube every 5 min. Digestion of cell walls takes place in this step.
3. Pass the moss material through a sieve with a 40-μm nylon mesh and collect the filtrate into a 50-ml glass centrifuge tube.
4. Centrifuge the filtrate in the glass tube for 2 min at 1,000 rpm (TOMY TS-7 rotor, 140 × *g*). Discard the supernatant using a 25-ml measuring pipette or Komagome pipette (keeping a small volume of the supernatant) and gently roll the glass tube between your hands.
5. Add 8 % mannitol solution (Subheading 2.4, **item 1**) slowly without disturbing the pellet up to 40 ml and centrifuge again for 2 min at 1,000 rpm and then discard the supernatant. Resuspend the pellet in 40 ml of 8 % mannitol solution and centrifuge again. This step is repeated once again.
6. Remove a 100 μl aliquot with a cut pipette tip and determine the protoplast number using a counting chamber. Meanwhile, centrifuge the protoplasts again for 2 min at 1,000 rpm.
7. Discard the supernatant and resuspend the pellet in MMM medium (Subheading 2.4, **item 3**), adjusting the density to 1.6×10^6 protoplasts/ml. Usually, 7–20 ml of protoplast solution is obtained.

3.2 Introduction of DNA into the Moss Protoplasts

1. Transfer the DNA solution (30 μl, 30 μg linearized DNA) into a 14-ml Falcon tube (#352006) and then carefully add 300 μl of the protoplast solution using a cut pipette tip.
2. Add 300 μl of PEG/T solution (Subheading 2.4, **item 4**) pre-incubated at 45 °C and mix gently by rolling the tube. Incubate the mixture at 45 °C for 5 min and then at 20 °C for 10 min.
3. Dilute the mixture with protoplast liquid solution (Subheading 2.4, **item 6**) every 3 min adding 0.3 ml (repeat five times) and then every 3 min adding 1 ml (repeat five times), respectively, and gently mix the solution after each step by rolling the tube.
4. Transfer the protoplast solution (total ~7 ml) into a 6-cm plastic dish. Seal the plate with Parafilm and cover it with aluminum foil, and incubate overnight (up to 24 h) at 25 °C. During this time protoplast regeneration is initiated.
5. Transfer the protoplast mixture using a P5000 Gilson Pipetman into a 15-ml Falcon tube (#35-2096); centrifuge at 1,000 rpm for 2 min at 4 °C. Discard the supernatant by decantation or using a P5000 Pipetman.
6. Slowly add 8 ml of PRM/T solution (Subheading 2.4, **item 7**) incubated at 45 °C. This step should be performed as quickly as possible before the medium solidifies.

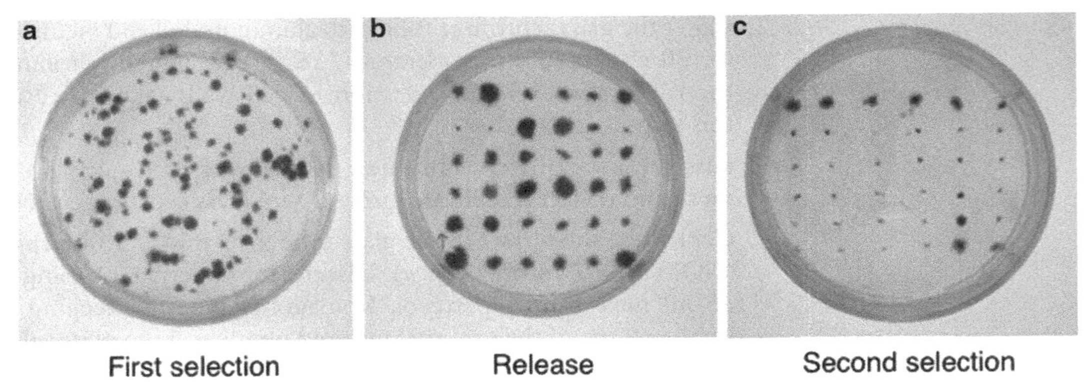

Fig. 1 Moss protonema colonies grown on selection medium. (**a**) Step 1 spec-resistant colonies grown for 3 weeks after culture on selection medium (first selection). Spec-resistant colonies are mixed with stable transformants and transient transformants. (**b**) The step 1 colonies are picked up and transferred to medium without spectinomycin and cultured for 1 week. This step is referred to as "release" and transient transformants can be excluded during second selection. (**c**) Step 2 spec-resistant colonies in "second selection"

7. Plate 2 ml each of the regenerating protoplasts onto 9-cm Petri dishes containing solid PRM/B medium (Subheading 2.4, **item 8**). The medium should be covered with a sterile cellophane sheet, which facilitates the transfer of the regenerating plants at subsequent stages.
8. Seal the Petri dishes with surgical tape (3M Micropore, #1530-0) or Parafilm and grow the cultures at 25 °C for 3 days under continuous light (30 μmol photons/m^2/s).

3.3 Screening Moss Plastid Transformants

After PEG-mediated transformation and regeneration of protoplasts, first selection, release, and second selection are performed on BCDAT medium plates with or without antibiotics to select for stable plastid transformed moss plants.

1. Transfer the cellophane with regenerated protoplasts using forceps to selection medium plates containing spectinomycin and culture for 3 weeks to generate "step 1 spec-resistant colonies" (Fig. 1a). This step is defined as the "first selection."
2. Pick up step 1 colonies using forceps (flamed) to medium without spectinomycin and incubated for one week (Fig. 1b). This step is referred to as "release."
3. A small portion of the colony is picked up and transferred back to selection medium. The remaining colonies on the medium without spectinomycin can be stored at 4 °C for several months.
4. The moss colonies transferred to the spectinomycin-containing medium are cultured for 2 weeks ("second selection"). The colonies that survive grow well within 1 week (Fig. 1c)

Table 1
Efficiency of plastid transformation in *P. patens*

	DNA1	DNA2	DNA3
No. of spec-resistant colonies			
Step 1	132	88	250
Step 2	30	28	44
Homoplasmic transplastomic lines	2	1	1

For plastid transformation 4.8×10^5 protoplasts were used

and are referred as "step 2 spec-resistant colonies." Most transient transformants can be excluded during second selection. Moss plants surviving the second round of selection are considered to be stable plastid transformants.

3.4 Analysis of Transformants

The transformants can be analyzed for stable integration of the transgene by PCR-based methods. Routinely a first PCR screen of the transformants is based on the detection of stable integration of the selection marker gene. To verify the integration of constructs containing an *aadA* selection, marker cassette can be used for PCR analysis and genomic Southern blot analysis. Two homoplasmic transplastomic lines (L1 and L4) were obtained for DNA 1 and one line (S26) for DNA2. Thus, efficiency of plastid transformation is quite reliable, and one or two homoplasmic transplastomic lines are yielded per plastid transformation experiment (Table 1).

3.4.1 One-Step PCR Analysis

Isolation of genomic DNA from small amounts of the moss colonies (5 mm) is performed according to an easy-to-handle protocol developed for the rapid screening of a large number of *P. patens* plants [12].

1. Place a small amount (1–5 mg) of the regenerated moss colonies into 50 μl of 1× DNA extraction buffer (Subheading 2.4, **item 10**) in a 250-μl tube for PCR.
2. Heat the moss samples at 60 °C for 20 min in a thermocycler. DNA is extracted in the buffer.
3. Use 2 ml of the moss-buffer mixture directly as template for PCR. The reaction is performed in the presence of 1 % polyvinylpyrrolidone (PVP-10, Sigma). Aliquots of the resulting extract can be immediately used for PCR analysis without further purification. Beside the detection of the selection marker gene, further analysis of the plants is dependent on the experimental approach.
4. Store samples at −20 °C and use as required.
5. PCR is performed under appropriate reaction conditions, depending on primers, target sequences, or quality of the

template DNA, etc. The thermocycler program is set to 95 °C for 3 min and is followed by 20 cycles (or 35–40 cycles) of 94 °C for 30 s, 55 °C for 30 s, and 72 °C for 2 min. Paq5000 DNA polymerase (Stratagene) is usually used for PCR. For PCR primers, *see* ref. 8.

6. Amplified products are run in a 1 % agarose gel, which is stained with ethidium bromide and photographed.

3.4.2 Southern Blot Analysis of Total Cellular DNA

To verify integration of the disruption construct by Southern blot analyses, total cellular DNA can be prepared.

1. Total cellular DNA is prepared from frozen powdered moss colonies (300 mg) according to a CTAB procedure [13].
2. The DNA (2 μg) is digested with *Bgl*II, separated on a 0.8 % agarose gel, and transferred to a nylon membrane (Hybond N^+, Amersham Pharmacia Biotech, USA).
3. DNA probes for detection of *rbcL* and *aadA* are prepared by PCR using appropriate primers and labeled with [α-^{32}P] dCTP by standard procedures [14].
4. Hybridization and washing of the membrane are performed at 65 °C using standard procedures [14].

An example of Southern blot analysis is shown in Fig. 2. Detected bands shift depending on the size of the inserted selection marker cassette indicating a homologous recombination event.

4 Notes

1. Before transformation of *Physcomitrella* protoplasts, it is recommended to linearize the DNA construct using appropriate restriction enzymes. The transformation of circular plasmids may result in extrachromosomal replication and the generation of unstable transformants [15].
2. Plastid transformation using a biolistic delivery system can be achieved for the *P. patens* protonemata. Transformation using particle bombardment is carried out by shooting the DNA into protonemal tissue growing on cellophane. After transformation, transgenic cells can be isolated by blending the tissue and subsequent plating onto selective medium. Particle bombardment is equally applicable to obtain stable transgenic *Physcomitrella* [16, 17]. Compared to the PEG-mediated transformation procedure, the application of the biolistic method requires less DNA (usually 5 μg).

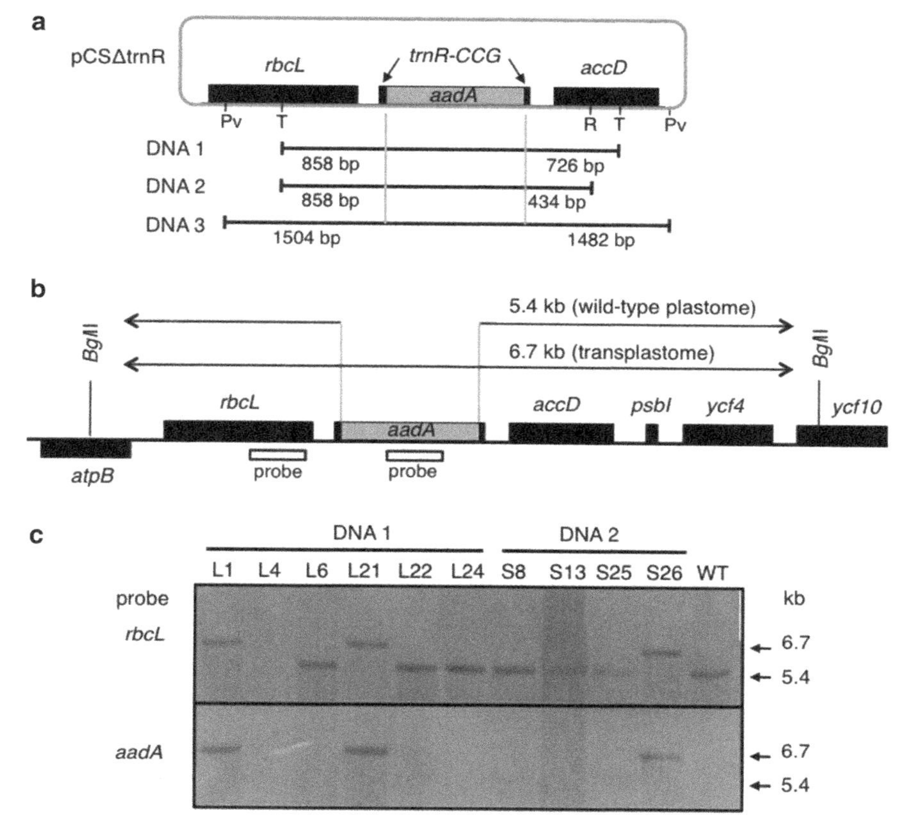

Fig. 2 Maps of plastid DNA and transforming plasmid. (**a**) Transforming plasmid pCSΔtrnR and the DNA fragments used for transformation are shown. Restriction enzyme-digested DNAs used for plastid transformation (DNAs 1–3) and the lengths of flanking plastid DNA sequences are shown below pCSΔtrnR. *Eco*RV (R), *Eco*T22I (T), and *Pvu*II (Pv) sites were used for fragmentation of the pCSΔtrnR plasmid before its introduction into protonemal protoplasts. (**b**) Partial map of wild-type plastome and transplastomes. Transplastomes were generated by the integration of *aadA* into the plastid genome via the flanking plastid DNA sequences, thus introducing the *trnR-CCG* gene disruption into the transformed lines. (**c**) Southern blot analysis of total cellular DNA from spec-resistant mosses (L1 to S26) and wild-type moss (WT) using *rbcL* or *aadA* DNA probe. L1 to L24 lines are derived from DNA1 and S8 to S26 lines transformed with DNA2

3. The most commonly used selection marker comprises the *aadA* gene encoding the enzyme aminoglycoside 3″-adenylyltransferase, which confers resistance to spectinomycin.
4. The plastid arginine tRNA gene, *trnR-CCG*, is present in bryophytes and gymnosperms, but not in angiosperms. The moss *trnR-CCG* is transcribed at negligible levels and the codon use of CGG is very low, thus suggesting a nonfunctional gene. Targeted gene disruption indicates that *trnR-CCG* is dispensable for *P. patens* plastid biogenesis.

5. *Physcomitrella* plants are provided from the International Moss Stock Center (IMSC), Freiburg [18]. Plant tissue cultures of gametophores can be maintained by subculturing the gametophores monthly. For routine use of *Physcomitrella* plants, it is recommended to keep plant cultures on solidified medium as a backup system or to keep spores which can be used to initiate new culture lines. Details of culture, growth, and maintenance; induction of sporophytes; and collection and storage of spores are publicly available online [19, 20].
6. The moss colony is composed of two types of protonemata (chloronema and caulonema cells). The addition of ammonium tartrate to the culture medium results in predominant growth of the moss protonema in the chloronema stage, which is thought to be the best source for protoplast isolation [21].

Acknowledgments

The author would like to thank Chika Sugiura Miyamoto for the development of plastid transformation of *P. patens*. This work was supported in part by a Grant-in-Aid from the Japan Society for the Promotion of Science (14340252).

References

1. Schaefer D, Zryd J-P (1997) Efficient gene targeting in the moss *Physcomitrella patens*. Plant J 11:1195–1206
2. Schaefer DG (2001) Gene targeting in *Physcomitrella patens*. Curr Opin Plant Biol 4:143–150
3. Cove D (2005) The moss *Physcomitrella patens*. Annu Rev Genet 39:339–358
4. Cove D, Bezanilla M, Harries P, Quatrano R (2006) Mosses as model systems for the study of metabolism and development. Annu Rev Plant Biol 57:497–520
5. Decker EL, Reski R (2004) The moss bioreactor. Curr Opin Plant Biol 7:166–170
6. Cho SH, Chung YS, Cho SK, Rim YW, Shin JS (1999) Particle bombardment mediated transformation and GFP expression in the moss *Physcomitrella patens*. Mol Cells 9:14–19
7. Sugiura C, Kobayashi Y, Aoki S, Sugita C, Sugita M (2003) Complete chloroplast DNA sequence of the moss *Physcomitrella patens*: evidence for the loss and relocation of *rpoA* from the chloroplast to the nucleus. Nucleic Acids Res 31:5324–5331
8. Sugiura C, Sugita M (2004) Plastid transformation reveals that moss $tRNA^{Arg}$-CCG is not essential for plastid function. Plant J 40: 314–321
9. Ashton NW, Cove DJ (1977) The isolation and preliminary characterisation of auxotrophic and analogue resistant mutants in the moss *Physcomitrella patens*. Mol Gen Genet 154:87–95
10. Reski R, Abel WQ (1985) Induction of budding on chloronemata and caulonemata of the moss, *Physcomitrella patens*, using isopentenyladenine. Planta 165:354–358
11. Svab Z, Maliga P (1993) High-frequency plastid transformation in tobacco by selection for a chimeric *aadA* gene. Proc Natl Acad Sci U S A 87:8526–8530
12. Schween G, Fleig S, Reski R (2002) High-throughput-PCR screen of 15,000 transgenic *Physcomitrella* plants. Plant Mol Biol Rep 20:43–47
13. Murray JM, Thompson WF (1980) Rapid isolation of high molecular weight plant DNA. Nucleic Acids Res 8:4321–4325
14. Sambrook J, Fritsch EF, Maniatis T (1989) Molecular cloning: a laboratory manual, 2nd edn. Cold Spring Harbor Laboratory Press, Cold Spring Harbor, NY

15. Ashton NW, Champagne CEM, Weiler T, Verkoczy LK (2000) The bryophyte *Physcomitrella patens* replicates extrachromosomal transgenic elements. New Phytol 146: 391–402
16. Bezanilla M, Pan A, Quatrno RS (2003) RNA interference in the moss *Physcomitrella patens*. Plant Physiol 133:470–474
17. Tasaki E, Hattori M, Sugita M (2010) The moss pentatricopeptide repeat protein with a DYW domain is responsible for RNA editing of mitochondrial *ccmFc* transcript. Plant J 62:560–570
18. Jenkins GI, Cove DJ (1983) Light requirements for regeneration of protoplasts of the moss *Physcomitrella patens*. Planta 157: 39–45
19. The International Moss Stock Center (IMSC Freiburg). http://www.moss-stock-center.org/
20. Home page of plant biotechnology (Prof. Dr. Ralf Reski lab.) The Freiburg standard methods for *Physcomitrella patens* cell culture, protoplastation and transformation. http://www.plant-biotech.net/
21. NIBB PHYSCObase (Prof. Dr. Mitsuyasu Hasebe lab.) Protocols. http://moss.nibb.ac.jp/

Chapter 30

Plastid Transformation of Sporelings and Suspension-Cultured Cells from the Liverwort *Marchantia polymorpha* L.

Shota Chiyoda, Katsuyuki T. Yamato, and Takayuki Kohchi

Abstract

We describe simple and efficient plastid transformation methods for suspension-cultured cells and sporelings of the liverwort, *Marchantia polymorpha* L. Use of rapidly proliferating cells such as suspension-cultured cells and sporelings, which are immature thalli developing from spores, as targets made plastid transformation by particle bombardment efficient. Selection on a sucrose-free medium and linearization of the transformation vector significantly improved the recovery rate of plastid transformants. With the methods described here, a few plastid transformants are obtained from a single bombardment of sporelings, while more efficient plastid transformation is expected in suspension-cultured cells, ~60 transformants from a single bombardment. Homoplasmic transformants of thalli are obtained immediately after primary selection, whereas homoplasmic transformants from suspension-cultured cells are obtained after 12–16 weeks of repeated subculture.

Key words Bryophyte, Chloroplast, Homologous recombination, Liverwort, *Marchantia polymorpha*, Particle bombardment, Plastid transformation, Suspension-cultured cell, Sporeling

1 Introduction

The liverwort, *Marchantia polymorpha* L., is an extant representative species of the first land plants. In spite of the substantial evolutionary divergence, the gene organization of the plastid genome is highly conserved between *M. polymorpha* and flowering plants [1, 2]. In particular, the nucleotide sequence of the 16S-23S rRNA operon shows a high degree of conservation [3]. The plastid genome size of *M. polymorpha* is 121,025 bp, which is relatively small compared to those of other land plants. Suspension culture cells of *M. polymorpha* are homogeneous, are stringently dependent on chloroplast functions, and proliferate vigorously [4, 5] (*see* **Note 1**). A sporeling is an immature thallus developing from a spore. Sporelings obtained by 5- to 7-day culture from sterilized spores [6] are also rich in actively proliferating cells, and thus

Pal Maliga (ed.), *Chloroplast Biotechnology: Methods and Protocols*, Methods in Molecular Biology, vol. 1132,
DOI 10.1007/978-1-62703-995-6_30, © Springer Science+Business Media New York 2014

suitable for *Agrobacterium*-mediated transformation of nuclear DNA [7].

We have developed an efficient plastid transformation method for suspension-cultured cells of *M. polymorpha* [8], a method, which is also applicable for sporelings with slight modifications. A transformation vector, pCS31, was constructed to integrate an *aadA* expression cassette encoding spectinomycin resistance into the *trnI–trnA* intergenic region of the *M. polymorpha* plastid DNA by homologous recombination. *M. polymorpha* suspension-cultured cells, or sporelings were bombarded with pCS31-coated gold projectiles and selected on a medium containing spectinomycin. Plastid transformants were efficiently isolated as spectinomycin-resistant calli or thalli on sucrose-free media. Although several rounds of dedifferentiation and regeneration are generally required to establish homoplasmic plastid transformants for flowering plants [9], this process is not required for *M. polymorpha*. Homoplasmicity in *M. polymorpha* plastid transformants can be readily achieved in suspension-cultured cells and sporelings by successive subculture and primary selection, respectively. This protocol is an update of our earlier Japanese paper on plastid transformation in *Marchantia polymorpha* [10].

2 Materials

2.1 Media and Buffer

1. B5 minor components (1,000× stock, 100 ml): 25 mg $NaMoO_4 \cdot 2H_2O$, 2.5 mg $CuSO_4 \cdot 5H_2O$, 2.5 mg $CoCl_2 \cdot 6H_2O$, 200 mg $ZnSO_4 \cdot 7H_2O$, 1 g $MnSO_4 \cdot 7H_2O$, and 300 mg H_3BO are dissolved in deionized water and stored at −30 °C.
2. B5 vitamin (1,000× stock, 100 ml): 10 g inositol, 100 mg nicotinic acid, 100 mg pyridoxine hydrochloride, and 1 g thiamine hydrochloride are dissolved in deionized water and stored at −30 °C.
3. 10× 0M51C stock solution (4 l): 80 g KNO_3, 16 g NH_4NO_3, 14.8 g $MgSO_4 \cdot 7H_2O$, 12 g $CaCl_2 \cdot H_2O$, 11 g KH_2PO_4, 1.6 g EDTA-NaFe (III), 40 ml B5 micro-components, 40 ml B5 vitamin, and 4 ml 0.75 % KI are dissolved in deionized water and stored at −30 °C. It is convenient to make 100 ml aliquots in sealable plastic bags for later use.
4. 0M51C/1M51C media (1 l): 100 ml 10× 0M51C stock solution, 20 g sucrose, 1.0 g casamino acids, and 0.3 g L-glutamate are dissolved in deionized water. 1 mg/l 2,4-dichlorophenoxyacetic acid is added for 1M51C medium. Adjust pH to 5.5 with 1 N KOH. Add 1.2 % agar for plates. Sterilize by autoclaving.
5. TE: 10 mM Tris–HCl (pH 8.0) and 1 mM EDTA (pH 8.0).

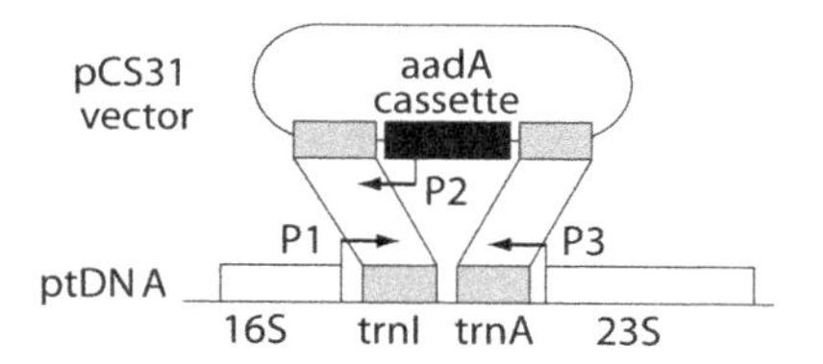

Fig. 1 The pCS31 transformation vector. Schematic illustration of pCS31 and its integration site on plastid DNA. Filled and gray boxes represent the *aadA* cassette and homologous sequences, respectively. *Arrows* indicate PCR primers for selection of plastid transformants and checking of homoplasmic state

2.2 Preparation of Surface-Sterilized Sporangia

1. Sterilization solution: 0.2 % sodium hypochlorite (final concentration of effective chloride is 0.017–0.027 %) and 0.1 % Triton X-100.
2. Tweezers.
3. Glass pipette.
4. Silica gel.

2.3 Transformation Vector

1. We use the intergenic region of the *trnI* and *trnA* genes as the site for targeted integration. A DNA fragment of interest can be placed between the two genes together with the *aadA* expression cassette. The *aadA* expression cassette consists of the promoter of the ribosomal RNA operon from the tobacco plastid genome, the ribosome binding site of the *M. polymorpha rbcL* gene, the *aadA* coding sequence, and the 3′-UTR of the tobacco *psbA* gene (Fig. 1). In the case of pCS31 [8], a 2 kb fragment carrying the *trnI* and *trnA* genes was first cloned into pBluescript II as homologous sequence (*see* **Note 2**), and then the *aadA* expression cassette alone was inserted between the *trnI* and *trnA* genes.
2. Prepare a sufficient amount of high-quality plasmid DNA for particle bombardment (use of a kit, such as QIAGEN plasmid midi kit, is recommended) and adjust DNA concentration to 1 μg/μl in TE and store at −30 °C. In this protocol, 25 μg of plasmid DNA is used to prepare enough DNA-coated particles for about ten bombardments.
3. (For transformation of sporelings) Linearize pCS31 by *Eco*RI and *Sca*I prior to bombardment.

2.4 Particle Bombardment

1. 2.5 M $CaCl_2$: aliquot 200 μl into 1.5 ml microtubes and store at −30 °C.
2. 0.1 M spermidine: aliquot 100 μl into 1.5 ml microtubes and store at −80 °C.
3. Filter (Nalgene: 134 mm × 230 mm, #300-4050) (*see* **Note 3**).
4. Filter disks (Whatman: No. 1φ55 mm) (*see* **Note 3**).

5. Cellophane: cut cellophane to fit into plates, and rinse in boiling water (*see* **Note 3**).
6. Aspirator.
7. Particle delivery system (Bio-Rad: PDS-1000/He Particle delivery system).
8. 0.6 μm gold particle (Bio-Rad: #165-2262): suspend in ethanol at 20 μg/μl, aliquot 110 μl into 1.5 ml microtubes, and store at −30 °C.
9. 900 psi rupture disk (Bio-Rad: #165-2328).
10. Macrocarrier (Bio-Rad, #165-2335).
11. Stopping screen (Bio-Rad, #165-2336).

2.5 Selection and Identification of Plastid Transformants

1. Selective media: make a solution of spectinomycin dihydrochloride hexahydrate (SIGMA, #S9007-5G) at a concentration of 50 mg/ml and filter through 0.20 μm syringe filter. Store in 50 ml plastic tube at −20 °C.
2. Sterilized water.
3. Tweezers.
4. 100 % ethanol.
5. DNA extraction buffer: 50 mM Tris–HCl (pH 8.0), 20 mM EDTA (pH 8.0), 0.3 M NaCl, 0.5 % SDS, 5 M urea, and 5 % (v/v) phenol (pH 7.0).
6. Tris–HCl with 10 μg/μl of RNase A.
7. Primer 1 (P1), 5′-ACCCATTAAATTATCCTTAGCATG-3′; Primer 2 (P2), 5′-TGGATCCCTCCCTACAACTG-3′; Primer 3 (P3), 3′-TGCCTAGGTATCCACCGTAA-3′.

3 Methods

3.1 Preparation of Suspension-Culture Cells for Bombardment

1. Maintain *M. polymorpha* suspension-cultured cells in 70 ml of 1M51C medium in a 300 ml flask on rotary shaker (130 rpm) under continuous white light (50–60 μmol photons/m^2/s) at 22 °C. Use cells from 7- to 10-day-old cultures for plastid transformation.
2. Layer suspension-cultured cells onto the center of a filter disk using vacuum filtration (1–2 mm thickness) (Fig. 2a).
3. Place the filter disk with the cells onto a 1M51C plate and incubate over night (22 °C, 50–60 μmol photons/m^2/s).

3.2 Preparation of Sporelings for Bombardment

1. Collect intact mature yellow sporangia (*see* **Note 4**) from archegoniophores with tweezers and transfer to microtubes.
2. Add sterilization solution to the microtubes and let stand for 1–2 min (do not break sporangia).

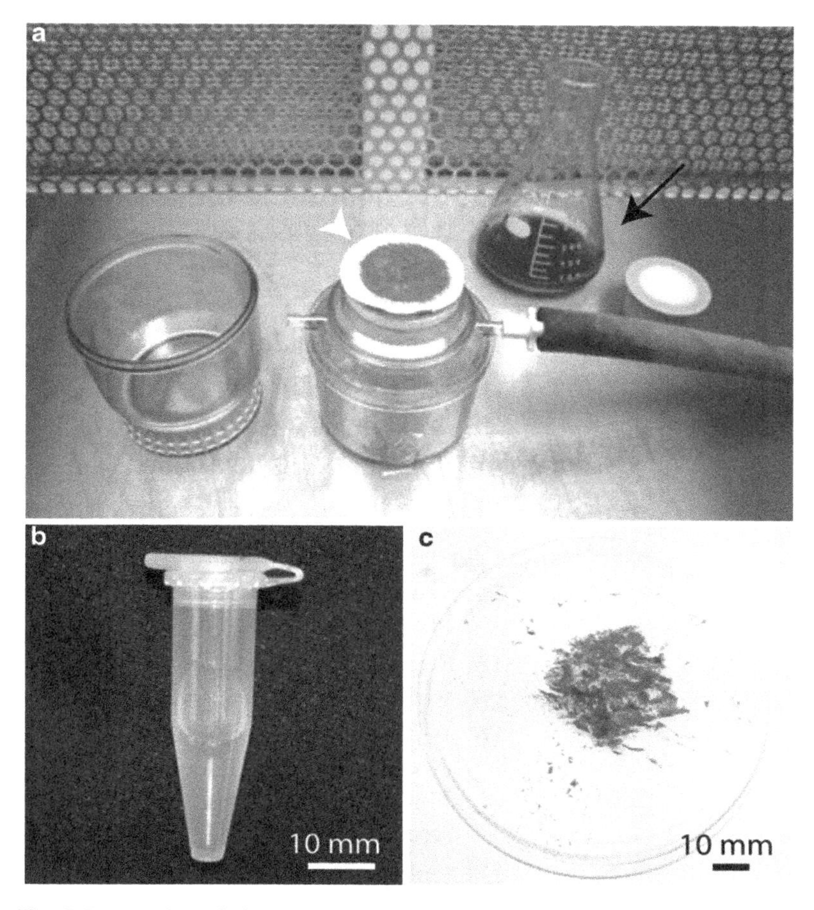

Fig. 2 Preparation of plant material for particle bombardment. (**a**) Suspension-cultured cells subcultured for 7 days in a flask (*arrow*) and layered on filter disk (*open arrowhead*) by aspirator. (**b**) Preparation of a suspension of sterilized spores. (**c**) Sporelings grown for 7 days from spores. Cellophane is layered onto medium to transfer cells to selective media easily after particle bombardment. Reproduced from ref. 10 with permission

3. Rinse sporangia three times with sterile water using a glass pipette.
4. Discard water, add a few beads of silica gel to the tubes, and store at 4 °C until use.
5. Suspend spores in sterile water (100 μl/sporangium) (Fig. 2b).
6. Drop the spore suspension onto the center of cellophane sheet on a 0M51C plate, and culture germinated spores for 1 week (22 °C, 50–60 μmol photons/m^2/s) (Fig. 2c).

3.3 Particle Bombardment

1. Centrifuge suspension of gold particles (20 μg/μl) at 6,000 × *g* for 1 min.
2. Discard supernatant.

3. Add 1 ml sterile water to pellet and vortex thoroughly.
4. Centrifuge at 2,000 × *g* for 3 min and discard supernatant.
5. Add 230 μl sterile water to pellet and vortex thoroughly.
6. Add in the following order: 250 μl of 2.5 M $CaCl_2$, 25 μl of 1 μg/μl vector DNA solution, and 50 μl of 0.1 M spermidine into microtube.
7. Incubate on ice for 10 min, vigorously mixing for 10 s once per minute.
8. Centrifuge at 2,000 × *g* for 3 min at 4 °C.
9. Discard supernatant.
10. Add 500 μl ethanol to pellet and vortex thoroughly.
11. Centrifuge at 2,000 × *g* for 3 min at 4 °C.
12. Repeat **steps 9–11**.
13. Suspend gold particles in 60 μl ethanol.
14. Use a 5.4 μl aliquot of the gold particle suspension for each bombardment.
15. Sterilize rupture disks, macrocarriers, and stopping screens with 70 % ethanol and air-dry in clean hood prior to use.
16. Perform particle bombardment according to the manufacturer's instructions. Parameters are as follows: vacuum, 28 in. Hg; the distance between the target stage and the stopping screen, 120 mm.
17. Culture bombarded cells overnight (22 °C, 50–60 μmol photons/m^2/s).
18. Divide cell culture into four and spread each portion evenly onto a selective 1M51C (for suspension-cultured cells) (Fig. 3a) or 0M51C (for sporelings) (Fig. 3b) plate containing 500 mg/l spectinomycin dihydrochloride without sucrose. For spreading, use tweezers and 2–3 ml of sterile water (*see* **Note 5**).
19. After 4 weeks of culture, transfer spectinomycin-resistant calli or thalli (Fig. 3c, d) to fresh selective media.

3.4 Identification of Homoplasmic Transformants by PCR

1. Transfer ~10 mg of cells or tissue (approx. 2 × 2 mm) into a 1.5 ml microtube (*see* **Note 6**).
2. Add 100 μl of DNA extraction buffer and grind cells by pestle.
3. Add 400 μl of DNA extraction buffer and mix vigorously.
4. Add 500 μl of phenol/chloroform, mix vigorously, and centrifuge (10,000 × *g*, 5 min).
5. Transfer 200 μl of supernatant to a new 1.5 ml microtube and add 500 μl of ethanol.

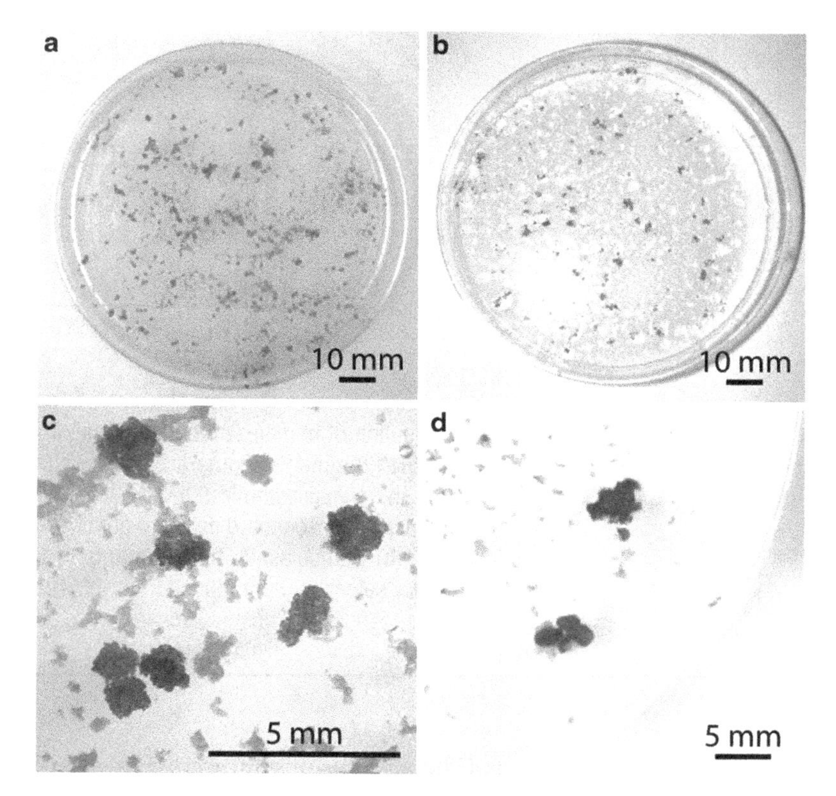

Fig. 3 Selection of spectinomycin-resistant lines. (**a**) Suspension-cultured cells and (**b**) sporelings spread onto selective media after particle bombardment. (**c**) Suspension-cultured cells and (**d**) thalli after 3–4 weeks of selection. Reproduced from ref. 10 with permission

6. Centrifuge (10,000 × *g*, 4 °C, 15 min) and discard supernatant.
7. Add 1 ml of 70 % ethanol and centrifuge (10,000 × *g*, 4 °C, 5 min).
8. Resuspend the pellet in 100–200 μl of Tris–HCl (pH 8.0) containing 10 μg/ml RNase A.
9. Select plastid transformants by PCR using primers P1 and P2 (Subheading 2.5, **item** 7) for the *aadA* cassette. For primer position, *see* Fig. 1.
10. Subculture the candidate plastid transformants on selective media.
11. Check the homoplasmic state of the transformants by PCR using primers P1 and P3, which anneal to the regions outside the homologous sequences (Figs. 1 and 4) (*see* **Note** 7).

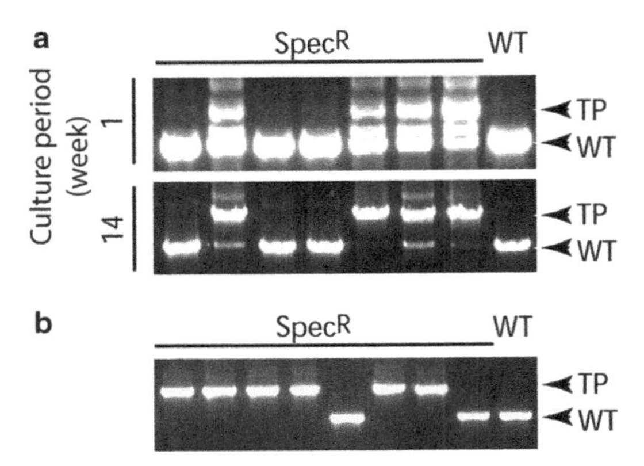

Fig. 4 Evaluation of homoplasmic state by agarose gel electrophoresis of PCR-amplified DNA fragments. Primers P1 and P3 (*see* Fig. 1) yield wild type (WT, 2,179 bp) and transplastomic (TP, 3,563 bp) DNA. (**a**) Homoplasmic transformants of suspension-cultured cells are obtained after 12–16 weeks of repeated subculture. (**b**) Homoplasmic transformants of thalli are obtained immediately after primary selection. Reproduced from ref. 10 with permission

4 Notes

1. The suspension-cultured cells used here have lost the totipotency to regenerate into intact plants due to prolonged subculture in liquid 1M51C medium.
2. A DNA fragment of approx. 1 kb is sufficient for homologous recombination.
3. Autoclave prior to use.
4. Sporangia can be obtained as previously described [6]. Here we use F1 spores produced by the genetic cross between the accessions Takaragaike-1 and Takaragaike-2.
5. High cell density at this step significantly increases the number of false-positive clones and thus should be avoided.
6. For DNA extraction, QIAGEN DNeasy Plant Mini Kit also works.
7. For sporelings, homoplasmic transformants are readily obtained after one (primary) selection cycle. For suspension-cultured cells, homoplasmic transformants are obtained after 12–16 weeks of repetitive subculture on selective medium.

Acknowledgements

We thank Dr. F. Sato for the use of a biolistic delivery system. This work was supported in part by the grant “Knowledge Cluster Initiative” from MEXT.

References

1. Ohyama K, Fukuzawa H, Kohchi T, Shirai H, Sano T, Sano S, Umesono K, Shiki Y, Takeuchi M, Chang Z, Aota S, Inokuchi H, Ozeki H (1986) Chloroplast gene organization deduced from complete sequence of liverwort *Marchantia polymorpha* chloroplast DNA. Nature 322:572–574
2. Shinozaki K, Ohme M, Tanaka M, Wakasugi T, Hayashida N, Matsubayashi T, Zaita N, Chunwongse J, Obokata J, Yamaguchi-Shinozaki K, Ohto C, Torazawa K, Meng BY, Sugita M, Deno H, Kamogashira T, Yamada K, Kusuda J, Takaiwa F, Kato A, Tohdoh N, Shimada H, Sugiura M (1986) The complete nucleotide sequence of the tobacco chloroplast genome: its gene organization and expression. EMBO J 5:2043–2049
3. Kohchi T, Shirai H, Fukuzawa H, Sano T, Komano T, Umesono K, Inokuchi H, Ozeki H, Ohyama K (1988) Structure and organization of *Marchantia polymorpha* chloroplast genome. IV. Inverted repeat and small single copy regions. J Mol Biol 203:353–372
4. Ono K, Ohyama K, Gamborg OL (1979) Regeneration of the liverwort *Marchantia polymorpha* L. from protoplasts isolated from cell suspension culture. Plant Sci Lett 14:225–229
5. Ohyama K, Wetter LR, Yamano Y, Fukuzawa H, Komano T (1982) A simple method for isolation of chloroplast DNA from *Marchantia polymorpha* L. cell suspension cultures. Agric Biol Chem 46:237–242
6. Chiyoda S, Ishizaki K, Kataoka H, Yamato KT, Kohchi T (2008) Direct transformation of the liverwort *Marchantia polymorpha* L. by particle bombardment using immature thalli developing from spores. Plant Cell Rep 27:1467–1473
7. Ishizaki K, Chiyoda S, Yamato KT, Kohchi T (2008) Agrobacterium-mediated transformation of the haploid liverwort *Marchantia polymorpha* L., an emerging model for plant biology. Plant Cell Physiol 49:1084–1091
8. Chiyoda S, Linley PJ, Yamato KT, Fukuzawa H, Yokota A, Kohchi T (2007) Simple and efficient plastid transformation system for the liverwort *Marchantia polymorpha* L. suspension-culture cells. Transgenic Res 16:41–49
9. Maliga P (2004) Plastid transformation in higher plants. Annu Rev Plant Biol 55:289–313
10. Chiyoda S, Kohchi T (2008) Plastid transformation system for suspension-culture cells and intact plants of liverwort *Marchantia polymorpha* L. Low Temp Sci 67:601–606 (in Japanese)

Index

Pal Maliga (ed.), *Chloroplast Biotechnology: Methods and Protocols*, Methods in Molecular Biology, vol. 1132,
DOI 10.1007/978-1-62703-995-6, © Springer Science+Business Media New York 2014

CPI Antony Rowe
Chippenham, UK
2017-03-21 21:40